Current Topics in Microbiology and Immunology

Volume 385

More information about this series at http://www.springer.com/series/82

Richard W. Compans · Michael B. A. Oldstone
Editors

Influenza Pathogenesis and Control - Volume I

Responsible Series Editors: Richard W. Compans and Michael B. A. Oldstone

Springer

Editors
Richard W. Compans
Department of Microbiology
and Immunology
Emory University
Atlanta, GA
USA

Michael B. A. Oldstone
Department of Immunology
and Microbial Science
The Scripps Research Institute
La Jolla, CA
USA

ISSN 0070-217X ISSN 2196-9965 (electronic)
ISBN 978-3-319-11154-4 ISBN 978-3-319-11155-1 (eBook)
DOI 10.1007/978-3-319-11155-1

Library of Congress Control Number: 2014951138

Springer Cham Heidelberg New York Dordrecht London

Printed on acid-free paper

Springer is part of Springer Science+Business Media (www.springer.com)

Preface

Perhaps first described in the Bible in Numbers 11:31–34 and later by the poet W.H. Auden:

Little birds with scarlet legs,
Sitting on their speckled eggs,
Eye each flu-infected city

Influenza has over the course of human experience been defined by the suffering, illness, and death rates it caused and causes of epidemics and pandemics worldwide.

Influenza virus belongs to the orthomyxovirus family and comes in three flavors, types A, B, and C, with type A and B viruses being important causes of disease in humans, with the majority of cases due to type A. These viruses have a negative strand sense, segmented RNA genome with eight genes that can code for up to 11 proteins. Because of the segmented genome and permissive infection of several animal species (primarily humans, birds, and pigs) influenza virus can easily re-assort differing RNA segments between human and animal viruses. Such an event leads to antigenic shift, and possible new pandemic strains. Further, as with other RNA viruses, influenza's RNA-dependent RNA polymerase is error-prone such that mutations occur frequently, and in the absence of a proofreading frame that eliminates and thus fails to control fit mutated viruses from evolving. The fitness of such mutated virus is suggested by their fidelity during replication. They undergo further selection because of antibody responses and immune escape, termed antigenic drift.

This work consisting of two volumes explores influenza pathogenesis and control from basic structure, binding, entry, replication, and release of influenza virus to its spread, the results of its interaction with animal models, the innate and adaptive immune systems and current epidemiologic efforts at rapid recognition and antiviral, anti-host cytokine storm therapies as well as vaccine strategies to control and prevent influenza virus infections.

Volume I provides overviews of current information on molecular determinants of viral pathogenicity, virus entry and cell tropism, pandemic risk assessment, transmission and pathogenesis in animal species, viral evolution, ecology, and antigenic variation. Leading investigators who provide current information on these topics include Hans-Dieter Klenk and colleagues on the influenza hemagglutinin, Guelsah Gabriel and Ervin Fodor on the polymerase complex, David Steinhauer and John McCauley on receptor binding, and Charles Russell on membrane fusion activity of the hemagglutinin protein. Nancy Cox and colleagues describe the influenza risk assessment tool (IRAT) to evaluate pandemic potential of novel influenza viruses. Yoshi Kawaoka, Ron Fouchier, Anice Lowen, John Steel, Hualan Chen, Juergen Richt, and their colleagues provide overviews of studies on viral transmission in animal hosts. Monique Franca, Jacqueline Katz, Ian York, Terrence Tumpey, Amy Vincent and colleagues review studies on pathogenesis in avian and mammalian hosts. Amber Smith and Jon McCullers describe the significance of secondary bacterial infection in viral pathogenesis. Viral ecology, evolution, and antigenic variation are discussed in chapters by Sun-Woo Yoon, Robert Webster, Richard Webby, Anice Lowen, John Steel, and Ruben Donis.

The second volume in this series, Volume II is concerned with innate immunity and adaptive immunity, vaccines, and antivirals. Experts in these various areas including John Teijaro, Adolfo Garcia-Sastre, Bali Pulendran, Stacey Shultz-Cherry, Paul Thomas, and colleagues have contributed chapters on specific aspects of innate immunity, while Tom Braciale, Rafi Ahmed, Donna Farber, and colleagues describe B and T cell adaptive immunity. Chapters on vaccines and vaccination include those contributed by Rino Rappuoli, Ian Wilson, Peter Palese, John Steel, Kanta Subbarao, Daniel Perez, Philip Dormitzer, Hongquan Wan, Maryna Eichelberger, Hiroshi Kida, Hideki Hasegawa, Richard Compans, Ioanna Skountzou, and their colleagues. Lastly, Ralph Tripp and S. Mark Tompkins discuss new antiviral discoveries, while John Teijaro writes about use of sphingosine-1-phosphate receptor 1 agonist to control the resultant cytokine storm caused by influenza virus infection.

La Jolla, USA — Michael B. A. Oldstone
Atlanta, USA — Richard W. Compans

Contents

Part I
Molecular Determinants of Pathogenicity

The Hemagglutinin: A Determinant of Pathogenicity

Eva Böttcher-Friebertshäuser, Wolfgang Garten, Mikhail Matrosovich and Hans Dieter Klenk

Abstract The hemagglutinin (HA) is a prime determinant of the pathogenicity of influenza A viruses. It initiates infection by binding to cell surface receptors and by inducing membrane fusion. The fusion capacity of HA depends on cleavage activation by host proteases, and it has long been known that highly pathogenic avian influenza viruses displaying a multibasic cleavage site differ in protease sensitivity from low pathogenic avian and mammalian influenza viruses with a monobasic cleavage site. Evidence is increasing that there are also variations in proteolytic activation among the viruses with a monobasic cleavage site, and several proteases have been identified recently that activate these viruses in a natural setting. Differences in protease sensitivity of HA and in tissue specificity of the enzymes are important determinants for virus tropism in the respiratory tract and for systemic spread of infection. Protease inhibitors that interfere with cleavage activation have the potential to be used for antiviral therapy and attenuated viruses have been generated by mutation of the cleavage site that can be used for the development of inactivated and live vaccines. It has long been known that human and avian influenza viruses differ in their specificity for sialic acid-containing cell receptors, and it is now clear that human tissues contain also receptors for avian viruses. Differences in receptor-binding specificity of seasonal and zoonotic viruses and differential expression of receptors for these viruses in the human respiratory tract account, at least partially, for the severity of disease. Receptor binding and fusion activation are modulated by HA glycosylation, and interaction of the glycans of HA with cellular lectins also affects virus infectivity. Interestingly, some of the mechanisms underlying pathogenicity are determinants of host range and transmissibility, as well.

E. Böttcher-Friebertshäuser · W. Garten · M. Matrosovich · H.D. Klenk (✉)
Institut für Virologie, Philipps-Universität, Hans-Meerwein-Str. 2,
35043 Marburg, Germany
e-mail: klenk@staff.uni-marburg.de

Current Topics in Microbiology and Immunology (2014) 385: 3–34
DOI: 10.1007/82_2014_384

Published Online: 17 July 2014

Contents

1 Introduction

It is generally agreed that most influenza A viruses originate from a large virus pool indigenous to aquatic birds. Since most of these agents cause no or only mild disease they are called low pathogenic avian influenza viruses (LPAIV). Because the host barrier is not an insurmountable obstacle for these viruses, they can occasionally be transmitted to terrestrial birds and mammals (Webster et al. 1992). Most of these transmissions are only transient. On rare occasions, however, the viruses adapt to a new species and give rise to a new lineage. Adaptation is facilitated by the high mutation rate of the viral polymerase and the segmented viral genome that allows reassortment after coinfection with another virus. Transmission and adaptation of LPAIV of subtypes H5 and H7 to domestic poultry may lead to the generation of highly pathogenic avian influenza viruses (HPAIV) (Roehm et al. 1995; Campitelli et al. 2004). Introduction and circulation of avian viruses in pigs may result in mutations allowing efficient growth in man. Moreover, the pig may serve as a mixing vessel in which viruses of different origin reassort to give rise to a new virus. If viruses with new surface antigens generated by these mechanisms are introduced into the human population they may cause a pandemic as has been the case in 1918, 1957, 1968, and 2009. So far, only influenza A viruses of subtypes H1N1, H2N2, and H3N2 have been observed to cause pandemics; influenza B viruses that co-cycle with influenza A viruses are not pandemic.

LPAIV have sporadically infected humans with mostly mild disease symptoms (Peiris 2009; Wei et al. 2013). LPAIV H7N9 that emerged recently (Gao et al. 2013) is an exception, since most of the human infections (ca. 400 between March 2013 and March 2014) have been severe with a high case-fatality rate. HPAIV have also been transmitted to humans, and the H5N1 infections, in particular, cause severe pulmonary disease with occasional dissemination to other organs (Peiris 2009). Although a few human H7N7 HPAIV infections have been reported before (Kurtz et al. 1996), there has been only one poultry outbreak in 2003 with a larger number of human infections, one of which was fatal (Koopmans et al. 2003; van Kolfschooten 2003; Fouchier et al. 2004).

Human influenza is a highly contagious disease that spreads by airborne transmission. Infection is usually confined to the epithelia of the respiratory tract. Severe cases typically develop primary influenza pneumonia or combined viral–bacterial pneumonia. Dissemination into other organs is rare (Kuiken and Taubenberger 2008). Infection of birds by avian influenza viruses is different. LPAIV cause local infections mostly confined to the intestinal tract. Virus is shed in the feces and transmitted through contaminated water. In contrast, HPAIV cause systemic infection. Virus can be isolated from many organs, and widely spread hemorrhages, edema, and cutaneous ischemia are major symptoms of the disease (Pantin-Jackwood and Swayne 2009).

Pathogenesis is the result of the complex interactions that the virus undergoes when it uses the biosynthetic machinery of the cell for replication and when it is exposed to the defence mechanisms of the host (Klenk et al. 2013). All viral proteins are involved in this interplay, but particularly prominent roles are played by the polymerase and the NS1 protein described in other chapters of these volumes and by the hemagglutinin (HA).

HA, the major viral surface glycoprotein, is inserted in the viral envelope as a type I membrane protein. More than 30,000 nucleotide sequences of the 18 HA subtypes identified to date have been determined. The three-dimensional structure of several serotypes has been analyzed by X-ray crystallography: H1 (Russell et al. 2004; Stevens et al. 2004), H2 (Liu et al. 2009), H3 (Wilson et al. 1981; Chen et al. 1998), H5 (Ha et al. 2002; Stevens et al. 2006), H7 (Russell et al. 2004), H9 (Ha et al. 2002) , H13 (Lu et al. 2013), H16 (Lu et al. 2012), H17 (Zhu et al. 2013), and H18 (Tong et al. 2013).

HA initiates infection by binding to cell surface receptors and by inducing membrane fusion. Receptor specificity and proteolytic activation of fusion capacity are major determinants of tissue tropism and virus dissemination in the organism. Receptor binding and fusion activation are modulated by HA glycosylation, and evidence is increasing that the carbohydrate side chains may also be directly involved in specific interactions with cell surface-bound and soluble lectins. In the following, the roles of these mechanisms in pathogenesis will be discussed.

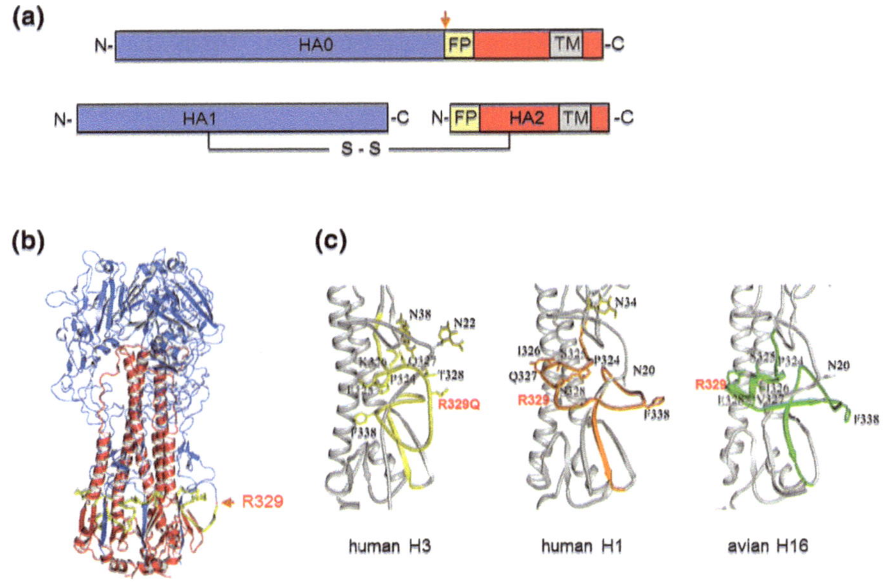

Fig. 1 Cleavage of HA0 into HA1 and HA2 at specific cleavage sites. **a** Schematic illustration of the HA0 precursor and the cleaved form consisting of the disulfide-linked subunits HA1 and HA2. The cleavage site is indicated by an *arrow*. *FP* Fusion peptide. *TM* transmembrane domain. **b** Structure of the trimeric H3 HA0 shown as a ribbon diagram (Galloway et al. 2013). The cleavage site (R329) is located in a surface loop and highlighted by an *arrow*. **c** The structures of noncleaved HA0 have been solved for subtypes H1, H3, and H16. Structural analysis of the H3 precursor (R329Q cleavage site mutant) demonstrated that the HA cleavage site is located in a prominent loop that protrudes from the surface (Chen et al. 1998). The cleavage site loop of HA0 of the 1918 pandemic H1N1 virus is less exposed and abuts the surface of the molecule (Stevens et al. 2004). Interestingly, structural analysis of noncleaved H16 revealed that the cleavage loop has an α-helix structure that hides the arginine residue (Lu et al. 2012). H16 has been shown to be resistant to cleavage by trypsin, but can be activated by TMPRSS2 in vitro (Lu et al. 2012; Galloway et al. 2013)

2 Proteolytic Cleavage of HA Is a Prerequisite for Membrane Fusion

Activation of HA by host proteases has been recognized nearly 40 years ago as a requirement for influenza virus infectivity (Klenk et al. 1975; Lazarowitz and Choppin 1975) and as a prime determinant of avian influenza virus pathogenicity (Bosch et al. 1979). Cleavage of the precursor protein HA0 into the subunits HA1 and HA2 occurs at a conserved arginine-glycine bond (R↓G) and activates fusion capacity (Huang et al. 1981; Maeda et al. 1981). The cleavage site is located in a loop that protrudes from the surface of HA. The amino acid sequence and the conformation of the loop vary with different viruses (Fig. 1, Table 1), and these variations determine cleavage of HA by different proteases (Klenk and Garten

Table 1 Hemagglutinin cleavage site variants of avian influenza A viruses

Virus strain	HA1 aa insertions		HA2	Pathogenicity
A/duck/Alberta/35/1976 (H1N1)	NVP---------SIQS**R**	↓	GLF	Low
A/chicken/Potsdam/4705/1984 (H2N2)	NVP---------QIES**R**	↓	GLF	Low
A/mallard/Alaska/715/2005 (H3N8)	NVP---------EKQT**R**	↓	GLF	Low
A/chicken/Germany/N/1949 (H10N7)	NVP--------EVVQG**R**	↓	GLF	Low
A/turkey/Wisconsin/1/1966 (H9N2)	NVP---------AVSS**R**	↓	GLF	Low
A/quail/Shantou/782/2000 (H9N2)	NVP---------A**R**SS**R**	↓	GLF	Low[a]
A/quail/Shantou/2061/2000 (H9N2)	NVP---------A**R**S**RR**	↓	GLF	Low[a]
A/chicken/Ibaraki/1/2005 (H5N2)	NVP---------QRET**R**	↓	GLF	Low
A/ostrich/SA/AI2114/2011 (H5N2)	NVP--------QR**RKKR**	↓	GLF	Low
A/chicken/HongKong/220/1997 (H5N1)	NTP-----QRERR**RKKR**	↓	GLF	High
A/chicken/FPV/Rostock/1934 (H7N1)	NVP-----EPSKK**R**E**KR**	↓	GLF	High
A/ostrich/Italy/984/2000 (H7N1)	NVP----EIPKGS**R**V**RR**	↓	GLF	High
A/chicken/Jalisco/CPA1/2012 (H7N3)	NVPENPKDRKSRH**RR**T**R**	↓	GLF	High

Amino acid sequences at the HA0 cleavage sites were aligned. The cleavage site between HA1 and HA2 is indicated by a downward arrow. Basic amino acids crucial for HA cleavage by relevant proteases are shown in boldface

[a] HA cleavage by matriptase may enhance the virulence of these viruses

1994; Steinhauer 1999). Cleavage of HA exposes a highly conserved fusion peptide consisting of hydrophobic amino acids at the N-terminus of the HA2 subunit. The high structural specificity of the fusion peptide is also indicated by the observation that the protease thermolysin cleaves HA0 C-terminal to the glycine residue (R-G↓L), resulting in fusion-incompetent HA and noninfectious virus (Garten et al. 1981). HA cleavage is a prerequisite for conformational changes at low pH that trigger fusion of the viral and endosomal membranes (Skehel and Wiley 2000) in order to release the viral RNP segments into the cytosol and further in the cell nucleus where transcription and replication of the viral genome takes place (Fig. 2).

2.1 *Cleavage Activation of HA: The Prime Pathogenicity Determinant of Avian Influenza Viruses*

LPAIV possess a single arginine (R) or rarely a single lysine (K) at the cleavage site and are activated by trypsin in vitro. The available evidence indicates that relevant trypsin-like proteases are restricted to the respiratory and the intestinal

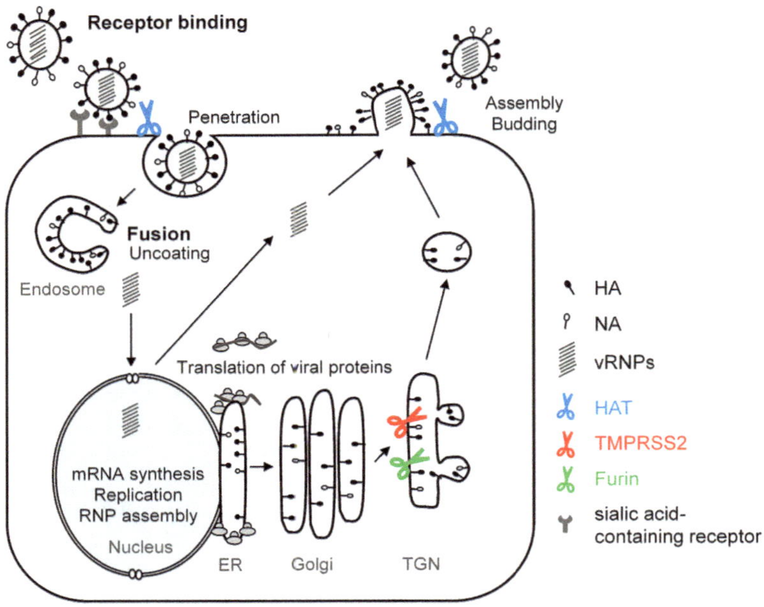

Fig. 2 The role of HA in influenza virus replication and the compartmentalization of the activating proteases. HA mediates binding to sialic acid-containing cell surface receptors and fusion of the viral and endosomal membrane following endocytosis at low pH in order to release the viral ribonucleoprotein complexes (RNPs) into the cell (uncoating). The vRNPs are imported into the nucleus, where transcription and replication occur. Translation of viral mRNAs is performed by the cellular machinery. HA, neuraminidase (NA), and the ion channel protein M2 are transported along the constitutive secretory pathway to the plasma membrane. New vRNPs are assembled in the nucleus and subsequently transported to the plasma membrane, too, where self-assembly of viral proteins leads to budding of new virions. NA cleaves sialic acid from carbohydrate moieties, facilitating release of progeny virus. HA is synthesized as fusion-incompetent precursor that requires posttranslational cleavage by host proteases. HA cleavage occurs by membrane-bound proteases and can take place in different compartments and at different time points during the viral life cycle and is indicated by scissors. HA with multibasic cleavage site is cleaved by furin in the trans Golgi-network (TGN). HA containing a monobasic cleavage site is cleaved by TMPRSS2 in the TGN or by HAT on the plasma membrane, either during assembly and budding of virions or during attachment and entry into the cell

tract of birds. LPAIV infection is therefore confined to these organs (Garten and Klenk 2008). A protease homologous to blood clotting factor Xa was identified activating LPAIV HA in embryonated chicken eggs (Gotoh et al. 1990), but the identity of HA-activating proteases in the intestinal and respiratory tracts of avian species is unknown.

HA of HPAIV is cleaved at the C-terminus of the multibasic consensus motif R-X-R/K-R by the ubiquitous eukaryotic subtilase furin or the closely related proprotein convertase 5/6 (PC5/6) (Stieneke-Gröber et al. 1992; Vey et al. 1992; Horimoto et al. 1994; Feldmann et al. 2000). Furin is a calcium-dependent serine endoprotease that activates pro-proteins and pro-hormones, such as pro-insulin receptor, pro-albumin and von Willebrand factor, at neutral pH during their transport along the secretory pathway to the plasma membrane (Seidah et al. 2008). Moreover, furin activates also the fusion proteins of a large number of other enveloped viruses (e.g., retroviruses including HIV, paramyxoviruses, flaviviruses), as well as bacterial toxins (e.g., anthrax toxin, botulinum toxin) at multibasic motifs (Klenk and Garten 1994; Thomas 2002). Cleavage of HA of HPAIV by furin takes place in the TGN (Fig. 2). The ubiquitous expression of furin supports proteolytic activation of HA in multiple organs and tissues, causing systemic infection and fatal disease.

It has been demonstrated that HPAIV emerge from LPAIV by insertion of the R-X-R/K-R motif into the cleavage site loop (Table 1). Why acquisition of a multibasic cleavage site susceptible to furin is confined to subtypes H5 and H7 is not known. Different mechanisms underlying the insertion of amino acids at the HA cleavage site have been observed: (1) successive addition of basic amino acids due to polymerase slippage at purine-rich arginine or lysine codons (Horimoto et al. 1995; Garcia et al. 1996; Perdue et al. 1997), (2) recombination of the HA gene with other viral gene segments (Orlich et al. 1994; Suarez et al. 2004; Pasick et al. 2005), or (3) recombination of the HA gene with ribosomal RNAs (Khatchikian et al. 1989; Maurer-Stroh et al. 2013). There are also examples where the insertion mechanism is unknown (Banks et al. 2001).

Introduction of a multibasic cleavage site in vitro conferred high cleavability to vector-expressed H3 HA (Ohuchi et al. 1991), but it did not lead to systemic infection and increased pathogenicity of a H3N2 virus in ferrets (Schrauwen et al. 2011). Furthermore, replacement of a monobasic HA cleavage site by a multibasic motif does not automatically confer high pathogenicity to a LPAIV, demonstrating that the emergence of a HPAIV is a multifactorial process (Stech et al. 2009). On the other hand, H5N1 reassortants with modified H2, H4, H8, and H14 HAs containing multibasic cleavage sites caused lethal infections in chickens (Veits et al. 2012). These findings demonstrated that a multibasic HA cleavage site can confer high pathogenicity to subtypes other than H5 and H7 in a suitable genetic background.

2.2 Proteolytic Activation of Mammalian influenza viruses

Like LPAIV, mammalian influenza viruses possess a single arginine at the cleavage site. Relevant host proteases show a restricted tissue distribution, and propagation in cell culture usually requires addition of exogenous trypsin. A number of trypsin-like proteases was isolated from rat and swine lung, including tryptase Clara, mini-plasmin and tryptase TC30, and shown to support proteolytic activation of influenza viruses with a monobasic HA cleavage site in vitro (Kido et al. 1992, 2007; Murakami et al. 2001; Sato et al. 2003). Because the genetic identity of these proteases is still unknown, it remains unclear whether they play a role in influenza virus infection in vivo. Moreover, plasmin, urokinase, or tissue kallikreins also have been reported to cleave HA with a monobasic cleavage site in vitro, but their role in in vivo infection is not clear, either (Lazarowitz and Choppin 1975; Hamilton and Whittaker 2013). Host proteases that activate influenza virus HA with a monobasic cleavage site in the human airways were unknown for a long time.

Cleavage of HA by soluble proteases such as tryptase Clara or by exogenous trypsin in cell culture occurs extracellularly during assembly and budding when HA is present on the plasma membrane or after virus is released from the infected cell. Hence, it was believed that HA with a monobasic cleavage site is activated extracellularly and, therefore, differs from HA of HPAIV, which is cleaved in the TGN by furin. However, in human airway epithelial cells and in human intestinal Caco-2 cells activation of HA with a monobasic cleavage site was shown to take place intracellularly, but the proteases remained unknown in these studies (Zhirnov et al. 2002, 2003).

In 2006, the proteases TMPRSS2 (transmembrane protease serine S1 member 2; also designated as epitheliasin) and HAT (human airway trypsin-like protease; also designated as TMPRSS11D) were identified as proteases present in the human airways that activate influenza virus HA with a monobasic cleavage site in vitro (Böttcher et al. 2006). Later on, the TMPRSS2-related protease TMPRSS4 was shown to activate HA with a monobasic cleavage site in vitro, too (Chaipan et al. 2009). TMPRSS2- or HAT-homologous proteases in swine and mouse have been identified and were shown to be capable of activating HA at a single arginine, suggesting that homologous proteases are involved in HA cleavage in different host species (Bertram et al. 2012; Peitsch et al. 2014; Tarnow et al. 2014).

TMPRSS2 and HAT belong to the family of type II transmembrane serine proteases (TTSPs) (Szabo and Bugge 2008). HA cleavage by TMPRSS2 and HAT takes place in different cellular compartments and at different time points during the viral life cycle (Fig. 2). HAT localizes to the plasma membrane as an active enzyme that can cleave newly synthesized HA0, probably during assembly and budding of progeny virus, and activates incoming virus at the stage of entry upon attachment to the cell (Böttcher-Friebertshäuser et al. 2010). In contrast, HA cleavage by TMPRSS2 takes place within the cell (Böttcher-Friebertshäuser et al. 2010). TMPRSS2 expressed at the cell surface does not activate incoming virus. Recent studies demonstrated that the protease accumulates in the TGN where it co-localizes with furin, suggesting that HA cleavage by TMPRSS2 and furin occurs in the same cellular compartment (Stieneke-Gröber et al. 1992; Schäfer et al. 1995; Böttcher-Friebertshäuser et al. 2013; Peitsch et al. 2014).

The catalytic domain of TTSPs may also be shed from the cell surface and soluble forms of TMPRSS2 and HAT have been described (Yasuoka et al. 1997; Afar et al. 2001). However, the soluble forms did not support proteolytic activation of influenza viruses, supporting the concept that HA cleavage in the airways occurs by cell-associated proteases (Böttcher-Friebertshäuser et al. 2010). Enhanced shedding of TMPRSS2 and HAT from differentiated human nasal epithelial cells and hence enhanced influenza virus replication has been observed upon exposure to ozone (Kesic et al. 2012). Thus, enhanced protease shedding under stress might play a role in influenza virus infection, but further investigations are needed to prove this hypothesis.

2.3 Variations in Cleavage Activation of LPAIV and Mammalian Influenza A Viruses

It has long been known that HAs with monobasic cleavage sites differ in protease sensitivity. Thus, A/WSN/33 (H1N1) HA is activated by plasmin (Lazarowitz et al. 1973), whereas A/chicken/Germany/49 (H10N7) HA depends on trypsin (Klenk et al. 1975). Furthermore, comparison of the sensitivity of 16 HA subtypes to TMPRSS2 and HAT as well as pancreatic trypsin in vitro demonstrated also that cleavage efficiency can vary significantly among the different subtypes (Galloway et al. 2013). The mechanisms underlying protease specificity of HA are still unknown and may be related to the structure and exposure of the cleavage site loop (Fig. 1) or steric hindrance by adjacent carbohydrate moieties (Kawaoka et al. 1984).

Although the monobasic HA cleavage site of the mammalian viruses is a major determinant for the confinement of infection to the respiratory tract, there are examples where the protease specificity promotes the spread of such viruses to other organs. A/WSN/33 H1N1 virus contains the unusual H1 HA cleavage site I-Q-Y-R instead of I-Q-S-R (cf. Table 1) that facilitates efficient cleavage by plasmin and has been shown to contribute to the neurotropism of the virus in mice (Goto and Kawaoka 1998; Sun et al. 2010). Another example are H9N2 viruses. Within the last years these viruses have attracted particular attention since they have become highly prevalent in poultry in many countries and are occasionally transmitted to humans and pigs, causing mild disease. Moreover, H9N2 viruses show a high genetic compatibility with other subtypes and have been involved in the emergence of H5N1 and H7N9 viruses (To et al. 2013). When compared to other subtypes, H9N2 viruses vary remarkably in their amino acid sequence at the HA cleavage site, and many isolates from Asia and the Middle East possess uncommon di- or tribasic cleavage site motifs R-S-S-R and R-S-R-R, respectively. In contrast to HPAIV of subtypes H5 and H7, the cleavage sites of H9N2 viruses evolved by substitution and not by insertion of basic amino acids (Table 1). They are not susceptible to cleavage by furin, but are activated by TMPRSS2 and HAT in vitro (Gohrbandt et al. 2011; Soda et al. 2011; Baron et al. 2013). Recent studies suggest that insertion of further amino acids to extend the cleavage loop or removal of a carbohydrate moiety in HA1 can render H9 susceptible to furin (Gohrbandt et al. 2011; Soda et al. 2011; Tse et al. 2014). Interestingly, the type II transmembrane protease matriptase was shown to cleave H9 at R-S-S-R and R-S-R-R motifs, too. Matriptase is expressed in epithelial cells of many tissues (Oberst et al. 2003) and has been suggested to contribute to H9N2 virus replication in the kidney of chickens (Baron et al. 2013). Whether matriptase is also involved in cleavage of other HA subtypes is under discussion (Baron et al. 2013; Beaulieu et al. 2013; Hamilton and Whittaker 2013).

Interestingly, bacterial proteases secreted by some strains of *Staphylococcus aureus* or *Aerococcus viridans* have also been shown to support proteolytic activation of influenza viruses with monobasic cleavage site in vitro. Coinfection of mice resulted in lethal infection with increased virus titers and extended lesions in the lung (Tashiro et al. 1987a, b; Scheiblauer et al. 1992). The susceptibility of HA to bacterial proteases also differed considerably and was specific for both virus and bacteria strains.

2.4 HA Cleavage Is Essential for Pneumotropism and Pathogenicity in Mammalian Hosts

Three recent studies demonstrated that knockout mice that lack expression of the protease TMPRSS2 are protected from pulmonary disease with lethal outcome when infected with influenza viruses of subtypes H7N9 and H1N1 (Hatesuer et al. 2013; Sakai et al. 2014; Tarnow et al. 2014). Intriguingly, H7N9 and H1N1 viruses were apathogenic in TMPRSS2-deficient mice, whereas wild-type mice developed severe infection with 100 and 20 % mortality, respectively. Virus growth kinetics in explants of murine respiratory tissues and in infected mice demonstrated that knockout of TMPRSS2 prevents spread of H7N9 and H1N1 viruses into the lung and consequently development of disease in mice (Tarnow et al. 2014) (Fig. 3a). In contrast, H3N2 virus replication in ex vivo airway models was only marginally affected by knockout of TMPRSS2 and both TMPRSS2-deficient and wild-type mice succumbed to infection. Analysis of the distribution of TMPRSS2 and HAT in the airways demonstrated that both proteases are present in the larynx, trachea and bronchi of mice, but only TMPRSS2 is expressed in the lung (Szabo and Bugge 2008; Sales et al. 2011; Tarnow et al. 2014) (Fig. 3a). Thus, HAT may support virus activation in upper airways and the tracheobronchial epithelium, but expression of TMPRSS2 is crucial for virus spread into the lung. Since, TMPRSS2-knockout mice lack expression of both TMPRSS2 and HAT in the lungs, it appears that multicycle replication of H3N2 virus in murine lung is due to another, not yet identified protease. Taken together, these studies demonstrated for the first time that expression of an appropriate HA-activating protease along the respiratory tract is essential for pneumotropism and pathogenicity of influenza viruses in mammals. Therefore, distribution of both receptors (see below) and virus-activating proteases is crucial for spread of influenza virus in the airways. The protease TMPRSS2 has been identified as a single host cell factor essential for H7N9 and H1N1 influenza virus pathogenicity in mice. It remains to be investigated what contributes to the differences in the protease specificity of H7N9 and H1N1 viruses, on one hand, and of H3N2 virus, on the other hand. Analysis of HAT and TMPRSS2 expression in the respiratory tracts of humans and swine suggests that protease distribution is similar to that in mice, indicating that also in man and pigs TMPRSS2 supports proteolytic activation of HA along the respiratory tract and virus spread into the lung, whereas HA cleavage by HAT seem to be confined to the upper airways,

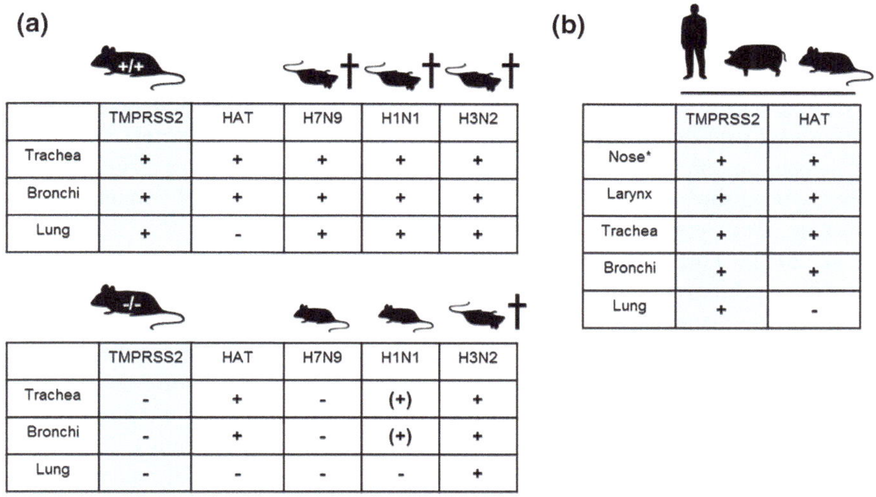

(a)

+/+

	TMPRSS2	HAT	H7N9	H1N1	H3N2
Trachea	+	+	+	+	+
Bronchi	+	+	+	+	+
Lung	+	-	+	+	+

-/-

	TMPRSS2	HAT	H7N9	H1N1	H3N2
Trachea	-	+	-	(+)	+
Bronchi	-	+	-	(+)	+
Lung	-	-	-	-	+

(b)

	TMPRSS2	HAT
Nose*	+	+
Larynx	+	+
Trachea	+	+
Bronchi	+	+
Lung	+	-

Fig. 3 Role of HA cleavage in pneumotropism and pathogenicity in mice. **a** TMPRSS2 is essential for pneumotropism and pathogenicity of H7N9 and H1N1 virus in mice. Replication and pathogenicity of H7N9, H1N1 and H3N2 influenza viruses, respectively, in TMPRSS2-knockout mice (*TMPRSS2*$^{-/-}$) and wild-type littermates (*TMPRSS2*$^{+/+}$) (Tarnow et al. 2014). Knockout of TMPRSS2 inhibits multicycle replication of H7N9 in murine respiratory tissues and prevents spread of H1N1 virus into the lung and thereby protects mice from pulmonary disease with lethal outcome. Low levels of H1N1 replication in trachea and bronchi of *TMPRSS2*$^{-/-}$ mice may be due to HA activation by HAT. In contrast, replication and pathogenicity of H3N2 virus is only marginally affected by knockout of TMPRSS2 and both wild-type and *TMPRSS2*$^{-/-}$ mice succumb to lethal infection. HAT might support multicycle replication of H3N2 virus in trachea and bronchi, but replication of H3N2 virus in the lung depends on a so far unknown protease. Presence (+) and absence (−) of protease expression and multicycle virus replication, respectively. Mice indicate apathogenic infection and lethal outcome of infection, respectively. **b** Expression of TMPRSS2 or HAT (+) and absence of expression (−) in respiratory tissues of human, swine or mouse. *Expression of HAT and TMPRSS2 in nasal epithelial cells of mice has not been determined

trachea and bronchi (Szabo and Bugge 2008; Sales et al. 2011; Bertram et al. 2012; Peitsch et al. 2014). Interestingly, TMPRSS2, but not HAT, has also been detected in human myocytes, indicating that the protease might contribute to influenza-associated myocarditis (Bertram et al. 2012).

2.5 *Protease Inhibitors as Drugs for Influenza Treatment*

It has long been known that inhibition of HA cleavage by aprotinin, a natural protease inhibitor from bovine lung, suppresses influenza virus replication and spread in mice and reduces symptoms of disease. Inhalation of aerosolized aprotinin by influenza patients markedly reduced the duration of symptoms without causing side effects (reviewed in Zhirnov et al. 2011). The resistance of TMPRSS2-knockout mice to pulmonary disease upon H7N9 or H1N1 virus infection identified TMPRSS2 as a potential target for drug development. Inhibition of TMPRSS2 activity by peptidomimetics or downregulation of TMPRSS2 expression by antisense peptide-conjugated phosphorodiamidate morpholino oligomers (PPMO) have already been shown to efficiently inhibit multicycle replication of human influenza virus in airway epithelial cells (Böttcher-Friebertshäuser et al. 2011, 2012; Meyer et al. 2013). Furthermore, the combination of protease inhibitors and the NA inhibitor oseltamivir carboxylate efficiently blocked influenza virus propagation in a synergistic manner.

Targeting host cell factors instead of viral proteins has the advantage to reduce or even prevent the emergence of drug resistant viruses. A major concern in targeting a host factor is side effects or toxicity due to inhibition of the physiological functions. The physiological roles of TMPRSS2 and HAT are still unknown, but mice deficient in expression of either protease are phenotypically asymptomatic (Kim et al. 2006; Sales et al. 2011), indicating functional redundancy or compensation of the function by other host proteases. Thus, inhibition of TMPRSS2 or HAT during an acute influenza infection seems feasible and the further development of above mentioned inhibitors may lead to novel drugs for influenza chemotherapy. Inhibition of HA cleavage may provide a promising approach also for the treatment of HPAIV infections in humans. Specific inhibitors of furin have been demonstrated to strongly suppress replication of HPAIV of subtype H7N1 in cell culture (Garten et al. 1989; Becker et al. 2010). The first approach for suppression of virus replication by furin inhibitors was performed with peptidomimetics using decanoylated basic tetrapeptide derivates, such as decRVKR chloromethylketone, that imitate the furin recognition motif (Hallenberger et al. 1992; Stieneke-Gröber et al. 1992; Garten et al. 1994). Recently, furin peptidomimetica with improved stability and efficacy have been designed (Becker et al. 2010, 2012).

2.6 The Use of Protease Activation Mutants for Vaccine Design

Modification of the HA cleavage site has been shown to provide a useful tool for vaccine development. Thus, replacement of the multibasic HA cleavage site of H5N1 and H7N1 viruses by a single arginine led to low pathogenic seed viruses that can be used for the generation of inactivated vaccines under biosafety-level-2 (BSL2) conditions (Webby et al. 2004).

Replacement of the conserved arginine at the HA cleavage site by an elastase motif (V↓G) has been demonstrated to attenuate influenza viruses in mice and to provide a promising strategy for the development of live vaccines (Stech et al. 2005). Such elastase-dependent mutant viruses can be efficiently propagated in the presence of elastase in vitro, but undergo restricted virus replication and spread in vivo due to missing HA activation. Generation of elastase-dependent virus variants of both influenza A virus [A/WSN/33 (H1N1)] and influenza B virus (B/Lee/40) has been shown to result in highly attenuated viruses that elicit protective immunity against lethal challenge with the respective wild-type virus after intranasal administration as live vaccine (Stech et al. 2005, 2011). Furthermore, when the multibasic HA cleavage site of the SC35M (H7N7) strain, which is highly pathogenic for mice, was replaced by an elastase motif, an attenuated mutant was obtained that was immunogenic in mice and protected the animals from lethal challenge (Gabriel et al. 2008). Thus, modification of the HA cleavage site may also be a promising strategy for the development of HPAIV live vaccines.

3 Receptor Specificity of HA

As pointed out in the introduction, human influenza viruses typically replicate in the epithelium of the upper respiratory tract (URT), trachea, and bronchi and cause transient uncomplicated rhinitis, pharyngitis, and tracheobronchitis. A major complication, infection of terminal bronchioles and alveoli, results in a severe, occasionally fatal, pneumonia (for review, see Kuiken and Taubenberger 2008; Taubenberger and Morens 2008). Seasonal human influenza viruses rarely cause severe disease, whereas it occurs unusually frequently in the case of zoonotic infections with H5N1 and H7N9 avian viruses (Beigel et al. 2005; Liu et al. 2013). These distinctive patterns of infection may depend, at least partially, on the differences in receptor-binding specificity of seasonal and zoonotic viruses and differential expression of receptors for these viruses in human target tissues.

3.1 Influenza Viruses Attach to Sia2-3Gal- and Sia2-6Gal-Containing Receptors

Influenza viruses attach to cells via HA interactions with sialic acids (Sia) expressed at the terminal positions of carbohydrate chains of cell surface glycoproteins and glycolipids. The Sia-binding site is a shallow pocket located in the membrane-distal part of HA (Fig. 4). Individual interactions of HA monomers with the sialyloligosaccharide moieties of the receptors are very weak (Kdiss $\sim$ 0.1 mM), and cooperative binding of multiple HA spikes to multiple copies of the receptor provides for the efficient virus attachment to cells. Attachment therefore depends on (i) HA affinity for the terminal Sia-containing receptor moiety, (ii) HA interactions with the penultimate sugar of the oligosaccharide chain and the

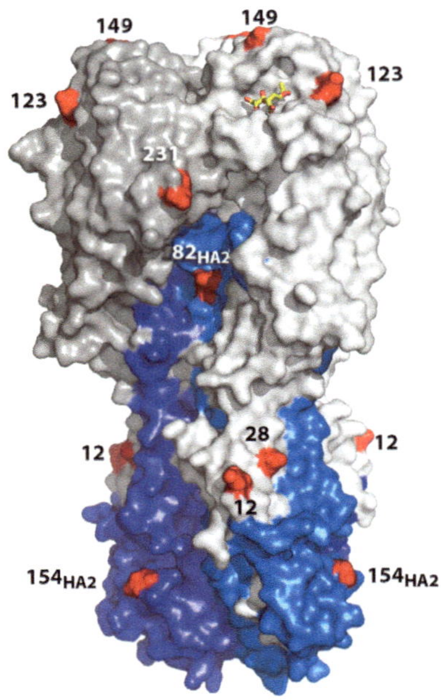

Fig. 4 Attachment sites and functions of the oligosaccharides of the H7 hemagglutinin. The structure of the H7 trimer is shown. HA1 (*gray*) with N-acetylneuraminic acid (*yellow*) attached to the receptor-binding site, HA2 (*blue*), and the asparagine residues at the glycosylation sites (*red*) are indicated. The oligosaccharide at position HA2-82 is of the oligomannosidic type, all other glycosylation sites contain complex or intermediate type glycans. The glycans at position 12, 28, and HA2-154 are conserved and essential for intracellular transport and stability of HA. The glycans at positions 123 and 149 vary with different H7 strains and modulate receptor binding. For references see text

protein/lipid part of the receptor, and (iii) ability of the virus particle to establish multivalent interactions with clusters of receptor molecules expressed on the cell surface. Thus, the receptor-binding properties of influenza viruses can be affected by amino acid substitutions inside the Sia-binding pocket, on the pocket rim, and by more distant mutations resulting in altered glycosylation or electrostatic charge of the HA globular head (for reviews, see Matrosovich et al. 2006, 2008; Gamblin and Skehel 2010; Viswanathan et al. 2010).

Sia species differ from each other by modifications at the amino group (N5) and at four hydroxyl groups (O4, O7, O8, and O9). Influenza A and B viruses typically bind to the most common species, N-acetylneuraminic acid (Neu5Ac), although some viruses can in addition bind to N-glycolylneuraminic acid (Neu5Gc) and 4-O-acetylated Sia (for review, see Matrosovich et al. 2006, 2008). Only viruses of the extinct equine H7N7 lineage bind to Neu5Gc much more strongly than to Neu5Ac (Gambaryan et al. 2012). The structure of natural Sia-containing oligosaccharide chains is highly variable and complex (Glycobiology 2009), and effects of their structural differences on influenza virus binding are still not well characterized. The best known effect is the ability of influenza viruses from different host species to discriminate between the two major natural sialylgalactosyl moieties, Sia2-3Gal and Sia2-6Gal. Thus, human and swine viruses preferentially bind to Sia2-6Gal-containing oligosaccharides (often called "6-linked" or "human-type" receptors); avian and equine viruses show the opposite specificity and bind to Sia2-3Gal-containing glycans ("3-linked" or "avian-type" receptors).

3.2 Expression of Receptors and Virus Tissue Tropism in the Human Respiratory Tract

The respiratory tract in humans contains a diverse population of cell types which vary along the respiratory tree (Tomashefski and Farver 2008). Two major cell types of the superficial cells of the air-conducting zone (nasal and tracheobronchial epithelium) are ciliated cells and nonciliated secretory cells. Secretory cells, together with the cells of sub-epithelial glands, secret mucins, and other molecules that become incorporated in the airway surface liquid and mucous blanket. Ciliary beating of ciliated cells moves the blanket toward the pharynx where mucus is swallowed together with trapped particles and microorganisms. The respiratory zone consists of respiratory bronchioles, alveolar ducts, and alveoli. Alveoli, which are responsible for 90 % of the gas exchange, are lined by type I and type II pneumocytes. These flat type I cells cover 90 % of the alveolar surface. They are the main components of the alveolar wall and the barrier between air and blood. The type II cells occupy around 5 % of the alveolar surface. They produce surfactant, resorb fluid from the alveolar lumen, and serve as reserve cells which differentiate into type I cells in response to alveolar damage.

Paulson and colleagues were the first who studied expression of receptors for influenza viruses in the human respiratory tract using fixed paraffin-embedded tissue sections of human trachea and labeled plant lectins and influenza viruses as Sia-binding histochemical probes (Baum and Paulson 1990; Couceiro et al. 1993). They found that Sia2-6Gal-specific lectin and human influenza virus efficiently bound to the apical surface of the epithelium and that binding of Sia2-3Gal-specific lectin and avian virus were restricted to intracellular mucin-containing secretory granules of goblet cells. The authors suggested that the preferential tropism to Sia2-6Gal-containing receptors, which is shared by various human viruses, can be determined by two simultaneous selective pressures, namely abundant expression of 6-linked Sia on airway epithelial cells and abundant expression of 3-linked Sia on decoy receptors of the mucous blanket.

More recently, several groups used the same methodological approach to study expression of receptors in all regions of the human respiratory epithelium from the nasal epithelium to alveoli (Shinya et al. 2006; van Riel et al. 2006, 2007, 2010, 2013; Nicholls et al. 2007a, b; Yao et al. 2008). Table 2 summarizes the main findings. Lectin binding data demonstrated that 6-linked Sia is present in high

Table 2 Expression of 2-3-linked and 2-6-linked sialic acids and virus attachment studied on histological sections of fixed human respiratory tissues

Tissue	Lectin binding[1, 2]			Virus attachment[3, 4]		
	SNA	MAA-1	MAA-2	Human H1N1, H3N2	H5N1	H7N9
Nasal	++	ND	–	++[a]	–	+/++[a]
Nasopharynx	++	++	–	++[a]	+/–	NT
Trachea	++	ND	–	++[a]	+/–	+[a]
Bronchus	++	++	–	++[a]	+/–[a]	++[a]
Bronchiole	+	+	++	++[a]	+[c]	++[a, c]
Alveolus	+	+	++	+[I]	+[e]	++[d, e]
Macrophages	+	+/++	–	–	+	+

[1] Shinya et al. (2006), Nicholls et al. (2007a), Chan et al. (2013b)
[2] SNA, *Sambucus nigra* agglutinin, preferentially binds to Sia2-6Gal. MAA-1, *Maackia amurensis* leukoagglutinin, preferentially binds to Sia2-3Gal1-4GlcNAc; MAA-2, *Maackia amurensis* hemagglutinin, preferentially binds to Sia2-3Gal1-3GalNAc and Sia2-3Gal1-3[Sia2-6]GalNAc (Geisler and Jarvis 2011)
[3] van Riel et al. (2006, 2007, 2010, 2013)
[4] Predominant cell type to which the virus attaches: [a] ciliated, [b] nonciliated, [c] Clara, [d] type I pneumocyte, [e] type II pneumocyte

amounts on the surface of epithelial cells of the URT, trachea and bronchi, and in lower but significant amounts on the surface of bronchiolar and alveolar cells. Expression of 3-linked Sia was studied using the lectins MAA-1 and MAA-2 which differ in their fine binding specificity. Sia2-3Gal1-3GalNAc moieties, which are typical for O-linked glycans, were only detected in bronchioles and alveoli. By contrast, Sia2-3Gal1-4GlcNAc-containing glycans were detected throughout the whole respiratory tract (Nicholls et al. 2007a; Walther et al. 2013). Unfortunately, lectin staining is not quantitative, precluding direct comparison of the concentrations in the tissues of different sialic acid types. Van Riel and colleagues studied attachment patterns of labeled human and avian influenza viruses to histological sections (van Riel et al. 2006, 2007, 2010, 2013). In their experiments, pandemic and seasonal human viruses of different subtype attached efficiently to the apical surface of the URT and the tracheobronchial epithelium. H5N1 (and a few other avian viruses tested) attached to these tissues weakly if at all. Both human and avian viruses attached to bronchiolar and alveolar epithelial cells, but displayed a different cell tropism. Human viruses preferentially bound to ciliated cells in the bronchioles and type I pneumocytes, whereas H5N1 virus bound to bronchiolar nonciliated Clara cells, type II pneumocytes and alveolar macrophages.

Avian H7N9 viruses responsible for zoonotic infections of humans since March 2013 were found to bind to both 3-linked and 6-linked Sia (Belser et al. 2013; Watanabe et al. 2013; Xiong et al. 2013; Zhou et al. 2013). In accordance with their dual receptor specificity, these viruses displayed attachment patterns of both human and avian viruses (van Riel et al. 2013). Similarly to human viruses, H7N9 viruses attached, albeit less efficiently, to the ciliated airway epithelium and were able to bind to type I cells in alveoli. As other avian viruses, H7N9 viruses displayed relatively strong attachment to bronchiolar Clara cells, type II pneumocytes, and alveolar macrophages.

The observed binding patterns (Table 2) correlate with viral receptor specificity and features of infection in humans. Preference of human influenza viruses for 6-linked Sia, abundant expression of 6-linked Sia and efficient attachment of human viruses to the URT, trachea and bronchi are consistent with preferential replication of human viruses in these tissues. This latter effect correlates with the common presentation of human influenza virus infection as tracheobronchitis (Kuiken and Taubenberger 2008) and seems to be essential for airborne transmission of infection via sneezing and coughing (Sorrell et al. 2011; Imai et al. 2013). Thus, mutagenesis of the receptor-binding site of the HA of pandemic H1N1/1918 and H2N2/57 viruses and human seasonal H3N2 virus to switch binding preference from 6-linked Sia to 3-linked Sia did not abolish infection of ferrets, but prevented transmission (Tumpey et al. 2007; Pappas et al. 2010; Roberts et al. 2011). Also, mutations in the HA that switched receptor preference of genetically engineered H5N1 viruses toward 6-linked Sia were critical to ensure their attachment to the epithelium of nasal turbinates in ferrets and render them transmissible in ferrets (Herfst et al. 2012; Imai et al. 2012). It should be noted, that despite a lack of binding of avian viruses to human airway epithelium in histochemical assays (van Riel et al. 2007, 2010, 2013), the presence of 3-linked Sia in the epithelium was confirmed by lectin staining and glycomic analysis, and various avian virus strains were able to infect and replicate in ex vivo explants of human bronchus (Nicholls et al. 2007a, b; Chan et al. 2013a; Walther et al. 2013). Thus, paucity of attachment sites for avian viruses in the human URT, trachea and bronchi does not prevent infection with high virus doses, but limits efficiency of natural infection to an extent incompatible with airborne transmission.

Relatively high concentrations of 3-linked Sia in bronchioles and alveoli (Shinya et al. 2006; Nicholls et al. 2007a; Chan et al. 2013a, b) and the ability of H5N1, H7N7, and H7N9 viruses to attach to cells in the lower respiratory tract suggested that the receptor-mediated pneumotropism of these viruses contributes to their high pathogenicity in humans. It remains to be determined why human viruses, which also can attach to bronchiolar and alveolar epithelial cells (Table 2) and efficiently replicate in human lung explants, only rarely cause severe disease (Walther et al. 2013).

3.3 *Cell Tropism of Viral Infection as Depending on Receptor Specificity*

Autopsy material obtained from fatal cases of pandemic H1N1, H5N1, and H7N9 influenza has also been analyzed. Viral antigen and/or RNA were typically found in type II pneumocytes, ciliated and nonciliated cells of tracheobronchial epithelium, alveolar macrophages, and several other cell types (Uiprasertkul et al. 2005;

Korteweg and Gu 2010; Zhou et al. 2013). To throw light on cell tropism at the early stages of infection, two useful experimental models were employed in recent years, differentiated cultures of human airway epithelial (HAE) cells and ex vivo cultures of human respiratory tissues removed by biopsy or surgical excision (for review, see Chan et al. 2013b).

Experiments in fully differentiated cultures of human nasal and tracheobronchial epithelial cells showed that early in infection human viruses preferentially infected nonciliated cells, whereas avian viruses mainly infected ciliated cells (Matrosovich et al. 2004). This pattern correlated with the lectin analyses in these cultures, i.e., strong binding of SNA to nonciliated cells and preferential binding of MAA-1 to ciliated cells. The cell tropism was not absolute: for example, human viruses infected all types of cells later in infection, and avian virus infection was not fully restricted to ciliated cells. These findings have been confirmed by other groups (Thompson et al. 2006; Wan and Perez 2007). Interestingly, the H3N2/1968 pandemic virus displayed less prominent tropism to nonciliated cells than seasonal H3N2 and H1N1 human viruses, which correlates with the less strict preference of the 1968 pandemic virus to 6-linked Sia and its weak but significant binding to 3-linked Sia (Matrosovich et al. 2004, 2007; Thompson et al. 2006).

The data on virus infection in human airway epithelial cell cultures suggested for the first time that differences in replication and pathogenicity of human and avian influenza viruses may be partially related to their receptor-mediated differential cell tropism. This concept was also tested in studies on receptor-binding variants of the H1N1/2009 pandemic influenza viruses. These viruses occasionally caused severe and fatal infections, and there was a strong association of the HA mutation D222G with severe disease (Rykkvin et al. 2013). Interestingly, the D222G mutation in the HA was also found in fatal cases from the 1918 influenza pandemic. The D222G mutants of the H1N1/2009 virus were shown to differ from the original human viruses by enhanced binding to 3-linked Sia, enhanced attachment to type II pneumocytes and alveolar macrophages on lung tissue sections and enhanced infection of ciliated cells in HAE cultures (Chutinimitkul et al. 2010; Liu et al. 2010). It is believed therefore that the avian virus-like receptor specificity and cell tropism of the D222G mutants seemed to have contributed to the disease severity.

Based on the differences in the virus attachment patterns in the alveolar tissues (Table 2), it was assumed that human viruses target type I and avian viruses target type II pneumocytes (van Riel et al. 2007, 2010). However, recent studies in human lung explants revealed that all human and avian viruses examined, including avian H5N1 and H7N9, almost exclusively infected type II pneumocytes (Weinheimer et al. 2012; Knepper et al. 2013). These findings agree with the abundance of infected type II pneumocytes in autopsy material of patients that died from severe infection with both H1N1 pandemic and H5N1 avian viruses (Uiprasertkul et al. 2005; Korteweg and Gu 2010) and the rare observation of antigen/RNA-positive type I cells (Liem et al. 2008; Shieh et al. 2010). Therefore, differences in pathogenicity of H5N1 and H7N9 avian viruses and human viruses cannot be explained by replication of human and avian viruses in different types of pneumocytes. It remains to be determined whether differences in receptor-binding specificity of these viruses could affect the efficiency of their replication in type II pneumocytes.

Antigen-positive alveolar macrophages (AMs) are commonly observed in autopsy material from fatal influenza cases. AMs and peripheral blood monocyte-derived macrophages (PBDMs) were found to express both 3-linked and 6-linked Sia (Yu et al. 2011). In the attachment assay, avian viruses and viruses with dual receptor specificity bound stronger to AMs than human viruses (van Riel et al. 2007; Chutinimitkul et al. 2010). Both AMs and PBDMs were more susceptible to infection with highly pathogenic H5N1 viruses than with human seasonal and pandemic viruses (van Riel et al. 2011; Yu et al. 2011). No infectious virus was released from infected AM suggesting that these cells do not significantly contribute to the spread of infection in human alveolae. As infection of AM likely impairs their protective scavenger function, a high susceptibility of AM to H5N1 viruses may contribute to their unusual high pathogenicity in humans.

In addition to pneumocytes, endothelial cells of alveolar capillaries represent the major type of cells in the lung. Microvascular endothelial cells are separated from pneumocytes by two thin basement membranes. Passive gas exchange in the lung occurs through this epithelio-endothelial barrier. Findings of infected endothelial cells in lung autopsies of patients with severe influenza are rare (Liem et al. 2008; Shieh et al. 2010). Studies of virus infection in cultures of primary human microvascular endothelial cells were recently described. Lectin-based analysis revealed the presence of 3-linked and 6-linked Sia (Chan et al. 2009) with an apparent predominance of the former (Zeng et al. 2012). Highly pathogenic H5N1 viruses infected endothelial cells significantly more efficiently and replicated in these cells to much higher titers than human seasonal and pandemic viruses (Ocana-Macchi et al. 2009; Zeng et al. 2012). Reverse genetics analysis and virus attachment studies demonstrated a major role for the viral HA in the observed tropism of H5N1 to endothelial cells (Ocana-Macchi et al. 2009). After infection, a pronounced release of cytokines and adhesion molecules was observed as well as an elevated rate of cell death. These observations are particularly interesting in view of recent findings suggesting a major role of endothelial cells in the production of cytokines during influenza infection (Teijaro et al. 2011). Thus, enhanced receptor-mediated tropism of H5N1 viruses to endothelial cells in the human lung could contribute to the disease caused by these viruses.

4 Modulation of HA Functions by Glycosylation

The carbohydrate moiety of HA consists of N-glycans synthesized by the cellular glycosylation machinery. The viral carbohydrates therefore show host-dependent variations. The only difference from host glycans is the absence of sialic acid that is removed by NA. There are complex, oligomannosidic, and intermediate type side chains, and there is usually microheterogeneity at a given site (Keil et al.

1985). The glycosylation sites and the functional significance of the carbohydrate side chains of the H7 HA are shown in Fig. 4. The glycosylation sites at Asn-12 and HA2-Asn154 (Asn-478, H7 numbering) are highly conserved, and the glycosylation site at Asn-28 is also present in many strains. The high conservation of these three sites, which are located in the stem region of the HA spike, suggests that they play important structural and functional roles. Indeed, loss of all three oligosaccharides resulted in a temperature-sensitive block of HA transport to the cell surface (Roberts et al. 1993). It was also shown that these carbohydrates stabilize HA in the metastable form susceptible to the conformational change necessary for fusion activation (Ohuchi et al. 1997a). They are therefore essential for the formation of replication-competent virus (Wagner et al. 2002a). In contrast to the oligosaccharides in the stem region, the glycans attached to the globular head of HA vary significantly in number and position with different virus strains (Klenk and Schwarz 1987). It is well known that the carbohydrate side chains modulate HA activities by masking antigenic epitopes and functional domains. On the other hand, evidence is increasing that the glycans may be directly involved in specific interactions with host components. Virus pathogenicity may be altered by both strategies.

Glycosylation of an antigenic epitope prevents antibody binding. It therefore contributes to immune escape and is an important mechanism underlying antigenic drift. This concept is supported by several observations. Thus, intraepidemic variation of HA1 glycosylation has been observed during an H3N2 outbreak in 1974/75, whereby the number of glycans appeared to increase at the end of the outbreak (Seidel et al. 1991). In fact, the H3 HA acquired four additional carbohydrate side chains between 1968 and 1985, and inhibition of glycosylation and site-directed mutagenesis have directly shown that at least one of these glycans interfered with antibody recognition (Skehel et al. 1984; Abe et al. 2004).

Receptor binding may also be affected by glycosylation. When glycans at Asn-123 and Asn-149 in the vicinity of the receptor-binding site of an H7 HA were removed by site-directed mutagenesis, vector-expressed HA showed drastically enhanced hemadsorbing activity, and virions containing HA lacking these glycans were unable to elute from receptors, although they retained NA activity (Ohuchi et al. 1997b). Comparison of human H1 viruses grown in MDCK cells and in eggs revealed that the virus obtained from MDCK cells contained a carbohydrate side chain in close proximity to the receptor-binding site that was not present in the egg-grown virus. The cell and egg-grown viruses differed by receptor-binding properties (Gambaryan et al. 1999). Glycosylation-dependent variations in receptor affinity of HA may have to be balanced by NA variations. Thus it has been shown that acquisition of a carbohydrate close to the receptor-binding site, supposed to interfere with receptor access and to increase virus release, was accompanied by a deletion in the NA stalk that reduced the enzymatic activity and, thus, compensated for the reduced receptor binding (Baigent and McCauley 2001). These observations led to the concept that a functional balance between HA and NA is required for optimal infection (Mitnaul et al. 2000; Wagner et al. 2002b).

It is well known that a carbohydrate can determine pathogenicity by modulating cleavage activation of HA. Apathogenic strains of A/chick/Penn/83(H5N2) virus were found to have a glycan masking the multibasic cleavage site of HA. Loss of the side chain resulted in high cleavability and high pathogenicity (Kawaoka et al. 1984; Deshpande et al. 1987; Kawaoka and Webster 1988). In vitro studies showed that cleavability of human H3 HA with a multibasic cleavage site can also be modulated by a masking oligosaccharide (Ohuchi et al. 1991).

Evidence is increasing that interactions of the glycans of HA and NA with soluble and membrane-bound lectins also play an important role in influenza virus infection. Glycans of the globular head of HA bind to collectins, collagenous Ca++- dependent multimeric lectins, including surfactant protein D (SP-D) present in respiratory secretions and mannose-binding lectin (MBL) present in serum. SP-D and MBL neutralize the virus by several mechanisms, such as steric hindrance of the receptor-binding site of HA, inhibition of the enzymatic activity of NA, aggregation of virions, and activation of complement-dependent pathways of the innate immune system (for review see Reading et al. 2007; Hartshorn 2010; Hillaire et al. 2013; Tate et al. 2014). Virus variants with reduced glycosylation obtained by mouse passages or generated by reversed genetics were less sensitive to inhibition by collectins in vitro and more virulent to mice than their fully glycosylated counterparts (Reading et al. 2007, 2009; Tate et al. 2011a, b). Furthermore, the HAs of the influenza viruses that caused the pandemics in 1918, 1957, 1968, and 2009 had only a few glycosylation sites on the globular head, which is a typical feature of LPAIV and LPAIV-derived swine viruses. HA glycosylation of these viruses increased during post-pandemic circulation in humans resulting in more collectin binding sites. Consequently, recombinant influenza viruses with the HAs of pandemic viruses showed lower binding to SP-D and significantly higher pathogenicity for mice than isogenic viruses with the HAs of seasonal human viruses (Vigerust et al. 2007; Qi et al. 2011). These findings indicate that reduced glycosylation and increased resistance to collectins in body fluids might contribute to enhanced pathogenicity of pandemic and zoonotic influenza viruses in humans.

C-type lectin receptors are type II membrane proteins that have also been found to bind to N-glycans of HA and NA. They include MMR (macrophage mannose receptor) and DC-SIGN also recognizing mannose as well as MGL (macrophage galactose-type lectin). All of these lectin receptors have been implicated in promoting infection of macrophages and dendritic cells of mice (Chu and Whittaker 2004; Hillaire et al. 2011; Londrigan et al. 2011). MGL that binds to both galactose and N-acetylgalactosamine has been proposed to function as a secondary receptor mediating endocytosis after virus attachment to the primary sialic acid receptor (Ng et al. 2014). Highly glycosylated virus strains that infected macrophages to high levels were recognized efficiently by MGL, whereas virus containing HA with few oligosaccharides was not, suggesting that differences in MGL-mediated recognition determine the susceptibility of murine macrophages to infection and that this modulates pathogenicity in mice (Upham et al. 2010).

5 Outlook

Future work is needed to focus on the identification of new activating proteases, particularly on enzymes cleaving at monobasic cleavage sites. It will be important to determine the tissue specific expression of the activating enzymes in epithelia, macrophages, and endothelia of the respiratory tract, as well as in nonrespiratory tissues, and to throw light on their role in organ tropism and spread of infection. To better understand the pathogenetic mechanisms involved in coinfection, HA-activating proteases expressed in bacteria and other microorganisms will have to be analyzed. Since evidence is increasing that there is considerable variation in proteolytic activation of mammalian and low pathogenic avian influenza viruses, the specificity of the activating proteases as depending on the structure of the respective cleavage sites will have to be elucidated. It is also desirable to design new inhibitors with a high specificity for these proteases.

It is well known that the linkage between N-acetyl-neuraminic acid and the adjacent galactose is an important determinant for the receptor specificity of the virus. However, little is known about the contribution of the proximal sugar residues and about the roles of the polypeptide and lipid moieties of the receptor molecules. The high variability of the glycan pattern of the respiratory tract is another challenge for structural analysis. Moreover, little is known about the glycans of the respiratory mucus and their roles as decoy receptors that interfere with infection. Thus, there is a need for more glycomics research to fully understand the role of HA in pathogenesis.

Finally, more information is needed about the functional significance of the oligosaccharide side chains of HA. It is fairly well known that they shield cleavage site, receptor-binding site, and antigenic epitopes. However, much has still to be learned about their interactions with lectins and the roles of lectins as secondary receptors and mediators in innate immunity.

Acknowledgments We thank Sabine Fischbach for expert secretarial help. Research of the authors was supported by grants from the Deutsche Forschungsgemeinschaft (SFB 593, SFB 1021), the European Union 7th Framework Programme (278433 Predemics, 258084 FLUPIG), the Bundesministerium für Bildung und Forschung (FluResearchNet, DZIF), and the von Behring-Roentgen-Stiftung.

References

Abe Y, Takashita E, Sugawara K, Matsuzaki Y, Muraki Y, Hongo S (2004) Effect of the addition of oligosaccharides on the biological activities and antigenicity of influenza A/H3N2 virus hemagglutinin. J Virol 78:9605–9611. doi:10.1128/JVI.78.18.9605-9611.2004

Afar DE, Vivanco I, Hubert RS et al (2001) Catalytic cleavage of the androgen-regulated TMPRSS2 protease results in its secretion by prostate and prostate cancer epithelia. Cancer Res 61:1686–1692

Baigent SJ, McCauley JW (2001) Glycosylation of haemagglutinin and stalk-length of neuraminidase combine to regulate the growth of avian influenza viruses in tissue culture. Virus Res 79:177–185

Banks J, Speidel ES, Moore E et al (2001) Changes in the haemagglutinin and the neuraminidase genes prior to the emergence of highly pathogenic H7N1 avian influenza viruses in Italy. Arch Virol 146:963–973

Baron J, Tarnow C, Mayoli-Nüssle D et al (2013) Matriptase, HAT, and TMPRSS2 activate the hemagglutinin of H9N2 influenza A viruses. J Virol 87:1811–1820. doi:10.1128/JVI.02320-12

Baum LG, Paulson JC (1990) Sialyloligosaccharides of the respiratory epithelium in the selection of human influenza virus receptor specificity. Acta Histochem 40:35–38

Beaulieu A, Gravel E, Cloutier A et al (2013) Matriptase proteolytically activates influenza virus and promotes multicycle replication in the human airway epithelium. J Virol 87:4237–4251. doi:10.1128/JVI.03005-12

Becker GL, Sielaff F, Than ME et al (2010) Potent inhibitors of furin and furin-like proprotein convertases containing decarboxylated P1 arginine mimetics. J Med Chem 53:1067–1075. doi:10.1021/jm9012455

Becker GL, Lu Y, Hardes K et al (2012) Highly potent inhibitors of proprotein convertase furin as potential drugs for treatment of infectious diseases. J Biol Chem 287:21992–22003. doi:10.1074/jbc.M111.332643

Beigel JH, Farrar J, Han AM et al (2005) Avian influenza A (H5N1) infection in humans. N Engl J Med 353:1374–1385

Belser JA, Gustin KM, Pearce MB et al (2013) Pathogenesis and transmission of avian influenza A (H7N9) virus in ferrets and mice. Nature 501:556–559

Bertram S, Heurich A, Lavender H et al (2012) Influenza and SARS-coronavirus activating proteases TMPRSS2 and HAT are expressed at multiple sites in human respiratory and gastrointestinal tracts. PLoS One 7:e35876. doi:10.1371/journal.pone.0035876

Bosch FX, Orlich M, Klenk HD, Rott R (1979) The structure of the hemagglutinin, a determinant for the pathogenicity of influenza viruses. Virology 95:197–207

Böttcher E, Matrosovich T, Beyerle M, Klenk HD, Garten W, Matrosovich M (2006) Proteolytic activation of influenza viruses by serine proteases TMPRSS2 and HAT from human airway epithelium. J Virol 80:9896–9898. doi:10.1128/JVI.01118-06

Böttcher-Friebertshäuser E, Freuer C, Sielaff F et al (2010) Cleavage of influenza virus hemagglutinin by airway proteases TMPRSS2 and HAT differs in subcellular localization and susceptibility to protease inhibitors. J Virol 84:5605–5614. doi:10.1128/JVI.00140-10

Böttcher-Friebertshäuser E, Stein DA, Klenk HD, Garten W (2011) Inhibition of influenza virus infection in human airway cell cultures by an antisense peptide-conjugated morpholino oligomer targeting the hemagglutinin-activating protease TMPRSS2. J Virol 85:1554–1562. doi:10.1128/JVI.01294-10

Böttcher-Friebertshäuser E, Lu Y, Meyer D, Sielaff F, Steinmetzer T, Klenk HD, Garten W (2012) Hemagglutinin activating host cell proteases provide promising drug targets for the treatment of influenza A and B virus infections. Vaccine 30:7374–7380. doi:10.1016/j.vaccine.2012.10.001

Böttcher-Friebertshäuser E, Klenk HD, Garten W (2013) Activation of influenza viruses by proteases from host cells and bacteria in the human airway epithelium. Pathog Dis 69:87–100. doi:10.1111/2049-632X.12053

Campitelli L, Mogavero E, De Marco MA et al (2004) Interspecies transmission of an H7N3 influenza virus from wild birds to intensively reared domestic poultry in Italy. Virology 323:24–36. doi:10.1016/j.virol.2004.02.015

Chaipan C, Kobasa D, Bertram S et al (2009) Proteolytic activation of the 1918 influenza virus hemagglutinin. J Virol 83:3200–3211. doi:10.1128/JVI.02205-08

Chan MC, Chan RW, Yu WC et al (2009) Influenza H5N1 virus infection of polarized human alveolar epithelial cells and lung microvascular endothelial cells. Respir Res 10:102

Chan MC, Chan RW, Chan LL et al (2013a) Tropism and innate host responses of a novel avian influenza A H7N9 virus: an analysis of ex vivo and in vitro cultures of the human respiratory tract. Lancet Respir Med 1:534–542

Chan RW, Chan MC, Nicholls JM, Malik Peiris JS (2013b) Use of ex vivo and in vitro cultures of the human respiratory tract to study the tropism and host responses of highly pathogenic avian influenza A (H5N1) and other influenza viruses. Virus Res 178:133–145

Chen J, Lee KH, Steinhauer DA, Stevens DJ, Skehel JJ, Wiley DC (1998) Structure of the hemagglutinin precursor cleavage site, a determinant of influenza pathogenicity and the origin of the labile conformation. Cell 95:409–417

Chu VC, Whittaker GR (2004) Influenza virus entry and infection require host cell N-linked glycoprotein. Proc Natl Acad Sci USA 101:18153–18158. doi:10.1073/pnas.0405172102

Chutinimitkul S, Herfst S, Steel J et al (2010) Virulence-associated substitution D222G in the hemagglutinin of 2009 pandemic influenza A(H1N1) virus affects receptor binding. J Virol 84:11802–11813

Couceiro JN, Paulson JC, Baum LG (1993) Influenza virus strains selectively recognize sialyloligosaccharides on human respiratory epithelium; the role of the host cell in selection of hemagglutinin receptor specificity. Virus Res 29:155–165

Deshpande KL, Fried VA, Ando M, Webster RG (1987) Glycosylation affects cleavage of an H5N2 influenza virus hemagglutinin and regulates virulence. Proc Natl Acad Sci USA 84:36–40

Feldmann A, Schäfer MK, Garten W, Klenk HD (2000) Targeted infection of endothelial cells by avian influenza virus A/FPV/Rostock/34 (H7N1) in chicken embryos. J Virol 74:8018–8027

Fouchier RA, Schneeberger PM, Rozendaal FW et al (2004) Avian influenza A virus (H7N7) associated with human conjunctivitis and a fatal case of acute respiratory distress syndrome. Proc Natl Acad Sci USA 101:1356–1361. doi:10.1073/pnas.0308352100

Gabriel G, Garn H, Wegmann M, Renz H, Herwig A, Klenk HD, Stech J (2008) The potential of a protease activation mutant of a highly pathogenic avian influenza virus for a pandemic live vaccine. Vaccine 26:956–965. doi:10.1016/j.vaccine.2007.11.052

Galloway SE, Reed ML, Russell CJ, Steinhauer DA (2013) Influenza HA subtypes demonstrate divergent phenotypes for cleavage activation and pH of fusion: implications for host range and adaptation. PLoS Pathog 9:e1003151. doi:10.1371/journal.ppat.1003151

Gambaryan AS, Robertson JS, Matrosovich MN (1999) Effects of egg-adaptation on the receptor-binding properties of human influenza A and B viruses. Virology 258:232–239. doi:10.1006/viro.1999.9732

Gambaryan AS, Matrosovich TY, Philipp J et al (2012) Receptor-binding profiles of H7 subtype influenza viruses in different host species. J Virol 86:4370–4379

Gamblin SJ, Skehel JJ (2010) Influenza hemagglutinin and neuraminidase membrane glycoproteins. J Biol Chem 285:28403–28409

Gao R, Cao B, Hu Y et al (2013) Human infection with a novel avian-origin influenza A (H7N9) virus. N Engl J Med 368:1888–1897. doi:10.1056/NEJMoa1304459

Garcia M, Crawford JM, Latimer JW, Rivera-Cruz E, Perdue ML (1996) Heterogeneity in the haemagglutinin gene and emergence of the highly pathogenic phenotype among recent H5N2 avian influenza viruses from Mexico. J Gen Virol 77(Pt 7):1493–1504

Garten W, Klenk HD (2008) Cleavage activation of the influenza virus hemagglutinin and its role in pathogenesis. Klenk HD, Matrosovich MN, Stech J (eds) Avian Influenza. Karger, Basel, pp 156–167

Garten W, Bosch FX, Linder D, Rott R, Klenk HD (1981) Proteolytic activation of the influenza virus hemagglutinin: the structure of the cleavage site and the enzymes involved in cleavage. Virology 115:361–374

Garten W, Stieneke A, Shaw E, Wikstrom P, Klenk HD (1989) Inhibition of proteolytic activation of influenza virus hemagglutinin by specific peptidyl chloroalkyl ketones. Virology 172:25–31

Garten W, Hallenberger S, Ortmann D et al (1994) Processing of viral glycoproteins by the subtilisin-like endoprotease furin and its inhibition by specific peptidylchloroalkylketones. Biochimie 76:217–225

Geisler C, Jarvis DL (2011) Effective glycoanalysis with Maackia amurensis lectins requires a clear understanding of their binding specificities. Glycobiology 21:988–993

Glycobiology (2009) Essentials of glycobiology, 2nd edn. In: Cold Spring Harbor Laboratory Press. Cold Spring Harbor, New York

Gohrbandt S, Veits J, Breithaupt A et al (2011) H9 avian influenza reassortant with engineered polybasic cleavage site displays a highly pathogenic phenotype in chicken. J Gen Virol 92:1843–1853. doi:10.1099/vir.0.031591-0

Goto H, Kawaoka Y (1998) A novel mechanism for the acquisition of virulence by a human influenza A virus. Proc Natl Acad Sci USA 95:10224–10228

Gotoh B, Ogasawara T, Toyoda T, Inocencio NM, Hamaguchi M, Nagai Y (1990) An endoprotease homologous to the blood clotting factor X as a determinant of viral tropism in chick embryo. EMBO J 9:4189–4195

Ha Y, Stevens DJ, Skehel JJ, Wiley DC (2002) H5 avian and H9 swine influenza virus haemagglutinin structures: possible origin of influenza subtypes. EMBO J 21:865–875. doi:10.1093/emboj/21.5.865

Hallenberger S, Bosch V, Angliker H, Shaw E, Klenk HD, Garten W (1992) Inhibition of furin-mediated cleavage activation of HIV-1 glycoprotein gp160. Nature 360:358–361. doi:10.1038/360358a0

Hamilton BS, Whittaker GR (2013) Cleavage activation of human-adapted influenza virus subtypes by kallikrein-related peptidases 5 and 12. J Biol Chem 288:17399–17407. doi:10.1074/jbc.M112.440362

Hartshorn KL (2010) Role of surfactant protein A and D (SP-A and SP-D) in human antiviral host defense. Front Biosci (Schol.Ed) 2:527–546

Hatesuer B, Bertram S, Mehnert N, Bahgat MM, Nelson PS, Pöhlmann S, Schughart K (2013) Tmprss2 is essential for influenza H1N1 virus pathogenesis in mice. PLoS Pathog 9:e1003774. doi:10.1371/journal.ppat.1003774

Herfst S, Schrauwen EJ, Linster M et al (2012) Airborne transmission of influenza A/H5N1 virus between ferrets. Science 336:1534–1541

Hillaire ML, van Eijk M, van Trierum SE et al (2011) Assessment of the antiviral properties of recombinant porcine SP-D against various influenza A viruses in vitro. PLoS One 6:e25005. doi:10.1371/journal.pone.0025005

Hillaire ML, Haagsman HP, Osterhaus AD, Rimmelzwaan GF, van EM (2013) Pulmonary surfactant protein D in first-line innate defence against influenza A virus infections. J Innate Immun 5:197–208

Horimoto T, Nakayama K, Smeekens SP, Kawaoka Y (1994) Proprotein-processing endoproteases PC6 and furin both activate hemagglutinin of virulent avian influenza viruses. J Virol 68:6074–6078

Horimoto T, Rivera E, Pearson J, Senne D, Krauss S, Kawaoka Y, Webster RG (1995) Origin and molecular changes associated with emergence of a highly pathogenic H5N2 influenza virus in Mexico. Virology 213:223–230. doi:10.1006/viro.1995.1562

Huang RT, Rott R, Klenk HD (1981) Influenza viruses cause hemolysis and fusion of cells. Virology 110:243–247

Imai M, Watanabe T, Hatta M et al (2012) Experimental adaptation of an influenza H5 HA confers respiratory droplet transmission to a reassortant H5 HA/H1N1 virus in ferrets. Nature 486:420–428

Imai M, Herfst S, Sorrell EM et al (2013) Transmission of influenza A/H5N1 viruses in mammals. Virus Res. 178:15–20

Kawaoka Y, Webster RG (1988) Sequence requirements for cleavage activation of influenza virus hemagglutinin expressed in mammalian cells. Proc Natl Acad Sci U S A 85:324–328
Kawaoka Y, Naeve CW, Webster RG (1984) Is virulence of H5N2 influenza viruses in chickens associated with loss of carbohydrate from the hemagglutinin? Virology 139:303–316
Keil W, Geyer R, Dabrowski J, Dabrowski U, Niemann H, Stirm S, Klenk HD (1985) Carbohydrates of influenza virus. Structural elucidation of the individual glycans of the FPV hemagglutinin by two-dimensional 1H n.m.r. and methylation analysis. EMBO J 4:2711–2720
Kesic MJ, Meyer M, Bauer R, Jaspers I (2012) Exposure to ozone modulates human airway protease/antiprotease balance contributing to increased influenza A infection. PLoS One 7:e35108. doi:10.1371/journal.pone.0035108
Khatchikian D, Orlich M, Rott R (1989) Increased viral pathogenicity after insertion of a 28S ribosomal RNA sequence into the haemagglutinin gene of an influenza virus. Nature 340:156–157. doi:10.1038/340156a0
Kido H, Yokogoshi Y, Sakai K, Tashiro M, Kishino Y, Fukutomi A, Katunuma N (1992) Isolation and characterization of a novel trypsin-like protease found in rat bronchiolar epithelial Clara cells. A possible activator of the viral fusion glycoprotein. J Biol Chem 267:13573–13579
Kido H, Okumura Y, Yamada H, Le TQ, Yano M (2007) Proteases essential for human influenza virus entry into cells and their inhibitors as potential therapeutic agents. Curr Pharm Des 13:405–414
Kim TS, Heinlein C, Hackman RC, Nelson PS (2006) Phenotypic analysis of mice lacking the Tmprss2-encoded protease. Mol Cell Biol 26:965–975. doi:10.1128/MCB.26.3.965-975.2006
Klenk HD, Garten W (1994) Host cell proteases controlling virus pathogenicity. Trends Microbiol 2:39–43
Klenk HD, Schwarz RT (1987) Variations in glycosylation of the influenza virus hemagglutinin of subtype H7. Brief report. Arch Virol 97:359–363
Klenk HD, Rott R, Orlich M, Blodorn J (1975) Activation of influenza A viruses by trypsin treatment. Virology 68:426–439
Klenk HD, Garten W, Matrosovich M (2013) Pathogenesis. In: Webster RG, Monto AS, Braciale TJ, Lamb RA (eds) Textbook of influenza. Wiley-Blackwell, New York, pp 157–171
Knepper J, Schierhorn KL, Becher A et al (2013) The novel human influenza A(H7N9) virus is naturally adapted to efficient growth in human lung tissue. MBio. 4:e00601–e00613
Koopmans M, Vennema H, Heersma H et al (2003) Early identification of common-source foodborne virus outbreaks in Europe. Emerg Infect Dis 9:1136–1142. doi:10.3201/eid0909.020766
Korteweg C, Gu J (2010) Pandemic influenza A (H1N1) virus infection and avian influenza A (H5N1) virus infection: a comparative analysis. Biochem. Cell Biol 88:575–587
Kuiken T, Taubenberger JK (2008) Pathology of human influenza revisited. Vaccine 26(Suppl 4):D59–D66
Kurtz J, Manvell RJ, Banks J (1996) Avian influenza virus isolated from a woman with conjunctivitis. Lancet 348:901–902. doi:10.1016/S0140-6736(05)64783-6
Lazarowitz SG, Choppin PW (1975) Enhancement of the infectivity of influenza A and B viruses by proteolytic cleavage of the hemagglutinin polypeptide. Virology 68:440–454
Lazarowitz SG, Goldberg AR, Choppin PW (1973) Proteolytic cleavage by plasmin of the HA polypeptide of influenza virus: host cell activation of serum plasminogen. Virology 56:172–180
Liem NT, Nakajima N, Phat LP et al (2008) H5N1-infected cells in lung with diffuse alveolar damage in exudative phase from a fatal case in Vietnam. Jpn J Infect Dis 61:157–160
Liu J, Stevens DJ, Haire LF et al (2009) Structures of receptor complexes formed by hemagglutinins from the Asian Influenza pandemic of 1957. Proc Natl Acad Sci USA 106:17175–17180. doi:10.1073/pnas.0906849106
Liu Y, Childs RA, Matrosovich T et al (2010) Altered receptor specificity and cell tropism of D222G hemagglutinin mutants isolated from fatal cases of pandemic A(H1N1) 2009 influenza virus. J Virol 84:12069–12074

Liu S, Sun J, Cai J et al (2013) Epidemiological, clinical and viral characteristics of fatal cases of human avian influenza A (H7N9) virus in Zhejiang Province. China J Infect 67:595–605

Londrigan SL, Turville SG, Tate MD, Deng YM, Brooks AG, Reading PC (2011) N-linked glycosylation facilitates sialic acid-independent attachment and entry of influenza A viruses into cells expressing DC-SIGN or L-SIGN. J Virol 85:2990–3000. doi:10.1128/JVI.01705-10

Lu X, Shi Y, Gao F, Xiao H, Wang M, Qi J, Gao GF (2012) Insights into avian influenza virus pathogenicity: the hemagglutinin precursor HA0 of subtype H16 has an alpha-helix structure in its cleavage site with inefficient HA1/HA2 cleavage. J Virol 86:12861–12870. doi:10.1128/JVI.01606-12

Lu X, Qi J, Shi Y et al (2013) Structure and receptor binding specificity of hemagglutinin H13 from avian influenza A virus H13N6. J Virol 87:9077–9085. doi:10.1128/JVI.00235-13

Maeda T, Kawasaki K, Ohnishi S (1981) Interaction of influenza virus hemagglutinin with target membrane lipids is a key step in virus-induced hemolysis and fusion at pH 5.2. Proc Natl Acad Sci USA 78:4133–4137

Matrosovich MN, Matrosovich TY, Gray T, Roberts NA, Klenk HD (2004) Human and avian influenza viruses target different cell types in cultures of human airway epithelium. Proc Natl Acad Sci USA 101:4620–4624

Matrosovich MN, Klenk HD, Kawaoka Y (2006) Receptor specificity, host range and pathogenicity of influenza viruses. In: Kawaoka Y (ed) Influenza virology: current topics. Caister Academic Press, Wymondham, pp 95–137

Matrosovich M, Matrosovich T, Uhlendorff J, Garten W, Klenk HD (2007) Avian-virus-like receptor specificity of the hemagglutinin impedes influenza virus replication in cultures of human airway epithelium. Virology 361:384–390

Matrosovich MN, Gambaryan AS, Klenk HD (2008) Receptor specificity of influenza viruses and its alteration during interspecies transmission. In: Klenk HD, Matrosovich MN, Stech J (eds) Avian influenza. Karger, Basel, pp 134–155

Maurer-Stroh S, Lee RT, Gunalan V, Eisenhaber F (2013) The highly pathogenic H7N3 avian influenza strain from July 2012 in Mexico acquired an extended cleavage site through recombination with host 28S rRNA. Virol J 10:139. doi:10.1186/1743-422X-10-139

Meyer D, Sielaff F, Hammami M, Böttcher-Friebertshäuser E, Garten W, Steinmetzer T (2013) Identification of the first synthetic inhibitors of the type II transmembrane serine protease TMPRSS2 suitable for inhibition of influenza virus activation. Biochem J 452:331–343. doi:10.1042/BJ20130101

Mitnaul LJ, Matrosovich MN, Castrucci MR, Tuzikov AB, Bovin NV, Kobasa D, Kawaoka Y (2000) Balanced hemagglutinin and neuraminidase activities are critical for efficient replication of influenza A virus. J Virol 74:6015–6020

Murakami M, Towatari T, Ohuchi M et al (2001) Mini-plasmin found in the epithelial cells of bronchioles triggers infection by broad-spectrum influenza A viruses and Sendai virus. Eur J Biochem 268:2847–2855

Ng WC, Liong S, Tate MD et al (2014) The macrophage galactose-type lectin can function as an attachment and entry receptor for influenza virus. J Virol 88:1659–1672. doi:10.1128/JVI.02014-13

Nicholls JM, Bourne AJ, Chen H, Guan Y, Peiris JS (2007a) Sialic acid receptor detection in the human respiratory tract: evidence for widespread distribution of potential binding sites for human and avian influenza viruses. Respir Res 8:73

Nicholls JM, Chan MC, Chan WY et al (2007b) Tropism of avian influenza A (H5N1) in the upper and lower respiratory tract. Nat. Med. 13:147–149

Oberst MD, Singh B, Ozdemirli M, Dickson RB, Johnson MD, Lin CY (2003) Characterization of matriptase expression in normal human tissues. J Histochem Cytochem 51:1017–1025

Ocana-Macchi M, Bel M, Guzylack-Piriou L et al (2009) Hemagglutinin-dependent tropism of H5N1 avian influenza virus for human endothelial cells. J Virol 83:12947–12955

Ohuchi R, Ohuchi M, Garten W, Klenk HD (1991) Human influenza virus hemagglutinin with high sensitivity to proteolytic activation. J Virol 65:3530–3537

Ohuchi M, Ohuchi R, Feldmann A, Klenk HD (1997a) Regulation of receptor binding affinity of influenza virus hemagglutinin by its carbohydrate moiety. J Virol 71:8377–8384

Ohuchi R, Ohuchi M, Garten W, Klenk HD (1997b) Oligosaccharides in the stem region maintain the influenza virus hemagglutinin in the metastable form required for fusion activity. J Virol 71:3719–3725

Orlich M, Gottwald H, Rott R (1994) Nonhomologous recombination between the hemagglutinin gene and the nucleoprotein gene of an influenza virus. Virology 204:462–465. doi:10.1006/viro.1994.1555

Pantin-Jackwood MJ, Swayne DE (2009) Pathogenesis and pathobiology of avian influenza virus infection in birds. Rev Sci Tech 28:113–136

Pappas C, Viswanathan K, Chandrasekaran A, Raman R, Katz JM, Sasisekharan R, Tumpey TM (2010) Receptor specificity and transmission of H2N2 subtype viruses isolated from the pandemic of 1957. PLoS One 5:e11158

Pasick J, Handel K, Robinson J et al (2005) Intersegmental recombination between the haemagglutinin and matrix genes was responsible for the emergence of a highly pathogenic H7N3 avian influenza virus in British Columbia. J Gen Virol 86:727–731. doi:10.1099/vir.0.80478-0

Peiris MJS (2009) Avian influenza viruses in humans. Rev Sci Tech 28:161–173

Peitsch C, Klenk HD, Garten W, Böttcher-Friebertshäuser E (2014) Activation of influenza A viruses by host proteases from swine airway epithelium. J Virol 88:282–291. doi:10.1128/JVI.01635-13

Perdue ML, Garcia M, Senne D, Fraire M (1997) Virulence-associated sequence duplication at the hemagglutinin cleavage site of avian influenza viruses. Virus Res 49:173–186

Qi L, Kash JC, Dugan VG et al (2011) The ability of pandemic influenza virus hemagglutinins to induce lower respiratory pathology is associated with decreased surfactant protein D binding. Virology 412:426–434. doi:10.1016/j.virol.2011.01.029

Reading PC, Tate MD, Pickett DL, Brooks AG (2007) Glycosylation as a target for recognition of influenza viruses by the innate immune system. Adv Exp Med Biol 598:279–292

Reading PC, Pickett DL, Tate MD, Whitney PG, Job ER, Brooks AG (2009) Loss of a single N-linked glycan from the hemagglutinin of influenza virus is associated with resistance to collectins and increased virulence in mice. Respir Res 10:117

Roberts PC, Garten W, Klenk HD (1993) Role of conserved glycosylation sites in maturation and transport of influenza A virus hemagglutinin. J Virol 67:3048–3060

Roberts KL, Shelton H, Scull M, Pickles R, Barclay WS (2011) Lack of transmission of a human influenza virus with avian receptor specificity between ferrets is not due to decreased virus shedding but rather a lower infectivity in vivo. J Gen Virol 92:1822–1831

Roehm C, Horimoto T, Kawaoka Y, Suss J, Webster RG (1995) Do hemagglutinin genes of highly pathogenic avian influenza viruses constitute unique phylogenetic lineages? Virology 209:664–670. doi:10.1006/viro.1995.1301

Russell RJ, Gamblin SJ, Haire LF, Stevens DJ, Xiao B, Ha Y, Skehel JJ (2004) H1 and H7 influenza haemagglutinin structures extend a structural classification of haemagglutinin subtypes. Virology 325:287–296

Rykkvin R, Kilander A, Dudman SG, Hungnes O (2013) Within-patient emergence of the influenza A (H1N1) pdm09 HA1 222G variant and clear association with severe disease, Norway. Euro Surveill 18:20369

Sakai K, Ami Y, Tahara M et al (2014) The host protease TMPRSS2 plays a major role for in vivo replication of emerging H7N9 and seasonal influenza viruses. J Virol. doi:10.1128/JVI.03677-13

Sales KU, Hobson JP, Wagenaar-Miller R et al (2011) Expression and genetic loss of function analysis of the HAT/DESC cluster proteases TMPRSS11A and HAT. PLoS One 6:e23261. doi:10.1371/journal.pone.0023261

Sato M, Yoshida S, Iida K, Tomozawa T, Kido H, Yamashita M (2003) A novel influenza A virus activating enzyme from porcine lung: purification and characterization. Biol Chem 384:219–227. doi:10.1515/BC.2003.024

Schäfer W, Stroh A, Berghofer S et al (1995) Two independent targeting signals in the cytoplasmic domain determine trans-Golgi network localization and endosomal trafficking of the proprotein convertase furin. EMBO J 14:2424–2435

Scheiblauer H, Reinacher M, Tashiro M, Rott R (1992) Interactions between bacteria and influenza A virus in the development of influenza pneumonia. J Infect Dis 166:783–791

Schrauwen EJ, Bestebroer TM, Munster VJ et al (2011) Insertion of a multibasic cleavage site in the haemagglutinin of human influenza H3N2 virus does not increase pathogenicity in ferrets. J Gen Virol 92:1410–1415. doi:10.1099/vir.0.030379-0

Seidah NG, Mayer G, Zaid A et al (2008) The activation and physiological functions of the proprotein convertases. Int J Biochem Cell Biol 40:1111–1125. doi:10.1016/j.biocel.2008.01.030

Seidel W, Kunkel F, Geisler B, Garten W, Herrmann B, Dohner L, Klenk HD (1991) Intraepidemic variants of influenza virus H3 hemagglutinin differing in the number of carbohydrate side chains. Arch Virol 120:289–296

Shieh WJ, Blau DM, Denison AM et al (2010) 2009 pandemic influenza A (H1N1): pathology and pathogenesis of 100 fatal cases in the United States. Am J Pathol 177:166–175

Shinya K, Ebina M, Yamada S, Ono M, Kasai N, Kawaoka Y (2006) Avian flu: influenza virus receptors in the human airway. Nature 440:435–436

Skehel JJ, Wiley DC (2000) Receptor binding and membrane fusion in virus entry: the influenza hemagglutinin. Annu Rev Biochem 69:531–569. doi:10.1146/annurev.biochem.69.1.531

Skehel JJ, Stevens DJ, Daniels RS, Douglas AR, Knossow M, Wilson IA, Wiley DC (1984) A carbohydrate side chain on hemagglutinins of Hong Kong influenza viruses inhibits recognition by a monoclonal antibody. Proc Natl Acad Sci USA 81:1779–1783

Soda K, Asakura S, Okamatsu M, Sakoda Y, Kida H (2011) H9N2 influenza virus acquires intravenous pathogenicity on the introduction of a pair of di-basic amino acid residues at the cleavage site of the hemagglutinin and consecutive passages in chickens. Virol J 8:64. doi:10.1186/1743-422X-8-64

Sorrell EM, Schrauwen EJ, Linster M, de GM, Herfst S, Fouchier RA (2011) Predicting ‘airborne’ influenza viruses: (trans-) mission impossible? Curr Opin Virol 1:635–642

Stech J, Garn H, Wegmann M, Wagner R, Klenk HD (2005) A new approach to an influenza live vaccine: modification of the cleavage site of hemagglutinin. Nat Med 11:683–689. doi:10.1038/nm1256

Stech O, Veits J, Weber S et al (2009) Acquisition of a polybasic hemagglutinin cleavage site by a low-pathogenic avian influenza virus is not sufficient for immediate transformation into a highly pathogenic strain. J Virol 83:5864–5868. doi:10.1128/JVI.02649-08

Stech J, Garn H, Herwig A et al (2011) Influenza B virus with modified hemagglutinin cleavage site as a novel attenuated live vaccine. J Infect Dis 204:1483–1490. doi:10.1093/infdis/jir613

Steinhauer DA (1999) Role of hemagglutinin cleavage for the pathogenicity of influenza virus. Virology 258:1–20. doi:10.1006/viro.1999.9716

Stevens J, Corper AL, Basler CF, Taubenberger JK, Palese P, Wilson IA (2004) Structure of the uncleaved human H1 hemagglutinin from the extinct 1918 influenza virus. Science 303:1866–1870

Stevens J, Blixt O, Tumpey TM, Taubenberger JK, Paulson JC, Wilson IA (2006) Structure and receptor specificity of the hemagglutinin from an H5N1 influenza virus. Science 312:404–410

Stieneke-Gröber A, Vey M, Angliker H et al (1992) Influenza virus hemagglutinin with multibasic cleavage site is activated by furin, a subtilisin-like endoprotease. EMBO J 11:2407–2414

Suarez DL, Senne DA, Banks J et al (2004) Recombination resulting in virulence shift in avian influenza outbreak, Chile. Emerg Infect Dis 10:693–699. doi:10.3201/eid1004.030396

Sun X, Tse LV, Ferguson AD, Whittaker GR (2010) Modifications to the hemagglutinin cleavage site control the virulence of a neurotropic H1N1 influenza virus. J Virol 84:8683–8690. doi:10.1128/JVI.00797-10

Szabo R, Bugge TH (2008) Type II transmembrane serine proteases in development and disease. Int J Biochem Cell Biol 40:1297–1316. doi:10.1016/j.biocel.2007.11.013

Tarnow C, Engels G, Arendt A et al (2014) TMPRSS2 is a host factor that is essential for pneumotropism and pathogenicity of H7N9 influenza A virus in mice. J Virol 88:4744–4751. doi:10.1128/JVI.03799-13

Tashiro M, Ciborowski P, Klenk HD, Pulverer G, Rott R (1987a) Role of Staphylococcus protease in the development of influenza pneumonia. Nature 325:536–537. doi:10.1038/325536a0

Tashiro M, Ciborowski P, Reinacher M, Pulverer G, Klenk HD, Rott R (1987b) Synergistic role of staphylococcal proteases in the induction of influenza virus pathogenicity. Virology 157:421–430

Tate MD, Brooks AG, Reading PC (2011a) Inhibition of lectin-mediated innate host defences in vivo modulates disease severity during influenza virus infection. Immunol Cell Biol 89:482–491

Tate MD, Brooks AG, Reading PC (2011b) Specific sites of N-linked glycosylation on the hemagglutinin of H1N1 subtype influenza A virus determine sensitivity to inhibitors of the innate immune system and virulence in mice. J Immunol 187:1884–1894

Tate MD, Job ER, Deng YM, Gunalan V, Maurer-Stroh S, Reading PC (2014) Playing hide and seek: how glycosylation of the influenza virus hemagglutinin can modulate the immune response to infection. Viruses 6:1294–1316. doi:10.3390/v6031294

Taubenberger JK, Morens DM (2008) The pathology of influenza virus infections. Annu Rev Pathol 3:499–522

Teijaro JR, Walsh KB, Cahalan S et al (2011) Endothelial cells are central orchestrators of cytokine amplification during influenza virus infection. Cell 146:980–991

Thomas G (2002) Furin at the cutting edge: from protein traffic to embryogenesis and disease. Nat Rev Mol Cell Biol 3:753–766. doi:10.1038/nrm934

Thompson CI, Barclay WS, Zambon MC, Pickles RJ (2006) Infection of human airway epithelium by human and avian strains of influenza a virus. J Virol 80:8060–8068

To KK, Chan JF, Chen H, Li L, Yuen KY (2013) The emergence of influenza A H7N9 in human beings 16 years after influenza A H5N1: a tale of two cities. Lancet Infect Dis 13:809–821. doi:10.1016/S1473-3099(13)70167-1

Tomashefski JF, Farver CF (2008) Anatomy and histology of the lung. In: Tomashefski JF, Cagle PT, Farver CF, Fraire AE (eds) Dail and Hammar's pulmonary pathology. Springer, New York, pp 20–48

Tong S, Zhu X, Li Y et al (2013) New world bats harbor diverse influenza A viruses. PLoS Pathog 9:e1003657. doi:10.1371/journal.ppat.1003657

Tse LV, Hamilton AM, Friling T, Whittaker GR (2014) A novel activation mechanism of avian influenza virus H9N2 by furin. J Virol 88:1673–1683. doi:10.1128/JVI.02648-13

Tumpey TM, Maines TR, Van HN et al (2007) A two-amino acid change in the hemagglutinin of the 1918 influenza virus abolishes transmission. Science 315:655–659

Uiprasertkul M, Puthavathana P, Sangsiriwut K et al (2005) Influenza A H5N1 replication sites in humans. Emerg Infect Dis 11:1036–1041

Upham JP, Pickett D, Irimura T, Anders EM, Reading PC (2010) Macrophage receptors for influenza A virus: role of the macrophage galactose-type lectin and mannose receptor in viral entry. J Virol 84:3730–3737. doi:10.1128/JVI.02148-09

van Kolfschooten F (2003) Dutch veterinarian becomes first victim of avian influenza. Lancet 361:1444. doi:10.1016/S0140-6736(03)13156-X

van Riel D, Munster VJ, de WE, Rimmelzwaan GF, Fouchier RA, Osterhaus AD, Kuiken T (2006) H5N1 Virus Attachment to Lower Respiratory Tract. Science 312:399

van Riel D, Munster VJ, de Wit E, Rimmelzwaan GF, Fouchier RA, Osterhaus AD, Kuiken T (2007) Human and avian influenza viruses target different cells in the lower respiratory tract of humans and other mammals. Am J Pathol 171:1215–1223 van Riel D, den Bakker MA, Leijten LM et al (2010) Seasonal and pandemic human influenza viruses attach better to human upper respiratory tract epithelium than avian influenza viruses. Am J Pathol 176:1614–1618

van Riel D, Leijten LM, van der Eerden M et al (2011) Highly pathogenic avian influenza virus H5N1 infects alveolar macrophages without virus production or excessive TNF-alpha induction. PLoS Pathog 7:e1002099

van Riel D, Leijten LM, de Graaf M et al (2013) Novel avian-origin influenza A (H7N9) virus attaches to epithelium in both upper and lower respiratory tract of humans. Am J Pathol 183:1137–1143

Vey M, Orlich M, Adler S, Klenk HD, Rott R, Garten W (1992) Hemagglutinin activation of pathogenic avian influenza viruses of serotype H7 requires the protease recognition motif R-X-K/R-R. Virology 188:408–413

Vigerust DJ, Ulett KB, Boyd KL, Madsen J, Hawgood S, McCullers JA (2007) N-linked glycosylation attenuates H3N2 influenza viruses. J Virol 81:8593–8600

Viswanathan K, Chandrasekaran A, Srinivasan A, Raman R, Sasisekharan V, Sasisekharan R (2010) Glycans as receptors for influenza pathogenesis. Glycoconj J 27:561–570

Veits J, Weber S, Stech O, Breithaupt A, Gräber M, Gohrbandt S, Bogs J, Hundt J, Teifke JP, Mettenleiter TC, Stech J (2012) Avian influenza hemagglutinins H2, H4, H8 and H14 support a highly pathogenic phenotype. Proc Natl Acad Sci USA 109:2579–2584

Wagner R, Heuer D, Wolff T, Herwig A, Klenk HD (2002a) N-Glycans attached to the stem domain of haemagglutinin efficiently regulate influenza A virus replication. J Gen Virol 83:601–609

Wagner R, Matrosovich M, Klenk HD (2002b) Functional balance between haemagglutinin and neuraminidase in influenza virus infections. Rev Med Virol 12:159–166. doi:10.1002/rmv.352

Walther T, Karamanska R, Chan RW et al (2013) Glycomic analysis of human respiratory tract tissues and correlation with influenza virus infection. PLoS Pathog 9:e1003223

Wan H, Perez DR (2007) Amino acid 226 in the hemagglutinin of H9N2 influenza viruses determines cell tropism and replication in human airway epithelial cells. J Virol 81:5181–5191

Watanabe T, Kiso M, Fukuyama S et al (2013) Characterization of H7N9 influenza A viruses isolated from humans. Nature 501:551–555

Webby RJ, Perez DR, Coleman JS et al (2004) Responsiveness to a pandemic alert: use of reverse genetics for rapid development of influenza vaccines. Lancet 363:1099–1103. doi:10.1016/S0140-6736(04)15892-3

Webster RG, Bean WJ, Gorman OT, Chambers TM, Kawaoka Y (1992) Evolution and ecology of influenza A viruses. Microbiol Rev 56:152–179

Wei SH, Yang JR, Wu HS et al (2013) Human infection with avian influenza A H6N1 virus: an epidemiological analysis. Lancet Respir Med 1:771–778. doi:10.1016/S2213-2600(13)70221-2

Weinheimer VK, Becher A, Tonnies M et al (2012) Influenza A viruses target type II pneumocytes in the human lung. J Infect Dis 206:1685–1694

Wilson IA, Skehel JJ, Wiley DC (1981) Structure of the haemagglutinin membrane glycoprotein of influenza virus at 3 A resolution. Nature 289:366–373

Xiong X, Martin SR, Haire LF et al (2013) Receptor binding by an H7N9 influenza virus from humans. Nature 499:496–499

Yao L, Korteweg C, Hsueh W, Gu J (2008) Avian influenza receptor expression in H5N1-infected and noninfected human tissues. FASEB J 22:733–740

Yasuoka S, Ohnishi T, Kawano S et al (1997) Purification, characterization, and localization of a novel trypsin-like protease found in the human airway. Am J Respir Cell Mol Biol 16:300–308. doi:10.1165/ajrcmb.16.3.9070615

Yu WC, Chan RW, Wang J et al (2011) Viral replication and innate host responses in primary human alveolar epithelial cells and alveolar macrophages infected with influenza H5N1 and H1N1 viruses. J Virol 85:6844–6855

Zeng H, Pappas C, Belser JA et al (2012) Human pulmonary microvascular endothelial cells support productive replication of highly pathogenic avian influenza viruses: possible involvement in the pathogenesis of human H5N1 virus infection. J Virol 86:667–678

Zhirnov OP, Ikizler MR, Wright PF (2002) Cleavage of influenza a virus hemagglutinin in human respiratory epithelium is cell associated and sensitive to exogenous antiproteases. J Virol 76:8682–8689
Zhirnov OP, Vorobjeva IV, Ovcharenko AV, Klenk HD (2003) Intracellular cleavage of human influenza a virus hemagglutinin and its inhibition. Biochemistry (Mosc) 68:1020–1026
Zhirnov OP, Klenk HD, Wright PF (2011) Aprotinin and similar protease inhibitors as drugs against influenza. Antiviral Res 92:27–36. doi:10.1016/j.antiviral.2011.07.014
Zhou J, Wang D, Gao R et al (2013) Biological features of novel avian influenza A (H7N9) virus. Nature 499:500–503
Zhu X, Yu W, McBride R et al (2013) Hemagglutinin homologue from H17N10 bat influenza virus exhibits divergent receptor-binding and pH-dependent fusion activities. Proc Natl Acad Sci USA 110:1458–1463. doi:10.1073/pnas.1218509110

Molecular Determinants of Pathogenicity in the Polymerase Complex

Gülsah Gabriel and Ervin Fodor

Abstract Viral pathogenesis involves numerous interactions between viral and cellular factors. In recent years, the influenza virus polymerase complex has emerged as a major determinant of interspecies transmission and pathogenicity. The viral RNA-dependent RNA polymerase, in concert with the nucleoprotein, mediates transcription and replication of the viral RNA genome in the nucleus of the infected cell. The activity by which the viral polymerase complex performs these processes in mammalian cells is considered to be a major contributor to viral pathogenicity in mammals. In this chapter, we summarise our current understanding on the pathogenicity determinants in the viral polymerase complex and highlight some of its cellular interaction partners. We particularly discuss the role of importin-α isoforms in host adaptation and pathogenesis as well as the role of the viral polymerase in regulating cellular responses to viral infection.

Abbreviations

PB1	Polymerase basic protein 1
PB2	Polymerase basic protein 2
PA	Polymerase acidic protein
NP	Nucleoprotein
NEP	Nuclear export protein
NS1	Non-structural protein 1
NS2	Non-structural protein 2
vRNA	Viral RNA
vRNP	Viral ribonucleoprotein
cRNA	Complementary RNA
CRM1	Chromosome region maintenance 1
NLS	Nuclear localisation signal

G. Gabriel (✉)
Heinrich-Pette-Institute, Leibniz Institute for Experimental Virology, Hamburg, Germany
e-mail: guelsah.gabriel@hpi.uni-hamburg.de

E. Fodor
Sir William Dunn School of Pathology, University of Oxford, Oxford, UK

Current Topics in Microbiology and Immunology (2014) 385: 35–60
DOI: 10.1007/82_2014_386

Published Online: 18 July 2014

Hsp90	Heat shock protein 90
IFN	Interferon
RIG-I	Retinoic acid-inducible gene 1
MAVS	Mitochondrial antiviral signalling protein
CPSF30	Cleavage and polyadenylation specificity factor 30
Pol II	RNA polymerase II

Contents

1 Introduction

Influenza viruses encode their own RNA-dependent RNA polymerase which consists of three subunits, PB1, PB2 and PA. The viral polymerase complex together with NP mediates transcription and replication of the viral genome in the nucleus of the infected host cell. The lack of proofreading activity results in high mutation rates during replication that facilitate viral adaptation to new host environments. Upon interspecies transmission, adaptive mutations in the viral polymerase complex and NP result in increased polymerase activity which may further facilitate enhanced pathogenesis and transmission in the mammalian host.

2 Transcription and Replication by the Viral Polymerase

2.1 Molecular Mechanisms of Transcription and Replication

The influenza A virus genome consists of eight single stranded RNA segments of negative polarity (reviewed in (Krug and Fodor 2013; Shaw and Palese 2013)). These RNA segments associate with the trimeric RNA-dependent RNA polymerase and oligomeric NP to form viral ribonucleoprotein (vRNP) complexes, which represent the minimal replication machinery of influenza virus (reviewed in (Fodor 2013; Resa-Infante et al. 2011)).

After virus internalisation by endocytosis and fusion of viral and endosomal membranes the vRNPs are released into the cytoplasm and are transported into the nucleus via the classical importin-α/β_1 pathway (O'Neill et al. 1995; O'Neill and Palese 1995). Initially, vRNPs act as transcription complexes producing viral mRNAs. Transcription is a primer-dependent event and it involves the use of 5′ capped RNA primers derived from host pre-mRNAs. These RNA primers are generated by the viral RNA polymerase. Association of the viral RNA polymerase with the cellular RNA polymerase II that synthesises capped pre-mRNAs is believed to facilitate the access of the viral RNA polymerase to cellular capped pre-mRNA (Engelhardt et al. 2005). Subsequently, the capped RNA primer is elongated by the viral polymerase using the vRNA as template. Transcription terminates by the addition of a poly(A) tail. Thus, viral transcription results in viral mRNAs that are structurally similar to cellular mRNAs containing a 5′ cap structure and a poly(A) tail at the 3′ end. This structural similarity enables viral mRNAs to be processed by cellular mRNA processing pathways and assembled into translation-competent mRNPs (reviewed in (York and Fodor 2013)). Translation of viral proteins is required for the onset of viral replication. vRNA is copied into a complementary RNA (cRNA) replication intermediate which then acts as a template for the synthesis of vRNA. Both these processes are primer-independent events resulting in RNAs containing 5′ triphosphates. Both cRNA and vRNA exist as ribonucleoprotein complexes in which the RNA is associated with RNA polymerase and oligomeric NP (York et al. 2013). The viral nuclear export protein (NEP), formerly known as non-structural protein 2 (NS2) is implicated in the regulation of viral RNA genome replication (reviewed in (Paterson and Fodor 2012)). The resulting vRNPs can act as templates for both transcription and replication or, after nuclear export and transport across the cytoplasm, they can be assembled into progeny virions. NEP acts as an export factor for progeny vRNPs mediating their interaction with the cellular CRM1-dependent nuclear export pathway (reviewed in (Paterson and Fodor 2012)).

2.2 *Structure and Function of the RNA Polymerase Complex*

PB1 is the catalytic subunit of the RNA polymerase directly involved in RNA synthesis (reviewed in (Mehle and McCullers 2013; Ruigrok et al. 2010)). Apart from short N-terminal and C-terminal fragments, which have been co-crystallised with a C-terminal fragment of PA and an N-terminal fragment of PB2, respectively, structural information is not available for PB1 (He et al. 2008; Obayashi et al. 2008; Sugiyama et al. 2009). Sequence alignments show that the central region of PB1 contains the classical polymerase motifs (Fig. 1). A nuclear

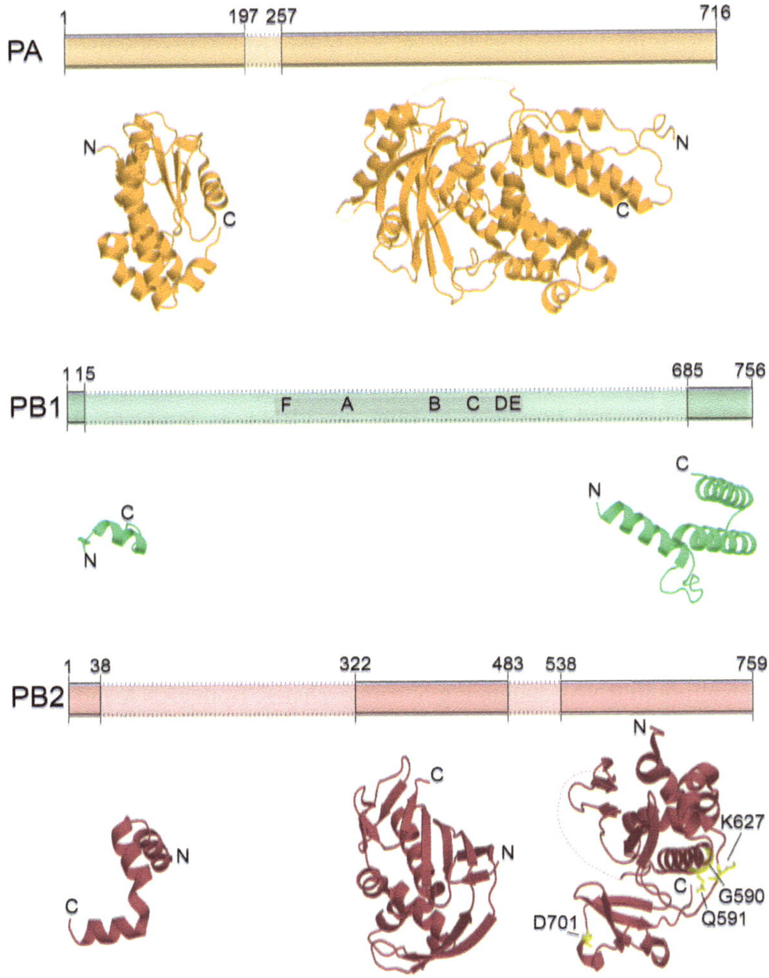

Fig. 1 Structure and functional domains of the influenza a virus polymerase. A diagram of the three subunits and domains with known three-dimensional structure is shown. These include the N-terminal endonuclease and C terminal domains of PA (Dias et al. 2009; He et al. 2008; Obayashi et al. 2008; Yuan et al. 2009) and the central cap-binding and C-terminal 627-NLS domain of PB2 (Guilligay et al. 2008; Kuzuhara et al. 2009a, b; Tarendeau et al. 2008; Yamada et al. 2010). Key adaptive positions are indicated in the 627-NLS domain of PB2. Structures of the N-terminal and C-terminal regions of PB1, as well as of the N-terminal region of PB2, are also shown (He et al. 2008; Obayashi et al. 2008; Sugiyama et al. 2009). The polymerase signature domains (A–F) are indicated in the central region of PB1 (Biswas and Nayak 1994; Poch et al. 1989). Structures were generated with PyMOL using the following PDB accession numbers: PA endonuclease domain, 3EBJ; PA C-terminal domain and PB1 N-terminus, 2ZNL; PB1 C-terminus and PB2 N-terminus, 2ZTT; PB2 cap-binding domain, 4CB4 and PB2 627-NLS domain, 2VY6

localisation signal (NLS) has been identified in the N-terminal third of PB1, although when PB1 is expressed in the absence of PB2 and PA, it predominantly accumulates in the cytoplasm (Fodor and Smith 2004). However, in the presence of co-expressed PA, it accumulates efficiently in the nucleus and RanBP5 and Hsp90 have been implicated in the nuclear import of a PB1-PA dimer (Deng et al. 2006; Naito et al. 2007). Hsp90 has also been implicated in the nuclear import of a PB1-PB2 dimer (Naito et al. 2007).

PB2 is a cap binding protein with the cap-binding domain located in the central region of the polypeptide as shown by structural analysis (Guilligay et al. 2008). Structural information is also available for the N-terminal fragment that interacts with PB1, the so called 627 domain, named after the prominent host-specific amino acid residue 627, as well as for the NLS domain (Fig. 1) (Kuzuhara et al. 2009a, b; Tarendeau et al. 2007, 2008; Yamada et al. 2010). Residue 627 is typically lysine in mammalian adapted influenza viruses, while in avian influenza viruses glutamic acid predominates. The replacement of the lysine with glutamic acid has little effect on the overall fold of the 627 domain. However, it disrupts a prominent basic patch on the protein surface. The PB2 of the 2009 pandemic H1N1 influenza viruses which unusually had a glutamic acid at this position acquired suppressor mutations at nearby sites (Fig. 2) (Mehle and Doudna 2009). The serine at position 590 and arginine at 591 were proposed to neutralise the glutamic acid at position 627 and partially restore the positively charged surface of the 627 domain. The function of the 627 domain remains obscure, although it has been implicated in RNA binding as well as in modulating the accessibility of the NLS domain (Tarendeau et al. 2007, 2008). The C-terminal bipartite NLS mediates nuclear import of PB2 via the importin-α/β_1 pathway (Gabriel et al. 2008; Tarendeau et al. 2007). The NLS domain has been co-crystallised with the cellular factor importin-α5. The structure shows that the extreme C-terminus containing the NLS assumes an extended conformation that positions the bipartite NLS into a superhelical groove on importin α5. However, in the absence of importin-α, the C-terminal extremity is tethered to the core of the domain. This interaction is mediated by a salt bridge between amino acid residues D701 and R753. PB2 D701N is another important host adaptive mutation observed upon avian influenza virus adaptation to mammals (Fig. 3). It was proposed that PB2 D701N mutation would disrupt the salt bridge between D701 and R753 and thus expose the NLS and lead to enhanced importin-α binding (Gabriel et al. 2008; Tarendeau et al. 2007). Individually expressed PB2 efficiently accumulates in the nucleus where it assembles with the PB1-PA dimer into a heterotrimer (Deng et al. 2006). However, it has been proposed that PB2 can also enter the nucleus as part of a PB1-PB2 dimer (Naito et al. 2007).

Structural analysis demonstrates that the PA subunit consists of two domains, the N-terminal endonuclease domain and a C-terminal domain that mediates the interaction with PB1 (Fig. 1) (Dias et al. 2009; He et al. 2008; Obayashi et al. 2008; Yuan et al. 2009). The two domains are separated by a flexible linker sequence. The N-terminal endonuclease domain is involved in the cleavage of cellular pre-mRNAs generating 5′ capped RNA fragments for priming transcription. The C-terminal

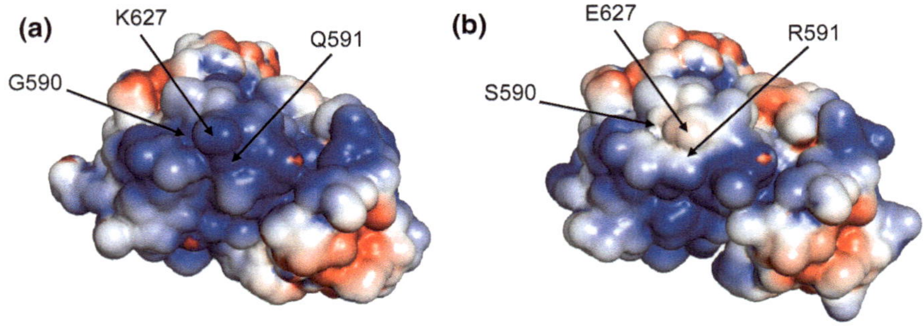

Fig. 2 Key adaptive mutations in the PB2 627-NLS domain of the influenza A virus RNA polymerase. Surface representation of the PB2 627-NLS domain of a human H5N1 isolate with PB2 627K mammalian signature (*panel A*) and of a 2009 pandemic H1N1 influenza virus with PB2 590S/591R mammalian signatures (*panel B*) (Yamada et al. 2010). Positions of key adaptive residues are highlighted. These are coloured according to the electrostatic surface potential calculated using APBS in PyMOL (Baker et al. 2001). *Red* and *blue* indicate negative and positive regions, respectively. Structures were generated using the following PDB accession numbers: 3KC6 for H5N1 and 3KHW for H1N1

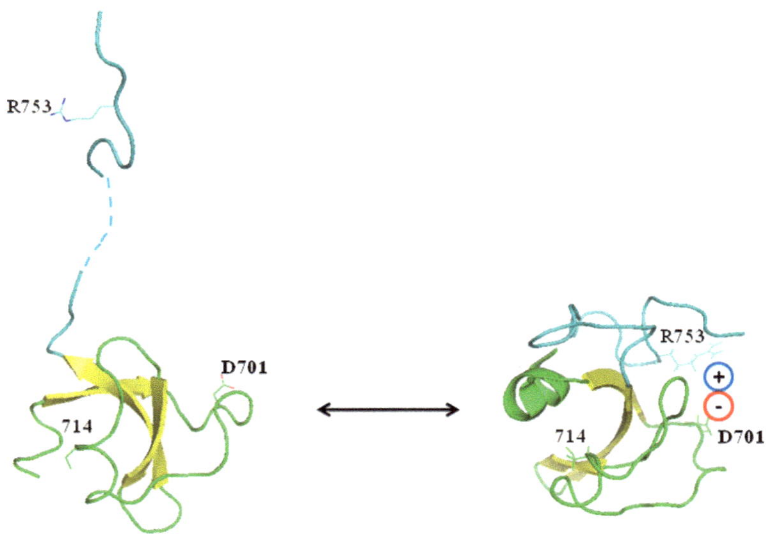

Fig. 3 Conformational change in the PB2 NLS domain. The C-terminus of PB2 shows two conformations. In the open form, the NLS at the extreme end is exposed to allow interaction with importin-α (*left side*). In the closed form the NLS is tethered to the main body of the domain through a salt bridge between R753 and D701 (*right side*). PB2 D701N adaptive mutation would facilitate exposure of the NLS. Structures were generated with PyMOL using published PDB accession number 2GMO and RCSB (Tarendeau et al. 2007). Modified from (Gabriel et al. 2013)

domain of PA that assumes a "dragon head" shape interacts with the N-terminus of PB1 through its "jaw" (He et al. 2008; Obayashi et al. 2008).

NP is an essential component of the influenza virus transcriptional machinery. It is a highly basic protein that consists of two major domains, the head and body domains (Ng et al. 2008; Ye et al. 2006). The two domains are separated by a negatively charged groove that is involved in RNA binding. Within the RNP, the NP occurs as a homo-oligomer; a tail loop of one NP molecule is inserted into a cavity in the body domain of a neighbouring NP (Chan et al. 2010). NP is responsible for maintaining the double helical structures of vRNPs and cRNAs through NP–NP interactions (Arranz et al. 2012; Moeller et al. 2012; York et al. 2013).

Several auxiliary viral proteins encoded by the polymerase gene segments have been identified, i.e. PB1-F2, PB1-N40, and the more recently described PA-X, PA-N155 and PA-N182 polypeptides which play a role in influenza virus pathogenesis (Chen et al. 2001; Jagger et al. 2012; Muramoto et al. 2013; Shi et al. 2012; Wise et al. 2009).

3 Host Adaptive Mutations in the Viral Polymerase Complex

After the first report of an H5N1 highly pathogenic avian influenza virus (HPAIV) transmission from infected poultry to humans in 1997 in Hong Kong, evidence steadily increased that the viral polymerase is a major determinant of interspecies transmission and pathogenesis (Subbarao et al. 1998). Many subsequent in vitro and in vivo studies identified host adaptive signatures in the viral polymerase complex and NP which are associated with enhanced polymerase activity, pathogenicity and transmission in mammalian species.

Most notably, the PB2 627K mammalian signature is highly prevalent in isolates obtained from humans infected with H5N1, H7N7, H9N2, H7N9 and recently H10N8 where it often correlates with fatal outcome (Chen et al. 2013, 2014; de Jong et al. 2006; Fouchier et al. 2004; Kageyama et al. 2013; Peiris et al. 1999; Subbarao et al. 1998). However, some isolates obtained from H5N1 and H7N9 avian influenza virus infected patients have maintained the avian signature PB2 627E but acquired PB2 701N as another mammalian signature which also correlates with severe disease in humans (Chen et al. 2013; de Jong et al. 2006; Kageyama et al. 2013; Le et al. 2009). In contrast, 2009 pandemic H1N1 strains do neither contain PB2 627K nor 701N (Garten et al. 2009). Instead, they have acquired another alternative signature in PB2 590S/591R which is absent from seasonal H1N1 viruses that circulated before 2009 (Mehle and Doudna 2009).

Besides these PB2 adaptive signatures, several other mutations in PB1, PA, NP and NEP were implicated in enhanced viral replication and pathogenesis in mammalian species (Table 1) (reviewed in (Cauldwell et al. 2014; Gabriel et al. 2013;

Table 1 Host adaptive signatures in the viral polymerase complex

vRNP	Host adaptive mutations	Function and pathogenicity in mammals	Reference
PB2	D9N	Mitochondrial PB2 localisation ↓ IFN-ß expression ↑ virulence in mice	(Graef et al. 2010; Kim et al. 2010)
	L89V, G309D, T33K, R477G, I495V, A676T	↑ virulence of H5 in mice	(Li et al. 2009)
	E158G	↑ virulence of H1 and H5 in mice	(Zhou et al. 2013a, b)
	D253N	↑ polymerase activity of H9	(Mok et al. 2011)
	T271A	↑ polymerase activity of H1	(Bussey et al. 2010)
	I504V, PA I550L	Cellular RNA-polymerase II degradation ↑ pathogenicity of H1 in mice	(Llompart et al. 2014)
	T588I	↑ MAVS binding ↓ IFN-ß expression	(Foeglein et al. 2011; Zhao et al. 2013)
	S590R, G591Q	Observed in human H1 isolates ↑ polymerase activity of H1	(Mehle and Doudna 2009)
	Q591K	↑ polymerase activity of H5 and H9 ↑ virulence of H5 and H9 in mice	(Mok et al. 2011; Yamada et al. 2010)
	E627K	Observed in human H1, H3, H5, H7, H9 and H10 isolates ↑ polymerase activity of H1, H3, H5, H7 and H9 ↑ virus replication requires NP from same viral background ↑ polymerase mobility in the nuclei ↑ binding of vRNPs to importin-α1 and -α7; ↓ reduced polymerase activity in importin-α1 and -α7 silenced cells; ↓ reduced viral replication in importin-α1 and -α7 silenced cells ↑ binding to NP ↑ recruitment of RNA promoter	(Bogs et al. 2011; Bortz et al. 2011; Chen et al. 2013, 2014; de Jong et al. 2006; Foeglein et al. 2011; Fouchier et al. 2004; Gabriel et al. 2005, 2008, 2011; Gao et al. 2009; Hatta et al. 2001, 2007; Herfst et al. 2010, 2012; Hudjetz and Gabriel 2012; Jagger et al. 2012; Kageyama et al. 2013; Labadie et al. 2007; Li et al. 2009; Mok et al. 2011; Paterson et al. 2014; Peiris et al. 1999; Shinya et al. 2004, 2009; Steel et al. 2009; Subbarao et al. 1993, 1998; Yamada et al. 2010)

(continued)

Table 1 (continued)

vRNP	Host adaptive mutations	Function and pathogenicity in mammals	Reference
		↑ virulence in mice; ↓virulence in importin-$\alpha7^{-/-}$ mice ↑ contact and airborne transmission of H1, H3 and H5 in guinea pigs ↑ airborne transmission of H5 in ferrets	
	D701N	Observed in human H5 and H7 isolates ↑ polymerase activity of H1, H3, H5 and H7 ↑ polymerase mobility in the nuclei ↑ binding to importin-α; ↓ reduced viral replication in importin-α1 and -α7 silenced cells ↑ virulence in mice; ↓ virulence in importin-$\alpha7^{-/-}$ mice ↑ contact and airborne transmission of H1, H3 and H5 in guinea pigs ↑ airborne transmission of H1 in ferrets	(Boivin and Hart 2011; Bortz et al. 2011; de Jong et al. 2006; Foeglein et al. 2011; Gabriel et al. 2005, 2008, 2011; Gao et al. 2009; Herfst et al. 2010; Kageyama et al. 2013; Resa-Infante et al. 2008; Shinya et al. 2009; Steel et al. 2009; Tarendeau et al. 2007; Zhou et al. 2013a, b)
PB1	L472V, L598P	↑ polymerase activity of H1	(Xu et al. 2012)
NP	A150R	↑ polymerase activity of H5 ↑ binding to NP	(Ng et al. 2012)
	N319K	↑ polymerase activity of H7 ↑ virulence of H7 in mice ↑ binding to importin-α	(Gabriel et al. 2005, 2008, 2011)
NEP	S7L, Y41C, E75G, X161M	↑ virus replication in the absence of PB2 627K	(Manz et al. 2012)

Effect of host adaptive mutations on viral polymerase activity, pathogenicity and transmission in mammals. ↑ enhancing effect; ↓ reducing effect

Manz et al. 2013)). Here, we will summarise our current knowledge on the key PB2 adaptive sites at positions 627, 701 and 590/591 and discuss their implications for viral replication, pathogenesis and transmission in mammals.

3.1 Polymerase Activity in Mammalian Cells

The PB2 E627K mutation was shown to mediate enhanced viral polymerase activity in a broad range of mammalian, including human cells. This ability generally correlated with avian-mammalian adaptation and pathogenesis in mice (Gabriel et al. 2005; Hatta et al. 2001). Interestingly, the enhancing effect of this host adaptive mutation on polymerase activity was not observed in avian cells (Hudjetz and Gabriel 2012; Labadie et al. 2007; Mehle and Doudna 2008; Paterson et al. 2014). Several studies have proposed potential mechanisms for how the PB2 E627K mutation might result in enhanced viral polymerase activity in mammalian cells. Avian influenza viruses preferentially replicate in the avian intestine where the temperature is around 40 °C (Murakami et al. 1988; Murphy et al. 1982a, b). In contrast, human influenza viruses have a replication optimum at lower temperatures (33–37 °C). This cold-sensitivity of avian influenza viruses is believed to restrict their ability to replicate efficiently in the upper respiratory tract of humans where the medium temperature is around 33 °C. The PB2 E627K mutation was shown to promote replication of H5N1 HPAIV in the upper respiratory tract of humans and mice (Hatta et al. 2007; Le et al. 2009; Massin et al. 2001). However, the mechanism by which the E627K mutation would promote higher polymerase activity at lower temperature remains unclear. Several studies proposed that the PB2 E627K mutation increased polymerase activity by promoting vRNP assembly in mammalian cells by increasing the affinity of PB2 for NP (Labadie et al. 2007; Mehle and Doudna 2008; Ng et al. 2012; Rameix-Welti et al. 2009). However, recent studies challenged this idea by suggesting that the observed apparent increase in PB2-NP affinity is the result of increased accumulation of vRNPs due to increased polymerase activity (Cauldwell et al. 2013; Paterson et al. 2014). In fact, more recently the PB2 627E mediated restriction of the viral polymerase was shown to be independent of NP but dependent on the length of the viral RNA template. Moreover, the restricted polymerase activity of PB2 627E complexes could be rescued by the introduction of specific mutations into the promoter region of viral RNAs (Crescenzo-Chaigne et al. 2002; Paterson et al. 2014). Therefore, it was proposed that PB2 627 might affect the recruitment of the viral RNA promoter by the viral RNA polymerase in mammalian cells. It was also hypothesised that an inhibitory factor must exist in human cells that restricts PB2 627E polymerase activity (Mehle and Doudna 2008). This concept was opposed by another study suggesting that instead, a positive factor is present in human cells which promotes PB2 627K polymerase activity (Moncorge et al. 2010). In agreement with the existence of a host factor that interacts with the viral polymerase and modulates its activity, it was found that the identity of 627 also affects the intranuclear mobility of the viral polymerase (Foeglein et al. 2011). In particular, low polymerase activity of PB2 627E correlated with significantly slower diffusion in human but not avian nuclei suggesting an interaction with a relatively large soluble inhibitory cellular factor and/or with an insoluble "static structure". Several host factors have been identified that differentially regulate H5N1 polymerase according to the identity of the PB2 627 residue, including the DEAD box RNA helicase DDX17/p72 that facilitated efficient transcription and replication of a human isolate of an H5N1 virus (627K) in mammalian cells (Bortz et al. 2011). In addition, importin-α isoforms show differential interaction with the viral polymerase and vRNPs depending on the identity of PB2 627 (see below). Several compensatory mutations in the viral polymerase subunits NP and NEP have been described that result in increased activity of avian influenza virus derived polymerases in mammalian cells (Table 1). The 2009 swine origin pandemic H1N1 virus acquired suppressor mutations at positions 590/591 of PB2 (Mehle and Doudna 2009), while adaptive mutations in the viral NEP were able to enhance polymerase activity in mammalian cells of an H5N1 polymerase containing the PB2 627E avian signature (Manz et al. 2012).

The host adaptive mutation PB2 D701N also mediates enhanced viral polymerase activity in mammalian, including human cells, and furthermore, correlates with avian-mammalian adaptation and pathogenesis in mice (Gabriel et al. 2005). Originally, this mutation was observed upon adaptation of H3N2 and H7N7 HPAIV to mice (Brown et al. 2001; Gabriel et al. 2005). The polymerase activity increasing effect of the PB2 D701N mutation was observed at a broad range of temperatures from 33 to 39 °C (Gabriel et al. 2005; Zhou et al. 2013a). Sequencing analysis of isolates obtained from humans infected with H5N1 revealed that PB2 701N, similar to 627K, contributes to efficient virus propagation in the human upper respiratory tract (Le et al. 2009).

It was suggested that PB2 590S/591R mammalian signatures specific for 2009 pandemic H1N1 influenza viruses contribute to elevated polymerase activity compared to seasonal H1N1 strains in which PB2 590G/591Q are present. The PB2 S590G/R591Q mutation reduced 2009 pandemic H1N1 viral polymerase activity in mammalian cells (Mehle and Doudna 2009). Interestingly, introduction of PB2 627K or 701N mammalian signatures into PB2 of 2009 pandemic H1N1 strains further elevated viral polymerase activity in human cells (Herfst et al. 2010; Song et al. 2011; Yamada et al. 2010).

In general, enhanced polymerase activity in mammalian cells seems to be predictive of elevated pathogenicity in mammalian species. However, it should be noted that none of these mammalian signatures (PB2 672K, 701N and 590S/591R) occur in combination in naturally circulating influenza virus strains. These observations suggest that during viral adaptation influenza viruses can acquire alternative mutations that enhance replicative fitness in mammalian cells.

3.2 Pathogenicity in Mammals

Both key PB2 627K and 701N mammalian signatures were shown to enhance not only viral polymerase activity but also pathogenicity of numerous influenza virus subtypes and strains in mice (Bogs et al. 2011; Fornek et al. 2009; Gabriel et al. 2005; Hatta et al. 2001; Salomon et al. 2006; Shinya et al. 2004). Furthermore, both signatures contributed individually to H5N1 virus propagation in the human upper respiratory tract (Le et al. 2009). In H5N1 infected patients in Vietnam, the presence of either PB2 627K or 701N mammalian signatures correlated with a high virus load in the blood. Disseminated virus replication and hypercytokinemia were hallmarks of fatal H5N1 and H7N9 infection in humans (de Jong et al. 2006; Zhou et al. 2013b). Interestingly, the presence of viral RNA in the blood was only observed together with very high pharyngeal virus loads. This suggests that after primary infection of the human respiratory tract, H5N1 HPAIV containing PB2 627K or 701N mammalian signatures may disseminate to other organs when a certain threshold in the lung is reached. Other studies performed in chickens further support the observation that these mammalian signatures specifically enhance pathogenicity in mammals. In contrast to mammalian species, PB2 E627K mutation did not change virulence in chickens (Bogs et al. 2011).

Whether the polymerase activity elevating effect of PB2 G590S/Q591R mutations (Mehle and Doudna 2009) contributes to pathogenicity in mammals remains unclear. However, a PB2 Q591K mutation was shown to enhance not only H5N1 polymerase activity, but also replication and virulence in mice (Yamada et al. 2010). On the other hand, introduction of PB2 627K or 701N into 2009 pandemic H1N1 influenza viruses did enhance viral polymerase activity but not virus replication or virulence in mice (Herfst et al. 2010; Song et al. 2011; Yamada et al. 2010). Future studies will be required to understand the role of PB2 590S/591R in 2009 pandemic H1N1 influenza virus pathogenicity in mammals.

3.3 Transmissibility in Mammals

There is ample evidence that mammalian signatures in the PB2 subunit of the viral polymerase are a prerequisite for the acquisition of sustained inter-host transmission. Host adaptive mutations in PB2 E627K or D701N have been repeatedly shown not only to enhance pathogenicity in mammals but also transmission of H3N2 or H5N1 influenza viruses between mammalian hosts (Gao et al. 2009; Steel et al. 2009). PB2 K627E mutation in human H3N2 or avian H5N1 strains reduced contact and aerosol transmission in guinea pigs, whereas introduction of PB2 701N was sufficient to recover transmissibility (Steel et al. 2009). However, the presence of these single mammalian signatures, while required, is generally not sufficient to confer sustained mammalian-to-mammalian transmission. Introduction of PB2 627K or 701N into various avian H5N1 virus strains showed that the presence of either of these signatures is a prerequisite for efficient contact transmission in guinea pigs but required additional mammalian signatures in the hemagglutinin (Gao et al. 2009; Steel et al. 2009). Studies performed in ferrets further highlighted that airborne transmissibility of H5 influenza viruses occured when mammalian signatures in the hemagglutinin as well as in PB2, such as 627K or 590S/591R were present (Herfst et al. 2012; Imai et al. 2012). This is in agreement with observations that recent H7N9 strains which contain the mammalian signatures PB2 627K or 701N show limited airborne transmissibility in ferrets (Richard et al. 2013). Thus, the presence of mammalian signatures seems to be required but not sufficient for sustained airborne transmission in mammals.

The 2009 pandemic H1N1 strains showed airborne transmission properties in the human population as well as in the ferret model. However, discrepancies were reported when their transmission efficiencies were compared to seasonal H1N1 influenza viruses. While some studies report that 2009 pandemic H1N1 influenza viruses have an increased potential for airborne transmission compared to seasonal H1N1 strains (Munster et al. 2009), others show decreased transmissibility (Maines et al. 2009) in the ferret model. Furthermore, it was also reported that PB2 R591Q reduced airborne transmission in ferrets suggesting that PB2 591R is required for replicative advantage and aerosol transmission in mammals (Yamada et al. 2010). On the other hand, introduction of neither PB2 627K nor 701N mammalian signatures into 2009 pandemic H1N1 influenza viruses enhanced airborne transmission in ferrets (Herfst et al. 2010; Yamada et al. 2010). However, a recent study reports that introduction of PB2 701N into a distinct 2009 pandemic H1N1 lineage was able to elevate transmissibility in ferrets (Zhou et al. 2013a).

Thus, transmission is a polygenic trait and the presence of key mammalian signatures in PB2 seems to be necessary but not always sufficient for sustained mammalian-to-mammalian transmission.

4 Cellular Interaction Partners of the RNA Polymerase

Genome-wide RNA interference (RNAi) screens (Brass et al. 2009; Hao et al. 2008; Karlas et al. 2010; Konig et al. 2010; Shapira et al. 2009; Sui et al. 2009) as well as proteomic analyses (Bradel-Tretheway et al. 2011; Jorba et al. 2008; Mayer et al. 2007) have identified numerous cellular factors involved in influenza virus replication. Thereby, cellular networks were identified which are involved in various steps of the viral life cycle, such as virus entry, nuclear import, transcription and replication, cell signalling and nuclear export (reviewed in (Gabriel et al. 2013; Shaw 2011; Watanabe et al. 2010)). Substantial body of evidence emerged in the last years that components of the cellular import machinery play a major role in influenza virus interspecies transmission and pathogenesis. Therefore, here we will focus on the role of the nuclear import factors, namely importin-α proteins, and discuss their implication in influenza virus replication and avian-mammalian host adaptation. Furthermore, we will highlight novel insights into the interactions of the viral polymerase with innate immune pathways and cellular

RNA polymerase II that were recently found to contribute to influenza virus pathogenicity in the mammalian host.

4.1 Importin-α Mediated Host Adaptation and Pathogenesis

Influenza virus vRNPs as well as the polymerase subunits utilise differential nuclear transport pathways to enter the nucleus of the host cell (reviewed in (Hutchinson and Fodor 2013; Resa-Infante and Gabriel 2013)). Transport of vRNPs is mediated by the classical importin-α/ß$_1$ pathway (O'Neill et al. 1995; O'Neill and Palese 1995). Thus, NP binds directly to the adaptor protein importin-α which then triggers the nuclear import of vRNPs together with the receptor protein importin-ß$_1$. The monomeric PB2 and NP proteins are also believed to be transported into the nucleus by the classical, importin-α dependent, pathway (Gabriel et al. 2008; Tarendeau et al. 2007). In the nucleus, PB2 assembles with PB1 and PA which are transported separately as PB1-PA dimers by the non-classical, importin-α independent, nuclear import pathway by direct interaction with RanBP5 (Fig. 4) (Deng et al. 2006). In addition, Hsp90 was proposed to be involved in the nuclear import of PB1-PB2 or PB1-PA heterodimers (Naito et al. 2007).

In recent years, evidence increased that key mammalian signatures, such as PB2 627K and 701N, utilise importin-α isoforms to promote viral replication in mammalian cells. It was proposed that divergence between avian and human homologues of importin-α isoforms (α1- α7) (Gabriel et al. 2011) but not RanBP5 might be sufficient to restrict virus replication in a host-dependent manner. Accordingly, interaction studies revealed that PB2 proteins from human influenza virus strains have higher binding affinities to human importin-α isoforms than PB2 proteins from avian strains (Resa-Infante et al. 2008). PB2 D701N as well as NP N319K mutations were shown to enhance PB2 and NP binding affinities to human but not avian importin-α isoforms. This correlated with enhanced nuclear localisation of PB2 and NP in mammalian cells (Boivin and Hart 2011; Gabriel et al. 2008). In contrast, the E627K mutation did neither affect monomeric PB2 binding to human importin-α isoforms nor alter PB2 nuclear localisation. In the context of vRNPs on the other hand, the PB2 E627K mutation led to enhanced binding of vRNPs to human importin-α1, -α5 and -α7 isoforms without affecting subcellular distribution of the viral polymerase subunits. This enhanced vRNP binding to these importin-α isoforms was proposed to be mediated by NP (Hudjetz and Gabriel 2012). This is in agreement with previous findings suggesting that importin-α isoforms may have additional, yet unknown, functions beyond nuclear transport (Resa-Infante et al. 2008). It should also be noted that single cargo proteins might interact with several importin-α isoforms but may show distinct specificities for certain isoforms when complexed with other proteins due to competition for endgoneous importin-α availability (Friedrich et al. 2006). Moreover, importin-α1 and -α7 were proposed to act as positive regulatory host factors for polymerases containing PB2 627K mammalian signatures (Hudjetz and Gabriel 2012). These differential importin-α

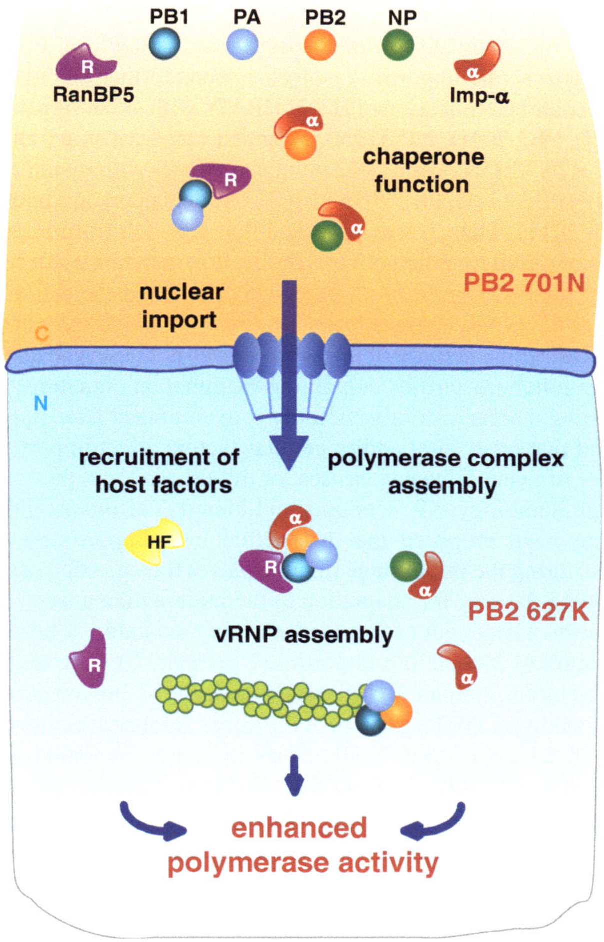

Fig. 4 Potential cytoplasmic and nuclear functions of importin-α proteins and their effect on influenza virus polymerase activity. Chaperone activities were proposed for RanBP5 and importin-α which might inhibit aggregation of PB1 and NP. Importin-α may also participate in the assembly of the viral polymerase complex contributing to the formation of functional vRNP complexes. Following nuclear transport, importin-α might be additionally required for viral transcription and/or replication. Moreover, importin-α might indirectly recruit other cellular host factors that additionally affect viral polymerase activity. All of these functions might contribute to enhanced polymerase activity mediated by host adaptive signatures in PB2. While PB2 D701N facilitates nuclear localisation determined by enhanced importin-α binding in the cytoplasm, PB2 E627K containing vRNPs utilise importin-α for promoting viral polymerase activity in the nucleus. *C* Cytoplasm; *N* Nucleoplasm; *R* RanBP5; α Importin-α proteins; *HF* Host factors. Modified from (Resa-Infante and Gabriel 2013)

interaction patterns with PB2 containing 701N and 627K mammalian signatures are consistent with structural reports that PB2 D701N but not E627K would modulate NLS exposure and accessibility for importin-α binding (Tarendeau et al. 2007, 2008). Whether this is linked to the exposed "basic patch" on PB2 627K and its potential biding to acidic importin-α isoforms needs further elucidation.

In the viral context, replication of H7N7 HPAIV with avian signatures in PB2 and NP (PB2 627E, PB2 701D, NP 319N) depended on importin-α1 and -α3, whereas replication of H7N7, H5N1 and H3N2 influenza viruses with mammalian signatures in PB2 and NP (PB2 627K, PB2 701N, NP 319K) depended on importin-α1 and -α7 (Gabriel et al. 2011). Thus, it was proposed that a switch from importin-α3 to -α7 dependency is required for efficient virus replication in mammalian cells. In contrast to importin-α1 and -α7, importin-α3 negatively regulated the activity of both PB2 627K as well as 627E polymerases in mammalian cells (Hudjetz and Gabriel 2012). However, the restricting activity of importin-α3 was not observed during replication of mammalian influenza viruses containing mammalian signatures (PB2 627K or 701N) suggesting that its restrictive activity is overcome at later stages during viral replication and that other viral and/or cellular factors affect importin-α specificity. The preference of avian influenza viruses for importin-α3 was particular interesting due to the high homology (99 % amino acid identity) of this isoform in birds and humans. It has been proposed that interaction with importin-α3 might play an important role during the initial stage of interspecies transmission of avian influenza viruses by paving the way for adaptation to the mammalian host.

With importin-α knockout ($\alpha^{-/-}$) mice becoming available, it became possible to confirm that animals lacking the importin-α7 gene ($\alpha 7^{-/-}$) were less susceptible to mammalian including human H1N1, H5N1 and H7N7 influenza virus infections compared to wildtype (WT), $\alpha 4^{-/-}$ or $\alpha 5^{-/-}$ mice. Reduced pathogenicity in $\alpha 7^{-/-}$ mice correlated with a restricted pulmonary infection compared to disseminated virus replication in WT animals (Gabriel et al. 2011). Interestingly, 2009 pandemic H1N1 influenza viruses with PB2 590S/591R mammalian signatures were exceptional in displaying importin-α3 as well as -α7 dependencies for their replication in human cells. This highlights that adaptive mutations other than PB2 E627K and D701N might additionally affect importin-α specificities.

Future studies will be required to understand how importin-α isoforms are utilised beyond nuclear transport and thereby contribute to enhanced viral replication and pathogenesis in the mammalian host.

4.2 *Interactions of the Viral Polymerase with Innate Immune Pathways*

The outcome of influenza virus infections is greatly influenced by the cellular innate immune responses, dominated by the type I interferon system. Influenza virus infection triggers the activation of innate immune responses through the recognition of pathogen associated molecular patterns by pathogen recognition receptors (reviewed in (Iwasaki and Peiris 2013)). The RNA helicase, RIG-I, is a key sensor of influenza virus RNA (Rehwinkel et al. 2010). Upon activation, RIG-I signals through the mitochondrial antiviral signalling (MAVS) protein that in turn activates kinases leading to the phosphorylation of transcription factors required for the expression type I IFN and other proinflammatory cytokines.

Influenza virus has evolved several strategies to counteract these host immune responses. The viral non-structural protein 1 (NS1) is an IFN antagonist that inhibits the expression of IFN by several independent mechanisms that include the specific inhibition of RIG-I activation. NS1 is also known to inhibit CPSF30, a cellular mRNA polyadenylation factor, leading to the general inhibition of host gene expression, including of antiviral genes (reviewed in (Hale et al. 2008; Krug and Garcia-Sastre 2013)). In addition to NS1, several other influenza virus proteins possess activities that lead to the inhibition of IFN expression. A systematic mapping of physical and regulatory interactions between influenza virus and the host cell resulted in the discovery of an important role of the RNA polymerase subunits and NP in regulating antiviral host responses (Shapira et al. 2009). Importantly, over-expression of PB1, PB2 and NP, individually or in combination, was found to inhibit cellular interferon responses to either vRNA transfection or viral infection, implicating non-NS1 viral proteins in modulating host responses. Further studies highlighted the viral polymerase subunit PB2 as a major player in mediating NS1-independent inhibition of type I interferon expression by associating with and inhibiting MAVS (Graef et al. 2010; Iwai et al. 2010). Although most PB2 in infected cells localises to the nucleus where transcription and replication of the viral genome occurs, PB2 of certain influenza viruses can also be detected at the mitochondria (Carr et al. 2006; Woodfin and Kazim 1993). Only PB2 proteins of seasonal H1N1 (pre-2009), H2N2 and H3N3 human influenza viruses associate with mitochondria while PB2 proteins of the 2009 H1N1 pandemic and avian influenza viruses, including H5N1 HPAIV, do not (Graef et al. 2010). A single amino acid polymorphism at amino acid residue 9 within the N-terminal mitochondrial targeting signal was found to be responsible for this differential localisation. An asparagine residue at position 9 in seasonal human influenza viruses results in mitochondrial localisation, while an aspartic acid in avian influenza viruses leads to non-mitochondrial localisation. Compared to a wild-type virus encoding mitochondrial PB2, an influenza virus expressing a non-mitochondrial PB2 N9D mutant induced higher levels of IFN-β in cell culture and it was attenuated in mice (Graef et al. 2010). In contrast, the D9N mutation in the PB2 protein of an avian H5N1 virus that allows mitochondrial localisation resulted in increased virulence in mice (Kim et al. 2010). Taken together, these studies suggest that the mitochondrial localisation of PB2 and consequently, its ability to inhibit IFN-β expression, are important determinants of virulence.

This conclusion has been reinforced by a more recent study reporting that a T588I mutation in PB2 that has been identified in swine isolates of the 2009 H1N1 pandemic virus enhanced their virulence in the mouse model. The T588I mutation was shown to enhance PB2 binding to MAVS and as a result, exacerbate PB2-mediated inhibition of IFN-β expression (Zhao et al. 2013).

It has been hypothesised that the non-mitochondrial PB2 expressed by highly pathogenic H5N1 viruses might, at least in part, be responsible for hypercytokinemia induced by these viruses in human hosts. In agreement with this, an H5N1 avian influenza virus PB2 was found to promote type I IFN inducing properties of a swine influenza virus strain in porcine dendritic cells (Ocana-Macchi et al. 2012). In particular, a reassortant swine H1N1 influenza virus with a PB2 segment derived from an H5N1 HPAIV induced higher levels of INF-β expression compared to the wild-type swine virus. Both the avian and swine virus PB2 proteins contain an aspartic acid at position 9, characteristic of non-mitochondrial PB2 proteins, suggesting that other differences between these two proteins, which differ at 17 amino acid positions, are responsible for their differential effects on IFN-β expression.

Thus, PB2 might act as a determinant of pathogenicity not only by contributing to polymerase activity as part of the trimeric viral polymerase complex, but also by acting independently of PB1 and PA as a regulator of innate immune responses.

4.3 Interactions of the Viral Polymerase with Cellular RNA Polymerase II

The influenza virus polymerase is also likely to contribute to the inhibition of cellular responses to viral infection through a mitochondria-independent mechanism. The trimeric influenza virus polymerase is known to bind to the C-terminal domain of the large subunit of cellular RNA polymerase II. It has been proposed that this binding inhibits elongation by Pol II during cellular transcription and it also triggers the proteolytic degradation of the large subunit of Pol II (reviewed in (Vreede and Fodor 2010)). General inhibition of host gene expression appears to be an efficient way of counteracting antiviral host mechanisms.

At early stages during infection, the influenza virus RNA polymerase associates with the cellular Pol II transcriptional machinery, presumably in order to facilitate access to 5′ capped RNA primers required for viral transcription (Engelhardt et al. 2005). However, later in infection, this interaction might serve as a means to induce the degradation of the large subunit of Pol II in order to prevent the expression of antiviral genes (Rodriguez et al. 2007; Vreede et al. 2010). The ability of influenza viruses to induce Pol II degradation has been correlated with virulence, and it has been proposed that this property might be an important factor in determining influenza virus pathogenicity (Rodriguez et al. 2009). In particular, amino acid residues 504 in PB2 and 550 in PA were found to be important for Pol II degradation and the pathogenicity of H1N1 as well as 2009 H1N1 pandemic influenza viruses in mice (Llompart et al. 2014).

These studies suggest that the viral polymerase might be involved in determining pathogenicity also by binding to cellular Pol II and inhibiting cellular gene expression. However, further studies are required to correlate the ability of influenza viruses to inhibit and degrade Pol II and elevated pathogenicity. Moreover, it remains unclear to what extent the binding of the viral polymerase to Pol II contributes to changes in general cellular gene expression and how this affects pathogenicity.

5 Conclusion and Future Perspectives

The viral polymerase complex is a major determinant of influenza virus interspecies transmission and pathogenesis in the mammalian host. However, viral pathogenesis is a polygenic trait which involves the interaction of multiple viral proteins with large cellular networks. Virus–host interactions may change during viral adaptation in order to ensure efficient virus replication in the new host cell. Thereby, viral polymerase proteins might switch from the use of one cellular factor to another to promote various steps in the viral life cycle such as nuclear import and transcription/replication of the viral genome.

We are still far from understanding the impact of these interactions on viral replication in the host cell. Thus, it remains a challenge for the future to elucidate pathways which affect interspecies adaptation, pathogenesis and transmission in mammals.

Acknowledgments We thank Edward Hutchinson and Anna Otte for critical reading of the manuscript and Narin Hengrung and Aartjan te Velthuis for their assistance with the figures. Work in the laboratory of the authors was supported by the German Research Foundation (Emmy-Noether Programme to GG), the European Union (Flu-Pharm to GG) and the Medical Research Council (programme grant MR/K000241/1 to EF).

References

Arranz R, Coloma R, Chichon FJ, Conesa JJ, Carrascosa JL, Valpuesta JM, Ortin J, Martin-Benito J (2012) The structure of native influenza virion ribonucleoproteins. Science 338(6114):1634–1637

Baker NA, Sept D, Joseph S, Holst MJ, McCammon JA (2001) Electrostatics of nanosystems: application to microtubules and the ribosome. Proc Natl Acad Sci U S A 98(18):10037–10041

Biswas SK, Nayak DP (1994) Mutational analysis of the conserved motifs of influenza a virus polymerase basic protein 1. J Virol 68(3):1819–1826

Bogs J, Kalthoff D, Veits J, Pavlova S, Schwemmle M, Manz B, Mettenleiter TC, Stech J (2011) Reversion of PB2-627E to -627K during replication of an H5N1 Clade 2.2 virus in mammalian hosts depends on the origin of the nucleoprotein. J Virol 85(20):10691–10698

Boivin S, Hart DJ (2011) Interaction of the influenza A virus polymerase PB2 C-terminal region with importin alpha isoforms provides insights into host adaptation and polymerase assembly. J Biol Chem 286(12):10439–10448

Bortz E, Westera L, Maamary J, Steel J, Albrecht RA, Manicassamy B, Chase G, Martinez-Sobrido L, Schwemmle M, Garcia-Sastre A (2011) Host- and strain-specific regulation of influenza virus polymerase activity by interacting cellular proteins. MBio 2(4):e00151–11

Bradel-Tretheway BG, Mattiacio JL, Krasnoselsky A, Stevenson C, Purdy D, Dewhurst S, Katze MG (2011) Comprehensive proteomic analysis of influenza virus polymerase complex reveals a novel association with mitochondrial proteins and RNA polymerase accessory factors. J Virol 85(17):8569–8581

Brass AL, Huang IC, Benita Y, John SP, Krishnan MN, Feeley EM, Ryan BJ, Weyer JL, van der Weyden L, Fikrig E, Adams DJ, Xavier RJ, Farzan M, Elledge SJ (2009) The IFITM proteins mediate cellular resistance to influenza A H1N1 virus, West Nile virus, and dengue virus. Cell 139(7):1243–1254

Brown EG, Liu H, Kit LC, Baird S, Nesrallah M (2001) Pattern of mutation in the genome of influenza a virus on adaptation to increased virulence in the mouse lung: identification of functional themes. Proc Natl Acad Sci U S A 98(12):6883–6888

Bussey KA, Bousse TL, Desmet EA, Kim B, Takimoto T (2010) PB2 residue 271 plays a key role in enhanced polymerase activity of influenza A viruses in mammalian host cells. J Virol 84(9):4395–4406

Carr SM, Carnero E, Garcia-Sastre A, Brownlee GG, Fodor E (2006) Characterization of a mitochondrial-targeting signal in the PB2 protein of influenza viruses. Virology 344(2): 492–508

Cauldwell AV, Moncorge O, Barclay WS (2013) Unstable polymerase-nucleoprotein interaction is not responsible for avian influenza virus polymerase restriction in human cells. J Virol 87(2):1278–1284

Cauldwell AV, Long JS, Moncorge O, Barclay WS (2014) Viral determinants of influenza a host range. J Gen Virol 95(Pt 6):1193–1210

Chan WH, Ng AK, Robb NC, Lam MK, Chan PK, Au SW, Wang JH, Fodor E, Shaw PC (2010) Functional analysis of the influenza virus H5N1 nucleoprotein tail loop reveals amino acids that are crucial for oligomerization and ribonucleoprotein activities. J Virol 84(14):7337–7345

Chen W, Calvo PA, Malide D, Gibbs J, Schubert U, Bacik I, Basta S, O'Neill R, Schickli J, Palese P, Henklein P, Bennink JR, Yewdell JW (2001) A novel influenza a virus mitochondrial protein that induces cell death. Nat Med 7(12):1306–1312

Chen Y, Liang W, Yang S, Wu N, Gao H, Sheng J, Yao H, Wo J, Fang Q, Cui D, Li Y, Yao X, Zhang Y, Wu H, Zheng S, Diao H, Xia S, Chan KH, Tsoi HW, Teng JL, Song W, Wang P, Lau SY, Zheng M, Chan JF, To KK, Chen H, Li L, Yuen KY (2013) Human infections with the emerging avian influenza A H7N9 virus from wet market poultry: clinical analysis and characterisation of viral genome. Lancet 381(9881):1916–1925

Chen H, Yuan H, Gao R, Zhang J, Wang D, Xiong Y, Fan G, Yang F, Li X, Zhou J, Zou S, Yang L, Chen T, Dong L, Bo H, Zhao X, Zhang Y, Lan Y, Bai T, Dong J, Li Q, Wang S, Li H, Gong T, Shi Y, Ni X, Li J, Fan J, Wu J, Zhou X, Hu M, Wan J, Yang W, Li D, Wu G, Feng Z, Gao GF, Wang Y, Jin Q, Liu M, Shu Y (2014) Clinical and epidemiological characteristics of a fatal case of avian influenza A H10N8 virus infection: a descriptive study. Lancet 383:714–721

Crescenzo-Chaigne B, van der Werf S, Naffakh N (2002) Differential effect of nucleotide substitutions in the 3' arm of the influenza A virus vRNA promoter on transcription/replication by avian and human polymerase complexes is related to the nature of PB2 amino acid 627. Virology 303(2):240–252

de Jong MD, Simmons CP, Thanh TT, Hien VM, Smith GJ, Chau TN, Hoang DM, Chau NV, Khanh TH, Dong VC, Qui PT, Cam BV, Ha do Q, Guan Y, Peiris JS, Chinh NT, Hien TT, Farrar J (2006) Fatal outcome of human influenza A (H5N1) is associated with high viral load and hypercytokinemia. Nat Med 12(10):1203–1207

Deng T, Engelhardt OG, Thomas B, Akoulitchev AV, Brownlee GG, Fodor E (2006) Role of ran binding protein 5 in nuclear import and assembly of the influenza virus RNA polymerase complex. J Virol 80(24):11911–11919

Dias A, Bouvier D, Crepin T, McCarthy AA, Hart DJ, Baudin F, Cusack S, Ruigrok RW (2009) The cap-snatching endonuclease of influenza virus polymerase resides in the PA subunit. Nature 458(7240):914–918

Engelhardt OG, Smith M, Fodor E (2005) Association of the influenza A virus RNA-dependent RNA polymerase with cellular RNA polymerase II. J Virol 79(9):5812–5818

Fodor E (2013) The RNA polymerase of influenza a virus: mechanisms of viral transcription and replication. Acta Virol 57(2):113–122
Fodor E, Smith M (2004) The PA subunit is required for efficient nuclear accumulation of the PB1 subunit of the influenza a virus RNA polymerase complex. J Virol 78(17):9144–9153
Foeglein A, Loucaides EM, Mura M, Wise HM, Barclay WS, Digard P (2011) Influence of PB2 host-range determinants on the intranuclear mobility of the influenza a virus polymerase. J Gen Virol 92(Pt 7):1650–1661
Fornek JL, Gillim-Ross L, Santos C, Carter V, Ward JM, Cheng LI, Proll S, Katze MG, Subbarao K (2009) A single-amino-acid substitution in a polymerase protein of an H5N1 influenza virus is associated with systemic infection and impaired T-cell activation in mice. J Virol 83(21):11102–11115
Fouchier RA, Schneeberger PM, Rozendaal FW, Broekman JM, Kemink SA, Munster V, Kuiken T, Rimmelzwaan GF, Schutten M, Van Doornum GJ, Koch G, Bosman A, Koopmans M, Osterhaus AD (2004) Avian influenza A virus (H7N7) associated with human conjunctivitis and a fatal case of acute respiratory distress syndrome. Proc Natl Acad Sci U S A 101(5):1356–1361
Friedrich B, Quensel C, Sommer T, Hartmann E, Kohler M (2006) Nuclear localization signal and protein context both mediate importin alpha specificity of nuclear import substrates. Mol Cell Biol 26(23):8697–8709
Gabriel G, Dauber B, Wolff T, Planz O, Klenk HD, Stech J (2005) The viral polymerase mediates adaptation of an avian influenza virus to a mammalian host. Proc Natl Acad Sci U S A 102(51):18590–18595
Gabriel G, Herwig A, Klenk HD (2008) Interaction of polymerase subunit PB2 and NP with importin alpha1 is a determinant of host range of influenza a virus. PLoS Pathog 4(2):e11
Gabriel G, Klingel K, Otte A, Thiele S, Hudjetz B, Arman-Kalcek G, Sauter M, Shmidt T, Rother F, Baumgarte S, Keiner B, Hartmann E, Bader M, Brownlee GG, Fodor E, Klenk HD (2011) Differential use of importin-alpha isoforms governs cell tropism and host adaptation of influenza virus. Nat Commun 2:156
Gabriel G, Czudai-Matwich V, Klenk HD (2013) Adaptive mutations in the H5N1 polymerase complex. Virus Res 178(1):53–62
Gao Y, Zhang Y, Shinya K, Deng G, Jiang Y, Li Z, Guan Y, Tian G, Li Y, Shi J, Liu L, Zeng X, Bu Z, Xia X, Kawaoka Y, Chen H (2009) Identification of amino acids in HA and PB2 critical for the transmission of H5N1 avian influenza viruses in a mammalian host. PLoS Pathog 5(12):e1000709
Garten RJ, Davis CT, Russell CA, Shu B, Lindstrom S, Balish A, Sessions WM, Xu X, Skepner E, Deyde V, Okomo-Adhiambo M, Gubareva L, Barnes J, Smith CB, Emery SL, Hillman MJ, Rivailler P, Smagala J, de Graaf M, Burke DF, Fouchier RA, Pappas C, Alpuche-Aranda CM, Lopez-Gatell H, Olivera H, Lopez I, Myers CA, Faix D, Blair PJ, Yu C, Keene KM, Dotson PD Jr, Boxrud D, Sambol AR, Abid SH, St George K, Bannerman T, Moore AL, Stringer DJ, Blevins P, Demmler-Harrison GJ, Ginsberg M, Kriner P, Waterman S, Smole S, Guevara HF, Belongia EA, Clark PA, Beatrice ST, Donis R, Katz J, Finelli L, Bridges CB, Shaw M, Jernigan DB, Uyeki TM, Smith DJ, Klimov AI, Cox NJ (2009) Antigenic and genetic characteristics of swine-origin 2009 A(H1N1) influenza viruses circulating in humans. Science 325(5937):197–201
Graef KM, Vreede FT, Lau YF, McCall AW, Carr SM, Subbarao K, Fodor E (2010) The PB2 subunit of the influenza virus RNA polymerase affects virulence by interacting with the mitochondrial antiviral signaling protein and inhibiting expression of beta interferon. J Virol 84(17):8433–8445
Guilligay D, Tarendeau F, Resa-Infante P, Coloma R, Crepin T, Sehr P, Lewis J, Ruigrok RW, Ortin J, Hart DJ, Cusack S (2008) The structural basis for cap binding by influenza virus polymerase subunit PB2. Nat Struct Mol Biol 15(5):500–506
Hale BG, Randall RE, Ortin J, Jackson D (2008) The multifunctional NS1 protein of influenza a viruses. J Gen Virol 89(Pt 10):2359–2376

Hao L, Sakurai A, Watanabe T, Sorensen E, Nidom CA, Newton MA, Ahlquist P, Kawaoka Y (2008) Drosophila RNAi screen identifies host genes important for influenza virus replication. Nature 454(7206):890–893

Hatta M, Gao P, Halfmann P, Kawaoka Y (2001) Molecular basis for high virulence of Hong Kong H5N1 influenza a viruses. Science 293(5536):1840–1842

Hatta M, Hatta Y, Kim JH, Watanabe S, Shinya K, Nguyen T, Lien PS, Le QM, Kawaoka Y (2007) Growth of H5N1 influenza a viruses in the upper respiratory tracts of mice. PLoS Pathog 3(10):1374–1379

He X, Zhou J, Bartlam M, Zhang R, Ma J, Lou Z, Li X, Li J, Joachimiak A, Zeng Z, Ge R, Rao Z, Liu Y (2008) Crystal structure of the polymerase PA(C)-PB1(N) complex from an avian influenza H5N1 virus. Nature 454(7208):1123–1126

Herfst S, Chutinimitkul S, Ye J, de Wit E, Munster VJ, Schrauwen EJ, Bestebroer TM, Jonges M, Meijer A, Koopmans M, Rimmelzwaan GF, Osterhaus AD, Perez DR, Fouchier RA (2010) Introduction of virulence markers in PB2 of pandemic swine-origin influenza virus does not result in enhanced virulence or transmission. J Virol 84(8):3752–3758

Herfst S, Schrauwen EJ, Linster M, Chutinimitkul S, de Wit E, Munster VJ, Sorrell EM, Bestebroer TM, Burke DF, Smith DJ, Rimmelzwaan GF, Osterhaus AD, Fouchier RA (2012) Airborne transmission of influenza A/H5N1 virus between ferrets. Science 336(6088): 1534–1541

Hudjetz B, Gabriel G (2012) Human-like PB2 627K influenza virus polymerase activity is regulated by importin-alpha1 and -alpha7. PLoS Pathog 8(1):e1002488

Hutchinson EC, Fodor E (2013) Transport of the influenza virus genome from nucleus to nucleus. Viruses 5(10):2424–2446

Imai M, Watanabe T, Hatta M, Das SC, Ozawa M, Shinya K, Zhong G, Hanson A, Katsura H, Watanabe S, Li C, Kawakami E, Yamada S, Kiso M, Suzuki Y, Maher EA, Neumann G, Kawaoka Y (2012) Experimental adaptation of an influenza H5 HA confers respiratory droplet transmission to a reassortant H5 HA/H1N1 virus in ferrets. Nature 486(7403):420–428

Iwai A, Shiozaki T, Kawai T, Akira S, Kawaoka Y, Takada A, Kida H, Miyazaki T (2010) Influenza a virus polymerase inhibits type I interferon induction by binding to interferon beta promoter stimulator 1. J Biol Chem 285(42):32064–32074

Iwasaki A, Peiris M (2013) Innate immunity. In: Webster R, Monto A, Braciale T, Lamb R (eds) Textbook of influenza, Wiley-Blackwell, UK, pp 269–282

Jagger BW, Wise HM, Kash JC, Walters KA, Wills NM, Xiao YL, Dunfee RL, Schwartzman LM, Ozinsky A, Bell GL, Dalton RM, Lo A, Efstathiou S, Atkins JF, Firth AE, Taubenberger JK, Digard P (2012) An overlapping protein-coding region in influenza a virus segment 3 modulates the host response. Science 337(6091):199–204

Jorba N, Juarez S, Torreira E, Gastaminza P, Zamarreno N, Albar JP, Ortin J (2008) Analysis of the interaction of influenza virus polymerase complex with human cell factors. Proteomics 8(10):2077–2088

Kageyama T, Fujisaki S, Takashita E, Xu H, Yamada S, Uchida Y, Neumann G, Saito T, Kawaoka Y, Tashiro M (2013) Genetic analysis of novel avian A(H7N9) influenza viruses isolated from patients in China, February to April 2013. Euro Surveill 18(15):20453

Karlas A, Machuy N, Shin Y, Pleissner KP, Artarini A, Heuer D, Becker D, Khalil H, Ogilvie LA, Hess S, Maurer AP, Muller E, Wolff T, Rudel T, Meyer TF (2010) Genome-wide RNAi screen identifies human host factors crucial for influenza virus replication. Nature 463(7282):818–822

Kim JH, Hatta M, Watanabe S, Neumann G, Watanabe T, Kawaoka Y (2010) Role of host-specific amino acids in the pathogenicity of avian H5N1 influenza viruses in mice. J Gen Virol 91(Pt 5):1284–1289

Konig R, Stertz S, Zhou Y, Inoue A, Hoffmann HH, Bhattacharyya S, Alamares JG, Tscherne DM, Ortigoza MB, Liang Y, Gao Q, Andrews SE, Bandyopadhyay S, De Jesus P, Tu BP, Pache L, Shih C, Orth A, Bonamy G, Miraglia L, Ideker T, Garcia-Sastre A, Young JA, Palese P, Shaw ML, Chanda SK (2010) Human host factors required for influenza virus replication. Nature 463(7282):813–817

Krug RM, Fodor E (2013) The virus genome and its replication, 2nd edn. In: Webster RG, Monto AS, Braciale TJ, and Lamb R (eds) Textbook of influenza, Wiley-Blackwell, UK, pp 57–66

Krug RM, Garcia-Sastre A (2013) The NS1 protein: a master regulator of host and viral functions, 2nd edn. In: Webster R, Monto A, Braciale T, Lamb R (eds) Textbook of influenza, Wiley-Blackwell, UK, pp 114–132

Kuzuhara T, Kise D, Yoshida H, Horita T, Murazaki Y, Nishimura A, Echigo N, Utsunomiya H, Tsuge H (2009a) Structural basis of the influenza a virus RNA polymerase PB2 RNA-binding domain containing the pathogenicity-determinant lysine 627 residue. J Biol Chem 284(11):6855–6860

Kuzuhara T, Kise D, Yoshida H, Horita T, Murazaki Y, Utsunomiya H, Tsuge H (2009b) Crystallization and X-ray diffraction analysis of the RNA primer/promoter-binding domain of influenza a virus RNA-dependent RNA polymerase PB2. Acta Crystallogr Sect F Struct Biol Cryst Commun 65(Pt 2):144–146

Labadie K, Dos Santos Afonso E, Rameix-Welti MA, van der Werf S, Naffakh N (2007) Host-range determinants on the PB2 protein of influenza a viruses control the interaction between the viral polymerase and nucleoprotein in human cells. Virology 362(2):271–282

Le QM, Sakai-Tagawa Y, Ozawa M, Ito M, Kawaoka Y (2009) Selection of H5N1 influenza virus PB2 during replication in humans. J Virol 83(10):5278–5281

Li J, Ishaq M, Prudence M, Xi X, Hu T, Liu Q, Guo D (2009) Single mutation at the amino acid position 627 of PB2 that leads to increased virulence of an H5N1 avian influenza virus during adaptation in mice can be compensated by multiple mutations at other sites of PB2. Virus Res 144(1–2):123–129

Llompart CM, Nieto A, Rodriguez-Frandsen A (2014) Specific residues of PB2 and PA influenza virus polymerase subunits confer the ability for RNA polymerase II degradation and virus pathogenicity in mice. J Virol 88:3443–3454

Maines TR, Jayaraman A, Belser JA, Wadford DA, Pappas C, Zeng H, Gustin KM, Pearce MB, Viswanathan K, Shriver ZH, Raman R, Cox NJ, Sasisekharan R, Katz JM, Tumpey TM (2009) Transmission and pathogenesis of swine-origin 2009 A(H1N1) influenza viruses in ferrets and mice. Science 325(5939):484–487

Manz B, Brunotte L, Reuther P, Schwemmle M (2012) Adaptive mutations in NEP compensate for defective H5N1 RNA replication in cultured human cells. Nat Commun 3:802

Manz B, Schwemmle M, Brunotte L (2013) Adaptation of avian influenza a virus polymerase in mammals to overcome the host species barrier. J Virol 87(13):7200–7209

Massin P, van der Werf S, Naffakh N (2001) Residue 627 of PB2 is a determinant of cold sensitivity in RNA replication of avian influenza viruses. J Virol 75(11):5398–5404

Mayer D, Molawi K, Martinez-Sobrido L, Ghanem A, Thomas S, Baginsky S, Grossmann J, Garcia-Sastre A, Schwemmle M (2007) Identification of cellular interaction partners of the influenza virus ribonucleoprotein complex and polymerase complex using proteomic-based approaches. J Proteome Res 6(2):672–682

Mehle A, Doudna JA (2008) An inhibitory activity in human cells restricts the function of an avian-like influenza virus polymerase. Cell Host Microbe 4(2):111–122

Mehle A, Doudna JA (2009) Adaptive strategies of the influenza virus polymerase for replication in humans. Proc Natl Acad Sci U S A 106(50):21312–21316

Mehle A, McCullers J (2013). Structure and function of the influenza virus replication machinery and PB1-F2. In: Webster R, Monto A, Braciale T, Lamb R (eds) Textbook of influenza, Wiley-Blackwell, UK, pp 133–145

Moeller A, Kirchdoerfer RN, Potter CS, Carragher B, Wilson IA (2012) Organization of the influenza virus replication machinery. Science 338(6114):1631–1634

Mok CK, Yen HL, Yu MY, Yuen KM, Sia SF, Chan MC, Qin G, Tu WW, Peiris JS (2011) Amino acid residues 253 and 591 of the PB2 protein of avian influenza virus A H9N2 contribute to mammalian pathogenesis. J Virol 85(18):9641–9645

Moncorge O, Mura M, Barclay WS (2010) Evidence for avian and human host cell factors that affect the activity of influenza virus polymerase. J Virol 84(19):9978–9986

Munster VJ, de Wit E, van den Brand JM, Herfst S, Schrauwen EJ, Bestebroer TM, van de Vijver D, Boucher CA, Koopmans M, Rimmelzwaan GF, Kuiken T, Osterhaus AD, Fouchier RA (2009) Pathogenesis and transmission of swine-origin 2009 A(H1N1) influenza virus in ferrets. Science 325(5939):481–483
Murakami Y, Nerome K, Yoshioka Y, Mizuno S, Oya A (1988) Difference in growth behavior of human, swine, equine, and avian influenza viruses at a high temperature. Arch Virol 100(3–4):231–244
Muramoto Y, Noda T, Kawakami E, Akkina R, Kawaoka Y (2013) Identification of novel influenza a virus proteins translated from PA mRNA. J Virol 87(5):2455–2462
Murphy BR, Hinshaw VS, Sly DL, London WT, Hosier NT, Wood FT, Webster RG, Chanock RM (1982a) Virulence of avian influenza a viruses for squirrel monkeys. Infect Immun 37(3):1119–1126
Murphy BR, Sly DL, Tierney EL, Hosier NT, Massicot JG, London WT, Chanock RM, Webster RG, Hinshaw VS (1982b) Reassortant virus derived from avian and human influenza a viruses is attenuated and immunogenic in monkeys. Science 218(4579):1330–1332
Naito T, Momose F, Kawaguchi A, Nagata K (2007) Involvement of Hsp90 in assembly and nuclear import of influenza virus RNA polymerase subunits. J Virol 81(3):1339–1349
Ng AK, Zhang H, Tan K, Li Z, Liu JH, Chan PK, Li SM, Chan WY, Au SW, Joachimiak A, Walz T, Wang JH, Shaw PC (2008) Structure of the influenza virus A H5N1 nucleoprotein: implications for RNA binding, oligomerization, and vaccine design. FASEB J 22(10):3638–3647
Ng AK, Chan WH, Choi ST, Lam MK, Lau KF, Chan PK, Au SW, Fodor E, Shaw PC (2012) Influenza polymerase activity correlates with the strength of interaction between nucleoprotein and PB2 through the host-specific residue K/E627. PLoS One 7(5):e36415
Obayashi E, Yoshida H, Kawai F, Shibayama N, Kawaguchi A, Nagata K, Tame JR, Park SY (2008) The structural basis for an essential subunit interaction in influenza virus RNA polymerase. Nature 454(7208):1127–1131
Ocana-Macchi M, Ricklin ME, Python S, Monika GA, Stech J, Stech O, Summerfield A (2012) Avian influenza a virus PB2 promotes interferon type I inducing properties of a swine strain in porcine dendritic cells. Virology 427(1):1–9
O'Neill RE, Palese P (1995) NPI-1, the human homolog of SRP-1, interacts with influenza virus nucleoprotein. Virology 206(1):116–125
O'Neill RE, Jaskunas R, Blobel G, Palese P, Moroianu J (1995) Nuclear import of influenza virus RNA can be mediated by viral nucleoprotein and transport factors required for protein import. J Biol Chem 270(39):22701–22704
Paterson D, Fodor E (2012) Emerging roles for the influenza a virus nuclear export protein (NEP). PLoS Pathog 8(12):e1003019
Paterson D, te Velthuis AJ, Vreede FT, Fodor E (2014) Host restriction of influenza virus polymerase activity by PB2 627E is diminished on short viral templates in a nucleoprotein-independent manner. J Virol 88(1):339–344
Peiris M, Yuen KY, Leung CW, Chan KH, Ip PL, Lai RW, Orr WK, Shortridge KF (1999) Human infection with influenza H9N2. Lancet 354(9182):916–917
Poch O, Sauvaget I, Delarue M, Tordo N (1989) Identification of four conserved motifs among the RNA-dependent polymerase encoding elements. EMBO J 8(12):3867–3874
Rameix-Welti MA, Tomoiu A, Dos Santos Afonso E, van der Werf S, Naffakh N (2009) Avian Influenza a virus polymerase association with nucleoprotein, but not polymerase assembly, is impaired in human cells during the course of infection. J Virol 83(3):1320–1331
Rehwinkel J, Tan CP, Goubau D, Schulz O, Pichlmair A, Bier K, Robb N, Vreede F, Barclay W, Fodor E, Reis e Sousa C (2010) RIG-I detects viral genomic RNA during negative-strand RNA virus infection. Cell 140(3):397–408
Resa-Infante P, Gabriel G (2013) The nuclear import machinery is a determinant of influenza virus host adaptation. Bioessays 35(1):23–27
Resa-Infante P, Jorba N, Zamarreno N, Fernandez Y, Juarez S, Ortin J (2008) The host-dependent interaction of alpha-importins with influenza PB2 polymerase subunit is required for virus RNA replication. PLoS One 3(12):e3904

Resa-Infante P, Jorba N, Coloma R, Ortin J (2011) The influenza virus RNA synthesis machine: advances in its structure and function. RNA Biol 8(2):207–215
Richard M, Schrauwen EJ, de Graaf M, Bestebroer TM, Spronken MI, van Boheemen S, de Meulder D, Lexmond P, Linster M, Herfst S, Smith DJ, van den Brand JM, Burke DF, Kuiken T, Rimmelzwaan GF, Osterhaus AD, Fouchier RA (2013) Limited airborne transmission of H7N9 influenza a virus between ferrets. Nature 501(7468):560–563
Rodriguez A, Perez-Gonzalez A, Nieto A (2007) Influenza virus infection causes specific degradation of the largest subunit of cellular RNA polymerase II. J Virol 81(10):5315–5324
Rodriguez A, Perez-Gonzalez A, Hossain MJ, Chen LM, Rolling T, Perez-Brena P, Donis R, Kochs G, Nieto A (2009) Attenuated strains of influenza a viruses do not induce degradation of RNA polymerase II. J Virol 83(21):11166–11174
Ruigrok RW, Crepin T, Hart DJ, Cusack S (2010) Towards an atomic resolution understanding of the influenza virus replication machinery. Curr Opin Struct Biol 20(1):104–113
Salomon R, Franks J, Govorkova EA, Ilyushina NA, Yen HL, Hulse-Post DJ, Humberd J, Trichet M, Rehg JE, Webby RJ, Webster RG, Hoffmann E (2006) The polymerase complex genes contribute to the high virulence of the human H5N1 influenza virus isolate A/Vietnam/1203/04. J Exp Med 203(3):689–697
Shapira SD, Gat-Viks I, Shum BO, Dricot A, de Grace MM, Wu L, Gupta PB, Hao T, Silver SJ, Root DE, Hill DE, Regev A, Hacohen N (2009) A physical and regulatory map of host-influenza interactions reveals pathways in H1N1 infection. Cell 139(7):1255–1267
Shaw ML (2011) The host interactome of influenza virus presents new potential targets for antiviral drugs. Rev Med Virol 21(6):358–369
Shaw M, Palese P (2013) Orthomyxoviridae, 6th edn. In: Knipe D, Howley P (eds) Fields virology, vol 1. Lippincott Williams and Wilkins, Philadelphia, pp 1151–1185
Shi M, Jagger BW, Wise HM, Digard P, Holmes EC, Taubenberger JK (2012) Evolutionary conservation of the PA-X open reading frame in segment 3 of influenza a virus. J Virol 86(22):12411–12413
Shinya K, Hamm S, Hatta M, Ito H, Ito T, Kawaoka Y (2004) PB2 amino acid at position 627 affects replicative efficiency, but not cell tropism, of Hong Kong H5N1 influenza A viruses in mice. Virology 320(2):258–266
Shinya K, Makino A, Ozawa M, Kim JH, Sakai-Tagawa Y, Ito M, Le QM, Kawaoka Y (2009) Ostrich involvement in the selection of H5N1 influenza virus possessing mammalian-type amino acids in the PB2 protein. J Virol 83(24):13015–13018
Song MS, Pascua PN, Lee JH, Baek YH, Park KJ, Kwon HI, Park SJ, Kim CJ, Kim H, Webby RJ, Webster RG, Choi YK (2011) Virulence and genetic compatibility of polymerase reassortant viruses derived from the pandemic (H1N1) 2009 influenza virus and circulating influenza a viruses. J Virol 85(13):6275–6286
Steel J, Lowen AC, Mubareka S, Palese P (2009) Transmission of influenza virus in a mammalian host is increased by PB2 amino acids 627K or 627E/701N. PLoS Pathog 5(1):e1000252
Subbarao K, Klimov A, Katz J, Regnery H, Lim W, Hall H, Perdue M, Swayne D, Bender C, Huang J, Hemphill M, Rowe T, Shaw M, Xu X, Fukuda K, Cox N (1998) Characterization of an avian influenza A (H5N1) virus isolated from a child with a fatal respiratory illness. Science 279(5349):393–396
Subbarao EK, London W, Murphy BR (1993) A single amino acid in the PB2 gene of influenza A virus is a determinant of host range. J Virol 67(4):1761–1764
Sugiyama K, Obayashi E, Kawaguchi A, Suzuki Y, Tame JR, Nagata K, Park SY (2009) Structural insight into the essential PB1-PB2 subunit contact of the influenza virus RNA polymerase. EMBO J 28(12):1803–1811
Sui B, Bamba D, Weng K, Ung H, Chang S, Van Dyke J, Goldblatt M, Duan R, Kinch MS, Li WB (2009) The use of random homozygous gene perturbation to identify novel host-oriented targets for influenza. Virology 387(2):473–481
Tarendeau F, Boudet J, Guilligay D, Mas PJ, Bougault CM, Boulo S, Baudin F, Ruigrok RW, Daigle N, Ellenberg J, Cusack S, Simorre JP, Hart DJ (2007) Structure and nuclear import

function of the C-terminal domain of influenza virus polymerase PB2 subunit. Nat Struct Mol Biol 14(3):229–233
Tarendeau F, Crepin T, Guilligay D, Ruigrok RW, Cusack S, Hart DJ (2008) Host determinant residue lysine 627 lies on the surface of a discrete, folded domain of influenza virus polymerase PB2 subunit. PLoS Pathog 4(8):e1000136
Vreede FT, Fodor E (2010) The role of the influenza virus RNA polymerase in host shut-off. Virulence 1(5):436–439
Vreede FT, Chan AY, Sharps J, Fodor E (2010) Mechanisms and functional implications of the degradation of host RNA polymerase II in influenza virus infected cells. Virology 396(1):125–134
Watanabe T, Watanabe S, Kawaoka Y (2010) Cellular networks involved in the influenza virus life cycle. Cell Host Microbe 7(6):427–439
Wise HM, Foeglein A, Sun J, Dalton RM, Patel S, Howard W, Anderson EC, Barclay WS, Digard P (2009) A complicated message: identification of a novel PB1-related protein translated from influenza a virus segment 2 mRNA. J Virol 83(16):8021–8031
Woodfin BM, Kazim AL (1993) Interaction of the amino-terminus of an influenza virus protein with mitochondria. Arch Biochem Biophys 306(2):427–430
Xu C, Hu WB, Xu K, He YX, Wang TY, Chen Z, Li TX, Liu JH, Buchy P, Sun B (2012) Amino acids 473V and 598P of PB1 from an avian-origin influenza A virus contribute to polymerase activity, especially in mammalian cells. J Gen Virol 93(Pt 3):531–540
Yamada S, Hatta M, Staker BL, Watanabe S, Imai M, Shinya K, Sakai-Tagawa Y, Ito M, Ozawa M, Watanabe T, Sakabe S, Li C, Kim JH, Myler PJ, Phan I, Raymond A, Smith E, Stacy R, Nidom CA, Lank SM, Wiseman RW, Bimber BN, O'Connor DH, Neumann G, Stewart LJ, Kawaoka Y (2010) Biological and structural characterization of a host-adapting amino acid in influenza virus. PLoS Pathog 6(8):e1001034
Ye Q, Krug RM, Tao YJ (2006) The mechanism by which influenza a virus nucleoprotein forms oligomers and binds RNA. Nature 444(7122):1078–1082
York A, Fodor E (2013) Biogenesis, assembly, and export of viral messenger ribonucleoproteins in the influenza a virus infected cell. RNA Biol 10(8):1274–1282
York A, Hengrung N, Vreede FT, Huiskonen JT, Fodor E (2013) Isolation and characterization of the positive-sense replicative intermediate of a negative-strand RNA virus. Proc Natl Acad Sci U S A 110(45):E4238–E4245
Yuan P, Bartlam M, Lou Z, Chen S, Zhou J, He X, Lv Z, Ge R, Li X, Deng T, Fodor E, Rao Z, Liu Y (2009) Crystal structure of an avian influenza polymerase PA(N) reveals an endonuclease active site. Nature 458(7240):909–913
Zhao Z, Yi C, Zhao L, Wang S, Zhou L, Hu Y, Zou W, Chen H, Jin M (2013) PB2 588I enhances the 2009 H1N1 pandemic influenza virus virulence by increasing viral replication and exacerbating PB2 inhibition of interferon-beta expression. J Virol 88(4):2260–2267
Zhou B, Pearce MB, Li Y, Wang J, Mason RJ, Tumpey TM, Wentworth DE (2013a) Asparagine substitution at PB2 residue 701 enhances the replication, pathogenicity, and transmission of the 2009 pandemic H1N1 influenza a virus. PLoS One 8(6):e67616
Zhou J, Wang D, Gao R, Zhao B, Song J, Qi X, Zhang Y, Shi Y, Yang L, Zhu W, Bai T, Qin K, Lan Y, Zou S, Guo J, Dong J, Dong L, Wei H, Li X, Lu J, Liu L, Zhao X, Huang W, Wen L, Bo H, Xin L, Chen Y, Xu C, Pei Y, Yang Y, Zhang X, Wang S, Feng Z, Han J, Yang W, Gao GF, Wu G, Li D, Wang Y, Shu Y (2013b) Biological features of novel avian influenza a (H7N9) virus. Nature 499(7459):500–503

Part II
Virus Entry and Cell Tropism

Receptor Binding Properties of the Influenza Virus Hemagglutinin as a Determinant of Host Range

Xiaoli Xiong, John W. McCauley and David A. Steinhauer

Abstract Host cell attachment by influenza A viruses is mediated by the hemagglutinin glycoprotein (HA), and the recognition of specific types of sialic acid-containing glycan receptors constitutes one of the major determinants of viral host range and transmission properties. Structural studies have elucidated some of the viral determinants involved in receptor recognition of avian-like and human-like receptors for various subtypes of influenza A viruses, and these provide clues relating to the mechanisms by which viruses evolve to adapt to human hosts. We discuss structural aspects of receptor binding by influenza HA, as well as the biological implications of functional interplay involving HA binding, NA sialidase functions, the effects of antigenic drift, and the inhibitory properties of natural glycans present on mucosal surfaces.

Abbreviations

HA	Hemagglutinin
NA	Neuraminidase
Sia	Sialic acid
Gal	Galactose
GlcNAc or NAG	*N*-acetylglucosamine
3-SLN	3′-Sialyl-*N*-acetyllactosamine
6-SLN	6′-Sialyl-*N*-acetyllactosamine
LSTc	LS-Tetrasaccharide C

X. Xiong · J.W. McCauley (✉)
Division of Virology, MRC National Institute for Medical Research,
The Ridgeway, Mill Hill, London NW7 1AA, UK
e-mail: jmccaul@nimr.mrc.ac.uk

X. Xiong
e-mail: axiong@nimr.mrc.ac.uk

D.A. Steinhauer (✉)
Department of Microbiology and Immunology, O. Wayne Rollins Research Center,
Emory University School of Medicine, 1510 Clifton Road, Atlanta, GA 30322, USA
e-mail: dsteinh@emory.edu

Current Topics in Microbiology and Immunology (2014) 385: 63–91
DOI: 10.1007/82_2014_423

Published Online: 31 July 2014

Contents

1 Introduction

Influenza A viruses present unique challenges for global human health and animal health due to their diversity in nature, and their propensity to adapt and become established in a wide variety of hosts. The genetic lineages of all influenza A viruses, excluding influenza viruses of bats, originate from strains that circulate in the waterfowl hosts that constitute their natural reservoir, but their ability to mutate and reassort their segmented genomes allows them to continuously probe the boundaries of tropism and host range. Mutation and reassortment facilitate transmission to alternative avian hosts including poultry, and mammalian hosts such as humans, pigs, horses, and most recently dogs. Numerous genetic factors are involved in transmission and host adaptation phenotypes, but the receptor binding properties of the viral hemagglutinin surface glycoprotein (HA) are a fundamental determinant. Contemporary molecular and structural biology approaches have allowed us to highlight mechanisms by which HA can evolve to recognize host cell receptors with specific properties, but a comprehensive appreciation of viral attachment, dissemination, and transmission amongst potential hosts, will require further understanding of a complex interplay involving viral surface proteins and host determinants.

Examples for the role of receptor binding in adaptation, and concurrent effects on additional phenotypic traits, derive from some of the earliest laboratory studies on influenza. In the wake of the description of erythrocyte agglutination properties by influenza and the inhibition of agglutination by specific sera in 1941 (Hirst 1941; McClelland and Hare 1941), Macfarlane Burnet and colleagues (Burnet and Bull 1943) published the classical O (original) to D (derivative) transition studies that demonstrated that human influenza isolates maintained their hemagglutination properties following replication in the amnion of the chicken embryo (O form), but had altered hemagglutination characteristics following passage and the selection of mutants in the allantoic cavity (D form). Not only were tissue tropism and binding properties altered during the O to D transition, but the mutations responsible also gave rise to detectable changes to their antigenic characteristics, as well as strain-specific constraints on replication by inhibitors present in human tears.

Early studies also highlighted the functional interplay between the HA binding properties, and those of the viral neuraminidase glycoprotein (NA). The observation that agglutinated erythrocytes could subsequently be "eluted" led to a series of studies revealing that the "receptor destroying" properties of bacterial neuraminidases displayed similarities with the virus-associated enzymatic activity, and resulted in the identification of the "split product" of the reaction as N-acetylneuraminic acid (Burnet and Stone 1947; Klenk et al. 1955; Gottschalk 1957). This led to the recognition that N-acetylneuraminic acid and its collective derivatives, commonly referred to as sialic acids, are the key components of influenza receptors. Comparative studies on the successive agglutination and elution of erythrocytes by different influenza viruses, as well as selected paramyxoviruses, demonstrated that the viruses "could be arranged in a linear series such that cells rendered resistant by treatment with a given virus failed to be agglutinated by virus strains earlier in the series but were still agglutinated by those succeeding it" (Burnet et al. 1946). These observations of "receptor gradients" for the opposing hemagglutinin and neuraminidase activities are now referred to as HA-NA functional balance, in which the HA receptor binding and the NA receptor-destroying properties of individual strains act in concert to fine-tune their activities and/or specificities for glycan receptors to achieve an optimal equilibrium between attachment and infectivity, and the capacity to disseminate and transmit.

Though we now have the tools to dissect specific features of HA binding and NA enzymatic functions, and have a knowledge of viral receptors, glycan inhibitors of binding, and immune mediators of antigenic drift, the challenge is to understand the interplay amongst these factors, and determine how they coordinately affect host range, transmission, and pathogenicity phenotypes. We will focus the majority of this chapter on the most fundamental of topics related to host range: the structural features of the HA involved in receptor binding, and the mechanisms by which HA discriminates amongst potential glycan receptors found at the sites of infection in different species, and we will discuss interrelationships between receptor binding properties and NA function, antigenic drift, and natural inhibitors.

2 HA Structure and a Description of the Receptor Binding Site

The influenza A envelope is embedded with HA and NA glycoprotein spikes that cover the viral surface (Fig. 1). The HA is generally more abundant, but the relative ratios of the HA and NA vary quite significantly depending on viral strain, and the NA spikes tend to cluster in "patches" on the viral surfaces (Calder et al. 2010). High-resolution X-ray crystal structures of HA and NA were first determined in the

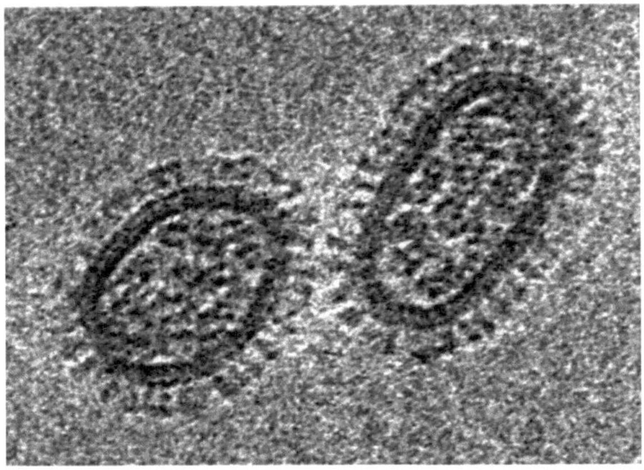

Fig. 1 Influenza A virions as visualized by electron cryotomography, with glycoprotein spikes protruding from the viral membranes. Courtesy of Lesley Calder and Peter Rosenthal, NIMR, UK

early 1980s (Wilson et al. 1981; Varghese et al. 1983), and we now have a great deal of structural information of these proteins. The HA is synthesized as precursor that, upon removal of the signal sequence, forms a noncovalently associated homotrimeric precursor (HA0). Proteolytic cleavage of each HA0 monomer into the disulfide-linked subunits, HA1 and HA2, is required to activate membrane fusion potential and, therefore, virus infectivity (Klenk et al. 1975; Chen et al. 1998). The first HA structure that was determined was that of a 1968 H3 subtype pandemic strain, and when residues that had evolved with antigenic drift were mapped to surface regions of the membrane distal HA1 head domains a pocket of conserved amino acids near the tip of each monomer was postulated as the sialic acid binding site (Wiley et al. 1981). This was confirmed by structures of HA in complexes with the receptor analog sialyllactose (Weis et al. 1988), and several structures for HA-receptor complexes have subsequently been determined and will be detailed below. The binding site is formed by noncontiguous residues of the HA1 subunit, and in this chapter all descriptions of the site will refer to specific residues using the H3 subtype numbering system. The major structural features of the binding site are shown in Fig. 2, and include: (i) a short helix at the membrane distal edge composed of HA1 residues 190–195 (190 helix), (ii) a loop region that forms a proximal boundary of the site when the trimer is viewed side-on with the membrane distal 190 helix at the top (130 loop), (iii) a 150 loop at the right side of the site in the same orientation, (iv) a 220 loop at the "left" side of the binding site, and (v) a hydrogen-bonded network of well-conserved residues that form the base of the binding site, which include Tyr98, Trp153, His183, and Tyr195.

There are now many structures of the HA of influenza A viruses with sialic acid receptor analogs in place and, regardless of the nature of receptor analogs in these complexes, the structure and orientation of sialic acid are essentially identical and exemplified in Fig. 2. Key conserved features of the sialic acid binding pocket of influenza A viruses include the residues Tyr98, Trp153, and His183 which form the floor of the site, where they form the base of a depression near the tip of each

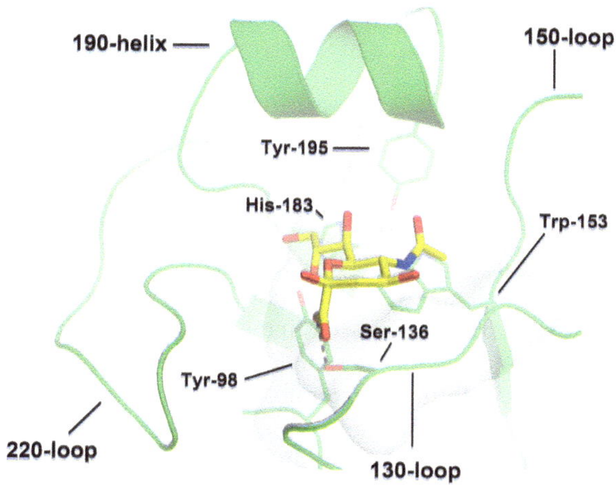

Fig. 2 The sialic acid binding pocket of the HA of A/Aichi/2/1968 (H3N2) illustrates four secondary structural elements of receptor binding site: the 130-loop, the 150-loop, the 190-helix and the 220-loop, which are labeled. Residues forming the shallow depression surrounding of the sialic acid are indicated and shown with a transparent surface. Structure from (Lin et al. 2012), [PDB: 2YPG]

monomer resembling a cirque. One side of the pyranose ring of the sialic acid sits on the base of the depression and the acetamido group on the C5 of sialic acid forms hydrogen bonds with the peptide backbone of residue 135 at the edge of the depression. The hydroxyl groups of C8 and C9 form hydrogen bonds with Tyr98 and, in certain cases, Ser228 and Glu190.

Though most of this chapter is devoted to influenza A viruses, we note that in influenza B viruses sialic acid sialic acid analogs have also been located in crystal structures (Wang et al. 2007). The residues of influenza B virus HA that form the base of the depression are similar, but not identical, to those of influenza A HAs. A notable exception is that the tyrosine residue at position 98 (H3 numbering) is a phenylalanine in influenza B HA (position 95, flu B numbering), which prohibits the ability of this residue to form a hydrogen bond with the glycerol side chain of sialic acid. It has been postulated that this difference might contribute to a different orientation of Trp153 (numbered 158, flu B) in the base of the depression. There are marked similarities on the right-hand edge of the depression, but the left-hand side is distinctly different between influenza A and influenza B HA molecules. A four residue insertion in the influenza B HA (residues 233–236) largely accounts for this difference and residues equivalent to residues Leu/Gln226 and Ser228 of the H3 HA are Pro (238) and Ser (240) and, strikingly, the side chain of influenza B HA residue Leu237 adopts a location analogous to residue Trp222 in the early H3 HA viruses.

3 Significance of Sialic Acid Glycosidic Linkage Stereochemistry

The observation that avian influenza viruses display a general preference for binding to receptors with terminal sialic acid with α2,3 glycosidic linkages to the penultimate sugar, while human viruses favor α2,6-linked receptors, forms one of the central doctrines of host specificity by influenza viruses (Rogers and Paulson 1983). The long-enduring explanation for this posits that avian hosts, which frequently transmit viruses via the fecal-oral route, express high levels of α2,3-linked receptors at sites of infection such as the intestine. On the other hand, in humans the upper respiratory tract contains a relatively high density of α2,6-linked glycans, and transmission is largely through respiratory droplets. Though these assumptions allow for logical interpretations of how influenza transmits among different hosts, the actual identity and distribution of influenza receptors in the tissues of their natural hosts is not well understood and will be discussed further below.

Direct evidence identifying the membrane distal pocket as the sialic acid binding site was first provided by the characterization of a glutamine to leucine substitution at HA1 position 226 that switched the receptor preference of 1968 H3 subtype pandemic strains from human α2,6-linkage to avian α2,3-linkage (Rogers et al. 1983a). This mutant virus was initially selected based on the capacity to replicate in the presence of horse serum, which contains an abundance of α-2-macroglobulin, a glycoprotein that is rich in sialosides with α2,6-linkages that inhibited the wild-type human strain. Interestingly, when proton-NMR binding affinity studies were carried out using the HAs of these 226 variants with α2,3- or α2,6-linked sialyllactose, all interactions were relatively weak (K_D values in the millimolar range), and the preferred receptor analogs displayed affinities that were only two- to threefold greater than the nonpreferred analogs (Sauter et al. 1989, 1992). The interpretation is that the relatively small differences observed for the binding of monovalent receptor analogs are amplified by the multiple HA-receptor interactions that are required for productive binding of viruses to host cells. Therefore, receptor density, presentation, and the fluid nature of both viral and host cell membranes all complicate interpretations of binding studies on influenza.

4 HA Mutations and Structural Alterations Associated with H1N1, H2N2, and H3N2 Human Pandemic Viruses

All or some of the genes of the three influenza pandemic viruses of the twentieth century evolved from those of influenza viruses of birds. The pandemic viruses arose either through the likely direct transfer of a virus from birds to humans in 1918, or by reassortment of an avian influenza virus with circulating human strains resulting in the reassortant pandemic viruses in 1957 and 1968. The evolution of the viruses from birds to humans is generally thought to be accompanied by a change in

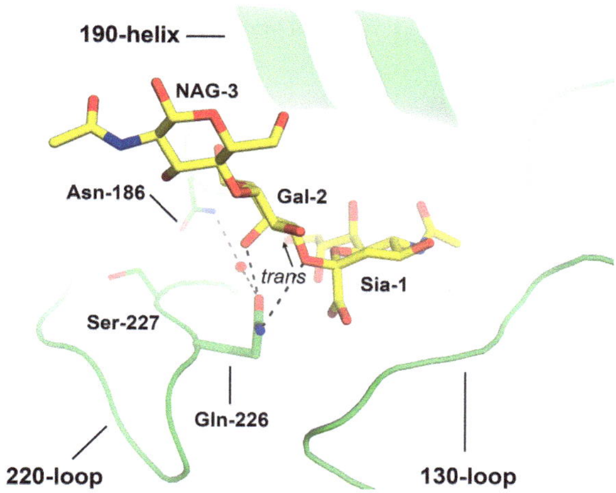

Fig. 3 Binding of an avian receptor analog by an avian HA. Binding of avian receptor analog (3-SLN) to HA of A/turkey/Turkey/1/2005 (H5N1). The avian receptor adopts the typical *trans* conformation when binding to avian HA. Gln226 interacts with glycosidic linkage and 4–OH of Gal-2 in the avian receptor (*dashed lines*) and it is also involved in a water mediated hydrogen bond with Asn-186 known to be important for this mode of receptor binding. Structure from (Xiong et al. 2013a), PDB: 4BH1

the recognition of the sialic acid receptors present on the upper respiratory tract of humans, with the HA gene acquiring a characteristic feature of preferring the human receptor over the avian receptor. Typically, avian influenza viruses recognize sialic acids linked in the α2,3 glycosidic linkage to galactose (Gal-2) in preference to those in the α2,6 linkage. Despite differences in the linkage between the second sugar, Gal-2, and GlcNAc-3, or modified GlcNAc-3, the avian receptor assumes a *trans* conformation between Sia-1 and Gal-2 that enables the receptor to adopt a trajectory in which the sugar chain exits the sialic acid binding site over the extremity of the 220 loop and in front of the left-hand side of the 190 helix (Fig. 3).

The preferential recognition of 2,6 linked sialic acid to galactose in human receptors by H3N2 viruses (Rogers and Paulson 1983; Connor et al. 1994) and by many H2N2 viruses (Connor et al. 1994) correlates in both cases with Q226L and G228S substitutions in the receptor binding site, though the position 226 substitution appears to be more critical for human receptor recognition (Matrosovich et al. 2000). In the structures of HAs complexed with receptor analogs, both the H3 HA and the H2 HA bind human α2,6 linked-receptor analogs such that the glycosidic bond between Sia-1 and Gal-2 adopts a *cis* conformation (Weis et al. 1988; Eisen et al. 1997; Ha et al. 2003; Liu et al. 2009), also see Fig. 4. The presence of leucine at residue 226 in both cases provides a hydrophobic environment with a 4 Å distance between Leu226 and the C6 of Gal-2. In the case of the H2N2 pandemic, the G228S substitution occurred subsequent to the Q226L change, and appeared to increase human receptor binding further, though it was probably not essential for

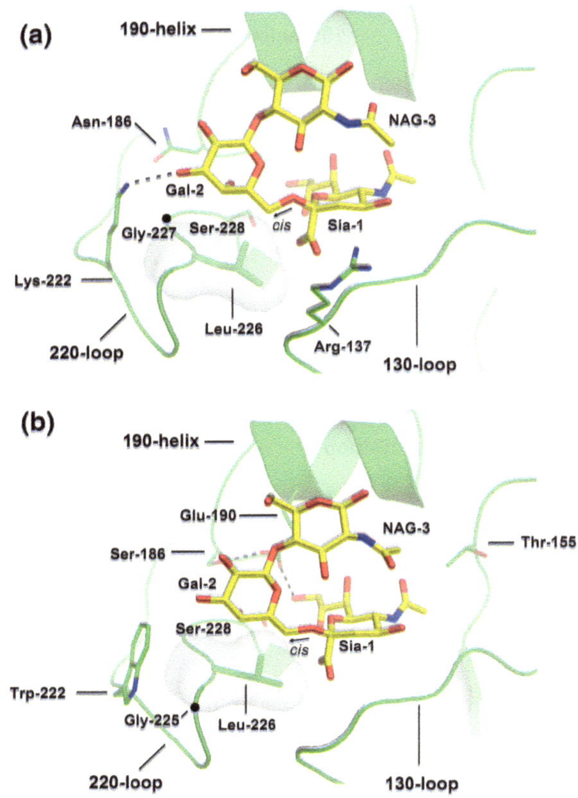

Fig. 4 Binding of human receptor analogs to the HAs of pandemic H2N2 and H3N2 viruses. A. Binding of human receptor analog (LSTc) to the HA of A/Singapore/1/1957 (H2N2). B. Binding of 6SLN to the HA of A/Aichi/2/1968 (H3N2). The human receptor bound to both HAs adopt *cis* conformation and both exit the receptor binding site vertically from the top of the HA. In both H2 and H3 human receptor complexes, the 6 carbon of the 2–6 glycosidic bond is in close contact with the hydrophobic Leu-226 (shown with transparent hydrophobic surface). Structures from (Liu et al. 2009), PDB: 2WR7; and (Lin et al. 2012), PDB: 2YPG

the initial spread of the H2N2 pandemic viruses in humans (Matrosovich et al. 2000).

When amino acid sequences of the HA of human 1968 H3 subtype strains were compared with avian strains such as A/Duck/Ukraine/1/63, thought to be representative of HA gene of the avian precursor for the 1968 human pandemic viruses, among the differences observed was the Gln to Leu substitution at HA1 position 226, that was subsequently identified in the horse serum selection experiments (see above), as well as a Gly to Ser mutation at nearby residue 228 (Fang et al. 1981; Ward and Dopheide 1981). Subsequent sequence comparisons of virus isolates from different hosts, as well as laboratory selection experiments, confirmed these residues as contributors to receptor binding specificity and host adaptation (Rogers et al. 1983a, 1985; Naeve et al. 1984; Kida et al. 1987; Bean et al. 1992; Connor et al. 1994). For H2 subtype viruses responsible for the 1957 H2N2 pandemic, residue 226 again appears to have played a role in altered receptor specificity and human adaptation, and the G228S change, as well as an N186I substitution, are thought potentially to contribute to adaptation to human transmission (Connor et al. 1994; Matrosovich et al. 2000). On the other hand, it appears that the emergence of

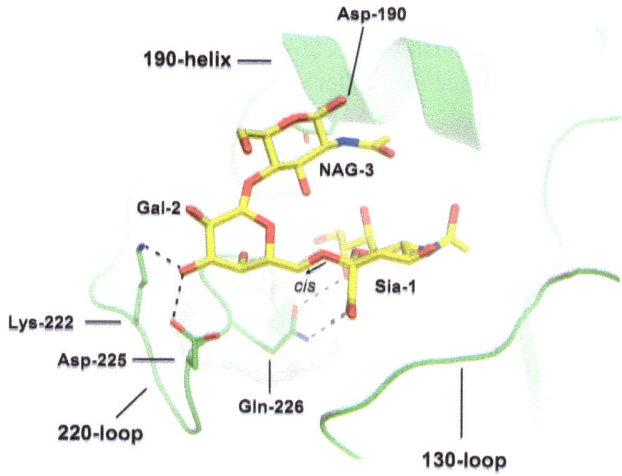

Fig. 5 Binding of a human receptor analog (LSTc) to the HA of the pandemic 1918 H1N1 virus (A/South Carolina/1/1918). Gal-2 of the human receptor interacts with Lys-222 and Asp-225 by hydrogen bonding and sits lower in the site comparing to that in equivalent complexes of H2 and H3 HAs from pandemic viruses. Gln-226, in contrast to H2 and H3 subtypes, remains unchanged in H1 HAs during pandemics and has the characteristic ability to hydrogen bond to the carboxylate and glycerol groups of sialic acid. Structure from (Liu et al. 2009); PDB: 2WRG

H1N1 human viruses responsible for the pandemic of 1918–1919 were facilitated by HA-receptor binding site substitutions at position 190 (E to D) and, to a lesser degree, position 225 (G to D).

In molecular terms, the preferred binding of the human receptor analog over the avian receptor in the H2 and H3 pandemic viruses conferred by the Q226L substitution is probably explained by the observation that the 226 side chain lies underneath the location of the glycosidic bond in an α-2, 3-linked avian receptor analog. The Q226L substitution abolishes the potential for Gln226 to interact with the 4–OH and the glycosidic bond of the avian receptor (Fig. 3). This prevents the avian receptor from binding in the *trans* conformation that is associated with tight binding and, as pointed out above, Leu226 provides a hydrophobic environment with a 4 Å distance between Leu226 and the C6 of Gal-2. In these ways, these pandemic viruses increased avidity to human receptors and reduced avidity to avian receptors.

The same explanation does not follow for the pandemic H1 HA. A key feature of the H1 HA is that a difference in location of the 130 and 220 loops results in Gln226 sitting lower in the receptor binding site (Fig. 5): for the human receptor Gal-2 is 2 Å lower in the binding site compared with its location in the H2 or H3 complex. Gamblin et al. (2004) highlighted the importance of the E190D amino acid substitution in the loss of recognition of the avian receptor. This observation was supported by the recognition that the same substitutions in the 1918 HA were shown to affect receptor binding specificity (Glaser et al. 2005). The importance of residue 225 in receptor specificity has also been highlighted (Xu et al. 2012a; Zhang et al. 2013a) in which the substitution D225G increases the binding of the HA to the avian receptor. It is striking that during the 2009 pandemic a D225G substitution was linked with increased virulence (Kilander et al. 2010; Mak et al. 2010; Wedde et al. 2013; Resende et al. 2014).

5 H5N1 Receptor Specificity and Potential for Human Adaptation

Zoonotic infections of humans by influenza viruses from birds or other species in which influenza viruses circulate is the likely first step in the evolution of a pandemic. Infection of humans with influenza A (H5N1) viruses was first reported in May 1997 (Claas et al. 1998) and since then highly pathogenic avian influenza H5N1 viruses have become widespread in poultry and wild birds in Eurasia and regions of Africa. More than 600 zoonotic human infections with a death rate approaching 60 % (WHO 2014) have now been reported but, as yet, there is no sustained human to human spread. However, transmission studies in ferrets showed that the H5 virus can adapt to aerosol spread. In part, a change in the binding preference of the H5 HA from avian receptor to human receptor was needed to gain aerosol transmissibility (Chen et al. 2012; Herfst et al. 2012; Imai et al. 2012; Xiong et al. 2013a; Zhang et al. 2013b) and this was associated with the substitution Q226L in the receptor binding site of the HA. Wild-type H5 virus binding to human and avian receptor has very strong avidity toward avian receptor but the transmissible-mutant virus has a preference ratio of about 200 in favor of human receptor binding (Xiong et al. 2013a). This is achieved by a moderate avidity for human receptor being offset by very weak binding to avian receptor. Structural studies showed that there is a strikingly different orientation of Gal-2 and GlcNAc-3 of the receptor in the transmissible-mutant HA complex compared with their arrangement in both wild-type HA complexes (Lu et al. 2013; Xiong et al. 2013a; de Vries et al. 2014). In the complex between the transmissible-mutant HA and the human receptor (Fig. 6a), Gal-2 adopts a *cis* configuration about the glycosidic bond with sialic acid, and GlcNAc-3 exits the receptor binding site towards the 190-helix, an orientation of Gal-2 and GlcNAc-3 very similar to that in complexes formed between human receptor and HAs from the 1918 (H1), 1957 (H2), and 1968 (H3) pandemics. In wild-type HA structures, Gal-2 of the human receptor also adopts a *cis* configuration about the glycosidic bond with sialic acid, but it is rotated by approximately 90° about its C6–C5 bond such that GlcNAc-3 exits from the side of the receptor binding site, over the 130-loop (Fig. 6b). This mode of human receptor binding has not been reported for any other avian or human HA.

To date, no isolates of H5N1 viruses from birds or humans have been reported to carry the Q226L substitution. Nevertheless, there has been considerable evolution of the H5 hemagglutinin (Shore et al. 2013; WHO/OIE/FAO 2014; Xiong et al. 2013c), and some of these, particularly the ones observed in human isolates,

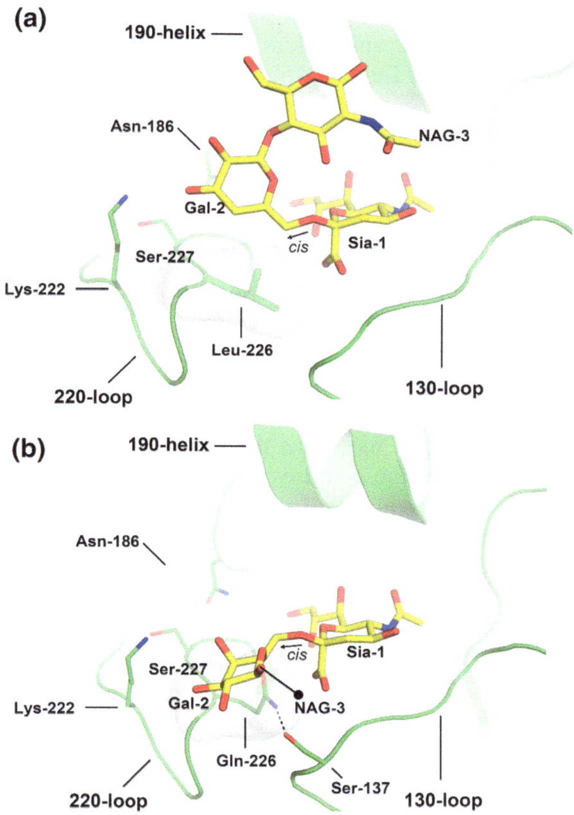

Fig. 6 Binding of human receptors by H5 HAs. **a** The ferret transmissible-mutant H5 HA derived from A/Vietnam/1203/2004 contacts Gal-2 on the edge, and the receptor exits from the top of the molecule with a vertical trajectory highly resembling those in human receptor complexes of HAs of pandemic viruses. **b** Gln-226 of the HA of A/turkey/Turkey/1/2005 contacts Gal-2 on the face of the sugar, the receptor adopts a novel conformation and exits the site above the 130-loop. The location of GlcNAc-3 is indicated by a black dot for clarity. Structures from (Xiong et al. 2013a); PDB: 4BH3 and PDB: 4BH0

show altered receptor avidity (Gambaryan et al. 2006; Yamada et al. 2006; Chutinimitkul et al. 2010; Watanabe et al. 2011). Human isolates with HAs containing S227N or N186K have been reported to display lower avidity for avian receptors (Xiong et al. 2014). Either substitution results in the loss of a hydrogen bond network that in wild-type HA links Gln-226 to Asn-186, and lifts Gln-226 about 1 Å higher than in the unliganded HA to form a hydrogen bond with the 4–OH of Gal-2 of the receptor (see Fig. 3). In contrast, in the mutants Gln-226 remains lower in the site and the interaction with Gal-2 is not observed. In addition, in these mutant HAs the α-2,3 glycosidic linkage between Sia-1 and Gal-2, adopts a *cis* configuration in the complex rather than the *trans* conformation that is characteristic of avian HA-avian receptor complexes. The selection for viruses with lowered avidity towards avian receptor may reflect the need for H5N1 virus to circumvent trapping by mucins that are densely decorated with avian receptors in the human respiratory tract as the first step in human adaptation.

Other H5N1 viruses can show increased avidity for both human and avian receptors. Examples include the Del133/Ile155Thr mutant HAs from recent

Egyptian isolates, which contain a more polar environment near the 130-and 150-loops resulting from the loss of the hydrophobic surface formed by the side-chains of Ala-133 and Ile-155. This polar environment results in better hydration of sialic acid upon receptor binding and higher avidity for both receptors, although the effect for human receptor binding is much more pronounced. Crystal structures show Del133/Ile155Thr mutant viruses bind both avian and human receptors in the same mode as wild-type (Xiong et al. 2014). The fact that Del133/Ile155Thr mutant viruses maintain the wild-type binding mode for avian receptor but increased avidity for human receptor is consistent with the observation that this mutant is able to infect humans at high frequency while maintaining its ability to infect poultry (Watanabe et al. 2011).

Although the substitutions seen in the evolution of the HA of H5N1 viruses increase the avidity of virus for the human receptor, none of the naturally occurring H5 mutants analyzed to date has changed receptor specificity sufficiently to preferentially recognize the human receptor over the avian receptor, a feature typical of pandemic viruses and the ferret transmissible H5N1 virus. Nevertheless, increased recognition of the human receptor is a concerning feature of the evolution of highly pathogenic H5N1 avian influenza viruses.

6 H7N9 Viruses with "Intermediate" Binding Specificity

The February 2013 emergence of avian H7N9 influenza viruses in China has led to a considerable number of infections with a mortality rate of about one in three of those infected. This has warranted a detailed examination of the receptor properties of these zoonotic H7 viruses and other more prototypical H7 avian influenza viruses. Binding of human H7N9 and avian H7N3 viruses to human and avian receptor analogs using a quantitative binding assay showed that the zoonotic H7N9 virus has a markedly increased avidity for human receptor, concomitant with a slightly decreased avidity for avian receptor over those of the H7N3 virus. Thus, the zoonotic H7N9 virus binds both receptor types but retains a small preference for the avian receptor (Shi et al. 2013; Xiong et al. 2013b), in contrast to the mammalian transmissible H5N1 mutant (see above). Studies using glycan arrays reached a similar conclusion (Xu et al. 2013). Thus, the zoonotic H7N9 virus does not have the characteristics of a human pandemic virus with a preference for the human receptor.

It is striking that two amino acid substitutions in the receptor binding site of the HAs of the avian and zoonotic H7N9 viruses are at residues 186 and 226. The glutamine residue at position 226 of the avian viruses is replaced by leucine, typical of the HAs of the H2 and H3 pandemic viruses, and a glycine residue at position 186 in the avian HA evolved to a valine residue in the zoonotic HA. The structure of the human receptor in a complex with HA of the zoonotic H7N9 has been determined and displays some unusual features (Fig. 7). In contrast to the human receptor complex with the avian exemplar HA, in which Gal-2 uses the face of the

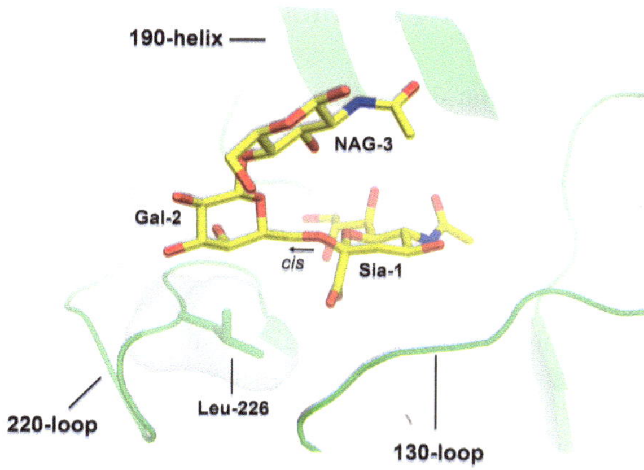

Fig. 7 Binding of human receptor by HA from an H7N9 virus (A/Anhui/1/2013). The substituted residue Leu-226 from glutamine in wild-type H7 HA is indicated and shown with a transparent molecular surface. The human receptor exits the site with a 45° bend towards the 130-loop in an off-vertical trajectory. Structure from (Xiong et al. 2013b); PDB: 4BSC

sugar ring to contact the HA, there is a 45° rotation around the C6–C5 bond of Gal-2 in the human receptor complex with the H7 HA from the zoonotic virus A/Anhui/1/2013. This results in the receptor taking an off-vertical trajectory out of the binding site. In both cases, the Sia-1-Gal-2 glycosidic bond of the human receptor analog takes on a *cis* conformation when bound to the HA. The part of the binding site where Gal-2 sits is significantly more hydrophobic in the HA from the zoonotic virus with a hydrophobic patch under the C6 of Gal-2. This orientation of Gal-2 and GlcNAc-3 in the complex is towards the top of the HA molecule, somewhat similar to that seen in complexes formed between human receptor and HA from all three human pandemic viruses but tilted towards the 130-loop because of the 45° rotation in Gal-2. The avian receptor adopts a *cis* configuration about the Sia-1–Gal-2 glycosidic bond when bound to zoonotic H7 HA, but a *trans* arrangement when bound to avian H7 HA. Avian receptors normally adopt only *cis* conformation when bound to human-adapted HAs, illustrating that the zoonotic H7N9 has already acquired features that promote human infection.

The receptor binding properties of the H7N9 viruses represent a possible intermediate in the process of adaptation of the virus to replication in humans. Currently, the H7N9 virus is able to cause zoonotic infection at an alarming frequency. The H7N9 virus has acquired the ability to recognize the human receptor efficiently, but, unlike human pandemic viruses, the zoonotic H7N9 virus retains the ability to recognize the avian receptor. Should this ability to recognize the avian receptor be lost, the virus could display receptor binding similar to human pandemic viruses.

7 Evolution of the H3 HA in Humans

Above, the focus has been on the evolution of the receptor binding properties of H5N1 viruses and H7N9 viruses, two viruses that show the potential to cause pandemics, and how these properties compare with viruses that have become established in humans. Influenza A(H3N2) viruses have circulated in the human population for over 45 years and have undergone significant levels of antigenic drift. The drift has warranted frequent changes in the composition of recommended viruses for inclusion in influenza vaccines. It is recognized that amino acid substitutions that prevent antibody binding can also result in altered receptor binding (Daniels et al. 1984, 1987; Underwood et al. 1987; Aytay and Schulze 1991; Hensley et al. 2009; Das et al. 2011; Koel et al. 2013). The evolution in receptor binding of the HA have been particularly marked for H3N2 viruses over the last 25 years. By the mid-1990s H3N2 viruses had a reduced ability to agglutinate chick red cells, associated with the substitution E190D (Nobusawa et al. 2000; Medeiros et al. 2001; Lin et al. 2012). From 1999 onwards, viruses had developed reduced avidity for both turkey red blood cells and human red blood cells. This evolution has resulted in several key changes that influence receptor binding: first W222R combined with G225D over the period 2001–2004, then from 2005 onwards, W222R combined with D225N, and in 2014 viruses carrying N225D have been in circulation. Since 2000 residue 226 changed twice, L226V in around 1999 and V226I in 2004.

The importance of the changes at residues 222 and 225 can be seen from quantitative assessment of the binding of virus to immobilized receptor analogs and from structures of HA-receptor complexes determined by X-ray crystallography (Fig. 8). Biolayer-interferometry showed that the avidity of H3N2 viruses for a human receptor analog decreased over time, with an approximate fourfold reduction between 1968 and 2001 and then a further estimated 200-fold reduction in binding over the period 2001–2004, with circulating viruses carrying W222R with G225D. By 2010, circulating viruses carrying W222R with G225N failed to bind to human receptor analog under standard assay conditions, although very limited binding could be detected using virus at a higher concentration (Lin et al. 2012). In the HA-receptor complex of the HA from a prototype 2004 virus, the human receptor analog adopts the conformation similar to that seen in its complex with the 1968 HA (Weis et al. 1988; Lin et al. 2012). However, unlike the 1968 HA, changes in the structure of the 220-loop of the 2004 HA occur upon receptor binding, which appear to facilitate the interaction. Notably, no significant conformational differences are observed between the unbound and receptor bound forms of the 1968 HA, but in the complex of the 2004 HA with the human receptor analog, Asp-225 forms a salt bridge with Arg-222 and is shifted closer to the bound receptor where it is able to form a hydrogen bond with the 3-hydroxyl of Gal-2 and enables Ile226 to contact the C6 of the sialic acid glycosidic bond. Moreover, the shorter side chain of D190 the in the 2004 HA does not reach into the receptor binding pocket sufficiently to interact with the glycerol substituent of sialic acid in the same manner as E190 seen in earlier 1968 HA.

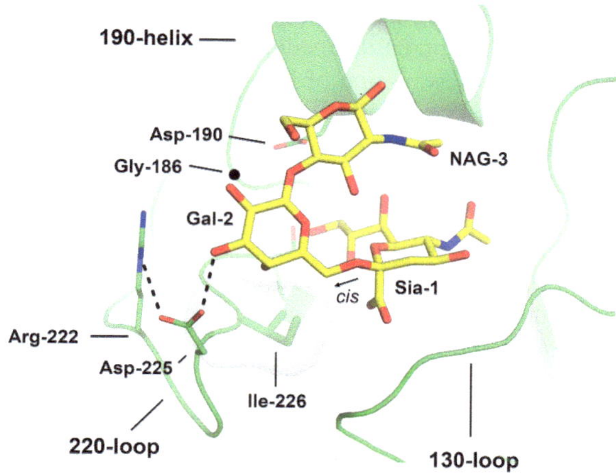

Fig. 8 Binding of human receptor by a 2004 H3 HA. The 2004 H3 HA differs significantly from 1968 H3 HA by three key substitutions - Trp222Arg, Gly225Asp, and Leu226Ile. The interaction between Asp-225 and Gal-2 of human receptor has been found to be important to facilitate the upward movement of the 220-loop to enable Ile-226 to contact the 2–6 glycosidic bond. Structure from (Lin et al. 2012); PDB: 2YP3

The subsequent substitution D225N reduces the ability of the residue 225 to form a hydrogen bond with Gal-2 and no HA structural change is observed upon binding receptor at concentrations of the receptor analog of up to 60 mM. Moreover, only electron density for the sialic acid of the sialyllactosamine was observed, and many interactions between sialic acid and the HA are made via intermediary water molecules, including interactions of the carboxylate group with Ser-136 and the glycerol group with Tyr-98. Less definitive information is available about the significance of the evolution of residue 226 from Leu through Val to Ile. There is no published structure of an HA-receptor complex of HA with Val-226. However, a comparison has revealed that in the 1968 HA the side chain of Leu226 is closer to C6 of Gal-2 (3.8 Å) than Ile226 (4.5 Å) in the corresponding 2004 HA-receptor complex. The change in aliphatic side chain length through Leu to Val to Ile are likely to affect the crucial hydrophobic interaction between residue 226 and the 2, 6 sialic acid linkage.

8 Altered Binding or Adaptation Properties Mediated by Mutations Outside the Binding Site

As detailed above, even amongst strains representing a range of HA subtypes, a select subset of HA positions are implicated as critical determinants of binding specificity on a reasonably consistent basis. However, residues not yet identified as binding determinants, including those at positions distal from the binding pocket, should not be dismissed as potential mediators of adaptation. For example, the ferret studies on airborne transmission of H5 subtype viruses indicated a potential role for HA stability (Herfst et al. 2012; Imai et al. 2012), and a comparison of membrane fusion pH phenotypes among avian and human viruses also suggest a role for HA stability in human adaptation (Galloway et al. 2013; Shelton et al. 2013). Mutations that affect HA stability can reside at residues positioned throughout the length of the HA trimer (Daniels et al. 1985), though there is little indication that these directly influence receptor binding. However, alternative mechanisms for modulating the binding phenotype mediated by residues outside the receptor sialic acid binding pocket have been characterized. For example, when a binding-impaired mutant virus with a Y98F substitution at the base of the binding pocket was used to infect mice intranasally, it displayed a highly attenuated phenotype, but adaptive changes were rapidly selected which partially or completely restored binding (Martin et al. 1998; Meisner et al. 2008; Bradley et al. 2011a). The structural locations of these included; (i) "second shell" mutations at residues buried in the structure near the binding pocket, (ii) residues that contact carbohydrates originating from the neighboring HA monomer, presumably affecting access to the site by receptor molecules, and (iii) residues that alter monomer–monomer interfaces, both near the site and at distances of up to 40 Å. Effects on the binding phenotype due to the gain or loss of carbohydrate attachment sites have been observed (Bradley et al. 2011a; Herfst et al. 2012; Imai et al. 2012), and can be reflected by differing transmission phenotypes in poultry (Iqbal et al. 2012), though the direct mechanisms involved are not known. In addition, HA contacts to sugars beyond those observed in the X-ray crystal structures may influence binding due to electrostatic interactions as exemplified by the certain recent H5N1 viruses that carry Q196R, G143R, and del133/I155T (Xiong et al. 2014).

Although it is prudent to monitor sites such as HA1 226 during surveillance for virus strains with pandemic potential, we need to be aware that other mechanisms exist for changing receptor binding characteristics, not least in H1N1 pandemic viruses, and we may not be able to predict the binding phenotype of given strains based on the presence of residues at any particular position.

9 Functional Balance and Compatibility Between the HA and NA Glycoproteins

The receptor gradient experiments by Burnet and colleagues provided early clues for functional compatibility between the HA-receptor binding properties and the NA enzymatic characteristics for given strains of influenza, or other viruses utilizing sialic acid-containing receptors (Burnet et al. 1946). These concepts were supported by the cosegregation of HA and NA during the emergence of human

H2N2 pandemic strains in 1957, the gradual evolution of NA sialidase specificity from α2,3 to dual α2,3/α2,6 specificity over the next 10 years, and the subsequent emergence of H3N2 human viruses with the same NA (Baum and Paulson 1991; Kobasa et al. 1999). A similar functional balance was observed in the 2009 H1N1 pandemic viruses, and may have been a crucial factor for the efficient human transmissibility displayed by these quadruple reassortant viruses (Xu et al. 2012b). HA–NA compatibility has also been demonstrated in multiple laboratory systems [for examples see (Yang et al. 1997; Hughes et al. 2000; Mitnaul et al. 2000)], demonstrating that if NA activity is compromised, the capacity for efficient multicycle replication can be restored by a reduction of HA binding such that progeny virions can be released from host cells and disseminate. The combination of the properties of the HA and the NA can determine the sensitivity of reassortant wild-type viruses to inhibition by zanamivir, reflecting their dependence on NA activity for spread in culture (Baigent et al. 1999; Baigent and McCauley 2001). Similar observations have been noted in mutant virus strains that are selected for resistance to the neuraminidase inhibitors zanamivir and oseltamivir (e.g. McKimm-Breschkin 2000). These mutant strains contain changes in the NA active site, or alternatively, changes that effect HA receptor binding properties. An extreme example of functional interplay is illustrated by H3N2 NA variants containing a D to G substitution at position 151, which displays agglutination properties that are sensitive to NA inhibitors (Lin et al. 2010; Zhu et al. 2012). When combined with HA substitutions that inhibit HA-mediated attachment, viruses with such NA variants are capable of multicycle growth in cell culture based on receptor binding function that has been completely subjugated by the NA glycoprotein (Hooper and Bloom 2013).

The NA functions in post assembly virus release are well documented based primarily on laboratory studies in cell culture; however, it also may act during early cycles of replication in vivo by assisting the navigation of relatively small numbers of virus particles through the mucus-rich environment of respiratory tissues, destroying decoy receptors and facilitating attachment to uninfected cells (Matrosovich et al. 2004). In this regard, it is known that NA inhibitors such as zanamivir and oseltamivir are most effective when administered within 48 h of the onset of symptoms. Following the initial infectious cycles in a region of tissue, copious levels of NA are expressed on infected-cell surfaces, and potentially modify the sialic acid distribution on adjacent cells or the proximal environment. Such factors might also provide selective pressure for the enrichment of viruses of filamentous morphology observed in clinical isolates (Mosley and Wyckoff 1946; Chu et al. 1949; Choppin et al. 1960; Itoh et al. 2009). More studies related to the activity and specificity of viral NAs for glycans that might act as potential receptors, or as inhibitors of binding, should provide further clues on HA–NA functional balance related to early infection.

10 Receptors for Influenza Virus

Studies of influenza virus binding properties or linkage specificity have always been limited due to the paucity of quantitative assays that reflect the multivalent nature of virus attachment, and aspects of membrane fluidity associated with virus binding to host cells. For decades, much of our fundamental understanding has been based on hemagglutination assays using erythrocytes from various species (Burnet and Bull 1943; Hoyle 1968; Ito et al. 1997), or neuraminidase-treated erythrocytes that were specifically resialylated using either α2,3- or α2,6-sialyltransferases (Paulson and Rogers 1987). Alternatively, assays have involved synthetically generated receptor analogs such as α2,3-, and α2,6-sialyllactose that were immobilized and probed for binding activity (Gambaryan and Matrosovich 1992; Gambaryan et al. 1997). In recent years, the technology for studying carbohydrate-binding microorganisms has advanced with the development of glycan microarray technology (Blixt et al. 2004). The most widely utilized glycan array is available through the Consortium for Functional Glycomics (CFG), and this array now contains 610 different chemically or enzymatically generated carbohydrates printed on glass slides, which can be interrogated for binding using fluorescence as a readout. The CFG arrays, and other glycan arrays based on similar technology, have been widely utilized to examine the binding properties of influenza viruses (Blixt et al. 2004; Stevens et al. 2006; Childs et al. 2009; Liu et al. 2010; Bradley et al. 2011b; Chen et al. 2011; Song et al. 2011b; Yen et al. 2011). Although most of the glycan microarray technology currently in use still suffers from the drawback that it involves immobilized substrates, such arrays have been valuable for broadly examining a variety of structures that can be recognized by viruses, and generally offer a useful snapshot of α2,3- or α2,6-recognition properties. However, they do not directly identify the actual receptors that influenza viruses utilize in natural hosts.

To date, our understanding of the distribution of influenza virus receptors in the cells and tissues of avian and mammalian hosts derives primarily from indirect methods. These include immunohistochemistry studies using plant lectins that recognize sialic acid-containing structures of specific linkage types. For example, the lectin SNA-1, derived from the inner bark of *Sambucus nigra* (elderberry), binds preferentially to α2,6-linked structures (Shibuya et al. 1987), whereas the lectins MAA-1 and MAA-2 derived from the seeds of the tree *Maackia amurensis*, recognize α2,3-linked structures (Wang and Cummings 1988). An alternative immunohistochemical method for identifying the distribution of receptors involves probing tissue sections with labeled viruses of known binding specificity, and these methods provide data that in general, support conclusions drawn using plant lectins. Initial studies on the distribution of α2,3- or α2,6-linked glycans in influenza host species indicated that the human respiratory tract was rich in α2,6-linked glycans, the avian intestinal tract expressed an abundance of α2,3-linkage types, and the respiratory tract of swine contained a mixture of the two (Ito et al. 1998); these results correlated with natural sites of infection in these hosts. The cells and tissues of these, and other hosts, have now been examined extensively using similar techniques, and a degree of variability is reported amongst these studies. There is reasonable agreement that tissues in the human upper respiratory tract are enriched for α2,6-linked glycans, whereas the lower respiratory tract contains a mixture of α2,3, and α2,6-linked glycans (Shinya et al. 2006; Nicholls et al. 2007). The respiratory tract of both swine (Nelli et al. 2010; Trebbien et al. 2011), and ferrets (Leigh et al. 1995; Jayaraman et al. 2012), appear similar to humans with respect to sialic acid linkage-type distribution. Data on receptor distribution in avian species demonstrate a wide range of α2,3- and α2,6-expression profiles, depending on the species, tissues examined, and the individual study (Kuchipudi et al. 2009; Costa et al. 2012; Franca et al. 2013).

Glycomic analysis of adult and pediatric human respiratory tract tissues by mass spectroscopy showed that a variety of both α2,3-, and α2,6-linked glycans were expressed in the bronchus and lungs (Walther et al. 2013). Analysis of the nasopharynx was limited due to tissue quantities, but appeared to contain a more restricted distribution of glycans. Glycome studies have also been performed on cultured primary swine respiratory epithelial cells isolated from trachea (Bateman et al. 2010), as well as ex vivo tracheal and lung tissues derived from Pigs (Chan et al. 2013), and mass spectrometry again revealed a complex distribution of glycans, with a greater prevalence of α2,6 glycans. These glycomic analyses reveal a diversity of glycan structures that are not comprehensively included on any of the currently available arrays, and highlight our lack of knowledge on the identity and distribution of *bona fide* natural receptors for influenza. To address this, a "shotgun glycomics" approach (Song et al. 2011a) has been developed to purify and interrogate the glycans present in pig lung tissue for binding to a panel of influenza viruses that include avian, swine, and human strains (Byrd-Leotis et al. 2014). A variety of α2,3-, and α2,6-linked glycans were present on the pig lung array, as determined by lectin binding profiles, and though avian viruses were generally α2,3 specific, and swine and human viruses α2,6 specific, a high degree of variability was observed in their binding profiles, even amongst isolates from the same species. In some cases, highly restricted subsets of glycans with the appropriate linkage type were utilized, and in many cases the preferential binding to di- and triantennary glycans highlight the relevance of valency in the binding interaction. It also highlighted N-glycans as preferred ligands in the pig lung, as opposed to O-linked glycans or glycolipids, supporting previous observations made with mutant CHO cell lines deficient in the N-acetylglucosaminyltransferase I (Chu and Whittaker 2004). The results demonstrate a fine-specificity of receptor recognition for particular strains, which would not be recognized using assays that only distinguish overall α2,3/α2,6 specificity or binding to synthetic receptor analogs. The challenge that remains involves the identification of natural binding substrates on a range of cells and tissues from natural hosts, and the elucidation of any receptors or binding profiles that might be critical determinants of host range or pathogenicity phenotypes. In turn, it might be possible to decipher in more detail the viral determinants of the HA and NA glycoproteins responsible for participating in such phenotypes.

11 Inhibitors

Glycan components with inhibitory properties for influenza have been isolated from the sera of various animals, as well as from secretions found in tears, airways, or other mucoid tissues. The inhibitory properties of such glycans appear to be related to a high content of sialic acid expressed in spacial arrangements that allow for multivalent engagement of viral surface HA binding sites, inhibiting viral attachment to host cells. As related above, inhibitory factors present in nonimmune horse serum have been particularly useful in driving our understanding of host specificity, and the role of sialic acid linkage recognition in defining this. Choppin and Tamm showed that horse serum was capable of inhibiting 1957 pandemic H2N2 viruses in a strain-specific fashion (Choppin and Tamm 1960a, b), and the active component was later identified as α-2 macroglobulin (Pritchett and Paulson 1989; Ryan-Poirier and Kawaoka 1991, 1993). Due to its high content of α2,6-linked SA it was particularly effective against human isolates, whereas α2,3-specific avian isolates tended to be less sensitive (Rogers et al. 1983b). These observations led to the well-documented selection studies, described above, that initially directed attention on residue 226 as a focal point governing the receptor binding specificity for many strains of influenza (Rogers et al. 1983a).

More relevant for a role in natural infections are the inhibitors present in mucus secretions (Burnet 1948). For humans, it has been shown that the glycans of bronchial mucins contain predominantly α2,3 sialic acid linkages (Breg et al. 1987; Lamblin and Roussel 1993), and such mucins can selectively inhibit viruses with α2,3-linkage specificity (Couceiro et al. 1993). This raises an important concept on the selective pressures operating for efficient human respiratory transmission. While it seems clear that human pandemic viruses should evolve a capacity to recognize the α2,6-linked glycans that are prevalent in human upper airways, such specificity alone might not provide the optimal transmission phenotype, and inefficient recognition of the α2,3-linkages present on secreted mucins might also be of significance. This has been noted in discussions of recent H7N9 viruses, which are reported to have dual specificity for both α2,3, and α2,6 linkages, but have not yet evolved to transmit efficiently in humans (Xiong et al. 2013b). On the other hand, analysis of mutant H5N1 viruses selected for aerosol transmission in ferrets showed that the mutant viruses display only a small increase in affinity for α2,6 linkages, but a marked decrease in recognition of α2,3-linked receptor analogs (Herfst et al. 2012; Xiong et al. 2013b). Should the analogy extend to humans, it would suggest that dual specificity is not optimal, and that even a moderate proficiency for binding to α2,6 receptors might be sufficient to support a human transmission phenotype (on the appropriate genetic background), as long as the recognition of α2,3 linkages is poor, thereby circumventing the inhibitory effects of mucins.

The number of licensed antiviral drugs for influenza is currently limited to the α-adamantane M2 inhibitors and the NA inhibitors described above. Such compounds could prove critical in the event of a new pandemic in the time interval while vaccine strains are identified, produced and made available to the public.

However, little is known about the potential effects of antivirals as selective forces that might influence HA binding properties. In the laboratory, we know that mutant viruses resistant to M2 inhibitors generally contain changes in the transmembrane channel domain of the target protein (Hay et al. 1985); however, mutants can also be selected with changes in the HA that lead to acid-stable phenotypes (Steinhauer et al. 1991; Ilyushina et al. 2007), and such phenotypes are now being recognized for their potential involvement in adaptation and pathogenicity (Herfst et al. 2012; Imai et al. 2012; Galloway et al. 2013; Shelton et al. 2013; Zaraket et al. 2013). As described above, NA inhibitors also have the potential for selecting HA mutants that alter binding phenotypes, but, like the M2 inhibitors, there is no evidence to date suggesting that such selective pressures have influenced the binding properties of naturally circulating viruses. In addition, novel drugs based on recombinant proteins have been designed to enzymatically destroy receptors using sialidase constructs (Fludase) (Belser et al. 2007; Triana-Baltzer et al. 2009), similar in nature to experiments carried out in the late 1940s (Stone 1948). More recent approaches using lectin domains to block access to receptors have yielded complete protection in mice from a lethal virus challenge (Connaris et al. 2014). Novel approaches such as these may provide useful alternatives to the available drugs, but effects on cell metabolism and signaling pathways have yet to be fully explored, and the potential consequences of indiscriminant action on the natural glycan inhibitors present in mucous secretions should probably be taken into consideration.

12 Conclusions and Future Perspectives

As genetic factors that play a role in the adaptation of influenza viruses to new hosts continue to be defined, it becomes increasingly clear that numerous factors and selective pressures are involved in driving the evolution of host range variants and partially adapted strains. These are often interrelated, and different genetic backgrounds can essentially provide a "clean slate" on which specific mutations may result in differing phenotypic effects. Nonetheless, we can justifiably concentrate on certain traits such as the HA-receptor binding properties, and even focus our attention on specific residues such as HA1 226 when trying to identify strains with pandemic potential. However, depending on the genetic background it will not always be possible to predict which residues might play a role in cross-species transmission events based on sequence analysis. Furthermore, numerous interrelated selective pressures exist to drive receptor binding phenotypes, including availability and distribution of potential receptors, effects of natural or synthetic inhibitors, immune pressure, HA-NA functional balance, and effects of environmental stability. As such, though we are generating improved techniques and approaches to identify viral strains with pandemic potential, it is likely that many surprises remain in store in our efforts to understand the unpredictable nature of influenza pandemics.

Acknowledgments The work by XX and JWM was funded by the Medical Research Council through programs U117512723 and U117584222. The work by DAS was supported by the U.S. Department of Health and Human Services contract HHSN272201400004C (NIAID Centers of Excellence for Influenza Research and Surveillance).

References

Aytay S, Schulze IT (1991) Single amino acid substitutions in the hemagglutinin can alter the host range and receptor binding properties of H1 strains of influenza A virus. J Virol 65:3022–3028

Baigent SJ, Bethell RC, McCauley JW (1999) Genetic analysis reveals that both haemagglutinin and neuraminidase determine the sensitivity of naturally occurring avian influenza viruses to zanamivir in vitro. Virology 263:323–338. doi:10.1006/viro.1999.9931

Baigent SJ, McCauley JW (2001) Glycosylation of haemagglutinin and stalk-length of neuraminidase combine to regulate the growth of avian influenza viruses in tissue culture. Virus Res 79:177–185

Bateman AC, Karamanska R, Busch MG, Dell A, Olsen CW, Haslam SM (2010) Glycan analysis and influenza a virus infection of primary swine respiratory epithelial cells: the importance of NeuAc{alpha}2–6 glycans. J Biol Chem 285:34016–34026. doi:10.1074/jbc.M110.115998

Baum LG, Paulson JC (1991) The N2 neuraminidase of human influenza virus has acquired a substrate specificity complementary to the hemagglutinin receptor specificity. Virology 180:10–15

Bean WJ, Schell M, Katz J, Kawaoka Y, Naeve C, Gorman O, Webster RG (1992) Evolution of the H3 influenza virus hemagglutinin from human and nonhuman hosts. J Virol 66:1129–1138

Belser JA, Lu X, Szretter KJ et al (2007) DAS181, a novel sialidase fusion protein, protects mice from lethal avian influenza H5N1 virus infection. J Infect Dis 196:1493–1499. doi:10.1086/522609

Blixt O, Head S, Mondala T et al (2004) Printed covalent glycan array for ligand profiling of diverse glycan binding proteins. Proc Natl Acad Sci USA 101:17033–17038. doi:10.1073/pnas.0407902101

Bradley KC, Galloway SE, Lasanajak Y et al (2011a) Analysis of influenza virus hemagglutinin receptor binding mutants with limited receptor recognition properties and conditional replication characteristics. J Virol 85:12387–12398. doi:10.1128/JVI.05570-11

Bradley KC, Jones CA, Tompkins SM et al (2011b) Comparison of the receptor binding properties of contemporary swine isolates and early human pandemic H1N1 isolates (Novel 2009 H1N1). Virology 413:169–182. doi:10.1016/j.virol.2011.01.027

Breg J, Van Halbeek H, Vliegenthart JF, Lamblin G, Houvenaghel MC, Roussel P (1987) Structure of sialyl-oligosaccharides isolated from bronchial mucus glycoproteins of patients (blood group O) suffering from cystic fibrosis. Eur J Biochem/FEBS 168:57–68

Burnet FM (1948) Mucins and mucoids in relation to influenza virus action. 4. Inhibition by purified mucoid of infection and haemagglutination with the virus strain WSE. Aust J exp Biol med Sci 26:381–387

Burnet FM, Bull DR (1943) Changes in influenza virus associated with adaptation to passage in chick embryos. Aust J Exp Biol Med Sci 21:55–69

Burnet FM, McCrea JF, Stone JD (1946) Modification of human red cells by virus action I. The receptor gradient for virus action in human red cells. Br J Exp Path 27:228–236

Burnet FM, Stone JD (1947) The receptor-destroying enzyme of V. cholera. Aust J Exp Biol Med Sci 25:227–233

Byrd-Leotis L, Liu R, Bradley KC et al (2014) Shotgun glycomics of pig lung identifies natural endogenous receptors for influenza viruses. Proc Natl Acad Sci USA. doi:10.1073/pnas.1323162111

Calder LJ, Wasilewski S, Berriman JA, Rosenthal PB (2010) Structural organization of a filamentous influenza a virus. Proc Natl Acad Sci USA 107:10685–10690. doi:10.1073/pnas.1002123107

Chan RW, Karamanska R, Van Poucke S et al (2013) Infection of swine ex vivo tissues with avian viruses including H7N9 and correlation with glycomic analysis. Influenza Other Respir Viruses 7:1269–1282. doi:10.1111/irv.12144

Chen J, Lee KH, Steinhauer DA, Stevens DJ, Skehel JJ, Wiley DC (1998) Structure of the hemagglutinin precursor cleavage site, a determinant of influenza pathogenicity and the origin of the labile conformation. Cell 95:409–417

Chen LM, Blixt O, Stevens J et al (2012) In vitro evolution of H5N1 avian influenza virus toward human-type receptor specificity. Virology 422:105–113. doi:10.1016/j.virol.2011.10.006

Chen LM, Rivailler P, Hossain J et al (2011) Receptor specificity of subtype H1 influenza A viruses isolated from swine and humans in the United States. Virology 412:401–410. doi:10.1016/j.virol.2011.01.015

Childs RA, Palma AS, Wharton S et al (2009) Receptor-binding specificity of pandemic influenza A (H1N1) 2009 virus determined by carbohydrate microarray. Nat Biotechnol 27:797–799. doi:10.1038/nbt0909-797

Choppin PW, Murphy JS, Tamm I (1960) Studies of two kinds of virus particles which comprise influenza A2 virus strains. III. Morphological characteristics: independence to morphological and functional traits. J Exp Med 112:945–952

Choppin PW, Tamm I (1960a) Studies of two kinds of virus particles which comprise influenza A2 Virus Strains : I. characterization of stable homogeneous substrains in reactions with specific antibody, mucoprotein inhibitors, and erythrocytes. J Exp Med 112:895–920

Choppin PW, Tamm I (1960b) Studies of two kinds of virus particles which comprise influenza A2 virus strains. II. Reactivity with virus inhibitors in normal sera. J Exp Med 112:921–944

Chu CM, Dawson IM, Elford WJ (1949) Filamentous forms associated with newly isolated influenza virus. Lancet 1:602

Chu VC, Whittaker GR (2004) Influenza virus entry and infection require host cell N-linked glycoprotein. Proc Natl Acad Sci USA 101:18153–18158. doi:10.1073/pnas.0405172102

Chutinimitkul S, van Riel D, Munster VJ et al (2010) In vitro assessment of attachment pattern and replication efficiency of H5N1 influenza a viruses with altered receptor specificity. J Virol 84:6825–6833 doi:10.1128/JVI.02737-09JVI.02737-09 [pii]

Claas EC, Osterhaus AD, van Beek R et al (1998) Human influenza A H5N1 virus related to a highly pathogenic avian influenza virus. Lancet 351:472–477. doi:10.1016/S0140-6736(97)11212-0

Connaris H, Govorkova EA, Ligertwood Y et al (2014) Prevention of influenza by targeting host receptors using engineered proteins. Proc Natl Acad Sci USA. doi:10.1073/pnas.1404205111

Connor RJ, Kawaoka Y, Webster RG, Paulson JC (1994) Receptor specificity in human, avian, and equine H2 and H3 influenza virus isolates. Virology 205:17–23. doi:10.1006/viro.1994.1615

Costa T, Chaves AJ, Valle R et al (2012) Distribution patterns of influenza virus receptors and viral attachment patterns in the respiratory and intestinal tracts of seven avian species. Vet Res 43:28. doi:10.1186/1297-9716-43-28

Couceiro JN, Paulson JC, Baum LG (1993) Influenza virus strains selectively recognize sialyloligosaccharides on human respiratory epithelium; the role of the host cell in selection of hemagglutinin receptor specificity. Virus Res 29:155–165

Daniels PS, Jeffries S, Yates P et al (1987) The receptor-binding and membrane-fusion properties of influenza virus variants selected using anti-haemagglutinin monoclonal antibodies. EMBO J 6:1459–1465

Daniels RS, Douglas AR, Skehel JJ et al (1984) Antigenic analyses of influenza virus haemagglutinins with different receptor-binding specificities. Virology 138:174–177

Daniels RS, Downie JC, Hay AJ, Knossow M, Skehel JJ, Wang ML, Wiley DC (1985) Fusion mutants of the influenza virus hemagglutinin glycoprotein. Cell 40:431–439

Das SR, Hensley SE, David A et al (2011) Fitness costs limit influenza a virus hemagglutinin glycosylation as an immune evasion strategy. Proc Natl Acad Sci USA 108:E1417–E1422. doi:10.1073/pnas.1108754108

de Vries RP, Zhu X, McBride R et al (2014) Hemagglutinin receptor specificity and structural analyses of respiratory droplet-transmissible H5N1 viruses. J Virol 88:768–773. doi:10.1128/JVI.02690-13

Eisen MB, Sabesan S, Skehel JJ, Wiley DC (1997) Binding of the influenza a virus to cell-surface receptors: structures of five hemagglutinin-sialyloligosaccharide complexes determined by X-ray crystallography. Virology 232:19–31. doi:10.1006/viro.1997.8526

Fang R, Min Jou W, Huylebroeck D, Devos R, Fiers W (1981) Complete structure of A/duck/Ukraine/63 influenza hemagglutinin gene: animal virus as progenitor of human H3 Hong Kong 1968 influenza hemagglutinin. Cell 25:315–323

Franca M, Stallknecht DE, Howerth EW (2013) Expression and distribution of sialic acid influenza virus receptors in wild birds. Avian pathol: J W.V.P.A 42:60–71 doi:10.1080/03079457.2012.759176

Galloway SE, Reed ML, Russell CJ, Steinhauer DA (2013) Influenza HA subtypes demonstrate divergent phenotypes for cleavage activation and pH of fusion: implications for host range and adaptation. PLoS Pathog 9:e1003151. doi:10.1371/journal.ppat.1003151

Gambaryan A, Tuzikov A, Pazynina G, Bovin N, Balish A, Klimov A (2006) Evolution of the receptor binding phenotype of influenza A (H5) viruses. Virology 344:432–438 doi:S0042-6822(05)00531-3 [pii] 10.1016/j.virol.2005.08.035

Gambaryan AS, Matrosovich MN (1992) A solid-phase enzyme-linked assay for influenza virus receptor-binding activity. J Virol Methods 39:111–123

Gambaryan AS, Tuzikov AB, Piskarev VE et al (1997) Specification of receptor-binding phenotypes of influenza virus isolates from different hosts using synthetic sialylglycopolymers: non-egg-adapted human H1 and H3 influenza A and influenza B viruses share a common high binding affinity for 6'-sialyl(N-acetyllactosamine). Virology 232:345–350. doi:10.1006/viro.1997.8572

Gamblin SJ, Haire LF, Russell RJ et al (2004) The structure and receptor binding properties of the 1918 influenza hemagglutinin. Science 303:1838–1842. doi:10.1126/science.1093155

Glaser L, Stevens J, Zamarin D et al (2005) A single amino acid substitution in 1918 influenza virus hemagglutinin changes receptor binding specificity. J Virol 79:11533–11536. doi:10.1128/JVI.79.17.11533-11536.2005

Gottschalk A (1957) Neuraminidase: the specific enzyme of influenza virus and Vibrio cholera. Biochim Biophys Acta 23:645

Ha Y, Stevens DJ, Skehel JJ, Wiley DC (2003) X-ray structure of the hemagglutinin of a potential H3 avian progenitor of the 1968 Hong Kong pandemic influenza virus. Virology 309:209–218

Hay AJ, Wolstenholme AJ, Skehel JJ, Smith MH (1985) The molecular basis of the specific anti-influenza action of amantadine. EMBO J 4:3021–3024

Hensley SE, Das SR, Bailey AL et al (2009) Hemagglutinin receptor binding avidity drives influenza a virus antigenic drift. Science 326:734–736 doi:10.1126/science.1178258326/5953/734 [pii]

Herfst S, Schrauwen EJ, Linster M et al (2012) Airborne transmission of influenza A/H5N1 virus between ferrets. Science 336:1534–1541. doi:10.1126/science.1213362

Hirst GK (1941) The agglutination of red cells by allantoic fluid of chick embryos infected with influenza virus. Science 94:22–23. doi:10.1126/science.94.2427.22

Hooper KA, Bloom JD (2013) A mutant influenza virus that uses an N1 neuraminidase as the receptor-binding protein. J Virol 87:12531–12540. doi:10.1128/JVI.01889-13

Hoyle L (1968) Virology monographs. Springer, Berlin

Hughes MT, Matrosovich M, Rodgers ME, McGregor M, Kawaoka Y (2000) Influenza a viruses lacking sialidase activity can undergo multiple cycles of replication in cell culture, eggs, or mice. J Virol 74:5206–5212

Ilyushina NA, Govorkova EA, Russell CJ, Hoffmann E, Webster RG (2007) Contribution of H7 haemagglutinin to amantadine resistance and infectivity of influenza virus. J Gen Virol 88:1266–1274. doi:10.1099/vir.0.82256-0

Imai M, Watanabe T, Hatta M et al (2012) Experimental adaptation of an influenza H5 HA confers respiratory droplet transmission to a reassortant H5 HA/H1N1 virus in ferrets. Nature 486:420–428. doi:10.1038/nature10831

Iqbal M, Essen SC, Xiao H, Brookes SM, Brown IH, McCauley JW (2012) Selection of variant viruses during replication and transmission of H7N1 viruses in chickens and turkeys. Virology 433:282–295. doi:10.1016/j.virol.2012.08.001

Ito T, Couceiro JN, Kelm S et al (1998) Molecular basis for the generation in pigs of influenza a viruses with pandemic potential. J Virol 72:7367–7373

Ito T, Suzuki Y, Mitnaul L, Vines A, Kida H, Kawaoka Y (1997) Receptor specificity of influenza a viruses correlates with the agglutination of erythrocytes from different animal species. Virology 227:493–499. doi:10.1006/viro.1996.8323

Itoh Y, Shinya K, Kiso M et al (2009) In vitro and in vivo characterization of new swine-origin H1N1 influenza viruses. Nature 460:1021–1025. doi:10.1038/nature08260

Jayaraman A, Chandrasekaran A, Viswanathan K, Raman R, Fox JG, Sasisekharan R (2012) Decoding the distribution of glycan receptors for human-adapted influenza a viruses in ferret respiratory tract. PLoS ONE 7:e27517. doi:10.1371/journal.pone.0027517

Kida H, Kawaoka Y, Naeve CW, Webster RG (1987) Antigenic and genetic conservation of H3 influenza virus in wild ducks. Virology 159:109–119

Kilander A, Rykkvin R, Dudman SG, Hungnes O (2010) Observed association between the HA1 mutation D222G in the 2009 pandemic influenza A(H1N1) virus and severe clinical outcome, Norway 2009–2010. Euro surveillance : bulletin Europeen sur les maladies transmissibles = European communicable disease bulletin 15

Klenk E, Faillard H, Lempfrid H (1955) Enzymatic effect of the influenza virus. Hoppe Seylers Z Physiol Chem 301:235–246

Klenk HD, Rott R, Orlich M, Blodorn J (1975) Activation of influenza a viruses by trypsin treatment. Virology 68:426–439

Kobasa D, Kodihalli S, Luo M et al (1999) Amino acid residues contributing to the substrate specificity of the influenza a virus neuraminidase. J Virol 73:6743–6751

Koel BF, Burke DF, Bestebroer TM et al (2013) Substitutions near the receptor binding site determine major antigenic change during influenza virus evolution. Science 342:976–979. doi:10.1126/science.1244730

Kuchipudi SV, Nelli R, White GA, Bain M, Chang KC, Dunham S (2009) Differences in influenza virus receptors in chickens and ducks: Implications for interspecies transmission. J Mol Gen Med Int J Bio Res 3:143–151

Lamblin G, Roussel P (1993) Airway mucins and their role in defence against micro-organisms. Respir Med 87:421–426

Leigh MW, Connor RJ, Kelm S, Baum LG, Paulson JC (1995) Receptor specificity of influenza virus influences severity of illness in ferrets. Vaccine 13:1468–1473

Lin YP, Gregory V, Collins P et al (2010) Neuraminidase receptor binding variants of human influenza A(H3N2) viruses resulting from substitution of aspartic acid 151 in the catalytic site: a role in virus attachment? J Virol 84:6769–6781. doi:10.1128/JVI.00458-10

Lin YP, Xiong X, Wharton SA et al (2012) Evolution of the receptor binding properties of the influenza A(H3N2) hemagglutinin. Proc Natl Acad Sci USA 109:21474–21479. doi:10.1073/pnas.1218841110

Liu J, Stevens DJ, Haire LF et al (2009) Structures of receptor complexes formed by hemagglutinins from the Asian Influenza pandemic of 1957. Proc Natl Acad Sci USA 106:17175–17180. doi:10.1073/pnas.0906849106

Liu Y, Childs RA, Matrosovich T et al (2010) Altered receptor specificity and cell tropism of D222G hemagglutinin mutants isolated from fatal cases of pandemic A(H1N1) 2009 influenza virus. J Virol 84:12069–12074. doi:10.1128/JVI.01639-10

Lu X, Shi Y, Zhang W, Zhang Y, Qi J, Gao GF (2013) Structure and receptor-binding properties of an airborne transmissible avian influenza a virus hemagglutinin H5 (VN1203mut). Protein Cell 4:502–511. doi:10.1007/s13238-013-3906-z
Mak GC, Au KW, Tai LS, Chuang KC, Cheng KC, Shiu TC, Lim W (2010) Association of D222G substitution in haemagglutinin of 2009 pandemic influenza A (H1N1) with severe disease. Euro surveillance: bulletin Europeen sur les maladies transmissibles = European communicable disease bulletin 15
Martin J, Wharton SA, Lin YP, Takemoto DK, Skehel JJ, Wiley DC, Steinhauer DA (1998) Studies of the binding properties of influenza hemagglutinin receptor-site mutants. Virology 241:101–111
Matrosovich M, Tuzikov A, Bovin N et al (2000) Early alterations of the receptor-binding properties of H1, H2, and H3 avian influenza virus hemagglutinins after their introduction into mammals. J Virol 74:8502–8512
Matrosovich MN, Matrosovich TY, Gray T, Roberts NA, Klenk HD (2004) Neuraminidase is important for the initiation of influenza virus infection in human airway epithelium. J Virol 78:12665–12667. doi:10.1128/JVI.78.22.12665-12667.2004
McClelland L, Hare R (1941) Adsorption of influenza virus by red cells and a new in vitro method of measuring antibodies for influenza virus. Can J Publ Health 32:530
McKimm-Breschkin JL (2000) Resistance of influenza viruses to neuraminidase inhibitors–a review. Antiviral Res 47:1–17
Medeiros R, Escriou N, Naffakh N, Manuguerra JC, van der Werf S (2001) Hemagglutinin residues of recent human A(H3N2) influenza viruses that contribute to the inability to agglutinate chicken erythrocytes. Virology 289:74–85. doi:10.1006/viro.2001.1121
Meisner J, Szretter KJ, Bradley KC et al (2008) Infectivity studies of influenza virus hemagglutinin receptor binding site mutants in mice. J Virol 82:5079–5083. doi:10.1128/JVI.01958-07
Mitnaul LJ, Matrosovich MN, Castrucci MR, Tuzikov AB, Bovin NV, Kobasa D, Kawaoka Y (2000) Balanced hemagglutinin and neuraminidase activities are critical for efficient replication of influenza a virus. J Virol 74:6015–6020
Mosley VM, Wyckoff RW (1946) Election micrography of the virus of influenza. Nature 157:263
Naeve CW, Hinshaw VS, Webster RG (1984) Mutations in the hemagglutinin receptor-binding site can change the biological properties of an influenza virus. J Virol 51:567–569
Nelli RK, Kuchipudi SV, White GA, Perez BB, Dunham SP, Chang KC (2010) Comparative distribution of human and avian type sialic acid influenza receptors in the pig. BMC Vet Res 6:4. doi:10.1186/1746-6148-6-4
Nicholls JM, Chan MC, Chan WY et al (2007) Tropism of avian influenza A (H5N1) in the upper and lower respiratory tract. Nat Med 13:147–149. doi:10.1038/nm1529
Nobusawa E, Ishihara H, Morishita T, Sato K, Nakajima K (2000) Change in receptor-binding specificity of recent human influenza a viruses (H3N2): a single amino acid change in hemagglutinin altered its recognition of sialyloligosaccharides. Virology 278:587–596. doi:10.1006/viro.2000.0679
Paulson JC, Rogers GN (1987) Resialylated erythrocytes for assessment of the specificity of sialyloligosaccharide binding proteins. Methods Enzymol 138:162–168
Pritchett TJ, Paulson JC (1989) Basis for the potent inhibition of influenza virus infection by equine and guinea pig alpha 2-macroglobulin. J Biol Chem 264:9850–9858
Resende PC, Motta FC, Oliveira Mde L et al (2014) Polymorphisms at residue 222 of the hemagglutinin of pandemic influenza A(H1N1)pdm09: association of quasi-species to morbidity and mortality in different risk categories. PloS one 9:e92789 doi:10.1371/journal.pone.0092789
Rogers GN, Daniels RS, Skehel JJ, Wiley DC, Wang XF, Higa HH, Paulson JC (1985) Host-mediated selection of influenza virus receptor variants. Sialic acid-alpha 2, 6Gal-specific clones of A/duck/Ukraine/1/63 revert to sialic acid-alpha 2, 3Gal-specific wild type in ovo. J Biol Chem 260:7362–7367

Rogers GN, Paulson JC (1983) Receptor determinants of human and animal influenza virus isolates: differences in receptor specificity of the H3 hemagglutinin based on species of origin. Virology 127:361–373
Rogers GN, Paulson JC, Daniels RS, Skehel JJ, Wilson IA, Wiley DC (1983a) Single amino acid substitutions in influenza haemagglutinin change receptor binding specificity. Nature 304:76–78
Rogers GN, Pritchett TJ, Lane JL, Paulson JC (1983b) Differential sensitivity of human, avian, and equine influenza a viruses to a glycoprotein inhibitor of infection: selection of receptor specific variants. Virology 131:394–408
Ryan-Poirier KA, Kawaoka Y (1991) Distinct glycoprotein inhibitors of influenza a virus in different animal sera. J Virol 65:389–395
Ryan-Poirier KA, Kawaoka Y (1993) Alpha 2-macroglobulin is the major neutralizing inhibitor of influenza a virus in pig serum. Virology 193:974–976. doi:10.1006/viro.1993.1208
Sauter NK, Bednarski MD, Wurzburg BA, Hanson JE, Whitesides GM, Skehel JJ, Wiley DC (1989) Hemagglutinins from two influenza virus variants bind to sialic acid derivatives with millimolar dissociation constants: a 500 MHz proton nuclear magnetic resonance study. Biochemistry 28:8388–8396
Sauter NK, Hanson JE, Glick GD et al (1992) Binding of influenza virus hemagglutinin to analogs of its cell-surface receptor, sialic acid: analysis by proton nuclear magnetic resonance spectroscopy and X-ray crystallography. Biochemistry 31:9609–9621
Shelton H, Roberts KL, Molesti E, Temperton N, Barclay WS (2013) Mutations in haemagglutinin that affect receptor binding and pH stability increase replication of a PR8 influenza virus with H5 HA in the upper respiratory tract of ferrets and may contribute to transmissibility. J Gen Virol 94:1220–1229. doi:10.1099/vir.0.050526-0
Shi Y, Zhang W, Wang F et al (2013) Structures and receptor binding of hemagglutinins from human-infecting H7N9 influenza viruses. Science 342:243–247. doi:10.1126/science.1242917
Shibuya N, Goldstein IJ, Broekaert WF, Nsimba-Lubaki M, Peeters B, Peumans WJ (1987) The elderberry (Sambucus nigra L.) bark lectin recognizes the Neu5Ac(alpha 2–6)Gal/GalNAc sequence. J Biol Chem 262:1596–1601
Shinya K, Ebina M, Yamada S, Ono M, Kasai N, Kawaoka Y (2006) Avian flu: influenza virus receptors in the human airway. Nature 440:435–436. doi:10.1038/440435a
Shore DA, Yang H, Balish AL et al (2013) Structural and antigenic variation among diverse clade 2 H5N1 viruses. PLoS ONE 8:e75209. doi:10.1371/journal.pone.0075209
Song X, Lasanajak Y, Xia B et al (2011a) Shotgun glycomics: a microarray strategy for functional glycomics. Nat Methods 8:85–90. doi:10.1038/nmeth.1540
Song X, Yu H, Chen X et al (2011b) A sialylated glycan microarray reveals novel interactions of modified sialic acids with proteins and viruses. J Biol Chem 286:31610–31622. doi:10.1074/jbc.M111.274217
Steinhauer DA, Wharton SA, Skehel JJ, Wiley DC, Hay AJ (1991) Amantadine selection of a mutant influenza virus containing an acid-stable hemagglutinin glycoprotein: evidence for virus-specific regulation of the pH of glycoprotein transport vesicles. Proc Natl Acad Sci USA 88:11525–11529
Stevens J, Blixt O, Tumpey TM, Taubenberger JK, Paulson JC, Wilson IA (2006) Structure and receptor specificity of the hemagglutinin from an H5N1 influenza virus. Science 312:404–410. doi:10.1126/science.1124513
Stone JD (1948) Prevention of virus infection with enzyme of V. cholerae, studies with influenza virus in mice. Aust J Exp Biol Med Sci 26:287–298
Trebbien R, Larsen LE, Viuff BM (2011) Distribution of sialic acid receptors and influenza a virus of avian and swine origin in experimentally infected pigs. Virol J 8:434. doi:10.1186/1743-422X-8-434
Triana-Baltzer GB, Gubareva LV, Nicholls JM et al (2009) Novel pandemic influenza A(H1N1) viruses are potently inhibited by DAS181, a sialidase fusion protein. PLoS ONE 4:e7788. doi:10.1371/journal.pone.0007788

Underwood PA, Skehel JJ, Wiley DC (1987) Receptor-binding characteristics of monoclonal antibody-selected antigenic variants of influenza virus. J Virol 61:206–208
Varghese JN, Laver WG, Colman PM (1983) Structure of the influenza virus glycoprotein antigen neuraminidase at 2.9 a resolution. Nature 303:35–40
Walther T, Karamanska R, Chan RW et al (2013) Glycomic analysis of human respiratory tract tissues and correlation with influenza virus infection. PLoS Pathog 9:e1003223. doi:10.1371/journal.ppat.1003223
Wang Q, Tian X, Chen X, Ma J (2007) Structural basis for receptor specificity of influenza B virus hemagglutinin. Proc Natl Acad Sci USA 104:16874–16879. doi:10.1073/pnas.0708363104
Wang WC, Cummings RD (1988) The immobilized leukoagglutinin from the seeds of Maackia amurensis binds with high affinity to complex-type Asn-linked oligosaccharides containing terminal sialic acid-linked alpha-2, 3 to penultimate galactose residues. J Biol Chem 263: 4576–4585
Ward CW, Dopheide TA (1981) Evolution of the Hong Kong influenza a sub-type. Structural relationships between the haemagglutinin from A/duck/Ukraine/1/63 (Hav 7) and the Hong Kong (H3) haemagglutinins. Biochem J 195:337–340
Watanabe Y, Ibrahim MS, Ellakany HF et al (2011) Acquisition of human-type receptor binding specificity by new H5N1 influenza virus sublineages during their emergence in birds in Egypt. PLoS Pathog 7:e1002068 doi:10.1371/journal.ppat.1002068 PPATHOGENS-D-10-00603 [pii]
Wedde M, Wahlisch S, Wolff T, Schweiger B (2013) Predominance of HA-222D/G polymorphism in influenza A(H1N1)pdm09 viruses associated with fatal and severe outcomes recently circulating in Germany. PLoS ONE 8:e57059. doi:10.1371/journal.pone.0057059
Weis W, Brown JH, Cusack S, Paulson JC, Skehel JJ, Wiley DC (1988) Structure of the influenza virus haemagglutinin complexed with its receptor, sialic acid. Nature 333:426–431. doi:10.1038/333426a0
WHO (2014) Monthly risk assessment. Influenza at the human-anima interface. Summary and assessment as of 27 June 2014 at (http://www.who.int/influenza/human_animal_interface/Influenza_Summary_IRA_HA_interface_27June14.pdf?ua=1)
WHO/OIE/FAO (2014) Revised and updated nomenclature for highly pathogenic avian influenza A (H5N1) viruses. Influenza Other Respir Viruses 8:384–388
Wiley DC, Wilson IA, Skehel JJ (1981) Structural identification of the antibody-binding sites of Hong Kong influenza haemagglutinin and their involvement in antigenic variation. Nature 289:373–378
Wilson IA, Skehel JJ, Wiley DC (1981) Structure of the haemagglutinin membrane glycoprotein of influenza virus at 3 a resolution. Nature 289:366–373
Xiong X, Coombs PJ, Martin SR et al (2013a) Receptor binding by a ferret-transmissible H5 avian influenza virus. Nature 497:392–396. doi:10.1038/nature12144
Xiong X, Martin SR, Haire LF et al (2013b) Receptor binding by an H7N9 influenza virus from humans. Nature 499:496–499. doi:10.1038/nature12372
Xiong X, Tuzikov A, Coombs PJ et al (2013c) Recognition of sulphated and fucosylated receptor sialosides by A/Vietnam/1194/2004 (H5N1) influenza virus. Virus Res 178:12–14. doi:10.1016/j.virusres.2013.08.007
Xiong X, Xiao H, Martin SR et al (2014) Enhanced human receptor binding by H5 haemagglutinins. Virology 456–457:179–187. doi:10.1016/j.virol.2014.03.008
Xu R, de Vries RP, Zhu X et al (2013) Preferential recognition of avian-like receptors in human influenza A H7N9 viruses. Science 342:1230–1235. doi:10.1126/science.1243761
Xu R, McBride R, Nycholat CM, Paulson JC, Wilson IA (2012a) Structural characterization of the hemagglutinin receptor specificity from the 2009 H1N1 influenza pandemic. J Virol 86:982–990. doi:10.1128/JVI.06322-11
Xu R, Zhu X, McBride R, Nycholat CM, Yu W, Paulson JC, Wilson IA (2012b) Functional balance of the hemagglutinin and neuraminidase activities accompanies the emergence of the 2009 H1N1 influenza pandemic. J Virol 86:9221–9232. doi:10.1128/JVI.00697-12

Yamada S, Suzuki Y, Suzuki T et al. (2006) Haemagglutinin mutations responsible for the binding of H5N1 influenza a viruses to human-type receptors. Nature 444:378–382 doi: nature05264 [pii] 10.1038/nature05264

Yang P, Bansal A, Liu C, Air GM (1997) Hemagglutinin specificity and neuraminidase coding capacity of neuraminidase-deficient influenza viruses. Virology 229:155–165. doi:10.1006/viro.1996.8421

Yen HL, Liang CH, Wu CY et al (2011) Hemagglutinin-neuraminidase balance confers respiratory-droplet transmissibility of the pandemic H1N1 influenza virus in ferrets. Proc Natl Acad Sci USA 108:14264–14269. doi:10.1073/pnas.1111000108

Zaraket H, Bridges OA, Russell CJ (2013) The pH of activation of the hemagglutinin protein regulates H5N1 influenza virus replication and pathogenesis in mice. J Virol 87:4826–4834. doi:10.1128/JVI.03110-12

Zhang Q, Shi J, Deng G et al (2013a) H7N9 influenza viruses are transmissible in ferrets by respiratory droplet. Science 341:410–414. doi:10.1126/science.1240532

Zhang W, Shi Y, Lu X, Shu Y, Qi J, Gao GF (2013b) An airborne transmissible avian influenza H5 hemagglutinin seen at the atomic level. Science 340:1463–1467 doi:10.1126/science.1236787 science.1236787 [pii]

Zhu X, McBride R, Nycholat CM, Yu W, Paulson JC, Wilson IA (2012) Influenza virus neuraminidases with reduced enzymatic activity that avidly bind sialic acid receptors. J Virol 86:13371–13383. doi:10.1128/JVI.01426-12

Acid-Induced Membrane Fusion by the Hemagglutinin Protein and Its Role in Influenza Virus Biology

Charles J. Russell

Abstract Membrane fusion is not spontaneous. Therefore, enveloped viruses have evolved membrane-fusion mediating glycoproteins that, once activated, refold, and release energy that fuses viral and cellular membranes. The influenza A virus hemagglutinin (HA) protein is a prototypic structural class I viral fusion glycoprotein that, once primed by proteolytic cleavage, is activated by endosomal low pH to form a fusogenic "leash-in-grooves" hairpin structure. Low-pH induced HA protein refolding is an irreversible process, so acid exposure in the absence of a target membrane leads to virus inactivation. The HA proteins of diverse influenza virus subtypes isolated from a variety of species differ in their acid stabilities, or pH values at which irreversible HA protein conformational changes are triggered. Recently, efficient replication of highly pathogenic avian influenza (HPAI) viruses such as H5N1 in avian species has been associated with a relatively high HA activation pH. In contrast, a decrease in H5N1 HA activation pH has been shown to enhance replication and airborne transmission in mammals. Mutations that alter the acid stabilities of H1 and H3 HA proteins have also been discovered that influence the amantadine susceptibilities, replication rates, and pathogenicities of human influenza viruses. An understanding of the role of HA acid stability in influenza virus biology is expected to aid in identifying emerging viruses with increased pandemic potential and assist in developing live attenuated virus vaccines. Acid-induced HA protein activation, which has provided a paradigm for protein-mediated membrane fusion, is now identified as a novel determinant of influenza virus biology.

Abbreviations

CT	Cytoplasmic tail
ER	Endoplasmic reticulum
HA	Hemagglutinin

C.J. Russell (✉)
Department of Infectious Diseases, St. Jude Children's Research Hospital, MS 330, 262 Danny Thomas Place, Memphis, TN 38105-3678, USA
e-mail: charles.russell@stjude.org

Current Topics in Microbiology and Immunology (2014) 385: 93–116
DOI: 10.1007/82_2014_393

Published Online: 11 July 2014

HAT	Human airway trypsin-like
HPAI	Highly pathogenic avian influenza
LPAI	Low pathogenic avian influenza
MSPL	Mosaic serine protease large-form
NA	Neuraminidase
NP	Nucleoprotein
RBD	Receptor-binding domain
R.B.P.	Receptor-binding pocket
RMSD	Root-mean-square deviation
SARS	Severe acute respiratory syndrome
TM	Transmembrane
TMPRSS	Transmembrane protease, serine

Contents

1 Introduction

The influenza A viruses are negative-strand RNA viruses of the *Orthomyxoviridae* family. They encode a variety of proteins from eight gene segments (Shaw and Palese 2013): PB2, PB1, PA, HA, NP, NA, M, and NS. The HA and neuraminidase (NA) envelope glycoproteins and the M2 ion channel are embedded in the host cell-derived viral membrane. Internal proteins include the M1 matrix protein, the polymerase proteins (PB1, PB2, and PA), the nucleoprotein (NP), and the NEP/NS2 protein. Nonstructural accessory proteins include NS1 and the more recently discovered proteins PB1-N40 (Wise et al. 2009), PA-X (Jagger et al. 2012), PB1-F2 (Chen et al. 2001), M42 (Wise et al. 2012), NS3 (Mohammed Selman et al. 2012), and PA-N155 and PA-N182 (Muramoto et al. 2013).

Diverse influenza A viruses circulate in a reservoir of wild aquatic birds including 16 HA and 9 NA subtypes (Krauss and Webster 2010). Two additional HA (H17 and H18) and NA (N10 and N11) subtypes have recently been discovered in influenza viruses isolated from bats (Tong et al. 2012, 2013). Avian influenza viruses occasionally infect non-avian species including humans, causing limited outbreaks or global pandemics (Russell and Webster 2005). The replication efficiency, transmissibility, and pathogenicity of a given influenza virus in a particular host depend on multiple viral and host factors (Salomon and Webster 2009; Belser et al. 2010; Fukuyama and Kawaoka 2011; Sorrell et al. 2011). The focus of this review is on the role of the HA protein in viral entry and influenza virus biology with a focus on the acid stability of the HA protein. The HA protein is a prototypic structural class I viral fusion glycoprotein that is triggered by low pH in the endosome to undergo irreversible conformational changes that cause membrane fusion. The HA proteins from circulating viruses differ in the pH required for fusion activation, and recent studies identify HA acid stability as one of several important factors in the interspecies adaptation and evolution of influenza A viruses.

2 Molecular and Cell Biology of the HA Protein

2.1 Hemagglutinin (HA) Protein Synthesis and Maturation

As with the other influenza viral genes, HA mRNA is transcribed in the nucleus and exported to the cytoplasm (Shaw and Palese 2013). The HA0 precursor protein is a type I integral membrane protein. HA0 contains an N-terminal signal sequence (which targets protein synthesis to the ER before being cleaved and released), an ~513-residue ectodomain, an ~27-residue transmembrane (TM) domain, and an ~11-residue cytoplasmic tail (CT) (Fig. 1). In the presence of chaperones in the ER, HA0 monomers are extensively folded and form intramolecular disulfide bonds before forming a noncovalently linked homotrimeric complex (Copeland et al. 1988). N-linked glycosylation contributes to protein folding in the ER (Hebert et al. 1995). As the HA protein trafficks intracellularly to the cell surface, N-linked glycosylation sites undergo maturation and free cysteine residues in the TM domain and CT are acylated with fatty acids (Naeve and Williams 1990; Veit et al. 1991; Steinhauer et al. 1991b).

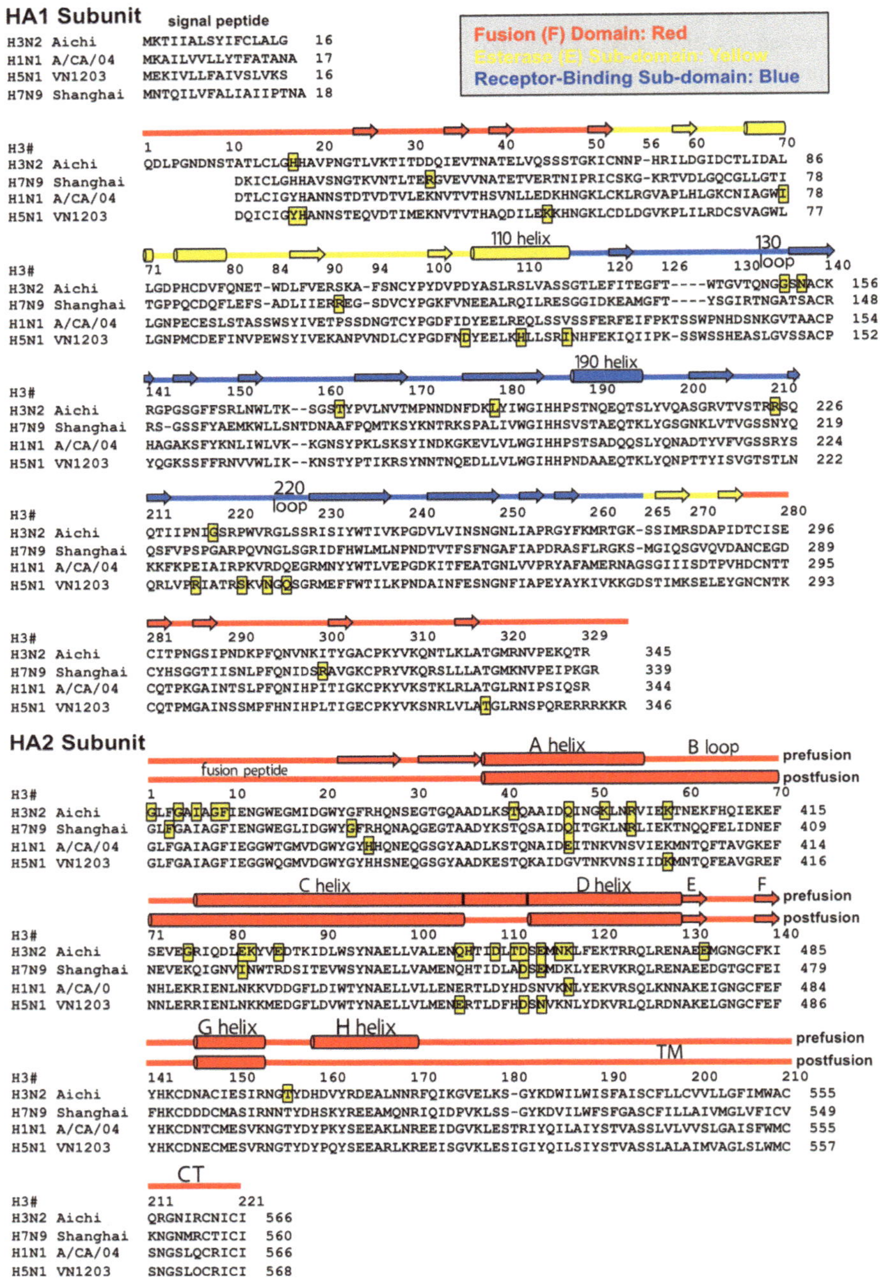
HA1 Subunit
signal peptide
H3N2 Aichi MKTIIALSYIFCLALG 16
H1N1 A/CA/04 MKAILVVLLYTFATANA 17
H5N1 VN1203 MEKIVLLFAIVSLVKS 16
H7N9 Shanghai MNTQILVFALIAIIPTNA 18
Fusion (F) Domain: Red
Esterase (E) Sub-domain: Yellow
Receptor-Binding Sub-domain: Blue
H3# 1 10 20 30 40 50 56 60 70
H3N2 Aichi QDLPGNDNSTATLCLGHHAVPNGTLVKTITDDQIEVTNATELVQSSSTGKICNNP-HRILDGIDCTLIDAL 86
H7N9 Shanghai DKICLGHHAVSNGTKVNTLTERGVEVVNATETVERTNIPRICSKG-KRTVDLGQCGLLGTI 78
H1N1 A/CA/04 DTLCIGYHANNSTDTVDTVLEKNVTVTHSVNLLEDKHNGKLCKLRGVAPLHLGKCNIAGWI 78
H5N1 VN1203 DQICIGYHANNSTEQVDTIMEKNVTVTHAQDILEKKHNGKLCDLDGVKPLILRDCSVAGWL 77
110 helix
130 loop
H3# 71 80 84 90 94 100 110 120 126 130 140
H3N2 Aichi LGDPHCDVFQNET-WDLFVERSKA-FSNCYPYDVPDYASLRSLVASSGTLEFITEGFT----WTGVTQNGGSNACK 156
H7N9 Shanghai TGPPQCDQFLEFS-ADLIIERREG-SDVCYPGKFVNEEALRQILRESGGIDKEAMGFT----YSGIRTNGATSACR 148
H1N1 A/CA/04 LGNPECESLSTASSWSYIVETPSSDNGTCYPGDFIDYEELREQLSSVSSFERFEIFPKTSSWPNHDSNKGVTAACP 154
H5N1 VN1203 LGNPMCDEFINVPEWSYIVEKANPVNDLCYPGDFNDYEELKHLLSRINHFEKIQIIPK-SSWSSHEASLGVSSACP 152
190 helix
H3# 141 150 160 170 180 190 200 210
H3N2 Aichi RGPGSGFFSRLNWLTK--SGSTYPVLNVTMPNNDNFDKLYIWGIHHPSTNQEQTSLYVQASGRVTVSTRRSQ 226
H7N9 Shanghai RS-GSSFYAEMKWLLSNTDNAAFPQMTKSYKNTRKSPALIVWGIHHSVSTAEQTKLYGSGNKLVTVGSSNYQ 219
H1N1 A/CA/04 HAGAKSFYKNLIWLVK--KGNSYPKLSKSYINDKGKEVLVLWGIHHPSTSADQQSLYQNADTYVFVGSSRYS 224
H5N1 VN1203 YQGKSSFFRNVVWLIK--KNSTYPTIKRSYNNTNQEDLLVLWGIHHPNDAAEQTKLYQNPTTYISVGTSTLN 222
220 loop
H3# 211 220 230 240 250 260 265 270 280
H3N2 Aichi QTIIPNIGSRPWVRGLSSRISIYWTIVKPGDVLVINSNGNLIAPRGYFKMRTGK-SSIMRSDAPIDTCISE 296
H7N9 Shanghai QSFVPSPGARPQVNGLSGRIDFHWLMLNPNDTVTFSFNGAFIAPDRASFLRGKS-MGIQSGVQVDANCEGD 289
H1N1 A/CA/04 KKFKPEIAIRPKVRDQEGRMNYYWTLVEPGDKITFEATGNLVVPRYAFAMERNAGSGIIISDTPVHDCNTT 295
H5N1 VN1203 QRLVPRIATRSKVNGQSGRMEFFWTILKPNDAINFESNGNFIAPEYAYKIVKKGDSTIMKSELEYGNCNTK 293
H3# 281 290 300 310 320 329
H3N2 Aichi CITPNGSIPNDKPFQNVNKITYGACPKYVKQNTLKLATGMRNVPEKQTR 345
H7N9 Shanghai CYHSGGTIISNLPFQNIDSRAVGKCPRYVKQRSLLLATGMKNVPEIPKGR 339
H1N1 A/CA/04 CQTPKGAINTSLPFQNIHPITIGKCPKYVKSTKLRLATGLRNIPSIQSR 344
H5N1 VN1203 CQTPMGAINSSMPFHNIHPLTIGECPKYVKSNRLVLATGLRNSPQRERRRKKR 346
HA2 Subunit
A helix
B loop
prefusion
fusion peptide
postfusion
H3# 1 10 20 30 40 50 60 70
H3N2 Aichi GLFGAIAGFIENGWEGMIDGWYGFRHQNSEGTGQAADLKSTQAAIDQINGKLNRVIEKTNEKFHQIEKEF 415
H7N9 Shanghai GLFGAIAGFIENGWEGLIDGWYGFRHQNAQGEGTAADYKSTQSAIDQITGKLNRLIEKTNQQFELIDNEF 409
H1N1 A/CA/04 GLFGAIAGFIEGGWTGMVDGWYGYHHQNEQGSGYAADLKSTQNAIDEITNKVNSVIEKMNTQFTAVGKEF 414
H5N1 VN1203 GLFGAIAGFIEGGWQGMVDGWYGYHHSNEQGSGYAADKESTQKAIDGVTNKVNSIIDKMNTQFEAVGREF 416
C helix
D helix
E
F
prefusion
postfusion
H3# 71 80 90 100 110 120 130 140
H3N2 Aichi SEVEGRIQDLEKYVEDTKIDLWSYNAELLVALENQHTIDLTDSEMNKLFEKTRRQLRENAEEMGNGCFKI 485
H7N9 Shanghai NEVEKQIGNVINWTRDSITEVWSYNAELLVAMENQHTIDLADSEMDKLYERVKRQLRENAEEDGTGCFEI 479
H1N1 A/CA/0 NHLEKRIENLNKKVDDGFLDIWTYNAELLVLLENERTLDYHDSNVKNLYEKVRSQLKNNAKEIGNGCFEF 484
H5N1 VN1203 NNLERRIENLNKKMEDGFLDVWTYNAELLVLMENERTLDFHDSNVKNLYDKVRLQLRDNAKELGNGCFEF 486
G helix
H helix
prefusion
TM
postfusion
H3# 141 150 160 170 180 190 200 210
H3N2 Aichi YHKCDNACIESIRNGTYDHDVYRDEALNNRFQIKGVELKS-GYKDWILWISFAISCFLLCVVLLGFIMWAC 555
H7N9 Shanghai FHKCDDDCMASIRNNTYDHSKYREEAMQNRIQIDPVKLSS-GYKDVILWFSFGASCFILLAIVMGLVFICV 549
H1N1 A/CA/04 YHKCDNTCMESVKNGTYDYPKYSEEAKLNREEIDGVKLESTRIYQILAIYSTVASSLVLVVSLGAISFWMC 555
H5N1 VN1203 YHKCDNECMESVRNGTYDYPQYSEEARLKREEISGVKLESIGIYQILSIYSTVASSLALAIMVAGLSLWMC 557
CT
H3# 211 221
H3N2 Aichi QRGNIRCNICI 566
H7N9 Shanghai KNGNMRCTICI 560
H1N1 A/CA/04 SNGSLQCRICI 566
H5N1 VN1203 SNGSLOCRICI 568

◀ **Fig. 1** Alignment of HA protein sequences with key features identified. Residues are identified by H3 numbering above the alignment and real numbering (starting with methionine 1) to the *right* of the sequences. Alpha helical—(*cylinders*) and beta strand—(*arrows*) secondary structures are included above the alignments. Both prefusion and postfusion secondary structures are shown in HA2 (they are equivalent in HA1). Fusion (*red*), esterase (*yellow*), and receptor-binding (*blue*) subdomains are color-coded in the secondary structures. Signal peptide, fusion peptide, transmembrane (TM), and cytoplasmic tail (CT) regions are also identified. Residues governing HA acid stability (Table 1) are identified in the primary sequence by yellow boxes. Sequences are from A/Aichi/2/68/X-31 (H3N2), A/Shanghai/02/2013 (H7N9), A/California/04/2009 (H1N1), and A/Vietnam/1203/04 (H5N1)

The HA0 precursor protein is incapable of causing membrane fusion, and proteolytic cleavage into an HA1/HA2 complex is required to prime the protein into a fusion-competent form (Steinhauer 1999). The HA0 proteins from HPAI viruses contain a polybasic (arginine and lysine) cleavage site with an R-X-R/K-R sequence that is recognized intracellularly in the trans-Golgi network by ubiquitously expressed subtilisin-like enzymes such as furin and PC6 (Garten et al. 1981; Webster and Rott 1987; Vey et al. 1992; Stieneke-Gröber et al. 1992). For human and low pathogenic avian influenza (LPAI) viruses, HA0 is cleaved extracellularly by soluble trypsin-like proteases including tryptase Clara (Kido et al. 1992), mini-plasmin (Murakami et al. 2001), and ectopic anionic trypsin I (Towatari et al. 2002). More recently, cell-associated serine proteases have also been discovered to cleave monobasic HA0 proteins including TMPRSS2, human airway trypsin-like (HAT) protease (Böttcher et al. 2006), and TMPRSS4 (Chaipan et al. 2009). MSPL and TMPRSS13 serine proteases have been found to cleave HA proteins from HPAI isolates (Okumura et al. 2010).

2.2 Virus Assembly and Budding

At the plasma membrane, the HA and NA glycoproteins associate with lipid rafts while the M2 ion channel is largely excluded (Takeda et al. 2003; Leser and Lamb 2005). Lipid rafts are variable-sized regions of the plasma membrane enriched with cholesterol and sphingolipid (Lingwood and Simons 2010) that serve as a platform for virus assembly and budding (Suomalainen 2002; Rossman and Lamb 2011) before M2-mediated scission (Rossman et al. 2010). The NA protein desialyates cellular and viral glycoconjugates, enabling efficient release of progeny virions from the host cell and preventing HA-mediated virion–virion aggregation (Compans et al. 1969; Palese et al. 1974). Efficient influenza virus replication appears to require an optimal balance between HA receptor-binding and NA receptor-destroying activities, which can be disturbed by reassortment, interspecies transmission, or NA-inhibitor treatment [reviewed in (Wagner et al. 2002)]. A role for NA enzymatic activity in HA-mediated membrane fusion has also been described (Huang et al. 1980). Inhibition of NA activity decreases the pH required to activate the HA proteins of some H5N1 isolates (Reed et al. 2010), and the absence of the NA protein results in decreased membrane fusion by an H1N1 HA protein (Su et al. 2009). The influence of NA activity on membrane fusion activity may be limited to specific isolates and its mechanism is unknown, although it may include destabilizing the HA protein via modification of HA glycosylation sites.

2.3 HA Protein Prefusion Structures

X-ray crystallographic structures of bromelain-released HA ectodomain from A/Aichi/68 (H3N2) in its HA0 uncleaved form (Chen et al. 1998) and its fusion-primed HA1/HA2 form (Wilson et al. 1981) reveal at atomic resolution the prefusion conformation of this prototypic structural class I viral fusion protein (Fig. 2a). Tethered to the viral envelope is the more highly conserved fusion (F) domain (also known as stalk or stem). The F domain is trapped in a high-energy conformation in part by the presence of the membrane-distal globular receptor-binding domain (RBD) head domains. The RBD consists of receptor-binding and vestigial esterase subdomains (identified in Fig. 1) that were most likely inserted into a primordial F protein during the virus' evolution (Rosenthal et al. 1998). The immunodominant, poorly conserved RBD head is directed toward target cells and contains the receptor-binding pocket (R.B.P.) and most of the major antigenic sites (Wiley et al. 1981). The long, fibrous F domain has at its core a 54-residue triple-stranded coiled coil (helices C and D) that is flanked by spring-loaded B loops, A helices, and membrane-proximal regions (Fig. 2b). The prefusion HA0 and HA1/HA2 structures are superimposable with the exception of five residues upstream of the cleavage site and the first 12 residues of the fusion peptide located near the base of the molecule. In uncleaved HA0, these 17 residues form a prominent loop that is protease-accessible (Chen et al. 1998); whereas, in cleaved HA1/HA2 the fusion peptide tucks into a pocket formed by residues from HA1 and the HA2 D and A helices (Wilson et al. 1981).

X-ray crystal structures have been determined for bromelain-released and baculovirus-expressed HA1/HA2 and HA0 prefusion structures from numerous influenza A viruses including those isolated from humans (H1N1, pH1N1, H2N2, H3N2, H5N1, H7N7, H7N9), avians (H1, H2, H3, H5, H7, H13, H16), swine (H1, H9), and bat (H17). Despite amino-acid sequence identity of less than 50 %, the structures across subtypes are highly conserved (RMSD values of ~ 1 Å within domains) and can be classified into two structural/antigenic groups (Air 1981; Nobusawa et al. 1991; Gamblin and Skehel 2010): group 1 (H1, H2, H5, H6, H8, H9, H11, H12, H13, H16, and H17) and group 2 (H3, H4, H7, H10, H14, and H15). The main differences between group 1 and 2 HA structures are (a) an ~20° axial rotation and ~4 Å displacement of the globular RBD head atop the central coiled coil in the stalk and (b) the sharpness of the turn of the B loop back toward the base of the C helix (Ha et al. 2002). These differences have been ascribed to differences in HA1 residue 107 (in the 110 helix of the vestigial esterase domain), residues 63 and 75 in the HA2 B loop, and residue 88 in the HA2 C helix (Ha et al. 2002).

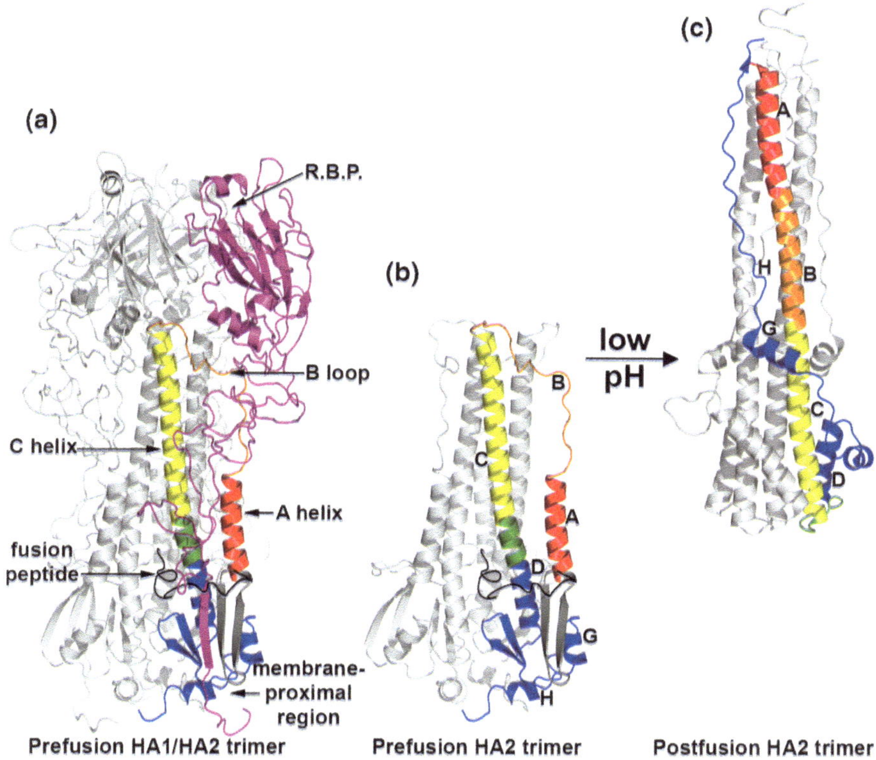

Fig. 2 Structural changes in the HA protein after low-pH induced activation. **a** Prefusion HA1/HA2 trimer with HA1 shown in magenta and HA2 secondary structural elements shown in multiple colors. The receptor-binding pocket (R.B.P.) is located in the membrane-distal head domain and the metastable fusion domain is located in the membrane-proximal stalk. **b** Prefusion HA2 trimer shown without HA1 residues. **c** Postfusion HA2 trimer showing changes in secondary and tertiary structure after acid-induced irreversible protein refolding. In each panel, two protomers are colored *gray*. Figure adapted from (Bullough et al. 1994) using A/Aichi/2/68/X-31 (H3N2) protein data bank structures 1HGF and 1QUI

2.4 H3 Numbering Convention

HA protein structure and function are highly conserved across subtypes despite substantial differences in primary amino-acid sequence and numbering. To facilitate the description of significant residues (e.g., residues in the R.B.P. or those altering HA acid stability), HA residues are conventionally identified using "H3 numbering." H3 numbering derived from the first HA structure of A/Aichi/68 (H3N2) (Wilson et al. 1981) is described in Fig. 1 and will be used throughout this review.

2.5 *Influenza Virus Internalization into a New Host Cell*

After virus egress and HA priming by cleavage, an infectious progeny virion is poised to invade a new host cell. During influenza A virus entry, the HA protein binds host cell receptors containing sialic acid, or N-acetyl-neuraminic acid (Weis et al. 1988; Chu and Whittaker 2004). It should be noted that A/WSN/33 (H1N1) has been shown to enter cells in the absence of sialyated N-glycans by a dynamin-dependent mechanism (de Vries et al. 2012) and the HA protein from a recently discovered H17N10 bat virus does not bind sialic acid-containing receptors (Tong et al. 2012; Sun et al. 2013). In general, the HA proteins from human-adapted viruses preferentially bind to α(2,6)-linked sialosides while those from avian viruses tend to prefer those with an α(2,3) linkage (Matrosovich et al. 2009). As few as one amino-acid mutation in the R.B.P. can switch receptor-binding specificity (Rogers et al. 1983; Matrosovich et al. 2000) and, in turn, can alter the tropism, host range, and pathogenicity of an influenza virus (Belser et al. 2010; Fukuyama and Kawaoka 2011; Sorrell et al. 2011). Receptor binding triggers internalization of the virion into the host cell (Rust et al. 2004), which can occur via clathrin-mediated or clathrin-independent endocytosis (Sieczkarski and Whittaker 2002; Lakadamyali et al. 2006) or by macropinocytosis (de Vries et al. 2011; De Conto et al. 2011).

2.6 *Acid-induced Activation of the HA Protein*

Once internalized, an influenza virus trafficks through the endosomal network (Lakadamyali et al. 2004; Sun and Whittaker 2013). Within 5 min, endocytosed material is exposed to pH 6.5–6.0 in early endosomes and then is slowly acidified (over 30–40 min) to pH 5.5–5.0 in late endosomes and pH 5.0–4.6 in lysosomes (Mellman et al. 1986; Cain et al. 1989; Grove and Marsh 2011). Exposure to low pH triggers fusion-primed HA1/HA2 to undergo irreversible conformational changes that cause membrane fusion in the presence of target membranes or virion inactivation in the absence of target membranes (Skehel and Wiley 2000). While the biological trigger for HA activation is low pH, the native protein is trapped in a metastable (high-energy) conformation and its activation energy barrier can also be overcome in vitro by other mild destabilizing agents such as heat and the denaturant urea (Scholtissek 1985; Ruigrok et al. 1986; Carr et al. 1997). Human H1, H2, and H3 HA proteins from twentieth century pandemics were found to be activated and cause membrane fusion at pH values of 5.0–5.2 (Hoekstra and Klappe 1993; Galloway et al. 2013), leading to the classical view that HA-mediated membrane fusion takes place in late endosomes (Grove and Marsh 2011; Sun and Whittaker 2013). However, broader surveys that include avian and swine isolates have shown that HA activation pH values range from pH 4.6 to 6.0 (Scholtissek 1985; Galloway et al. 2013). This raises the possibility that isolates with relatively high activation pH values may cause membrane fusion in early endosomes, as has been proposed for HPAI H5N1 viruses (Reed et al. 2010; DuBois et al. 2011b). The pH values of endocytic compartments may also vary between host species, tissues, and individuals.

2.7 Low-pH-induced Structural Changes

Several large-scale structural changes occur during HA protein refolding (Fig. 2). Upon reaching a threshold pH, protonation appears to induce deformations in the RBD and relax the B loops at the top of the stalk (Xu and Wilson 2011). Next, the RBD protomers dissociate from each other and the stalk (Bottcher et al. 1999; Huang et al. 2002) with the overall fold within the RBD domains remaining intact (Bizebard et al. 1995; DuBois et al. 2011a). Solvent penetration into the stalk induces conformational changes within the central helices (Xu and Wilson 2011). Meanwhile, dissociation of the RBDs from the stalk enables the B loops to spring out and extend the central coiled coil from helix C through helix A, an energetically favorable process (Carr and Kim 1993) that propels the fusion peptide harpoon toward the target membrane (Bullough et al. 1994; Huang et al. 2003). This prehairpin intermediate is topologically reminiscent of those formed by fellow structural class I viral fusion glycoproteins from HIV gp41 (Chan and Kim 1998) and the paramyxovirus F protein (Russell et al. 2001). Subsequently, HA2 residues 106–112 switch from a coiled coil to a reverse turn (Wilson et al. 1981; Bullough et al. 1994). This allows the membrane-proximal "leash" region to zipper up in the grooves of the central coiled coil in an antiparallel orientation, juxtaposing the TM domains and fusion peptides (Fig. 2c).

2.8 Membrane Fusion

Membrane fusion is not spontaneous and requires energy. Energy barriers that must be overcome for membrane fusion include (a) bringing together two membranes containing repulsive charges; (b) inducing lipid curvatures and deformations to form a hemifusion stalk; (c) in some cases, forming a hemifusion diaphragm; and (d) forming and expanding a fusion pore to allow mixing of contents (Chernomordik and Kozlov 2008; Martens and McMahon 2008). The two fusing membranes are brought into proximity first by high-avidity binding of multiple trimeric HA proteins to the host cell. Additionally, HA protein refolding juxtaposes target-membrane interacting fusion peptides with viral-envelope spanning TM domains (Durrer et al. 1996; Chen et al. 1999). Formation of the leash-in-grooves hairpin by the HA protein helps drive lipid mixing and hemifusion (Borrego-Diaz et al. 2003; Park et al. 2003). Similarly, the energy released by six-helix bundle formation in HIV gp41 and paramyxovirus F is coupled to the work of membrane fusion for those other class I fusion proteins (Melikyan et al. 2000; Russell et al. 2001).

Lipid packing is also disturbed by penetration into the host cell membrane by the fusion peptide, which adopts a bent "boomerang" structure (Han et al. 2001). The presence of the TM domains helps to break the hemifusion stalk or diaphragm (Melikyan et al. 1995; Armstrong et al. 2000). Finally, continued HA protein refolding outside the initial contact zone is necessary for fusion pore expansion that is sufficient for delivery of the viral RNA and internal structural proteins into the cytosol of the host cell (Leikina et al. 2004).

2.9 Amino-acid Residues Influencing HA Acid Stability

Over 70 mutations in H1, H2, H3, H5, and H7 HA proteins have been discovered that alter the pH at which the HA protein is triggered to undergo its irreversible conformational changes and cause membrane fusion (Table 1). Residues determining HA acid stability are located throughout the sequence (Fig. 1) and structure (Fig. 3). In general, the mutations are located in regions of the HA molecule that undergo dramatic changes in secondary and tertiary structure during the prefusion to postfusion transition (Wilson et al. 1981; Daniels et al. 1985; Bullough et al. 1994; Thoennes et al. 2008). As activation pH-altering mutations have not altered substantially the backbone structures of human H3N2 (Weis et al. 1990), avian H5N1 (DuBois et al. 2011b; Xiong et al. 2013; Zhang et al. 2013; de Vries et al. 2014), and human H2N2 (Xu and Wilson 2011) HA proteins in the prefusion conformation, acid stability mutations may in general exact their greatest influence on protein folding intermediates.

Pioneering work to identify HA acid stability mutants involved the selection of amantadine-resistant viruses that had elevated activation pH values (Daniels et al. 1985). These mutations were later mapped to four structural regions (Bullough et al. 1994): (1) the fusion peptide and its pocket; (2) helix A and loop B packing against the spring-loaded long CD helix; (3) the HA2-HA1 interface between the RBD esterase subdomain, helix C, and loop B; and (4) the HA1-HA1 interface between RBD protomers. Mapping the locations of the expanded list of acid stability mutations adds three more structural regions (Fig. 3): (5) near and in the R.B.P.; (6) at the interface of RBD and esterase subdomains; and (7) in the membrane-proximal region. Furthermore, it is likely that mutations in the TM domain (Zhou et al. 2013) and cytoplasmic tails may also influence the energy required for HA protein activation, just as has been reported for other structural class I viral fusion proteins from retroviruses and paramyxoviruses (Li et al. 2001; Tong et al. 2002; Waning et al. 2004). Some combinations of acid stability mutations are additive while others or not (Steinhauer et al. 1996), consistent with multiple structural transitions occurring during HA protein refolding. A great variety of acid stability mutations have been shown to be functionally equivalent with respect to amantadine resistance (Daniels et al. 1985). It has been hypothesized, but has not yet been demonstrated, that acid stability mutations are also functionally equivalent with respect to other biological properties (Zaraket et al. 2013a, b).

Table 1 Mutations altering the activation pH of the influenza A virus HA protein

H3 #[a]	Sub-type	Mutation and change in activation pH	Region[b]	Reference
17_1	H3N2	H17R (+0.6), H17A (0.4), H17Q (+0.25), H17Y (-0.3)	1	Daniels et al. (1985), Steinhauer et al. (1996), Lin et al. (1997), Thoennes (2008)
17_1	H5N1	Y17H (+0.4)	1	Reed et al. (2009), Zaraket et al. (2013a)
18_1	H5N1	H18Q (−0.3)	1	Reed et al. (2009), Zaraket et al. (2013a)
32_1	H7N1	R32G (+0.2)	2	Daniels et al. (1985)
45_1	H5N1	K45D (−0.1)	1	Reed et al. (2009)
70_1/ 25_2	H1N1	L70_1P/Q25_2H (−0.2)	6/7	Koerner et al. (2012)
91_1	H7N1	R91L (+0.3), R91Q (+0.1)	3	Daniels et al. (1985)
104_1/ 115_1	H5N1	D104N/I115T (−0.3)	3	Hulse et al. (2004), DuBois et al. (2011b)
110_1	H5N1	H110Y (−0.4)	3	Herfst et al. (2012), Zhang et al. (2013)
135_1/ 4_2	H3N2	G135_1E/G4_2E (+0.4)	5/1	Lin et al. (1997)
137_1/ 82_2	H3N2	N137D/K82T (+0.4)	5/2	Lin et al. (1997)
162_1	H3N2	P162S (+0.2)	5	Keleta et al. (2008)
179_1	H3N2	L179P (+0.5)	5	Nakowitsch et al. (2011)
210_1	H3N2	Q210R (+0.15)	4	Keleta et al. (2008)
216_1	H5N1	E216K (−0.4), K216E (+0.4)	4	Hulse et al. (2004), DuBois et al. (2011b)
218_1	H3N2	G218E,W(+0.4)	4	Steinhauer et al. (1996), Lin et al. (1997), Narasaraju et al. (2009), Keleta et al. (2008)
221_1	H5N1	S221P (−0.15)	4	Hulse et al. (2004), DuBois et al. (2011b)
224_1/ 226_1	H5N1	N224K/Q226L (+0.2)	5	Imai et al. (2012)
300_1	H7N1	R300S (+0.3)	2	Daniels et al. (1985)
318_1	H5N1	T318I (−0.2)	2	Imai et al. (2012)
1_2	H3N2	G1F,H,I,L (+0.3)	1	Steinhauer et al. (1995)
3_2	H7N1	F3L (+0.4)	1	Daniels et al. (1985)
4_2	H3N2	G4A (+0.4)	1	Steinhauer et al. (1995)
6_2	H3N2	I6M (+0.3)	1	Daniels et al. (1985)
8_2	H3N2	G8A (+0.5)	1	Steinhauer et al. (1995)
9_2	H3N2	F9L (+0.6)	1	Daniels et al. (1985)

(continued)

Table 1 (continued)

H3 #[a]	Sub-type	Mutation and change in activation pH	Region[b]	Reference
23_2	H7N7	G23C (−1.0)	1	Ilyushina et al. (2007)
41_2/ 85_2	H3N2	T41A/E85D (+0.4)	1/2	Lin et al. (1997)
47_2	H7N1	Q47L (+0.45)	2	Daniels et al. (1985)
47_2	H3N2	Q47R (+0.35)	2	Daniels et al. (1985)
47_2	H1N1	E47K (−0.4)	2	Cotter et al. (2014)
51_2	H3N2	K51A,E (+0.2)	2	Thoennes et al. (2008)
54_2	H3N2	R54E (+0.2)	2	Steinhauer et al. (1996)
54_2	H7N1	R54K (+0.3), R54G,S (+0.1)	2	Daniels et al. (1985)
58_2	H3N2	K58I (−0.7)	2	Steinhauer et al. (1991a, 1996)
58_2	H5N1	K58I (−0.5)	2	Reed et al. (2009, 2010), Zaraket et al. (2013a, b)
75_2	H3N2	G75R (+0.4)	2	Nakowitsch et al. (2011)
81_2	H3N2	E81G (+0.3)	2	Daniels et al. (1985)
81_2	H7N1	I81S (+0.1)	2	Daniels et al. (1985)
105_2	H3N2	Q105K (+0.3), Q105A (-0.2)	2	Daniels et al. (1985), Thoennes (2008)
105_2	H5N1	E105K (−0.2)	2	Reed et al. (2009)
106_2	H3N2	H106A (+0.1), H106F (-0.1)	2	Thoennes et al. (2008)
106_2	H2N2	R106H (−1.0)	2	Xu and Wilson (2011)
112_2	H3N2	D112A (+0.5), D112G (+0.4), D112N (+0.35) D112E (+0.25)	1	Daniels et al. (1985), Steinhauer et al. (1996), Thoennes et al. (2008)
112_2	H7N1	D112G (+0.4)	1	Daniels et al. (1985)
112_2	H5N1	D112G (+0.3)	1	Reed et al. (2009)
114_2	H7N1	E114K (+0.5)	1	Daniels et al. (1985)
114_2	H5N1	N114K (+0.5)	1	Reed et al. (2009, 2010)
114_2	H3N2	E114K (+0.6)	1	Daniels et al. (1985)
116_2	H3N2	N116D (+0.4)	1	Lin et al. (1997)
117_2	H3N2	K117R (+0.4)	1	Lin et al. (1997)
117_2	H1N1	N117D (+0.4)	1	Murakami et al. (2012)
132_2	H3N2	D132N (+0.2)	7	Doms et al. (1986)
156_2	H3N2	T156N (−0.1)	7	Keleta et al. (2008)

[a] H3 numbering is described in Fig. 1 and subscripts refer to HA1 and HA2 subunits

[b] Structural regions are defined as: (1) Fusion peptide and its surrounding pocket, (2) Spring-loaded coiled coil, (3) Esterase-stalk interface, (4) HA1-HA1 protomer interface, (5) Near or in receptor-binding pocket, (6) Esterase-RBD interface, and (7) Membrane proximal region

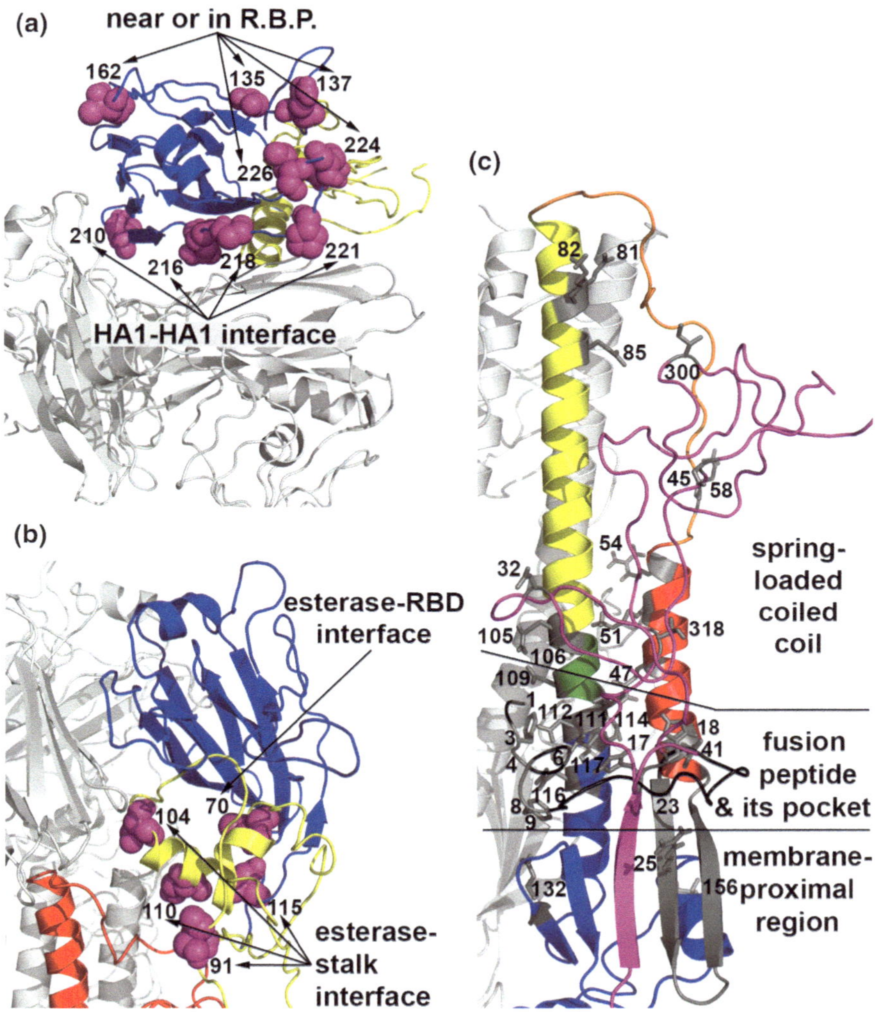

Fig. 3 Locations of HA acid stability mutations in the prefusion conformation. **a** RBD residues near or in the receptor-binding pocket (R.B.P.) or at the interface of HA1-HA1 protomers. **b** Esterase sub-domain residues at the interface of the receptor-binding subdomain and the stalk region. **c** HA1 and HA2 residues in the spring-loaded coiled coil region, in and around the fusion peptide pocket, and in the membrane-proximal region. In panels A and B, the receptor-binding subdomain is colored blue and the esterase subdomain is colored yellow. In panel C, HA1 residues in the stalk are colored magenta and HA2 secondary structural elements are colored as in Fig. 2. Residues governing HA acid stability are identified using H3 numbering on the A/Aichi/2/68/X-31 (H3N2) protein data bank structure 1HGF. Acid stability mutations are described in detail in Table 1

3 Impact of HA Acid Stability on Influenza Virus Biology

3.1 Adaptation of HA Acid Stability In Vitro

The acid stability of the HA protein has been shown to play a role in amantadine susceptibility and interspecies adaptation in vitro. Mutations that increase the pH of activation of HA proteins from H3N2 and H7N1 viruses have been shown to enhance viral fitness in the presence of ammonium chloride and high concentrations of amantadine, compounds that raise endosomal pH (Daniels et al. 1985; Doms et al. 1986; Steinhauer et al. 1995, 1996). Alternatively, HPAI H7N1 and H7N7 viruses containing mutations that decrease the HA activation pH are resistant to lower concentrations of amantadine that prevent M2-mediated neutralization of the secretory pathway (Steinhauer et al. 1991a; Ilyushina et al. 2007). The adaptation of egg-adapted A/Aichi/2/68 (H3N2) to mammalian MDCK and MDBK cells (Lin et al. 1997) and A/PR/8/34 (H1N1) to Vero cells (Murakami et al. 2012) resulted in increases in HA activation pH from pH 5.2 to 5.6 for the H3N2 virus and from pH 5.2 to 5.4 for the H1N1 virus.

3.2 Decrease in HA Activation pH Promotes HPAI Virus Adaptation to Mammals

Mutations that alter HA acid stability have also been implicated in the adaptation of avian influenza viruses from wild aquatic birds, the reservoir of influenza A viruses (Kim et al. 2009), to land-based poultry and mammals. A comparison of LPAI H7N3 viruses found that a virus isolated from a turkey had a lower HA activation pH than a related virus previously isolated from a duck, suggesting the adaptation of avian H7 viruses from wild aquatic birds to terrestrial poultry may be associated with enhanced HA acid stability (Giannecchini et al. 2006). The HPAI virus A/chicken/Vietnam/C58/2004 (H5N1), with an HA activation pH of 5.9 (Reed et al. 2009), was rendered unfit for growth and transmission in mallards by mutations that substantially raised (HA1-Y17H, activation pH 6.3) or lowered (HA2-K58I, activation pH 5.4) the HA activation pH. While the acid-stabilizing K58I mutation attenuated H5N1 replication in ducks, this same mutation increased replication and pathogenesis in mice (Zaraket et al. 2013b) and enhanced early growth of A/Vietnam/1203/04 (H5N1) in the upper respiratory tracts of ferrets (Zaraket et al. 2013a). Thus, a decrease in the HA activation pH has been shown to assist in the adaptation of H5N1 to mammals. On the other hand, an increase in HA activation pH (in the range of pH 5.2–6.0) has been associated with increased H5N1 virus growth and virulence in chickens (Hulse et al. 2004; DuBois et al. 2011b). This agrees with observations that the HA proteins from HPAI influenza viruses tend to have relatively high (pH 5.6–6.0) activation pH values while those from human-adapted seasonal influenza viruses are lower (pH 5.0–5.2) (Scholtissek 1985; Reed et al. 2010; DuBois et al. 2011b; Galloway et al. 2013).

Two recent studies provide compelling evidence for a role of HA acid stability in the adaptation of avian H5-containing viruses to mammals (Imai et al. 2012; Herfst et al. 2012). Two H5 viruses were adapted to transmit in ferrets, either A/Indonesia/5/2005 (H5N1) (Herfst et al. 2012) or a reassortant virus containing the HA gene from A/Vietnam/1203/04 (H5N1) and the internal genes of A/California/04/09 (H1N1) (Imai et al. 2012). In both studies, three analogous rounds of mutations occurred. First, two mutations were introduced into the R.B.P. that switched receptor-binding specificity from a preference of α(2,3) to α(2,6). Second, a mutation atop the RBD head removed a glycosylation site, increasing the accessibility of the R.B.P. The final mutation necessary for airborne transmissibility in ferrets in both cases was an HA1 mutation adjacent to (H110Y) or in (T318I) the fusion-mediating stalk domain that decreased the HA activation pH to ~5.6. While an acid-stabilizing mutation was required for airborne transmissibility in ferrets, such a mutation was not sufficient in the absence of α(2,6) receptor-binding specificity and/or glycosylation site deletion (Shelton et al. 2013; Zaraket et al. 2013a).

3.3 *Adaptation of Human Influenza Viruses to Murine Lungs*

The adaptation of human influenza viruses for replication in the murine lung has been associated with increases in HA activation pH (Narasaraju et al. 2009). An increase in HA activation pH from 5.2 to 5.6 resulted in increased growth and virulence of A/Hong Kong/1/68 (H3N2) in mice (Keleta et al. 2008). Similarly, an increase in HA activation pH from 5.3 to 5.8 enhanced A/PR/8/34 (H1N1) growth and virulence in mice (Koerner et al. 2012). Mouse-adapted A/Philippines/82 (H3N2) was also found to have increases in HA activation pH and virulence in mice (Hartley et al. 1997). The effects of activation-pH altering mutations on influenza A virus growth in the murine lung may not extend to growth in ferrets, as the acid-stabilized HA2-K58I mutant of A/Vietnam/1203/04 (H5N1) had increased growth in the lungs of mice but reduced growth in the lungs of ferrets (Zaraket et al. 2013a).

3.4 *HA Acid Stability in Live Attenuated Vaccine Development*

Just as knowledge of optimal HA activation pH values in various species may enhance risk assessment of emerging influenza viruses with increased pandemic potential, such an understanding may also enhance the development of live attenuated vaccines. Acid-destabilizing mutations that increase the HA activation pH from 5.8 to 6.2–6.3 have been shown to impair the immunogenicity of live attenuated, NS1-deleted H3N2 vaccine candidates in ferrets (Nakowitsch et al. 2011). In contrast, the acid-stabilizing HA2-K58I mutation that enhances upper respiratory tract growth of A/Vietnam/1203/04 in ferrets (Zaraket et al. 2013a) has been shown also to increase the infectivity and immunogenicity of a live attenuated, NS1-deleted H5N1 vaccine candidate in mice (Krenn et al. 2011). Similarly, a decrease in the pH of activation of the HA protein from 5.4 to 5.0 has been shown to enhance the stability and infectivity of p2009 H1N1 vaccine candidates (Cotter et al. 2014). Overall, HA acid-stabilizing mutations may enhance live attenuated influenza virus vaccines.

3.5 Hypothesis for Role of HA Acid Stability in Influenza Virus Biology

While further studies are needed to define the mechanism by which the pH of activation of the HA protein influences the interspecies adaptation of influenza A viruses, a working hypothesis is emerging. Intracellular pH is determined by the presence and activity of cellular vacuolar H^+-ATPases (V-ATPases) (Mellman et al. 1986; Jefferies et al. 2008), which may differ between different cell types, tissues, hosts, and metabolic states. For example, increased glucose exposure in MDCK cells leads to increased vATPase activity and decreased endosomal pH, thereby enhancing the replication of A/PR/8/34 (H1N1) (Kohio and Adamson 2013), which has a relatively low HA activation pH of 5.0–5.1 (Galloway et al. 2013). For avian H5N1 viruses, a relatively high HA activation pH enhances replication in the enteric and respiratory tracts of ducks and chickens (Reed et al. 2010; DuBois et al. 2011b), presumably because a destabilized HA protein is more readily triggered for membrane fusion during viral entry. In contrast, a high HA activation pH appears to be a liability for growth in the mammalian upper respiratory tract (Imai et al. 2012; Herfst et al. 2012; Shelton et al. 2013; Zaraket et al. 2013a), most likely because it enhances the susceptibility of pre-cleaved H5 HA trimers to irreversible inactivation in airway tissues. Mammalian airway tissue is acidic (pH 5.5–6.9) (Washington et al. 2000; Fischer and Widdicombe 2006) and acid secretion into the airway is enhanced upon infection by influenza viruses, lowering the nasal pH to 5.2 (Jacoby et al. 1988; Fischer et al. 2002). Human-adapted influenza viruses with monobasic cleavage sites may also benefit by having a relatively low HA activation pH value, which may help prevent premature inactivation in the mammalian respiratory tract. Human-adapted H1N1, H2N2, and H3N2 viruses from the twentieth century pandemics have HA activation pH values of pH 5.0–5.2 (Scholtissek 1985; Galloway et al. 2013), and a naturally occurring mutation in circulating p2009 H1N1 viruses has been shown to decrease the HA activation pH from 5.4 to 5.0 (Cotter et al. 2014).

4 Conclusions and Future Perspectives

Membrane fusion is not a spontaneous process but instead is driven by the formation of hairpin structures that bring together the viral TM domain (embedded in the viral envelope) and the viral fusion peptide (inserted into the host cell membrane) so as to juxtapose the viral and cellular membranes. HA-mediated membrane fusion has been a paradigm for understanding viral membrane fusion and is topographically similar to intracellular membrane fusion driven by SNARE proteins (Söllner 2004; Wickner and Schekman 2008). Recent studies add HA protein acid stability, or the pH at which the HA is triggered to undergo irreversible structural changes to cause membrane, to the list of molecular properties that contribute to the interspecies adaptation, pathogenesis, and transmissibility of influenza A viruses. Influenza viruses are internalized by endocytosis, and the HA protein is activated for membrane fusion by low pH. Many other enveloped viruses invade host cells similarly including hepatitis C virus, Epstein-Barr virus, vesicular stomatitis virus, avian leukemia virus, human rhinovirus, dengue virus, and severe acute respiratory syndrome (SARS) coronavirus (Grove and Marsh 2011). These other viruses may also regulate their host range and tropism by adaptive mutations that alter the acid stabilities of their fusion glycoproteins.

HA acid stability is just one of multiple simultaneous factors that dictate influenza virus biology, and many questions remain with respect to its mechanism and breadth of impact. It is not yet clear the extent to which this one molecular property interacts with others long known for influencing the replication, interspecies adaptation, pathogenicity, and transmissibility of influenza viruses (Salomon and Webster 2009; Belser et al. 2010; Fukuyama and Kawaoka 2011; Sorrell et al. 2011). For example, the mechanism by which NA activity influences HA acid stability is unknown (Huang et al. 1980; Su et al. 2009; Reed et al. 2010). It is also unknown whether mutations that alter the acid stabilities of monobasic HA proteins exert as strong of effects as those altering intracellularly cleaved (and more acid-neutralization sensitive) polybasic HA proteins. Compared to HPAI viruses, LPAI viruses are reported to have a broader and lower range of HA activation pH values (5.3–6.0 in ducks and 5.1–5.5 in chickens and turkeys) (Galloway et al. 2013). Avian influenza viruses disseminate throughout the enteric tracts of birds, including environments of extreme low pH, yet it is not clear how these viruses avoid inactivation. The pH values in endocytic vesicles have been measured in vitro in cell monolayers, but it is unknown how the pH values of early and late endosomes vary in vivo in the tissues of various living species and individuals with altered metabolic states. Perhaps a knowledge of species-specific and tissue-specific endosomal pH values and extracellular pH may better explain why a relatively low HA activation pH is disfavored by HPAI viruses in avian species yet favored in mammals. The preferred range of HA activation pH in swine and whether swine may serve as a mixing vessel during the acquisition of human-preferred HA acid stability is not yet known. Finally, the long-term viability of HA stalk-binding antiviral drugs and antibodies, in addition to the development of stalk-based universal vaccines, will also depend on the extent to which such treatments and prophylactics are prone to altering HA acid stability and the concomitant effects of such changes on influenza virus replication and the host response.

References

Air GM (1981) Sequence relationships among the hemagglutinin genes of 12 subtypes of influenza A virus. Proc Natl Acad Sci USA 78:7639–7643

Armstrong RT, Kushnir AS, White JM (2000) The transmembrane domain of influenza hemagglutinin exhibits a stringent length requirement to support the hemifusion to fusion transition. J Cell Biol 151:425–437

Belser JA, Maines TR, Tumpey TM, Katz JM (2010) Influenza A virus transmission: contributing factors and clinical implications. Expert Rev Mol Med 12:e39. doi:10.1017/S1462399410001705

Bizebard T, Gigant B, Rigolet P et al (1995) Structure of influenza virus haemagglutinin complexed with a neutralizing antibody. Nature 376:92–94. doi:10.1038/376092a0

Borrego-Diaz E, Peeples ME, Markosyan RM et al (2003) Completion of trimeric hairpin formation of influenza virus hemagglutinin promotes fusion pore opening and enlargement. Virology 316:234–244

Bottcher C, Ludwig K, Herrmann A et al (1999) Structure of influenza haemagglutinin at neutral and at fusogenic pH by electron cryo-microscopy. FEBS Lett 463:255–259

Böttcher E, Matrosovich T, Beyerle M et al (2006) Proteolytic activation of influenza viruses by serine proteases TMPRSS2 and HAT from human airway epithelium. J Virol 80:9896–9898. doi:10.1128/JVI.01118-06

Bullough PA, Hughson FM, Skehel JJ, Wiley DC (1994) Structure of influenza haemagglutinin at the pH of membrane fusion. Nature 371:37–43

Cain CC, Sipe DM, Murphy RF (1989) Regulation of endocytic pH by the Na+, K+-ATPase in living cells. Proc Natl Acad Sci USA 86:544–548

Carr CM, Chaudhry C, Kim PS (1997) Influenza hemagglutinin is spring-loaded by a metastable native conformation. Proc Natl Acad Sci USA 94:14306–14313

Carr CM, Kim PS (1993) A spring-loaded mechanism for the conformational change of influenza hemagglutinin. Cell 73:823–832

Chaipan C, Kobasa D, Bertram S et al (2009) Proteolytic activation of the 1918 influenza virus hemagglutinin. J Virol 83:3200–3211. doi:10.1128/JVI.02205-08

Chan DC, Kim PS (1998) HIV entry and its inhibition. Cell 93:681–684

Chen J, Lee KH, Steinhauer DA et al (1998) Structure of the hemagglutinin precursor cleavage site, a determinant of influenza pathogenicity and the origin of the labile conformation. Cell 95:409–417

Chen J, Skehel JJ, Wiley DC (1999) N- and C-terminal residues combine in the fusion-pH influenza hemagglutinin HA(2) subunit to form an N cap that terminates the triple-stranded coiled coil. Proc Natl Acad Sci USA 96:8967–8972

Chen W, Calvo PA, Malide D et al (2001) A novel influenza A virus mitochondrial protein that induces cell death. Nat Med 7:1306–1312. doi:10.1038/nm1201-1306

Chernomordik LV, Kozlov MM (2008) Mechanics of membrane fusion. Nat Struct Mol Biol 15:675–683. doi:10.1038/nsmb.1455

Chu VC, Whittaker GR (2004) Influenza virus entry and infection require host cell N-linked glycoprotein. Proc Natl Acad Sci USA 101:18153–18158. doi:10.1073/pnas.0405172102

Compans RW, Dimmock NJ, Meier-Ewert H (1969) Effect of antibody to neuraminidase on the maturation and hemagglutinating activity of an influenza A2 virus. J Virol 4:528–534

Copeland CS, Zimmer KP, Wagner KR et al (1988) Folding, trimerization, and transport are sequential events in the biogenesis of influenza virus hemagglutinin. Cell 53:197–209

Cotter CR, Jin H, Chen Z (2014) A single amino acid in the stalk region of the H1N1pdm influenza virus HA protein affects viral fusion, stability and infectivity. PLoS Pathog 10:e1003831

Daniels RS, Downie JC, Hay AJ et al (1985) Fusion mutants of the influenza virus hemagglutinin glycoprotein. Cell 40:431–439

De Conto F, Covan S, Arcangeletti MC et al (2011) Differential infectious entry of human influenza A/NWS/33 virus (H1N1) in mammalian kidney cells. Virus Res 155:221–230. doi:10.1016/j.virusres.2010.10.008

de Vries E, de Vries RP, Wienholts MJ et al (2012) Influenza A virus entry into cells lacking sialylated N-glycans. Proc Natl Acad Sci USA 109:7457–7462. doi:10.1073/pnas.1200987109

de Vries E, Tscherne DM, Wienholts MJ et al (2011) Dissection of the influenza a virus endocytic routes reveals macropinocytosis as an alternative entry pathway. PLoS Pathog 7:e1001329. doi:10.1371/journal.ppat.1001329

de Vries RP, Zhu X, McBride R et al (2014) Hemagglutinin receptor specificity and structural analyses of respiratory droplet-transmissible H5N1 viruses. J Virol 88:768–773. doi:10.1128/JVI.02690-13

Doms RW, Gething MJ, Henneberry J et al (1986) Variant influenza virus hemagglutinin that induces fusion at elevated pH. J Virol 57:603–613

DuBois RM, Aguilar-Yañez JM, Mendoza-Ochoa GI et al (2011a) The receptor-binding domain of influenza virus hemagglutinin produced in *Escherichia coli* folds into its native, immunogenic structure. J Virol 85:865–872. doi:10.1128/JVI.01412-10

DuBois RM, Zaraket H, Reddivari M et al (2011b) Acid stability of the hemagglutinin protein regulates H5N1 influenza virus pathogenicity. PLoS Pathog 7:e1002398. doi:10.1371/journal.ppat.1002398

Durrer P, Galli C, Hoenke S et al (1996) H+-induced membrane insertion of influenza virus hemagglutinin involves the HA2 amino-terminal fusion peptide but not the coiled coil region. J Biol Chem 271:13417–13421

Fischer H, Widdicombe JH (2006) Mechanisms of acid and base secretion by the airway epithelium. J Membr Biol 211:139–150. doi:10.1007/s00232-006-0861-0

Fischer H, Widdicombe JH, Illek B (2002) Acid secretion and proton conductance in human airway epithelium. Am J Physiol Cell Physiol 282:C736–C743. doi:10.1152/ajpcell.00369.2001

Fukuyama S, Kawaoka Y (2011) The pathogenesis of influenza virus infections: the contributions of virus and host factors. Curr Opin Immunol 23:481–486. doi:10.1016/j.coi.2011.07.016

Galloway SE, Reed ML, Russell CJ, Steinhauer DA (2013) Influenza HA subtypes demonstrate divergent phenotypes for cleavage activation and pH of fusion: implications for host range and adaptation. PLoS Pathog 9:e1003151. doi:10.1371/journal.ppat.1003151

Gamblin SJ, Skehel JJ (2010) Influenza hemagglutinin and neuraminidase membrane glycoproteins. J Biol Chem 285:28403–28409. doi:10.1074/jbc.R110.129809

Garten W, Bosch FX, Linder D et al (1981) Proteolytic activation of the influenza virus hemagglutinin: the structure of the cleavage site and the enzymes involved in cleavage. Virology 115:361–374

Giannecchini S, Campitelli L, Calzoletti L et al (2006) Comparison of in vitro replication features of H7N3 influenza viruses from wild ducks and turkeys: potential implications for interspecies transmission. J Gen Virol 87:171–175

Grove J, Marsh M (2011) Host-pathogen interactions: the cell biology of receptor-mediated virus entry. J Cell Biol 195:1071–1082. doi:10.1083/jcb.201108131

Ha Y, Stevens DJ, Skehel JJ, Wiley DC (2002) H5 avian and H9 swine influenza virus haemagglutinin structures: possible origin of influenza subtypes. EMBO J 21:865–875

Han X, Bushweller JH, Cafiso DS, Tamm LK (2001) Membrane structure and fusion-triggering conformational change of the fusion domain from influenza hemagglutinin. Nat Struct Biol 8:715–720

Hartley CA, Reading PC, Ward AC, Anders EM (1997) Changes in the hemagglutinin molecule of influenza type A (H3N2) virus associated with increased virulence for mice. Arch Virol 142:75–88

Hebert DN, Foellmer B, Helenius A (1995) Glucose trimming and reglucosylation determine glycoprotein association with calnexin in the endoplasmic reticulum. Cell 81:425–433

Herfst S, Schrauwen EJA, Linster M et al (2012) Airborne transmission of influenza A/H5N1 virus between ferrets. Science 336:1534–1541. doi:10.1126/science.1213362

Hoekstra D, Klappe K (1993) Fluorescence assays to monitor fusion of enveloped viruses. Meth Enzymol 220:261–276

Huang Q, Opitz R, Knapp EW, Herrmann A (2002) Protonation and stability of the globular domain of influenza virus hemagglutinin. Biophys J 82:1050–1058

Huang Q, Sivaramakrishna RP, Ludwig K et al (2003) Early steps of the conformational change of influenza virus hemagglutinin to a fusion active state: stability and energetics of the hemagglutinin. Biochim Biophys Acta 1614:3–13

Huang RT, Rott R, Wahn K et al (1980) The function of the neuraminidase in membrane fusion induced by myxoviruses. Virology 107:313–319

Hulse DJ, Webster RG, Russell RJ, Perez DR (2004) Molecular determinants within the surface proteins involved in the pathogenicity of H5N1 influenza viruses in chickens. J Virol 78:9954–9964

Ilyushina NA, Govorkova EA, Russell CJ et al (2007) Contribution of H7 haemagglutinin to amantadine resistance and infectivity of influenza virus. J Gen Virol 88:1266–1274. doi:10.1099/vir.0.82256-0

Imai M, Watanabe T, Hatta M et al (2012) Experimental adaptation of an influenza H5 HA confers respiratory droplet transmission to a reassortant H5 HA/H1N1 virus in ferrets. Nature. doi:10.1038/nature10831

Jacoby DB, Tamaoki J, Borson DB, Nadel JA (1988) Influenza infection causes airway hyperresponsiveness by decreasing enkephalinase. J Appl Physiol 64:2653–2658

Jagger BW, Wise HM, Kash JC et al (2012) An overlapping protein-coding region in influenza A virus segment 3 modulates the host response. Science 337:199–204. doi:10.1126/science.1222213

Jefferies KC, Cipriano DJ, Forgac M (2008) Function, structure and regulation of the vacuolar (H+)-ATPases. Arch Biochem Biophys 476:33–42. doi:10.1016/j.abb.2008.03.025

Keleta L, Ibricevic A, Bovin NV et al (2008) Experimental evolution of human influenza virus H3 hemagglutinin in the mouse lung identifies adaptive regions in HA1 and HA2. J Virol 82:11599–11608. doi:10.1128/JVI.01393-08

Kido H, Yokogoshi Y, Sakai K et al (1992) Isolation and characterization of a novel trypsin-like protease found in rat bronchiolar epithelial Clara cells. A possible activator of the viral fusion glycoprotein. J Biochem 267:13573–13579

Kim JK, Negovetich NJ, Forrest HL, Webster RG (2009) Ducks: the "Trojan horses" of H5N1 influenza. Influenza Other Respir Viruses 3(IRV084):121–128. doi:10.1111/j.1750-2659.2009.00084.x

Koerner I, Matrosovich MN, Haller O et al (2012) Altered receptor specificity and fusion activity of the haemagglutinin contribute to high virulence of a mouse-adapted influenza A virus. J Gen Virol 93:970–979. doi:10.1099/vir.0.035782-0

Kohio HP, Adamson AL (2013) Glycolytic control of vacuolar-type ATPase activity: a mechanism to regulate influenza viral infection. Virology 444:301–309. doi:10.1016/j.virol.2013.06.026

Krauss S, Webster RG (2010) Avian influenza virus surveillance and wild birds: past and present. Avian Dis 54:394–398

Krenn BM, Egorov A, Romanovskaya-Romanko E et al (2011) Single HA2 mutation increases the infectivity and immunogenicity of a live attenuated H5N1 intranasal influenza vaccine candidate lacking NS1. PLoS One 6:e18577. doi:10.1371/journal.pone.0018577

Lakadamyali M, Rust MJ, Zhuang X (2006) Ligands for clathrin-mediated endocytosis are differentially sorted into distinct populations of early endosomes. Cell 124:997–1009. doi:10.1016/j.cell.2005.12.038

Lakadamyali M, Rust MJ, Zhuang X (2004) Endocytosis of influenza viruses. Microbes Infect 6:929–936. doi:10.1016/j.micinf.2004.05.002

Leikina E, Mittal A, Cho M-S et al (2004) Influenza hemagglutinins outside of the contact zone are necessary for fusion pore expansion. J Biochem 279:26526–26532. doi:10.1074/jbc.M401883200

Leser GP, Lamb RA (2005) Influenza virus assembly and budding in raft-derived microdomains: a quantitative analysis of the surface distribution of HA, NA and M2 proteins. Virology 342:215–227. doi:10.1016/j.virol.2005.09.049

Li M, Yang C, Compans RW (2001) Mutations in the cytoplasmic tail of murine leukemia virus envelope protein suppress fusion inhibition by R peptide. J Virol 75:2337–2344

Lin YP, Wharton SA, Martin J et al (1997) Adaptation of egg-grown and transfectant influenza viruses for growth in mammalian cells: selection of hemagglutinin mutants with elevated pH of membrane fusion. Virology 233:402–410

Lingwood D, Simons K (2010) Lipid rafts as a membrane-organizing principle. Science 327:46–50. doi:10.1126/science.1174621

Martens S, McMahon HT (2008) Mechanisms of membrane fusion: disparate players and common principles. Nat Rev Mol Cell Biol 9:543–556. doi:10.1038/nrm2417

Matrosovich M, Stech J, Klenk HD (2009) Influenza receptors, polymerase and host range. Rev Off Int Epizoot 28:203–217

Matrosovich M, Tuzikov A, Bovin N et al (2000) Early alterations of the receptor-binding properties of H1, H2, and H3 avian influenza virus hemagglutinins after their introduction into mammals. J Virol 74:8502–8512

Melikyan GB, Markosyan RM, Hemmati H et al (2000) Evidence that the transition of HIV-1 gp41 into a six-helix bundle, not the bundle configuration, induces membrane fusion. J Cell Biol 151:413–423

Melikyan GB, White JM, Cohen FS (1995) GPI-anchored influenza hemagglutinin induces hemifusion to both red blood cell and planar bilayer membranes. J Cell Biol 131:679–691

Mellman I, Fuchs R, Helenius A (1986) Acidification of the endocytic and exocytic pathways. Annu Rev Biochem 55:663–700. doi:10.1146/annurev.bi.55.070186.003311

Selman M, Dankar SK, Forbes NE et al (2012) Adaptive mutation in influenza A virus non-structural gene is linked to host switching and induces a novel protein by alternative splicing. Emerg Microbes Infect 1:e42. doi:10.1038/emi.2012.38

Murakami M, Towatari T, Ohuchi M et al (2001) Mini-plasmin found in the epithelial cells of bronchioles triggers infection by broad-spectrum influenza A viruses and Sendai virus. Eur J Biochem 268:2847–2855

Murakami S, Horimoto T, Ito M et al (2012) Enhanced growth of influenza vaccine seed viruses in vero cells mediated by broadening the optimal pH range for virus membrane fusion. J Virol 86:1405–1410. doi:10.1128/JVI.06009-11

Muramoto Y, Noda T, Kawakami E et al (2013) Identification of novel influenza A virus proteins translated from PA mRNA. J Virol 87:2455–2462. doi:10.1128/JVI.02656-12

Naeve CW, Williams D (1990) Fatty acids on the A/Japan/305/57 influenza virus hemagglutinin have a role in membrane fusion. EMBO J 9:3857–3866

Nakowitsch S, Wolschek M, Morokutti A, Ruthsatz T (2011) Mutations affecting the stability of the haemagglutinin molecule impair the immunogenicity of live attenuated H3N2 intranasal influenza vaccine candidates lacking NS1. Vaccine 29:3517–3524

Narasaraju T, Sim MK, Ng HH et al (2009) Adaptation of human influenza H3N2 virus in a mouse pneumonitis model: insights into viral virulence, tissue tropism and host pathogenesis. Microbes Infect 11:2–11. doi:10.1016/j.micinf.2008.09.013

Nobusawa E, Aoyama T, Kato H et al (1991) Comparison of complete amino acid sequences and receptor-binding properties among 13 serotypes of hemagglutinins of influenza A viruses. Virology 182:475–485

Okumura Y, Takahashi E, Yano M et al (2010) Novel type II transmembrane serine proteases, MSPL and TMPRSS13, proteolytically activate membrane fusion activity of the hemagglutinin of highly pathogenic avian influenza viruses and induce their multicycle replication. J Virol 84:5089–5096. doi:10.1128/JVI.02605-09
Palese P, Tobita K, Ueda M, Compans RW (1974) Characterization of temperature sensitive influenza virus mutants defective in neuraminidase. Virology 61:397–410
Park HE, Gruenke JA, White JM (2003) Leash in the groove mechanism of membrane fusion. Nat Struct Biol 10:1048–1053
Reed ML, Bridges OA, Seiler P et al (2010) The pH of activation of the hemagglutinin protein regulates H5N1 influenza virus pathogenicity and transmissibility in ducks. J Virol 84:1527–1535. doi:10.1128/JVI.02069-09
Reed ML, Yen H-L, DuBois RM et al (2009) Amino acid residues in the fusion peptide pocket regulate the pH of activation of the H5N1 influenza virus hemagglutinin protein. J Virol 83:3568–3580. doi:10.1128/JVI.02238-08
Rogers GN, Paulson JC, Daniels RS et al (1983) Single amino acid substitutions in influenza haemagglutinin change receptor binding specificity. Nature 304:76–78
Rosenthal PB, Zhang X, Formanowski F et al (1998) Structure of the haemagglutinin-esterase-fusion glycoprotein of influenza C virus. Nature 396:92–96
Rossman JS, Jing X, Leser GP, Lamb RA (2010) Influenza virus M2 protein mediates ESCRT-independent membrane scission. Cell 142:902–913. doi:10.1016/j.cell.2010.08.029
Rossman JS, Lamb RA (2011) Influenza virus assembly and budding. Virology 411:229–236. doi:10.1016/j.virol.2010.12.003
Ruigrok RW, Martin SR, Wharton SA et al (1986) Conformational changes in the hemagglutinin of influenza virus which accompany heat-induced fusion of virus with liposomes. Virology 155:484–497
Russell CJ, Jardetzky TS, Lamb RA (2001) Membrane fusion machines of paramyxoviruses: capture of intermediates of fusion. EMBO J 20:4024–4034. doi:10.1093/emboj/20.15.4024
Russell CJ, Webster RG (2005) The genesis of a pandemic influenza virus. Cell 123:368–371. doi:10.1016/j.cell.2005.10.019
Rust MJ, Lakadamyali M, Zhang F, Zhuang X (2004) Assembly of endocytic machinery around individual influenza viruses during viral entry. Nat Struct Mol Biol 11:567–573. doi:10.1038/nsmb769
Salomon R, Webster RG (2009) The influenza virus enigma. Cell 136:402–410. doi:10.1016/j.cell.S0092-8674(09)00077-4, 10.1016/j.cell.2009.01.029
Scholtissek C (1985) Stability of infectious influenza A viruses at low pH and at elevated temperature. Vaccine 3:215–218
Shaw ML, Palese P (2013) Orthomyxoviridae. In: Knipe DM, Howley PM (eds) Fields virology, 6th edn. Lippincott Williams and Wilkins, Philadelphia, pp 1151–1185
Shelton H, Roberts KL, Molesti E et al (2013) Mutations in hemagglutinin that affect receptor binding and pH stability increase replication of a PR8 influenza virus with H5 HA in the upper respiratory tract of ferrets and may contribute to transmissibility. J Gen Virol. doi:10.1099/vir.0.050526-0
Sieczkarski SB, Whittaker GR (2002) Influenza virus can enter and infect cells in the absence of clathrin-mediated endocytosis. J Virol 76:10455–10464
Skehel JJ, Wiley DC (2000) Receptor binding and membrane fusion in virus entry: the influenza hemagglutinin. Annu Rev Biochem 69:531–569
Sorrell EM, Schrauwen EJA, Linster M et al (2011) Predicting "airborne" influenza viruses: (trans-) mission impossible? Curr Opin Virol 1:635–642. doi:10.1016/j.coviro.2011.07.003
Söllner TH (2004) Intracellular and viral membrane fusion: a uniting mechanism. Curr Opin Cell Biol 16:429–435. doi:10.1016/j.ceb.2004.06.015
Steinhauer DA (1999) Role of hemagglutinin cleavage for the pathogenicity of influenza virus. Virology 258:1–20

Steinhauer DA, Martin J, Lin YP et al (1996) Studies using double mutants of the conformational transitions in influenza hemagglutinin required for its membrane fusion activity. Proc Natl Acad Sci USA 93:12873–12878

Steinhauer DA, Wharton SA, Skehel JJ et al (1991a) Amantadine selection of a mutant influenza virus containing an acid-stable hemagglutinin glycoprotein: evidence for virus-specific regulation of the pH of glycoprotein transport vesicles. Proc Natl Acad Sci USA 88: 11525–11529

Steinhauer DA, Wharton SA, Skehel JJ, Wiley DC (1995) Studies of the membrane fusion activities of fusion peptide mutants of influenza virus hemagglutinin. J Virol 69:6643–6651

Steinhauer DA, Wharton SA, Wiley DC, Skehel JJ (1991b) Deacylation of the hemagglutinin of influenza A/Aichi/2/68 has no effect on membrane fusion properties. Virology 184:445–448

Stieneke-Gröber A, Vey M, Angliker H et al (1992) Influenza virus hemagglutinin with multibasic cleavage site is activated by furin, a subtilisin-like endoprotease. EMBO J 11:2407–2414

Su B, Wurtzer S, Rameix-Welti M-A et al (2009) Enhancement of the influenza A hemagglutinin (HA)-mediated cell-cell fusion and virus entry by the viral neuraminidase (NA). PLoS One 4:e8495. doi:10.1371/journal.pone.0008495

Sun X, Shi Y, Lu X et al (2013) Bat-derived influenza hemagglutinin h17does not bind canonical avian or human receptors and most likely uses a unique entry mechanism. Cell Rep 3:769–778. doi:10.1016/j.celrep.2013.01.025

Sun X, Whittaker GR (2013) Entry of influenza virus. Adv Exp Med Biol 790:72–82. doi:10. 1007/978-1-4614-7651-1_4

Suomalainen M (2002) Lipid rafts and assembly of enveloped viruses. Traffic 3:705–709

Takeda M, Leser GP, Russell CJ, Lamb RA (2003) Influenza virus hemagglutinin concentrates in lipid raft microdomains for efficient viral fusion. Proc Natl Acad Sci USA 100:14610–14617. doi:10.1073/pnas.2235620100

Thoennes S, Li Z-N, Lee B-J et al (2008) Analysis of residues near the fusion peptide in the influenza hemagglutinin structure for roles in triggering membrane fusion. Virology 370:403–414. doi:10.1016/j.virol.2007.08.035

Tong S, Li M, Vincent A et al (2002) Regulation of fusion activity by the cytoplasmic domain of a paramyxovirus F protein. Virology 301:322–333

Tong S, Li Y, Rivailler P et al (2012) A distinct lineage of influenza A virus from bats. Proc Natl Acad Sci USA 109:4269–4274. doi:10.1073/pnas.1116200109

Tong S, Zhu X, Li Y et al (2013) New world bats harbor diverse influenza A viruses. PLoS Pathog 9:e1003657. doi:10.1371/journal.ppat.1003657

Towatari T, Ide M, Ohba K et al (2002) Identification of ectopic anionic trypsin I in rat lungs potentiating pneumotropic virus infectivity and increased enzyme level after virus infection. Eur J Biochem 269:2613–2621

Veit M, Kretzschmar E, Kuroda K et al (1991) Site-specific mutagenesis identifies three cysteine residues in the cytoplasmic tail as acylation sites of influenza virus hemagglutinin. J Virol 65:2491–2500

Vey M, Orlich M, Adler S et al (1992) Hemagglutinin activation of pathogenic avian influenza viruses of serotype H7 requires the protease recognition motif R-X-K/R-R. Virology 188:408–413

Wagner R, Matrosovich M, Klenk H-D (2002) Functional balance between haemagglutinin and neuraminidase in influenza virus infections. Rev Med Virol 12:159–166. doi:10.1002/rmv.352

Waning DL, Russell CJ, Jardetzky TS, Lamb RA (2004) Activation of a paramyxovirus fusion protein is modulated by inside-out signaling from the cytoplasmic tail. Proc Natl Acad Sci USA 101:9217–9222. doi:10.1073/pnas.0403339101

Washington N, Steele RJ, Jackson SJ et al (2000) Determination of baseline human nasal pH and the effect of intranasally administered buffers. Int J Pharm 198:139–146

Webster RG, Rott R (1987) Influenza virus A pathogenicity: the pivotal role of hemagglutinin. Cell 50:665–666

Weis W, Brown JH, Cusack S et al (1988) Structure of the influenza virus haemagglutinin complexed with its receptor, sialic acid. Nature 333:426–431. doi:10.1038/333426a0
Weis WI, Cusack SC, Brown JH et al (1990) The structure of a membrane fusion mutant of the influenza virus haemagglutinin. EMBO J 9:17–24
Wickner W, Schekman R (2008) Membrane fusion. Nat Struct Mol Biol 15:658–664. doi:10.1038/nsmb.1451
Wiley DC, Wilson IA, Skehel JJ (1981) Structural identification of the antibody-binding sites of Hong Kong influenza haemagglutinin and their involvement in antigenic variation. Nature 289:373–378
Wilson IA, Skehel JJ, Wiley DC (1981) Structure of the haemagglutinin membrane glycoprotein of influenza virus at 3 A resolution. Nature 289:366–373
Wise HM, Foeglein A, Sun J et al (2009) A complicated message: Identification of a novel PB1-related protein translated from influenza A virus segment 2 mRNA. J Virol 83:8021–8031. doi:10.1128/JVI.00826-09
Wise HM, Hutchinson EC, Jagger BW et al (2012) Identification of a novel splice variant form of the influenza A virus M2 ion channel with an antigenically distinct ectodomain. PLoS Pathog 8:e1002998. doi:10.1371/journal.ppat.1002998
Xiong X, Coombs PJ, Martin SR et al (2013) Receptor binding by a ferret-transmissible H5 avian influenza virus. Nature 497:392–396. doi:10.1038/nature12144
Xu R, Wilson IA (2011) Structural characterization of an early fusion intermediate of influenza virus hemagglutinin. J Virol 85:5172–5182. doi:10.1128/JVI.02430-10
Zaraket H, Bridges OA, Duan S et al (2013a) Increased acid stability of the hemagglutinin protein enhances H5N1 influenza virus growth in the upper respiratory tract but is insufficient for transmission in ferrets. J Virol 87:9911–9922. doi:10.1128/JVI.01175-13
Zaraket H, Bridges OA, Russell CJ (2013b) The pH of activation of the hemagglutinin protein regulates H5N1 influenza virus replication and pathogenesis in mice. J Virol 87:4826–4834. doi:10.1128/JVI.03110-12
Zhang W, Shi Y, Lu X et al (2013) An airborne transmissible avian influenza H5 hemagglutinin seen at the atomic level. Science 340:1463–1467. doi:10.1126/science.1236787
Zhou J, Xu S, Ma J et al (2013) Recombinant influenza A H3N2 viruses with mutations of HA transmembrane cysteines exhibited altered virological characteristics. Virus Genes. doi:10.1007/s11262-013-1011-2

Part III
Transmission and Pandemics

Pandemic Preparedness and the Influenza Risk Assessment Tool (IRAT)

Nancy J. Cox, Susan C. Trock and Stephen A. Burke

Abstract Influenza infections have resulted in millions of deaths and untold millions of illnesses throughout history. Influenza vaccines are the cornerstone of influenza prevention and control. Recommendations are made by the World Health Organization (WHO) 6–9 months in advance of the influenza season regarding what changes, if any, should be made in the formulation of seasonal influenza vaccines. This allows time to manufacture, test, distribute, and administer vaccine prior to the beginning of the influenza season. At the same time experts also consider which viruses not currently circulating in the human population, but with pandemic potential, pose the greatest risk to public health. Experts may conclude that one or more of these viruses are of enough concern to warrant development of a high-growth reassortant candidate vaccine virus. Subsequently, national authorities may determine that a vaccine should be manufactured, tested in clinical trials, and even stockpiled in some circumstances. The Influenza Risk Assessment Tool (IRAT) was created in an effort to develop a standardized set of elements that could be applied for decision making when evaluating pre-pandemic viruses. The tool is a simple, additive model, based on multi-attribute decision analysis. The ultimate goal is to identify an appropriate candidate vaccine virus and prepare a human vaccine before the virus adapts to infect and efficiently transmit in susceptible human populations. This pre-pandemic preparation allows production of vaccine—a strategy that could save lives and mitigate illness during a pandemic.

N.J. Cox (✉) · S.C. Trock
Influenza Division, National Center for Immunization and Respiratory Diseases, Centers for Disease Control and Prevention, Atlanta, GA, USA
e-mail: njc1@cdc.gov

S.A. Burke
Influenza Division, National Center for Immunization and Respiratory Diseases, Centers for Disease Control and Prevention and Battelle, Atlanta, GA, USA

Current Topics in Microbiology and Immunology (2014) 385: 119–136
DOI: 10.1007/82_2014_419

Published Online: 2 August 2014

Abbreviations

CDC	Centers for Disease Control and Prevention
HPAI	Highly pathogenic avian influenza
IRAT	Influenza Risk Assessment Tool
RVF	Rift Valley fever
SMEs	Subject matter experts
WHO	World Health Organization

Contents

1 Introduction

An influenza pandemic, which occurs when a novel influenza virus becomes efficiently transmissible in human populations with limited immunity, poses a challenging and unpredictable threat to human health by causing global, community-wide outbreaks. Global influenza pandemics have been virologically confirmed to have occurred in 1918, 1957, 1968, and 2009. The 1918–1919 pandemic was caused by an influenza A/H1N1 virus and was responsible for over 500,000 deaths in the U.S. (Glezen 1996; Simonsen et al. 1998) and an estimated 50–100 million deaths worldwide (Morans and Fauci 2007). During the Asian flu pandemic in 1957–1958, an influenza A/H2N2 virus was responsible for approximately 70,000 deaths in the U.S. (Glezen 1996; Simonsen et al. 1998) and approximately 2 million globally. In 1968–1969, the Hong Kong flu pandemic, caused by an influenza A/H3N2 virus, was responsible for ~30,000 U.S. deaths (Simonsen et al. 1998) and 1 million global deaths. The most recent pandemic occurred in 2009 when a heretofore novel H1N1 virus was first identified in the U.S. in April and rapidly spread across the globe. It was estimated that approximately 12,000 people died in the U.S. with an estimated 280,000 global deaths (Dawood et al. 2012). At this time it is not possible to predict the timing of the next influenza pandemic event or the virus that will cause it. Nevertheless, the charge for public health is to prepare for a pandemic with the intent to limit the spread of disease, suffering, and death as well as potential societal impacts (National Strategy for Pandemic Influenza 2005).

Advanced planning for influenza pandemics began in the United States in the 1970s with the formation of a governmental interagency group whose efforts culminated in a published pandemic influenza plan that acknowledged the need for surveillance, annual influenza immunization for high risk individuals, and increased support for influenza research. The plan also incorporated a methodology directed to planning and policy across agencies (Interagency Work Group on Pandemic Influenza 1978). This early effort was subsequently revised and served as a model for other plans (Iskander et al. 2013). Progress and investment in pandemic preparedness have increased over time both domestically and internationally (Schuchat et al. 2014). Contemporary planning efforts have incorporated risk-based approaches in guiding comprehensive planning and response efforts (National Strategy for Pandemic Influenza 2005; Pandemic Influenza Risk Management WHO Interim Guidance 2013). Critical to these plans are the objectives of early recognition of and rapid response to the emergence of a virus with pandemic potential. However, without advanced knowledge of the specific influenza virus that will cause the next pandemic, the ability to prioritize efforts and allocate resources is still a significant challenge and establishes a critical role for pre-pandemic risk assessment.

Risk assessment frameworks have been developed to anticipate and mitigate outbreaks of other infectious diseases. For example, Rift Valley fever (RVF) is a disease that occurs in Africa and impacts both animal and human populations. This disease causes economic losses in affected domestic animal populations through high mortality and abortion losses and also can have a significant negative impact on trade (Metwally 2008). In 2006–2008, the countries of Kenya, Somalia and Tanzania together reported over 1,000 human cases of RVF with over 300 deaths (Anyamba et al. 2010b). Risk-based prediction models for these nations in sub-Saharan Africa have been developed based upon environmental factors, weather patterns, vector habitats, and distribution of domestic animal and human populations (Anyamba et al. 2010a, b; Britch et al. 2013). Models developed to forecast malaria outbreaks also have taken advantage of climatological data, vector density, and advances in weather forecasting (Hoshen and Morse 2004; Briet et al. 2008).

While risk assessments for both RVF and malaria are based primarily on an evaluation of environmental factors affecting arthropod vectors necessary for transmission of the infectious agent, risk assessment for influenza poses a very different set of challenges. Although influenza viruses exhibit seasonal periodicity in many areas of the world, climatic changes affecting vegetation or vector density cannot be used as surrogate measures to predict outbreaks. And while environmental and host factors can influence the emergence of a novel influenza virus, it is primarily the virus' lack of fidelity during replication and its ability to exchange gene segments with other influenza viruses through reassortment that make it virtually impossible to accurately predict emergence and impact of a new virus.

Consequently, our approach to pre-pandemic influenza risk assessment focuses on the rapid characterization of newly emerging influenza viruses with pandemic potential in an effort to determine the relative risk posed by a range of such viruses in accord with our current understanding of factors that contribute to allowing an influenza virus to spread globally in humans. A relatively small expenditure of resources is usually required for the preliminary risk assessment of a novel influenza virus. If it is deemed that the risk to the population posed by a particular virus is great or eminent, then a decision can be made to advance pre-pandemic preparations to include one or more of the following escalating steps: developing a candidate vaccine virus; conducting pre-clinical vaccine studies in animals; manufacturing clinical trial lots of vaccine; carrying out human clinical trials, stockpiling vaccine; and priming the population against the novel influenza virus. Each of these steps requires an escalating expenditure of resources. Yet because not all novel influenza viruses pose the same level of risk, it is important to have a clear, concise method to convey the relative risk posed by a newly emerging influenza virus as compared with the risks for other potentially pandemic viruses to those who make decisions about the allocation of resources. This approach to risk assessment relies on global awareness of the importance of newly emerging viruses when they are detected during surveillance in animal or human populations. This approach also relies on rapid sharing of viruses and data about newly detected viruses among the influenza research and public health communities. This sharing of information and viruses must be coupled with development of diagnostic tests and vaccines along with examining these viruses for antigenic, antiviral susceptibility, and transmission properties. Additionally, past pandemics have shown that our understanding of influenza at the animal–human interface is critical for detecting events leading to the emergence of pandemic influenza viruses (Swayne and Perdue 2005; Zimmer and Burke 2009).

What goes on in animal hosts matters. Wild waterfowl and shore birds had been identified as the reservoir for 16 subtypes of influenza A viruses (Webster et al. 1992; Swayne 2008). More recently, influenza A/H17 and A/H18 viruses were recovered from New World bats, (Tong et al. 2012, 2013) however, these two new subtypes have not demonstrated the ability to infect humans or mammalian species other than bats. In contrast, avian origin influenza viruses have infected a variety of aquatic and terrestrial mammals on many occasions. In fact, human pandemic influenza viruses can all trace their lineage back to an avian origin (Swayne 2008). Indeed avian origin influenza viruses have demonstrated the ability to infect swine and these animals have been proposed as a likely 'mixing vessel' for pandemic influenza viruses, be they of avian or mammalian origin (Webster et al. 1992). It is therefore important to consider both avian and swine populations as potential sources for new influenza viruses with human pandemic potential.

Unfortunately, there is little or no surveillance for animal influenza viruses in many countries, often due to absent or insufficient resources. In other countries, often those with greater resources, programs to conduct influenza surveillance among animal populations have been well established (USDA 2007, 2010). While it is possible to synthetically reconstruct viruses based upon sequence information posted in publically accessible data banks, this information initially is often incomplete. Therefore, it is usually necessary to isolate and share the virus to be able to fully assess its human pandemic potential. Ideally influenza viruses recovered from animal hosts and information about them would be accessible simultaneously to both global animal health and public health for risk assessment. In addition, it is necessary to understand the context associated with the recovery of these viruses—such as the source species along with the date and location of sample collection. Such transparency will only be possible if those responsible for the identification and recovery of the virus are appropriately involved and acknowledged. Furthermore, partnerships and mutual respect between public health and animal health can leverage limited resources in ways that are of benefit to both and more effectively involve individuals from animal health with expertise in influenza viruses who are essential for conducting risk assessments of a novel influenza virus.

What if the 2009 H1N1 virus had been recognized while in animals? In late April 2009, a novel H1N1 influenza A virus was isolated from two human cases in California (CDC 2009). Genetic analysis indicated these viruses originated from viruses that were known to be present in swine (Garten et al. 2009). Within two weeks of receipt of the sample, the CDC had developed, validated, and was able to distribute a pH1N1 real-time RT-PCR diagnostic assay (Jernigan et al. 2011). During 2009, there were two distinct peaks of influenza activity in the U.S., an initial smaller peak in June followed by a significant increase in activity in August, with peak activity occurring during the week of October 24, 2009 (Kasowski et al. 2011; Jhung et al. 2011). Although vaccine development was initiated soon after the identification of the virus, a licensed monovalent H1N1 2009 vaccine was not available for distribution in the U.S. until the week of October 5, 2009 and remained in limited supply through the end of the year (CDC, Jan 22 2010). Had this virus been recovered from humans or its animal host and then shared with global public health experts even four weeks earlier than it was, it is likely that vaccine would have been more widely available before the peak of the second wave. Certainly the laboratory assessment would have been available and diagnostic test development could have proceeded earlier, without the added pressure of testing an onslaught of clinical specimens arriving during the early stages of the pandemic.

2 Creating a Framework for Influenza Risk Assessment

Recent identification of emerging influenza viruses including the 2009 H1N1virus (CDC 2009), the re-emergence and spread of highly pathogenic avian influenza (HPAI) H5N1viruses in 2003–2006 (WHO 2014), and the unanticipated appearance of low pathogenic H7N9, in China during 2013 and 2014 (WHO 2014a, b; Parry 2013; Gao et al. 2013) has reinforced the necessity for a systematic, transparent method to evaluate novel influenza viruses. However, before developing a framework for risk assessment it was necessary to specify what questions

Table 1 IRAT goals and requirements

Goal/requirement	Description/benefit
Objectivity	Consistent application of criteria; minimization of bias from one input or field of expertise
Transparancy	Open processes with clear definitions; documentation of data collection and calculations
Adaptability	Flexible to accommodate new data or criteria as new information or advancement in influenza virology become available
Priority setting	Enable comparison between influenza viruses; use weighted criteria based on specific risk questions
Uncertainty estimate	Incorporate estimates of uncertainty; communicate level of confidence associated with risk estimates
Clear communication	Provide clear, concise description of risk assessment output to inform decision-makers
Expert involvement	Involves subject matter experts in development and use of the tool; promotes collaboration, data sharing, and acceptance of risk assessments

would be addressed. In the context of pre-pandemic preparedness, the two most pressing public health questions are:

(1) *'What is the risk that a virus not currently circulating in the human population has the potential for sustained human-to-human transmission?' and,*
(2) *'If the virus were to achieve sustained human-to-human transmission, what is the risk that a virus not currently circulating in the human population has the potential for significant impact on public health?'*

The goals and requirements of the influenza risk assessment are summarized in Table 1 and in combination with the two public health questions created the framework for developing an influenza risk assessment tool (IRAT) (Trock et al. 2012).

2.1 Involvement of Subject Matter Experts (SMEs) in Developing the IRAT

The IRAT is based on input from SMEs in the areas of epidemiology, virology, human and veterinary medicine, animal ecology, and risk assessment. While various methods have been applied to risk assessments, the available information associated with a novel influenza virus is initially often incomplete, and thus it has been very useful to involve SMEs that were responsible for the initial isolation, identification, and recognition of an emerging influenza virus with pandemic potential. An ensuing collaboration among SMEs often takes place prior to formal publication and SMEs then contribute their insight applied both to published and unpublished scientific data; observations with negative findings also provide useful data points for the evaluation of emerging influenza viruses. These information exchanges and collaborations provide early data to incorporate into the IRAT and also provide the basis for incorporation of real-time findings into the decision-making process. Although it is not possible to eliminate all bias in the scoring process, the IRAT provides a systematic method designed to minimize bias while exploiting the experience and vision of influenza SMEs.

2.2 Identification and Definition of Risk Elements

Initial discussions with a relatively small number of SMEs focused on identifying a preliminary set of core elements that would be useful for risk assessment of influenza viruses that are not currently circulating in the human population, but which are deemed to have pandemic potential. These experts initially considered many parameters that were ultimately consolidated into ten elements, each with a "preliminary" definition. This concept for an IRAT along with defined draft elements then became the focus of a larger meeting of global influenza SMEs from academia and various national and global institutions representing the fields of influenza virology, diagnosis, epidemiology, ecology, and laboratory research in animal and human influenza; risk modelers were also included. Meeting participants were provided with an overview of the IRAT, including its intended purpose and the draft elements and definitions. Next small work groups were asked to consider these draft elements along with their definitions and were offered opportunities to provide suggestions and edits. Participants were also invited to propose additional elements to include and/or existing elements that could be omitted for the purpose of developing a sound risk assessment framework. The consensus was that, with edits, the ten elements effectively captured the types of data necessary for influenza risk assessment (Table 2). The elements were placed into one of three categories with the first category containing four elements that apply to the properties of the virus and are primarily assessed in a laboratory setting. These elements are as follows: viral genomic variation; receptor binding properties; transmission properties in laboratory animals; and antiviral susceptibility. The second category of elements considers the field and epidemiological findings and includes: global distribution of the virus in animals; the animal species that are known to be infected by the virus; and extent of human infections. The final category incorporates attributes of the human population and considers the following elements: existing population immunity; the severity of human disease; and the antigenic distance between the novel virus and existing influenza vaccines. The IRAT thus has been formulated to include those elements that SMEs consider essential for risk assessment and it integrates information in an understandable, transparent manner which can contribute to decision-making and allocation of resources when appropriate (Dolan 2010; Phillips 1984). To this end, the IRAT strives to provide a framework to differentiate among novel influenza viruses with pandemic potential through the application of standardized descriptions or definitions.

Table 2 Ten elements incorporated into the IRAT

Antigenic relatedness (to vaccines)	Human infections
Antiviral and treatment options	Infection in animal species
Disease severity and pathogenesis	Population immunity
Genomic variation	Receptor binding
Global distribution in animals	Transmission in laboratory animals

Each element was then defined carefully. For example, the element concerning population immunity states that *population immunity in humans for the purposes of this risk assessment tool is defined as the detection of pre-existing cross-reactive serum antibodies acquired through prior infection and/or vaccination in all age groups.* Risk assessment for this element included low, moderate, and high risk scoring categories, recognizing that different age groups may have differing levels of cross-reactive antibodies to emerging influenza viruses. Experts scoring this element also were asked to assign a risk score between 1 and 10, with 1 representing the lowest and 10 the highest risk. Further guidance was provided to SMEs to assign a low risk score (1–3) for evidence of cross-reactive antibodies in at least 30 % of the population in all age groups, except for children ≤17 years of age. A moderate risk score (4–7) was defined as evidence of cross-reactive antibodies in at least 30 % of the population only among persons ≥50 years of age and a high risk score (8–10) was defined as ≤10 % of all age groups having evidence of cross-reactive antibodies. Other elements of the IRAT were defined and similar guidance for scoring categories of low, moderate, and high risk was provided to the SMEs scoring particular risk elements. In each case a low risk score was between one and three; moderate risk was between four and seven, while a high risk rated a numerical score between eight and ten for a particular element. As an example, the H5N1 HPAI viruses currently circulating in birds in at least six countries (OIE 2014) scored in the high risk category when considering existing population immunity since there is little to no preexisting cross-reactive antibody in the population.

Another parameter included in the risk assessment elements was antiviral treatment options. This element was defined for risk assessment as the *predicted or demonstrated efficacy of available antiviral agents against animal influenza viruses.* A low risk was defined as no evidence of clinically relevant resistance to any of the antiviral drugs approved for human use (neuraminidase inhibitors and M2 blockers). Moderate risk would include sensitivity to all neuraminidase inhibitors but resistance to M2 blockers. While a high risk score would be defined as those viruses demonstrating resistance to one or more neuraminidase inhibitor antiviral drugs. HPAI H5N1 viruses score in the moderate risk range while the A/H1N1 [A/duck/NY/96] recovered in 1996 scored in the low risk range for this element because the former viruses are already resistant to adamantanes, but the H1N1 [A/duck/NY/96] is not resistant to either class of antiviral drug.

SMEs were asked to provide risk scores based strictly on the agreed upon definition of the element. For example, if it is known that a particular influenza virus such as an A/H5N1 virus is responsible for human infections and severe illness, this information should not impact the receptor binding score. Conversely, the receptor binding score is to be based only upon the definition for that element and is independent of scores for other elements. For viruses such as the HPAI A/H5N1 viruses, the overall elevated risk is captured in the risk score by elements that address human infections and disease severity.

2.3 Weighting of Risk Elements

The IRAT was designed to address two important public health questions, i.e., (1) "What is the risk that a virus not currently circulating in the human population has potential for sustained human-to-human transmission?" and (2) "If the virus were to achieve sustained human-to-human transmission, what is the risk that a virus not currently circulating in the human population has the potential for significant impact on public health?" The former question addresses the risk of emergence, while the latter characterizes the expected impact if the virus did achieve successful transmission among humans. The ten elements remain constant when considering how to answer each of these two questions as do the definitions of each element. This means that each virus is given a risk score for each element and that score is then incorporated into the final risk score for the virus. However, because the two questions are quite different, the specific elements that provide the greatest insight into answering each question must be ranked in order of importance relative to each other.

When faced with answering the question of emergence, SMEs were asked what one piece of information would be most useful in order to answer the emergence question. General consensus indicated that knowing whether or not there was evidence of human infections with the virus in question would provide the most useful information for this question. This element was followed in importance by the element that takes into account results of laboratory studies examining transmission of the virus using accepted animal models. This process of ranking the relative importance of each element in the context of answering the specific question continued until all 10 elements were ranked relative to each other for each of the two questions (Table 3). While knowing whether or not human infections have occurred is considered one of the most important considerations, other elements such as antiviral treatment options were judged to have little impact on the risk of emergence of a novel influenza virus and thus received a much lower weight.

The second question, referred to as the impact question, deals with the impact that a novel influenza virus will have on global public health if it achieves the ability to transmit efficiently from person-to-person. Again, SMEs were requested to identify which of the ten elements provide the most pertinent information to answer this question. Information regarding disease severity and population

Table 3 Relative importance of information as associated with elements listed from most important to least

Question 1: What is the risk that a virus not currently circulating in the human population has potential for sustained human-to-human transmission?	Question 2: If the virus were to achieve sustained human-to-human transmission, what is the risk that a virus not currently circulating in the human population has the potential for significant impact on public health?
Human infections	Disease severity
Transmission in laboratory animals	Population immunity
Receptor binding	Human infections
Population immunity	Antiviral and treatment options
Infections in animal species	Antigenic relatedness (to vaccines)
Genomic variation	Receptor binding
Antigenic relatedness (to vaccines)	Genomic variation
Global distribution in animals	Transmission in laboratory animals
Disease severity	Global distribution in animals
Antiviral and treatment options	Infections in animals

immunity ranked number one and two in importance for answering this question, while information regarding human infections ranked third and information regarding antiviral treatment options ranked fourth. Clearly, while this fourth element may not greatly impact the summary risk score for the question of emergence, it does influence the assessment of the impact a novel influenza virus could have on the public's health.

As demonstrated, not all elements have equal weights for each of the two questions, but rather the score for each element depends upon which of the two specific risk questions is being posed. Furthermore, the information provided by some elements drive the final risk score more than others. In recognition of this consideration, a weighting system was applied through a process by which surrogate weights were assigned based upon ranking of the ten elements in an effort to capture the relative importance of each specific element for each of the two questions (Edwards and Barron 1994). This approach also attempted to mitigate bias which could be introduced from any one research area. Weights were then calculated based upon the number of elements considered in the risk assessment such that the weights sum to 1. Use of weighting conveys the concept of a relative importance of one factor or element in relationship to others. The weighting process assigns a higher value or relative weight to the element which the SMEs deemed most important to answering that question. The weights decrease in the order that the remaining elements are ranked such that the elements of least importance are assigned the smallest values. The weights are then multiplied by the risk score assigned by SMEs to each element and the products of each are then totaled (Tables 4, 5). The resultant summary scores are the risk scores for that virus for each of the two questions. It is important to note that although the weight value is given to several decimal points, there is no inference that this degree of accuracy is intended. In order for the last element to carry a weight greater than

Table 4 Risk scores for Question 1 (Emergence) as applied to three influenza viruses—H1N1 North American Mallard, H7N9 [Anhui/1/2013] and H3N2(v) 2013

Element	N. Am. H1N1			H7N9		H3N2	
	Weight (W)	Risk score (RS)	W × RS	RS	W × RS	RS	W × RS
Human infections	0.2929	2.33	0.68	5	1.465	5.8	1.70
Transmission in laboratory animals	0.1929	2	0.39	7	1.350	7.5	1.45
Receptor binding	0.1429	2	0.29	6.3	0.900	8.6	1.23
Population immunity	0.1096	3	0.33	9	0.986	4.1	0.45
Infections in animals	0.0846	2	0.17	4.7	0.398	7.4	0.63
Genomic variation	0.0646	3	0.19	8.6	0.556	7.6	0.49
Antigenic relatedness	0.0479	2	0.10	3.7	0.177	5.5	0.26
Global distribution (animals)	0.0336	2.5	0.08	4.7	0.158	5.8	0.19
Disease severity	0.0211	2.25	0.05	8.5	0.179	4.8	0.10
Antiviral and treatment options	0.001	2.25	0.00	5.8	0.006	3.9	0.00
Summary risk score			2.28		6.18		6.50

Table 5 Risk scores for Question 2 (Impact) as applied to three influenza viruses—H1N1 North American Mallard, H7N9 [Anhui/1/2013] and H3N2(v) 2013

Element	N. Am. H1N1			H7N9		H3N2	
	Weight (W)	Risk score (RS)	W × RS	RS	W × RS	RS	W × RS
Disease severity	0.2929	2.25	0.66	8.5	2.490	4.8	1.41
Population immunity	0.1929	3	0.58	9	1.736	4.1	0.79
Human infections	0.1429	2.33	0.33	5	0.715	5.8	0.83
Antiviral and treatment options	0.1096	2.25	0.25	5.8	0.636	3.9	0.43
Antigenic relatedness	0.0846	2	0.17	3.7	0.313	5.5	0.47
Receptor binding	0.0646	2	0.13	6.3	0.407	8.6	0.56
Genomic variation	0.0479	3	0.14	8.6	0.412	7.6	0.36
Transmission in laboratory animals	0.0336	2	0.07	7	0.235	7.5	0.25
Global distribution (animals)	0.0211	2.5	0.05	4.7	0.099	5.8	0.12
Infections in animals	0.001	2	0.00	4.7	0.005	7.4	0.01
Summary risk score			2.38		7.05		5.22

zero, a three decimal value is calculated. The same is true of the other weights which are intended to provide relative comparison values, and not to imply an absolute degree of certainty.

2.4 Addressing Uncertainty

The risk score for each element of the IRAT is an average of risk scores provided by a group of individually surveyed SMEs. It is expected that risk scores provided by SMEs for an element will fall into general alignment, but there is no expectation that SMEs will all agree on exactly the same risk scores for a particular element and virus. Experts are provided with the working definition of the particular element that they have agreed to score and reminded that the score must be reflective of the components as defined by the element. Then each SME is asked to provide a point estimate that best captures the risk score for a particular influenza virus.

While some elements can be evaluated using more quantitative criteria, others are more qualitative in nature. For example, a relatively quantitative definition was developed for the element Receptor Binding (Table 6) but, in addition to the point risk score, experts were asked to provide an upper and lower boundary of risk scores that they would be willing to accept for each element. This request attempts to capture a range of uncertainty around the point estimates. For example, when scoring the Receptor Binding element for the H3N2 variant [A/Indiana/08/11] virus the point estimates from surveyed SMEs were all in the high risk range (i.e. 8–9) while the lowest acceptable score was 7.0; a high risk score was selected because H3N2v viruses have receptor binding profiles typical of influenza viruses that are transmissible in humans. All SMEs agreed that the highest acceptable score was 10. In contrast, the element describing Infections in Animals is more qualitative. Eleven SMEs provided a risk score for this element for HPAI A/H7N7 [A/NL/219/03] (definition provided in Table 7) and there was good agreement with nine raters providing a point estimate risk score of four or five and two experts providing risk scores of two and three. The average risk score for this element was 4.1. The same experts provided an upper end risk score ranging from three to nine, with eight of the experts providing an upper risk score of five, six or seven indicating that the consensus risk score for this virus fell into the low to moderate risk category. SMEs are asked to provide another input that speaks to uncertainty. As well as providing a point score, an upper and low acceptable risk score, the experts are requested to express their level of confidence in their scoring (Table 8). In the examples given above, the mean of the confidence score for the Receptor Binding element for the H3N2 variant [A/Indiana/08/11] virus was 2.75, while the mean confidence score for the Infections in Animals element was 2.8. In addition to point scores, upper and lower boundary scores, and confidence scores, SMEs were provided an open comment section and encouraged to provide specifics regarding publications or other materials that supported their scores and were prompted to provide additional comments and observations relevant to scoring the element.

After one round, the scores were summarized and sent back to the respective subject matter expert groups along with information regarding the variance and the range of scores. The comments and citations were collated and circulated to each of the SME groups, but kept anonymous. This approach allows those most familiar with data for a particular element the opportunity to incorporate comments and

Table 6 Example of an element that is quantitative

Receptor binding	For the purposes of the IRAT, Receptor Binding preference as a risk element for pandemic emergence is defined as a virus binding to glycans with sialic acid in α2,6 linkage to galactose
Low risk 1–3	Aquatic bird-like receptor binding specificity; i.e. glycans with α2,3 galactose-linked sialic acid and lack of binding to α2,6 galactose-linked sialic acids
Moderate risk 4–7	Dual receptor binding specificity; i.e. glycans with α2,3 galactose and α2,6 galactose linked sialic acids
High risk 8–10	Human-like receptor specificity; i.e. binding primarily to α2,6 galactose-linked sialic acid receptors

Table 7 Example of an element that is qualitative

Infections in animals	The element of Infections in Animals is defined as the ability of the virus to naturally infect animal species, the number and diversity of those species, and the ability to maintain sustained natural transmission in those populations, and the potential extent of exposure between humans and those species
Low risk 1–3	Sustained transmission in wild species (increased risk associate with species whose behavior patterns bring them into frequent contact with humans; ex: mallards/ducks)
Moderate risk 4–7	Limited outbreaks or sporadic disease in poultry or mammals Infected mammals, but without potential for mass exposure of humans (ex: feral mammals, dogs, cats) Sustained transmission in a limited number of host species (increased risk associated with infection in multiple species)
High risk 8–10	Endemicity established in an animal species Endemicity established in species with close contact with many humans (ex: infected animals at agricultural events [eg. Fairs, live markets, swap meets, etc.] zoos and other animal collections, households) Sustained transmission in multiple number of host species (increased risk associated with infection in multiple mammalian species)

Table 8 Confidence scoring guide

Score	
0	Low confidence; crude speculation only, due to lack of data or conflicting data
1	Weak correlation; educated guess, preliminary results of unknown reliability
2	Fair correlation; acceptable method, limited consensus on reliability
3	Good fit; small sample size, reliable method, independent verification of closely related variable
4	Exact measure; large sample set, independent verification of same variable

input from colleagues and adjust their scores if appropriate. Experts were then asked to reconsider their scores in light of this information and submit a second round of scoring to include a point risk score, an acceptable high and low score, their level of confidence for the score, and additional comments. This second round of scoring is used to generate a final virus risk score.

2.5 Addressing Missing Data

Early in the discovery of an emerging influenza virus there often is limited information available. In some instances there is a clearly recognized need to fill the information void, especially if human illnesses have occurred. An example is the HPAI H5N1 viruses which have caused 650 laboratory-confirmed human cases, 386 of whom have died between 2003 and January 2014 according to WHO(www.whi.int/influenza/human_animal_interface/EN_GIP_20140124 CumulativeNumberH5N1cases). In other instances, newly discovered viruses may not require a formal risk assessment from SMEs. Recent publications reported the discovery of H17 and H18 influenza viruses from New World bats (Tong et al. 2012, 2013). These recently identified viruses represent a distinct lineage that diverged from the influenza viruses that are known to have infected mammals including humans. Enough information is available to conclude that these viruses pose little risk to human health and that there is no evidence to date which indicates that a formal risk assessment should be conducted.

There will always be a need for additional data to assess the level of risk posed by a novel influenza virus to the human population. As with other risk assessments, this is also a limitation of the IRAT. However, it is not possible to defer some decisions or judgments until all information gaps are completely filled. In fact, at times additional information may be contradictory or offer little insight into key parameters associated with the novel influenza virus. When data gaps are associated with the elements of the IRAT that are of lower rank, the missing information is less likely to impact the overall risk score since these elements are associated with a lower weighting factor. On the other hand, filling information gaps associated with highly ranked elements is often essential to generate a meaningful risk score because of their higher proportional weights and consequent greater impact on the final score. Therefore, resource expenditure can be prioritized to gather information associated with some elements, while accepting a delay in filling all data gaps. In such a situation, the IRAT can still provide an understanding of the potential risk associated with a particular influenza virus when there is available information related to those elements of higher rank and weight.

2.6 Communication of the IRAT Scoring

The IRAT can be used to provide a uniform method to generate a risk score to compare the relative risks posed by influenza viruses that are not currently circulating in the human population, but cannot provide an exact quantification of risk. Inherent in the summary risk score for each virus is the need to incorporate uncertainty. While uncertainty is expressed through providing an upper and lower margin of acceptable risk scores as assigned by SMEs, it is also captured by assessing the level of confidence each expert assigns to their point score (Sluijs et al. 2005).

This information should be conveyed as part of the summary report so that those who used the IRAT scoring for decision-making have perspective regarding the interpretation of the final risk score. In fact a range of risk scores along with an explanation can be provided in the summary document.

3 Summary

The IRAT was designed to capture information that would answer two questions: (1) "What is the risk that a virus not currently circulating in the human population has potential for sustained human-to-human transmission?" and (2) "If the virus were to achieve sustained human-to-human transmission, what is the risk that a virus not currently circulating in the human population has the potential for significant impact on public health?" Ten elements were identified that can be used to differentiate among influenza viruses and provide information needed to answer these questions. Consensus working definitions for each element were developed by SMEs. The risk score for each of these elements provide the basis for development of a multi-attribute, additive approach to generate a summary risk score using the framework of the IRAT.

The IRAT draws upon subject matter expertise as well as published studies. SMEs for each risk element are asked to provide a point risk score between one and ten for influenza viruses of interest. Uncertainty is addressed by asking the SMEs to also provide a lower and upper acceptable score and an expression of confidence with their score. They are also encouraged to provide comments and citations which support their risk score for the particular element they are scoring.

SMEs were consulted and through consensus provided a rank ordering of elements from highest to lowest importance and information necessary to answer the two questions above using the derived definitions for each element. Depending upon the question, the order of importance of the elements varied. Information pertinent to some elements is more important than other information. Therefore the IRAT does not assign an equal weight to each element, but rather attaches surrogate weights to each element based upon rank order of importance. These weights are multiplied with the risk score (1–10) assigned by the SMEs for each element. The final risk score for a virus is the sum of these products and can be used during the decision-making process. One limitation of the IRAT is missing, incorrect, or ambiguous data; however, the IRAT can identify critical data gaps so that resources can be directed to fill them. It is anticipated that the IRAT will continue to evolve as scientific breakthroughs occur and provide insight into how the various elements can be refined, added to, or deleted as necessary. Additionally, as computational methods are incorporated into risk assessment, this information can help to refine risk scores. The IRAT is not intended to predict which virus will cause the next pandemic; rather it addresses the need for systematic, transparent, evidence-based prioritization of influenza viruses for pandemic preparedness with the ultimate goal of mitigating the impact should an influenza virus circulating in the animal population successfully jump the species barrier and move into humans.

References

Anyamba A, Chretien JP, Small J, Tucker CJ, Formenty PB, Richardson JH, Britch SC, Schnabel DC, Erickson RL, Linthicum KJ (2010a) Prediction of a Rift Valley fever outbreak. PNAS 106(3):955–959

Anyamba A, Linthicum KJ, Small J, Britch SC, Pak E, de La Rocque S, Formenty P, Hightower AW, Breiman RF, Chretien JP, Tucker CJ, Schnabel D, Sang R, Haagsma K, Lathem M, Lewandowski HB, Magdi SO, Mohamed MA, Nguku PM, Reynes JM, Swanepoel R (2010b) Prediction, assessment of the Rift Valley fever activity in east and southern Africa 2006–2008 and possible vector control strategies. Am J Trop Med Hyg 83(Sup 2):43–51

Briet OJT, Vounatsou P, Gunawardena DM, Galappaththy GNL, Amerasinghe PH (2008) Models for short term malaria prediction in Sri Lanka. http://www.malariajournal.com/content/7/1/76

Britch SC, Binepal YS, Ruder MG, Kariithi HM, Linthicum KJ, Anyamba A, Small JL, Tucker CJ, Ateya LO, Oriko AA, Gacheru S, Wilson WC (2013) Rift Valley fever risk map model and seroprevalence in selected wild ungulates and camels from Kenya. PLoS One 8(6):e66626

Centers for Disease Control and Prevention (2010) Interim results: Influenza A (H1N1) 2009 monovalent vaccination coverage—United States, October—December 2009. MMWR Morb Mortal Wkly Rep 59(02):44–48

Centers for Disease Control and Prevention (2009) Swine Influenza A (H1N1) infection in two children-Southern California, March-April 2009. MMWR Morb Mortal Wkly Rep 58(15):400–402

Dawood FS, Juliano AD, Reed C, Meltzer MI, Shay KD, Cheng PY, Dandaranayake D, Breiman RF, Brooks WA, Buchy P, Feikin DR, Fowler KB, Gordon A, Hien NT, Horby P, Huang QS, Katz MA, Krishnan A, Lal R, Montgomery JM, Molabak K, Pebody R, Presanis AM, Razuri H, Steens A, Tinoco YO, Wallinga J, Yu H, Vong S, Bresee J, Widdowson MA (2012) Estimated global mortality associated with the first 12 months of 2009 pandemic influenza A H1N1 virus circulation: a modelling study. Lancet 12:651–653

Department of Health, Education and Welfare (1978) Interagency work group on pandemic influenza. A plan for pandemic influenza

Dolan JG (2010) Multi-criteria clinical decision support: a primer on the use of multiple criteria decision making methods to promote evidence based, patient centered healthcare. Patient 4:229–248

Edwards W, Barron FH (1994) SMARTS and SMARTER: improved simple methods for multi-attribute utility measurement. Organ Behav Hum Decis Process 60:306–325

Gao R, Cao B, Hu Y, Feng Z, Wang D, Hu W, Chen J, Jie Z, Qiu H, Xu K, Xu X, Lu H, Zhu W, Gao Z, Xiang N, Shen Y, He Z, Gu Y, Zhang Z, Yang Y, Zhao X, Zhou L, Li X, Zou S, Zhang Y, Li X, Yang L, Guo J, Dong J, Li Q, Dong L, Zhu Y, Bai T, Wang S, Hao P, Yang W, Zhang Y, Han J, Yu H, Li D, Gao GF, Guizhen W, Wang Y, Yuan Z, Shu Y (2013) Human infection with a novel avian-origin influenza A (H7N9) virus. NEJM 368(20):1888–1897

Garten RJ, Davis CT, Russell CA, Shu B, Lindstrom S, Balish A, Sessions WM, Xu X, Skepner E, Deyde V, Okomo-Adhiambo M, Gubareva L, Barnes J, Smith CB, Emery SL, Hillman MJ, Rivailler P, Smagala J, de Graff M, Burke DF, Fouchier RAM, Pappas C, Alpuche-Aranda CM, Lopez-Gatell H, Olivera H, Lopez I, Myers CA, Faix D, Blair PJ, Yu C, Keene KM, Dotson PD, Boxrud D, Sambol AR, Abid SH, St. George K, Bannerman T, Moore AL, Stringer DJ, Blevins P, Demmler-Harrison GJ, Ginsberg M, Kriner P, Waterman S, Smole S,

Guevara HF, Belongia EA, Clark PA, Beatrice ST, Donis R, Katz J, Finelli L, Bridges CB, Shaw M, Jernigan DB, Uyeki TM, Smith DJ, Klimov AI, Cox NJ (2009) Antigenic and genetic characteristics of swine-origin 2009 A(H1N1) influenza viruses circulating in humans. Science 325:197–201

Glezen WP (1996) Emerging infections: Pandemic influenza. Epidemio Rev 18(1):64–76

Homeland Security Council (2005) National strategy for pandemic influenza. http://www.flu.gov/planning-preparedness/federal/pandemic-influenza.pdf

Hoshen MB, Morse AP (2004) A weather-driven model of malaria transmission. http://www.malariajournal.com/content/3/1/32

OIE World Organisation for Animal Health (2014) Update on highly pathogenic avian influenza in animals (type H5 and H7) http://www.oie.int/animal-health-in-the-world/update-on-avian-influenza/2014/

Iskander J, Strikas RA, Gensheimer KF, Cox NJ, Redd SC (2013) Pandemic influenza planning, United States, 1978–2008. EID 19(6):879–885

Jernigan DB, Lindstrom SL, Johnson JR, Miller JD, Hoelscher M, Humes R, Shively R, Brammer L, Burke SA, Villanueva JM, Balish A, Uyeki T, Mustaquim D, Bishop A, Handsfield JH, Astles R, Xu X, Klimov AI, Cox NJ, Shaw MW (2011) Detecting 2009 pandemic influenza A(H1N1) virus infection: availability of diagnostic testing led to rapid pandemic response. CID 52(Suppl 1):S36–S43

Jhung MA, Swerdlow D, Olson SJ, Jernigan D, Biggerstaff M, Kamimoto L, Kniss K, Reed C, Fry A, Brammer L, Gindler J, Gregg WJ, Bresee J, Finelli L (2011) Epidemiology of 2009 pandemic influenza A (H1N1) in the United States. CID 52(Suppl 1):S13–S26

Kasowski EJ, Garten RJ, Bridges CB (2011) Influenza pandemic epidemiologic and virologic diversity: reminding ourselves of the possibilities. CID 52(Suppl 1):S44–S49

Metwally S (2008) Rift Valley fever. In; Brown C, Torres A (eds) Foreign animal diseases, 7th edn. USAHA, Boca Raton, pp 369–375

Morans DM, Fauci AS (2007) The 1918 influenza pandemic: insights for the 21st century. JID 195:1018–1028

Parry J (2013) H7N9 avian flu infects humans for the first time. BMJ 346:f2151

Phillips LD (1984) A theory of requisite decision models. Acta Psychologica 56:29–48

Schuchat A, Tappero J, Blandford J (2014) Global health and the US Centers for Disease Control and Prevention. Lancet (accepted for publication)

Simonsen L, Clarke MJ, Schonberger LB, Arden NH, Cox NJ, Fududa K (1998) Pandemic versus epidemic influenza mortality: a pattern of changing age distribution. JID 178:53–60

Sluijs JP, Craye M, Funtowicz S, Kloprogge P, Ravetz J, Risbey J (2005) Combining quantitative and qualitative measures of uncertainty in model-based environmental assessment: The NUSAP system. Risk Anal 25(2):481–492

Swayne DE, Perdue ML (2005) Public health risk from avian influenza viruses. Av Dis 49:317–327

Swayne DE (ed) (2008) Avian influenza. Blackwell Publishing, Ames

Tong S, Zhu X, Li Y, Shi M, Zhang J, Bourgeois M, Yang H, Chen X, Rescuenco S, Gomez J, Chen LM, Johnson A, Tao Y, Dreyfus C, Yu W, McBride R, Carney PJ, Bilbert AT, Chang J, Guo Z, Davis CT, Paulson JC, Stevens J, Rupprecht CE, Holmes EC, Wilson IA, Donis RO (2013) New world bats harbor diverse influenza a viruses. Plos Pathogens 9(10):e1003657

Tong S, Li Y, Rivailler P, Conrardy C, Castillo DAA, Chen LM, Recuenco S, Ellison JA, Davis CT, York IA, Turmelle AS, Moran D, Rogers S, Shi M, Tao Y, Weil MR, Tang K, Rowe LA, Sammons S, Xu X, Frace M, Lindblade KA, Cox NJ, Anderson LJ, Rupprecht CE, Donis RO (2012) A distinct lineage of influenza a virus from bats. www.pnas.org/cgi/doi/10.1073/pnas.11162000109

Trock SC, Burke SA, Cox NJ (2012) Development of an influenza virologic risk assessment tool. Av Dis 56:1058–1061

USDA, National Poultry Improvement Plan (2007) http://www.aphis.usda.gov/vs/nahss/poultry/ai/avian_influenza_surveillance_plan_062907.pdf

USDA, National Surveillance Plan for Swine Influenza Virus in Pigs. 14 Jul 2010. http://www.aphis.usda.gov/animal_health/animal_dis_spec/swine/downloads/appendix_a_national_surv_plan.pdf
Webster RG, Bean WJ, Gorman OT, Chambers TM, Kawaok Y (1992) Evolution and ecology of influenza a viruses. Microbiol Rev 56(1):152–179
WHO/OIE/FAO H5N1 Evolution Working Group (2014) Revised and updated nomenclature for highly pathogenic avian influenza A (H5N1). Influenza Other Respir Viruses 8(3):384–388
World Health Organization (2014a) Human infection with influenza A(H7N) virus. http://who.int/influenza/human_animal_interface/influenza_h7n9/140225_H7N9RA_for_web_20140306FM.pdf?ua=1
World Health Organization (2014b) Influenza at the human-animal interface, monthly risk assessment summary. http://www.who.int/influenza/human_animal_interface/Influenza_Summary_IRA_HA_interface_5May14.pdf
World Health Organization (2013) Pandemic influenza risk management. http://www.who.int/influenza/preparedness/pandemic/GIP_PandemicInfluenzaRiskManagementInterimGuidance_Jun2013.pdf
Zimmer SM, Burke DS (2009) Historical perspective—emergence of influenza A (H1N1) viruses. NEJM 361(3):279–285

Avian Influenza Virus Transmission to Mammals

S. Herfst, M. Imai, Y. Kawaoka and R. A. M. Fouchier

Abstract Influenza A viruses cause yearly epidemics and occasional pandemics. In addition, zoonotic influenza A viruses sporadically infect humans and may cause severe respiratory disease and fatalities. Fortunately, most of these viruses do not have the ability to be efficiently spread among humans via aerosols or respiratory droplets (airborne transmission) and to subsequently cause a pandemic. However, adaptation of these zoonotic viruses to humans by mutation or reassortment with human influenza A viruses may result in airborne transmissible viruses with pandemic potential. Although our knowledge of factors that affect mammalian adaptation and transmissibility of influenza viruses is still limited, we are beginning to understand some of the biological traits that drive airborne transmission of influenza viruses among mammals. Increased understanding of the determinants and mechanisms of airborne transmission may aid in assessing the risks posed by avian

S. Herfst · R. A. M. Fouchier (✉)
Department of Viroscience, Postgraduate School Molecular Medicine,
Erasmus Medical Center, P.O. Box 2040, 3000 CA Rotterdam, The Netherlands
e-mail: r.fouchier@erasmusmc.nl

M. Imai
Department of Veterinary Medicine, Faculty of Agriculture, Iwate University,
Iwate 020-8550, Japan

Y. Kawaoka
Department of Pathobiological Sciences, School of Veterinary Medicine,
University of Wisconsin-Madison, 575 Science Drive, Madison WI, USA

Y. Kawaoka
Division of Virology, Department of Microbiology and Immunology,
Institute of Medical Science, University of Tokyo, Tokyo 108-8639, Japan

Y. Kawaoka
Department of Special Pathogens, International Research Center for Infectious Diseases,
Institute of Medical Science, University of Tokyo, Minato-ku, Tokyo 108-8639, Japan

Y. Kawaoka
ERATO Infection-Induced Host Responses Project, Japan Science and Technology Agency,
Kawaguchi-shi, Saitama 332-0012, Japan

Current Topics in Microbiology and Immunology (2014) 385: 137–155
DOI: 10.1007/82_2014_387

Published Online: 22 July 2014

influenza viruses to human health, and preparedness for such risks. This chapter summarizes recent discoveries on the genetic and phenotypic traits required for avian influenza viruses to become airborne transmissible between mammals.

Contents

1 Introduction

Influenza A viruses are classified into subtypes, based on the antigenic properties of the major surface glycoproteins hemagglutinin (HA) and neuraminidase (NA). To date, 16 major antigenic variants of HA and 9 of NA have been described in wild aquatic birds. The recently discovered viruses of the H17N10 and H18N11 subtypes have only been detected in bats (Fouchier et al. 2005; Tong et al. 2012, 2013). More than 20 other mammalian host species, including humans, pigs, dogs, horses, and marine mammals, as well as various poultry species can be infected by influenza A viruses as well (Webster et al. 1992). Domestic ducks and geese may serve as "bridge species" for influenza A virus transmission to poultry (Fig. 1). Both wild and domestic birds are suitable influenza virus hosts that mostly experience asymptomatic infections, despite high virus replication. Low pathogenic avian influenza viruses (LPAI) of subtypes H5 and H7 are an important exception, since they can evolve in poultry to become highly pathogenic avian influenza (HPAI) viruses, causing severe disease and high mortality.

Occasionally, avian influenza viruses that circulate in poultry (e.g., those of subtypes H5, H6, H7, H9, and H10) cross the species barrier and infect humans. Usually, such zoonotic events are restricted to individual cases and do not pose a public health threat. However, on rare occasions, avian influenza viruses acquire the ability to transmit between humans and start a pandemic. In the last century, four pandemics were recorded: the 1918 H1N1 Spanish pandemic, the 1957 H2N2 Asian pandemic, the 1968 H3N2 Hong Kong pandemic, and the 2009 H1N1 pandemic (pH1N1). The question of which influenza A virus will most likely cause the next pandemic is a highly debated topic in virology on which influenza experts disagree. It has been postulated that only influenza A viruses of the H1, H2, and

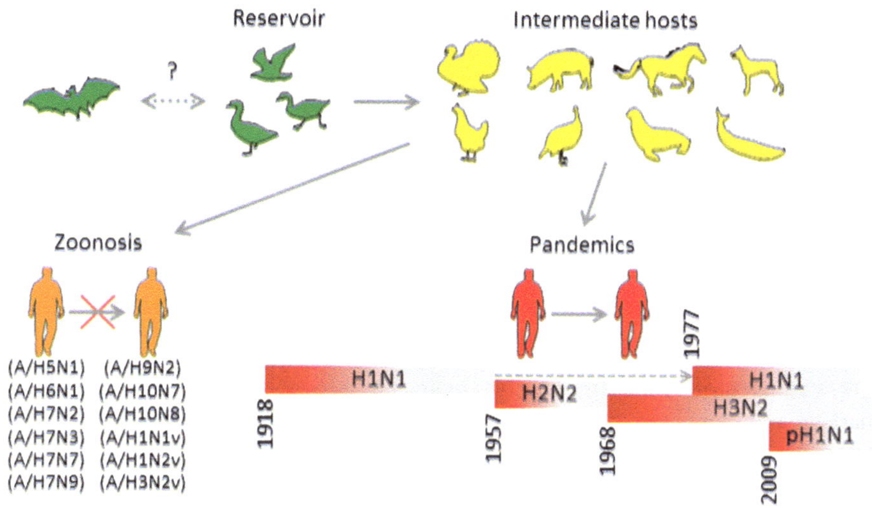

Fig. 1 Zoonotic and pandemic threats. Wild aquatic birds are thought to form the influenza virus reservoir in nature (*green*). Recently influenza viruses have also been detected in bats (*green*). Humans can be infected by avian influenza viruses via intermediate hosts such as various poultry species, pigs, dogs, horses and marine mammals (*yellow*). These zoonotic events usually result in isolated cases of human infections (*orange*). Occasionally, upon mutation or reassortment, these zoonotic viruses adapt to the new human host and may start a pandemic (*red*). In the last century, four major pandemics have been recorded

H3 subtypes can cause pandemics, as influenza viruses of no other subtypes have ever caused pandemics in the past. This would mean that we would not need to be worried about the pandemic potential of zoonotic infections caused by avian influenza viruses that were associated with poultry outbreaks. On the other hand, these viruses should be considered to represent pandemic threats since they have shown the ability to cross the species barrier and infect humans. Moreover, current virologic and serologic approaches can only determine the subtypes of influenza viruses that circulated during the last 150 years; there is no way to know what subtypes of viruses were circulating in humans more than 150 years ago.

The viruses that caused three of the four major pandemics of the last century emerged upon reassortment of animal and human influenza viruses. Pigs may serve as a 'mixing vessel' to allow reassortment of avian and mammalian influenza viruses and initial mammalian adaptation. However, a future pandemic may also be triggered by an entirely avian influenza virus without the need of reassortment. Early studies on transmissibility between mammals have shown that reassortment events between avian and human influenza viruses do not yield viruses that are readily transmitted in mammalian airborne transmission models (Maines et al. 2006, 2011; Yen et al. 2007; Jackson et al. 2009; Schrauwen et al. 2013), however, recent studies have demonstrated that such reassortants harboring an (un-)modified avian H5 or H9 subtype HA gene can gain the ability to be transmitted (Sorrell et al. 2009; Kimble et al. 2011, 2014; Chen et al. 2012; Imai et al. 2012; Zhang et al. 2013c). Similar research showed that a fully avian H5N1 virus can become airborne transmissible without the requirement for reassortment (Herfst et al. 2012).

In this review, we focus on recent discoveries on the genetic and phenotypic traits required for avian influenza viruses to become airborne transmissible between mammals.

2 Avian Influenza A Virus Zoonoses

The first zoonotic events with the HPAI H5N1 virus were recorded in Hong Kong in 1997, and led to the infection of 18 humans, of whom 6 died (de Jong et al. 1997). This virus was temporarily eradicated through depopulation of poultry markets and chicken farms in Hong Kong in late 1997. In 2001 and 2002, other outbreaks of HPAI H5N1 viruses in poultry occurred, but no new cases of human infection or disease were reported (Sims et al. 2003). Since 2003, HPAI H5N1 viruses have caused severe outbreaks in poultry in numerous countries throughout Asia, Europe and Africa, and sporadically the virus was transmitted to humans, mostly upon direct or indirect contact with infected poultry. As of 24 January 2014, the World Health Organization (WHO) has been informed of 650 laboratory-confirmed human cases of infection with H5N1 viruses, of which 386 were fatal (WHO 2014). Due to the enzootic circulation of the virus in poultry, the high incidence of zoonotic events, and the severity of disease in humans, the H5N1 virus is considered to pose a serious human health threat. Although human-to-human transmission between family members has been reported on several occasions, no sustained human-to-human transmission has occurred (Ungchusak et al. 2005; Kandun et al. 2006; Wang et al. 2008).

In spring 2013, a major zoonotic outbreak of a novel avian-origin H7N9 virus started in China, resulting in 370 laboratory-confirmed human cases of infection and 114 deaths (as of 24 February 2014) (CIDRAP 2014). Although some mild cases of infection have been reported, the H7N9 virus has frequently caused severe illness, characterized by severe pulmonary disease and acute respiratory distress syndrome (Gao et al. 2013a, b). A limited number of family clusters of H7N9 cases has been reported, but thus far no sustained transmission between humans has occurred (Qi et al. 2013). The 2013 outbreak already peaked in April, and only two new human cases of infection were reported during the summer months. However, additional human infections occurred in early October, resulting in a second wave of human H7N9 cases that has already exceeded the first wave in number of cases and duration (CIDRAP 2014). The timely detection of H7N9 in poultry, which may prevent transmission to humans, is hampered by the lack of clinical signs in poultry, which makes it more difficult to observe if poultry is affected by this LPAI virus.

Poultry outbreaks with other viruses of the H7 subtype have been reported as well (e.g., in the United Kingdom, United States, Canada, Mexico, the Netherlands and Italy (CDC 2004; Fouchier et al. 2004; Tweed et al. 2004; Puzelli et al. 2005; Lopez-Martinez et al. 2013)). These H7 viruses have also sporadically infected humans. However, the 2013 H7N9 virus appeared to jump the species barrier more easily and was generally associated with more serious disease in humans. Interestingly, the internal genes of the H7N9 virus belong to the same genetic lineage as the internal genes of the H5N1 virus; both are derived from H9N2 viruses (Guan et al. 1999; Lam et al. 2013).

H9N2 viruses began to be isolated frequently from poultry in the late 1980s, and soon became enzootic in domestic birds in many countries of the Eastern Hemisphere. Zoonotic events of H9N2 viruses have been reported upon human contact with poultry since 1999 (Peiris et al. 1999), usually resulting in relatively mild human infections. The most recent human cases of infection were reported in December 2013 in Hong Kong. Importantly, viruses of this subtype have been isolated from pigs, and numerous reassortment events between H9N2 virus and other influenza virus subtypes have been reported (Liu et al. 2013).

Although multiple introductions of H6 subtype influenza viruses into poultry have occurred and H6 viruses circulate extensively in poultry (Cheung et al. 2007), only one human case of avian H6N1 virus infection has been reported (Yuan et al. 2013). However, the ability of avian H6 viruses to replicate in mice and ferrets without prior adaptation (Gillim-Ross et al. 2008) and its high frequency of reassortment and high prevalence in birds have raised concerns about the threat posed by H6 viruses (Munster et al. 2007).

In 2004, human infections with influenza viruses of subtype H10N7 were reported in Egypt (Pan American Health Organization 2004). In March 2010, an outbreak of LPAI H10N7 in chickens was responsible for the infection of abattoir workers. In November 2013, an immune-compromised patient with underlying illnesses who had visited a local poultry market had succumbed from an infection with H10N8 influenza virus (ProMED-mail 2013). Reported outbreaks of avian H10 subtype influenza viruses are uncommon, but these human cases highlight the ability of these avian viruses to cross the species barrier (Arzey et al. 2012).

Over the past 15 years, there has been an increase in the number of zoonotic events that have been detected, however, this may actually be the result of increased diagnostics and pandemic preparedness, rather than a real increase in zoonotic events.

3 Genetic Determinants of Mammalian Adaptation

The recurring outbreaks of avian influenza viruses in poultry and the associated human cases of infection keep us alert to that these viruses pose threats to public health. Although avian influenza can infect mammals without prior reassortment or mutation, subsequent adaptation to the new host may facilitate improved replication in their new host.

The influenza virus replication cycle is initiated when the HA protein binds to sialic acid (SA) receptors on susceptible host cells. The specificity of influenza viruses for these receptors varies according to the species from which they are isolated. Avian influenza viruses preferentially bind oligosaccharides that terminate with a SA linked to galactose by α2,3-linkages (Rogers and Paulson 1983; Connor et al. 1994; Stevens et al. 2006), whereas human influenza viruses preferably recognize α2,6-linked SA receptors (Shinya et al. 2006; van Riel et al. 2006). The receptor distribution throughout the respiratory tract of ferrets resembles that of humans in that the α2,6 SA receptors are predominantly present on ciliated cells in the upper respiratory tract (URT), and the α2,3 SA receptors are mainly present on non-ciliated cells of the lower respiratory tract (LRT) and type II pneumocytes. In chickens, quail, and other land-based birds, α2,3 SAs predominate, but both α2,3 SA and α2,6 SA are present throughout the respiratory and enteric tracts. The host species restriction and cell and tissue tropism of influenza viruses are thought to be primarily determined by the differences in receptor distribution and specificity. Structural studies have shown that receptor-specificity is primarily determined by the receptor-binding site (RBS) in HA. For the H2N2 and H3N2 pandemic viruses, amino acid substitutions and G228S in the RBS (numbering throughout the manuscript is based on H3 HA) were shown to change the binding preference from avian α2,3 SA to human α2,6 SA receptors (Connor et al. 1994; Matrosovich et al. 2000). For the H1N1 virus subtype, substitutions E190D and D225G were important for the change in preference from avian to human receptors (Matrosovich et al. 2000; Glaser et al. 2005). For the 2009 pH1N1 virus, the D225G residue was associated with increased disease severity, presumably because viruses with this substitution acquired dual receptor specificity for α2,3 SA and α2,6 SA (Chutinimitkul et al. 2010; Kilander et al. 2010). Furthermore, for all three virus subtypes it was previously demonstrated that these RBS mutations that change the receptor-binding preference are also critical determinants for airborne transmissibility of these viruses between ferrets (Tumpey et al. 2007; Pappas et al. 2010; Roberts et al. 2011).

Upon infection of the host cell and subsequent release of the viral genome in the nucleus, the viral RNA-dependent RNA polymerase complex transcribes the viral RNA to produce copy RNA that serves as a template to produce more viral RNA. Furthermore, viral RNA is transcribed into messenger RNA that is subsequently translated into viral proteins. The viral polymerase complex consists of the basic polymerase protein 1 (PB1), basic polymerase protein 2 (PB2) and the acidic polymerase (PA) protein, of which the PB2 protein was shown to be a major determinant of host range restriction. The viral nucleoprotein (NP) binds to the viral RNA, and together with the polymerase proteins forms the ribonucleoprotein complexes (RNPs). Whereas avian influenza viruses in principle replicate at temperatures around 41 °C, the temperature in the intestinal tract of birds, for replication in humans the viruses need to adapt to 33 °C, the temperature of the human URT (Subbarao et al. 1993; Gao et al. 2009; Yamada et al. 2010). Key changes in PB2 have been associated with increased virus replication in mammalian cells at such lower temperatures. These substitutions at positions 627, 701, and 591 of PB2 have been shown to support the transmission of multiple virus subtypes in mammals (Subbarao et al. 1993; Mehle and Doudna 2009; Steel et al. 2009; Van Hoeven et al. 2009). It was shown that the function of influenza polymerase containing an avian-like PB2 (627E) in human cells was restricted by an inhibitory activity that prevented assembly of polymerase into the RNP. As a result of this assembly defect, genomic replication and transcription of viral RNAs was reduced (Mehle and Doudna 2008).

4 Modes of Influenza A Virus Transmission

The pandemic potential of an avian influenza virus depends on its ability to adapt to replication in the human respiratory tract and its ability to become transmissible. Human-to-human transmission of influenza viruses can occur through direct or indirect contact (via fomites), or via the airborne route (via aerosols or respiratory droplets). However, based on the transmission properties of animal-origin influenza viruses that have caused pandemics and epidemics in the past and were all transmissible via the airborne route, transmission via aerosol or respiratory droplets appears to be crucial for their pandemic potential.

Although there is no exact definition of particle size cut-off at which transmission changes from aerosols to (large) respiratory droplets, it is generally accepted that respiratory droplets, >5 μm, do not remain suspended in air and travel less than 1 m before settling on the mucosa of close contacts or environmental surfaces. On the contrary, small aerosols, <5 μm, have a slow settling velocity, thus remain suspended in the air longer than the larger respiratory droplets and can thus travel further. The majority of aerosol particles expelled by breathing, coughing or sneezing of humans were measured at <1 μm (Lindsley et al. 2010; Fabian et al. 2008). The importance of the airborne route for influenza virus transmission is supported by experimental evidence showing transmission to volunteers by aerosols, thereby reproducing the disease at doses much less than required by intranasal infections (Alford et al. 1966).

In recent years, a lot of effort was put on investigating the routes of influenza virus transmission between animals and the determinants of airborne transmission. To study airborne transmission of influenza viruses between mammals under high containment laboratory conditions, ferret and guinea pig transmission models were developed. Ferrets have been used in influenza research since 1933, and are still considered a suitable animal model for influenza virus research. Ferrets are susceptible to infection with human and avian influenza viruses and they develop respiratory disease and lung pathology similar to that observed in humans (Smith et al. 1933). More recently, pigs and guinea pigs are also used to study transmission properties of influenza viruses (Lowen et al. 2006). Using these mammalian transmission models, pandemic and epidemic viruses isolated from humans are generally transmitted efficiently via the airborne route, whereas avian viruses are generally not airborne transmissible (Belser et al. 2013b). Although the experimental setup may vary between the different laboratories, the basic setup consists of an inoculated donor animal in a cage that is adjacent to a cage housing a naive recipient animal. To allow airflow from the donor to the recipient animal, but to prevent direct contact or fomite transmission, the two cages are separated from each other by steel grids and/or a physical distance (Lowen et al. 2006; Maines et al. 2006).

5 Airborne Transmission of H5N1 Influenza Between Mammals

HPAI H5N1 viruses have previously shown their ability to infect mammals, including humans. Whereas mammalian adaptation markers have been detected in various human and avian HPAI H5N1 isolates, the genetic traits that result in efficient human-to-human transmissibility have obviously not been acquired by these viruses yet. The lack of airborne transmissibility of HPAI H5N1 virus has been demonstrated in several mammalian models including mice, guinea pigs, and ferrets. To study the requirements for airborne transmission of HPAI H5N1 virus, quite a few studies were conducted with wild type (wt) H5N1 viruses, reassortants between H5N1 and contemporary human influenza viruses (H3N2 or pH1N1), H5N1 viruses adapted via repeated passage, and genetically modified H5N1 viruses with substitutions known to alter receptor binding preference and/or replication. Although direct contact transmission was achieved with some of the candidate viruses, none of the early attempts resulted in H5 viruses that were airborne transmissible (Lowen et al. 2006; Maines et al. 2006, 2011; Yen et al. 2007; Jackson et al. 2009; Steel et al. 2009; Schrauwen et al. 2013). These studies confirm the complexity of the genetic and phenotypic requirements of influenza virus transmission and underline that significant adaptations are required for H5N1 viruses to become pandemic.

Recently, important discoveries towards understanding influenza A virus transmission via the airborne route were reported, using H5N1 viruses and ferret and guinea pig transmission models (Chen et al. 2012; Herfst et al. 2012; Imai et al. 2012; Zhang et al. 2013c). These discoveries have contributed to increased understanding of the genetic and phenotypic viral factors and host cell factors that drive transmission between mammals (Fig. 2).

A recent study by Herfst et al. demonstrated that a fully avian highly pathogenic H5N1 virus acquired the ability of efficient airborne transmission in the ferret model (Herfst et al. 2012). After the initial introduction of mammalian adaptation substitutions Q226L and G228S in HA and E627K in PB2 (see above) of A/Indonesia/5/2005 (Indo05), no airborne transmission was observed when this mutant virus was evaluated in the ferret transmission model. This indicated that a change in binding preference to α2,6 SAs, and adaptation of the polymerase complex alone is not enough to confer airborne transmissibility to a fully avian H5N1 virus. To allow further adaptation to replication in a mammalian host, this mutant virus was passaged ten times in the URT of ferrets, resulting in the accumulation of potential adaptation substitutions. When the virus mixture from

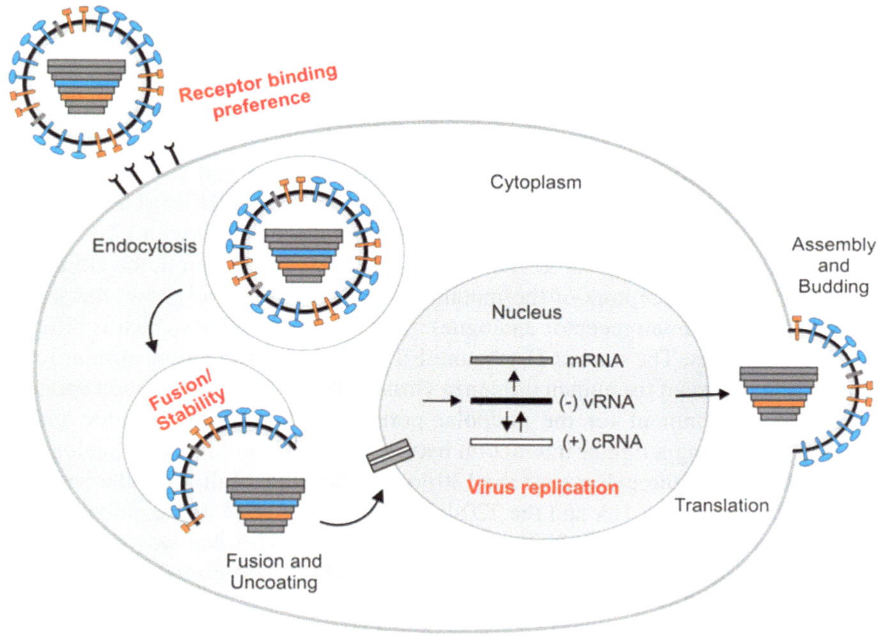

Fig. 2 Determinants of airborne transmission of influenza A viruses between mammals. To facilitate efficient airborne transmission of influenza A viruses, zoonotic viruses need to attach to appropriate cells of the mammalian airways, which is facilitated by a change in receptor binding preference of HA. The binding of HA to the host cell receptors can subsequently be increased by deletion of a potential N-linked glycosylation site adjacent to the RBS. Other mutations stabilize the HA protein to compensate for the instability induced by the change in receptor binding preference, and influence the pH of fusion between the endosomal and viral membranes. Mutations in the viral polymerase complex proteins increased replication by enhancing transcription of vRNA, mRNA and cRNA. Phenotypic traits that affect airborne transmissibility are highlighted in red

the passage ten ferret was evaluated for its ability to be transmitted between ferrets, three out of four recipient ferrets became infected via the airborne route. Subsequent consensus sequence analysis showed that five mutations were consistently found in all airborne-transmitted viruses. In addition to the three substitutions that were introduced on purpose (Q226L, G228S in HA and E627K in PB2), two other substitutions in HA were found: H110Y, located at the interface between the two monomers of the HA trimer; and T160A, that results in the loss of a potential N-linked glycosylation site at residues 158–160, which is proximal but not immediately adjacent to the receptor binding site. Subsequent extensive analysis of the critical genetic changes that confer airborne-transmissibility to Indo05, showed that two sets of five substitution determine the airborne transmission phenotype (Linster et al. 2014). The two mutations that changed the receptor binding preference to α2,6 SAs (Q226L and G2228) were independently capable of changing the binding preference sufficiently for airborne transmission.

In addition to the 'consistently found' five mutations described above, a PB1 H99Y mutation was identified as a new key adaptation marker required for airborne transmission of H5N1. In concert with PB2 E627K, this PB1 H99Y substitution enhanced transcription of vRNA, mRNA and cRNA and increased virus replication of Indo/05.

The effect of the HA substitutions on receptor binding and HA structure was investigated using surface plasmon resonance (Zhang et al. 2013b). The Indo/5/05 HA containing all four substitutions had a ~9-fold reduced affinity for α2,3 SA and a >3-fold increased affinity for α2,6 SA compared to wt HA. Overall, the affinity for avian and human receptors of the mutant HA was still relatively low: the wt HA bound LSTc (a human receptor analogue) in trans conformation, which is different from human HAs. The mutant HA bound LSTc in cis conformation, similar to the orientation observed for human influenza viruses. The Q226L substitution created a favorable environment for the nonpolar portion of the human receptor analog, thereby facilitating a tighter interaction between the RBS and receptor analog. The RBS is formed by three domains: the 130-loop at the edge of the globular head, the 190-helix at the top of HA and the 220-loop at the other edge of the globular head. Substitutions Q226L and G228S together resulted in a RBS that was ~1 Å wider between the 130 and 220 loops compared to that of wt HA, as reported for human HAs (Zhang et al. 2013b).

Three other recent studies demonstrated that H5N1 viruses can become airborne transmissible upon genetic reassortment with contemporary human influenza viruses, with or without mammalian adaptation substitutions.

Imai et al. introduced random mutations in the globular head (amino acids 120–259) of A/Vietnam/1203/2004 (VN1203), which includes the RBS. After screening of many mutants, it was found that Q226L and N224K in HA resulted in α2,6 SA receptor binding preference, allowing subsequent binding of H5 HA to human tracheal epithelium cells (Imai et al. 2012). Because of the high genetic compatibility of pH1N1 and H5N1 viruses (Octaviani et al. 2010) and the coexistence of these viruses in nature, which provides an opportunity for the generation of reassortants, a pH1N1 reassortant virus harboring this mutant H5 HA was generated. When evaluated in the ferret transmission model, this virus was not transmitted via the airborne route to recipient ferrets. However, passaging of this virus in ferrets resulted in the acquisition of two additional HA mutations (N158D and T318I) that conferred airborne transmissibility to this avian-human reassortant virus. The N158D substitution led to the loss of the same N-linked glycosylation site that was lost with the introduction of T160A as identified by Herfst et al. (2012). The other substitution, T318I, is located proximally to the fusion peptide that plays a role in the process of membrane fusion by HA. The effect of the four HA substitutions (N158D, N224K, Q226L and T318I) identified by Imai et al. on receptor binding and HA structure were investigated in the context of autologous VN1203 HA and the closely related A/Vietnam/1194/2004 (VN1194) HA (Lu et al. 2013; Xiong et al. 2013a). Binding of the mutated HA's to α2,6 SA was slightly increased, whereas the affinity for α2,3 SA was decreased substantially. Overall, the affinity of the mutant H5 HA for human and avian receptors was 5–10 fold lower than for human H3 HA (Xiong et al. 2013a). Similar to the structural change of the airborne Indo05 HA, it was demonstrated that the RBS between the 130 and 220 loops of both the mutant VN1203 and VN1194 HA, had increased in size by ~1–1.5 Å.

Influenza viruses attach to receptors on the host cell surface with their HA, followed by receptor-mediated endocytosis. A low-pH-triggered conformational change of HA mediates fusion of the viral and endosomal membranes to release the virus genome in the cytoplasm (Shaw and Palese 2013). Consequently, premature exposure of virus to low pH in the extracellular environment results in an irreversible conformational change in HA, leading to a considerable loss of viral infectivity. Similarly, heat treatment at neutral pH also induces the conformational change of HA and therefore serves as a surrogate for HA stability (Haywood and Boyer 1986; Carr et al. 1997). Both Imai et al. and Linster et al. showed that introduction of the substitutions responsible for the α2,6 SA binding preference increase the activation pH of fusion and decrease the associated thermal stability of HA (Imai et al. 2012; Linster et al. 2014). The ferret adaptation substitutions T318I and H110Y that were identified in the two transmission studies, were shown to increase HA stability upon acquisition of the α2,6 SA binding substitutions. These results suggest that an H5 HA-containing virus with α2,6 SA binding preference requires additional substitutions in HA to become more stable in the environment, thereby possibly enabling airborne transmissibility between ferrets. However, it is important to note that the increased acid stability and thermostability of HA, as studied in laboratory experiments, may only be a surrogate marker for another phenotype, such as stability in mucus, aerosols or respiratory droplets, or upon exposure to air. Collectively, these findings suggest that a fine balance of mutations associated with different functions in HA (such as receptor-binding specificity and HA protein stability) is likely to be required for efficient airborne transmission of H5N1 viruses between mammals.

Chen et al. have demonstrated that a reassortant virus containing an HPAI H5N1 virus HA protein with Q196R, Q226L and G228S substitutions (and already possessing N158D and T160A, resulting in the loss of a glycosylation site), a human H3N2 virus NA protein and the internal genes from an H5N1 virus could be transmitted via the airborne route among ferrets; viral shedding was detected in the nasal washes of one of two exposed ferrets (Chen et al. 2012).

Whereas earlier studies on reassortment of H5N1 virus with contemporary human influenza viruses did not yield viruses that were airborne transmissible between mammals (Maines et al. 2006; Jackson et al. 2009; Schrauwen et al. 2013), it was recently shown that several reassortant H5N1 virus variants were transmitted via the airborne route between guinea pigs without the need for mammalian adaptation substitutions (Zhang et al. 2013c). This H5 virus (A/duck/Guangxi/35/2001), which revealed dual specificity for α2,3 SA and α2,6 SA receptors and which already possessed the T160A substitution in HA and D701N in PB2, required the PA and NS genes of the 2009 pH1N1 virus to acquire airborne transmissibility between guinea pigs. In addition, the NP, M, and NA genes from 2009 pH1N1 virus were shown to be important in enhancing airborne transmissibility between mammals. These results appear to be in contrast to the study of Imai et al. in which a reassortant virus with a mutant H5 HA with human receptor specificity in a 2009 pH1N1 virus backbone failed to transmit via the airborne route between ferrets (Imai et al. 2012). This difference in transmissibility may be explained by the use of different lineages of H5 HAs. In addition, it is unclear how the guinea pig and ferret models compare with respect to H5 virus transmission studies. However, as mentioned above, the avian-human reassortant influenza virus in the study of Imai et al. transmitted via the airborne route after passaging in ferrets, when a few additional genetic changes in HA were acquired. These two studies thus demonstrated that different airborne transmissible H5N1 strains can emerge upon reassortment between human and avian influenza viruses (with or without the need for additional substitutions), as was previously also shown for avian H9N2 virus (see below).

6 Airborne Transmission of H7N9 Influenza Between Mammals

Soon after the emergence of the avian-origin H7N9 in humans, several studies were published on airborne transmissibility of H7N9 viruses between ferrets (Belser et al. 2013a; Richard et al. 2013; Watanabe et al. 2013; Zhang et al. 2013a; Zhu et al. 2013). Although different human isolates were used in the different studies (A/Shanghai/1/2013, A/Shanghai/2/2013 and A/Anhui/1/2013), similar results were obtained. It can be concluded that the novel H7N9 virus demonstrated limited airborne transmission between ferrets: transmission was more efficient than for other avian influenza viruses, which are not airborne transmitted in ferrets, but less robust—with fewer animals becoming infected, and less and delayed virus shedding—as compared to seasonal and pandemic A(H1N1) virus transmission (Itoh et al. 2009; Munster et al. 2009). It should be noted that mammalian adaptation markers that change receptor binding preference as A138S (known to enhance binding to α2,6 SA of swine H5N1 viruses), G186V and Q226L in HA as well as E627K / D701N in PB2 that affect replication at lower temperature, were present in most human H7N9 virus isolates (Yang et al. 2010, 2013; Gambaryan et al. 2012; Srinivasan et al. 2013). However, none of the H7N9 viruses contained G228S. Yang et al. have shown that introduction of G228S in H7N9 HA that already contained Q226L resulted in overall increased binding to α2,3 and α2,6 SA, but not in a switch towards human-type receptor preference (Yang et al. 2013). When attachment patterns to formalin-fixed human respiratory tract tissues were studied, visualized by histochemical methods (virus histochemistry), G228S resulted in increased binding to the apical surface of tracheal and alveolar tissues of humans (Tharakaraman et al. 2013). However, it is still unclear whether other amino acid substitutions could improve the receptor binding preference of the H7N9 viruses, thereby potentially increasing the airborne transmissibility of H7N9 viruses.

H7N9 viruses were found to have dual receptor specificity for both α2,3 SA and α2,6 SA receptors (Lam et al. 2013; Shi et al. 2013; van Riel et al. 2013; Watanabe et al. 2013; Xiong et al. 2013b). Overall, A/Anhui/1/2013 with the Q226L mutation in its HA was found to bind more efficient to α2,6 SA than A/Shanghai/1/2013 containing Q226 in its HA. In glycan arrays, A/Anhui/1/2013 bound to bi-antennary structures as well as structures containing long N-acetyllactosamine repeats (Belser et al. 2013a; Watanabe et al. 2013). H7N9 influenza viruses in general had higher affinity for α2,6 SA compared to other H7 influenza viruses. However, in most studies the H7N9 viruses maintained preference for α2,3 SA over α2,6 SA receptors, which is in contrast to human influenza viruses and airborne transmissible H5 viruses. The dual receptor specificity of H7N9 was also reflected by their attachment patterns using virus histochemistry; H7N9 viruses were shown to attach to cells of both the URT and LRT of humans, and to both type I and II pneumocytes (van Riel et al. 2013).

As described above, studies on H5N1 viruses suggest that a more stable HA is required for increased airborne transmissibility. The H7N9 viruses that showed limited airborne transmission between ferrets have a relatively unstable HA: the conformational change of HA can already be induced at low temperatures and high threshold pH for fusion as compared to the airborne H5 viruses (Richard et al. 2013). Research on airborne transmissibility of H7N9 viruses should therefore focus on the identification of substitutions in the H7N9 HA that increase the stability of HA and improve receptor binding to α2,6 SA receptors.

The H7N9 HA structure of A/Anhui/1/2013 in complex with human and avian receptor analogues was solved and compared with those of avian H7 HAs. Both LSTc and LSTa (an avian receptor analogue) were bound to HA in a *cis* conformation (Shi et al. 2013; Xiong et al. 2013b). However, LSTc bound to H7N9 HA was shown to exit the RBS in a different direction to that seen in HA-receptor complexes formed by HA's from human influenza viruses and the airborne transmissible H5 HA described by Imai et al. The H7 HAs have a subtype-specific insertion in the 150-loop, causing this loop to protrude closer to the RBS. However, the function of this 150-loop is not yet known (Russell et al. 2006). The human receptor-binding properties of human H7 viruses probably arise from the introduction of two bulky hydrophobic residues by the substitutions Q226L and G186V. The well-documented Q226L widens the separation between the 220-loop and the 130-loop by about 1 Å (as described above for human influenza viruses and airborne H5). However, in the case of the H7N9 HA there is only a slight increase in the 220-loop/130-loop separation compared to avian H7 HA (Xiong et al. 2013b).

7 Airborne Transmission of H9 Subtype Influenza Between Mammals

H9N2 viruses with either human or dual receptor specificity are now prevalent in many Eurasian countries, thereby increasing the possibility of these viruses to infect humans (Matrosovich et al. 2001). It was previously shown that a wt avian

H9N2 virus, or a reassortant virus carrying the H9N2 HA and NA surface proteins with the internal genes derived from a human H3N2 virus was not transmissible via the airborne route between ferrets. However, upon 10 passages of this reassortant in ferrets and the acquisition of ferret adaptation mutations Q226L, T189A and R192G in HA, this reassortant acquired airborne transmissibility (Sorrell et al. 2009). Furthermore, when this mutant H9 HA was evaluated on a pH1N1 backbone, with or without its own NA, this virus was also airborne transmissible (Kimble et al. 2011). Interestingly, whereas a reassortant containing the wt avian H9N2 HA and NA with the internal genes of a human H3N2 was not airborne transmissible between ferrets, the same HA and NA on a pH1N1 backbone was. However, when only the wt H9N2 HA was used in a pH1N1 backbone, airborne transmission was abolished, suggesting that an optimal balance between HA and NA activity is required for efficient airborne transmission.

8 Future Perspectives

In recent years, gain-of-function research has accelerated the accumulation of knowledge about phenotypic traits that are required for (highly pathogenic) avian influenza viruses to cross the species barrier and to adapt to replication and transmission in mammals. For specific A/H5N1 lineages there is experimental evidence that multiple adaptations are required: (1) The zoonotic viruses need to attach to appropriate cells of the mammalian airways, which is facilitated by a change in receptor binding preference of HA; (2) The binding of HA to the host cell receptors can subsequently be increased by deletion of a potential N-linked glycosylation site adjacent to the RBS; (3) Stabilization of the HA protein to compensate for the instability induced by the change in receptor binding preference; (4) Increased replication by the viral polymerase complex (by enhancing transcription of vRNA, mRNA and cRNA). A hypothesized fifth requirement for airborne transmissibility, for which no experimental evidence has been provided yet, is virus shedding as single particles rather than virus aggregates, which may be achieved by balanced HA and NA activities (de Wit et al. 2010; Sorrell et al. 2011).

The five phenotypic requirements described in this review should form the basis for future research on (airborne) transmissibility of avian influenza viruses. So far, these phenotypes have been described for specific H5N1 virus strains, but it should be investigated if these are common mechanisms for airborne transmissibility of all influenza virus subtypes. Although gain-of-function research should be continued on influenza viruses representing public health threats (like H5, H7, H9 and H10), additional evidence for the importance of these phenotypes may come from loss-of-function research on viruses that caused the previous pandemics.

Further identification of the genetic and phenotypic changes required for host adaptation and transmission of influenza viruses remains an important research topic to facilitate risk assessment for the emergence of zoonotic and pandemic influenza. If we succeed to identify common phenotypic properties, than simple phenotypic assays may complement ongoing surveillance activities rather than looking for specific adaptation mutations in sequences, since such mutations may be dependent on the genetic backbone or other functional equivalent mutations may exist.

References

Alford RH, Kasel JA, Gerone PJ, Knight V (1966) Human influenza resulting from aerosol inhalation. Proc Soc Exp Biol Med 122:800–804

Arzey GG, Kirkland PD, Arzey KE, Frost M, Maywood P, Conaty S et al (2012) Influenza virus A (H10N7) in chickens and poultry abattoir workers, Australia. Emerg Infect Dis 18:814–816

Belser JA, Gustin KM, Pearce MB, Maines TR, Zeng H, Pappas C et al (2013a) Pathogenesis and transmission of avian influenza A (H7N9) virus in ferrets and mice. Nature 501:556–559

Belser JA, Maines TR, Katz JM, Tumpey TM (2013b) Considerations regarding appropriate sample size for conducting ferret transmission experiments. Future Microbiol 8:961–965

Carr CM, Chaudhry C, Kim PS (1997) Influenza hemagglutinin is spring-loaded by a metastable native conformation. Proc Natl Acad Sci U S A 94:14306–14313

CDC (2004) Update: influenza activity–United States, 2003-04 season. MMWR Morb Mortal Wkly Rep 53:284–287

Chen LM, Blixt O, Stevens J, Lipatov AS, Davis CT, Collins BE et al (2012) In vitro evolution of H5N1 avian influenza virus toward human-type receptor specificity. Virology 422:105–113

Cheung CL, Vijaykrishna D, Smith GJ, Fan XH, Zhang JX, Bahl J et al (2007) Establishment of influenza A virus (H6N1) in minor poultry species in southern China. J Virol 81:10402–10412

Chutinimitkul S, Herfst S, Steel J, Lowen AC, Ye J, van Riel D et al (2010) Virulence-associated substitution D222G in the hemagglutinin of 2009 pandemic influenza A(H1N1) virus affects receptor binding. J Virol 84:11802–11813

Center for Infectious Disease Research and Policy (CIDRAP) (2014) Cases of H7N9 influenza in China by week of http://www.cidrap.umn.edu/sites/default/files/public/downloads/topics/mceirs_h7n9_update_022414_0.pdf

Connor RJ, Kawaoka Y, Webster RG, Paulson JC (1994) Receptor specificity in human, avian, and equine H2 and H3 influenza virus isolates. Virology 205:17–23

de Jong JC, Claas EC, Osterhaus AD, Webster RG, Lim WL (1997) A pandemic warning? Nature 389:554

de Wit E, Munster VJ, van Riel D, Beyer WE, Rimmelzwaan GF, Kuiken T et al (2010) Molecular determinants of adaptation of highly pathogenic avian influenza H7N7 viruses to efficient replication in the human host. J Virol 84:1597–1606

Fabian P, McDevitt JJ, DeHaan WH, Fung RO, Cowling BJ, Chan KH et al (2008) Influenza virus in human exhaled breath: an observational study. PLoS One 3:e2691

Fouchier RA, Munster V, Wallensten A, Bestebroer TM, Herfst S, Smith D et al (2005) Characterization of a novel influenza A virus hemagglutinin subtype (H16) obtained from black-headed gulls. J Virol 79:2814–2822

Fouchier RA, Schneeberger PM, Rozendaal FW, Broekman JM, Kemink SA, Munster V et al (2004) Avian influenza A virus (H7N7) associated with human conjunctivitis and a fatal case of acute respiratory distress syndrome. Proc Natl Acad Sci USA 101:1356–1361

Gambaryan AS, Matrosovich TY, Philipp J, Munster VJ, Fouchier RA, Cattoli G et al (2012) Receptor-binding profiles of H7 subtype influenza viruses in different host species. J Virol 86:4370–4379

Gao HN, Lu HZ, Cao B, Du B, Shang H, Gan JH et al (2013a) Clinical findings in 111 cases of influenza A (H7N9) virus infection. N Engl J Med 368:2277–2285

Gao R, Cao B, Hu Y, Feng Z, Wang D, Hu W et al (2013b) Human infection with a novel avian-origin influenza A (H7N9) virus. N Engl J Med 368:1888–1897
Gao Y, Zhang Y, Shinya K, Deng G, Jiang Y, Li Z et al (2009) Identification of amino acids in HA and PB2 critical for the transmission of H5N1 avian influenza viruses in a mammalian host. PLoS Pathog 5:e1000709
Gillim-Ross L, Santos C, Chen Z, Aspelund A, Yang CF, Ye D et al (2008) Avian influenza h6 viruses productively infect and cause illness in mice and ferrets. J Virol 82:10854–10863
Glaser L, Stevens J, Zamarin D, Wilson IA, Garcia-Sastre A, Tumpey TM et al (2005) A single amino acid substitution in 1918 influenza virus hemagglutinin changes receptor binding specificity. J Virol 79:11533–11536
Guan Y, Shortridge KF, Krauss S, Webster RG (1999) Molecular characterization of H9N2 influenza viruses: were they the donors of the "internal" genes of H5N1 viruses in Hong Kong? Proc Natl Acad Sci USA 96:9363–9367
Haywood AM, Boyer BP (1986) Time and temperature dependence of influenza virus membrane fusion at neutral pH. J Gen Virol 67(Pt 12):2813–2817
Herfst S, Schrauwen EJ, Linster M, Chutinimitkul S, de Wit E, Munster VJ et al (2012) Airborne transmission of influenza A/H5N1 virus between ferrets. Science 336:1534–1541
Imai M, Watanabe T, Hatta M, Das SC, Ozawa M, Shinya K et al (2012) Experimental adaptation of an influenza H5 HA confers respiratory droplet transmission to a reassortant H5 HA/H1N1 virus in ferrets. Nature 486:420–428
Itoh Y, Shinya K, Kiso M, Watanabe T, Sakoda Y, Hatta M et al (2009) In vitro and in vivo characterization of new swine-origin H1N1 influenza viruses. Nature 460:1021–1025
Jackson S, Van Hoeven N, Chen LM, Maines TR, Cox NJ, Katz JM et al (2009) Reassortment between avian H5N1 and human H3N2 influenza viruses in ferrets: a public health risk assessment. J Virol 83:8131–8140
Kandun IN, Wibisono H, Sedyaningsih ER, Yusharmen Hadisoedarsuno W, Purba W et al (2006) Three Indonesian clusters of H5N1 virus infection in 2005. N Engl J Med 355:2186–2194
Kilander A, Rykkvin R, Dudman SG, Hungnes O (2010) Observed association between the HA1 mutation D222G in the 2009 pandemic influenza A(H1N1) virus and severe clinical outcome, Norway 2009–2010. Euro Surveill 15(19):19498
Kimble JB, Angel M, Wan H, Sutton TC, Finch C, Perez DR (2014) Alternative reassortment events leading to transmissible H9N1 influenza viruses in the ferret model. J Virol 88:66–71
Kimble JB, Sorrell E, Shao H, Martin PL, Perez DR (2011) Compatibility of H9N2 avian influenza surface genes and 2009 pandemic H1N1 internal genes for transmission in the ferret model. Proc Natl Acad Sci USA 108:12084–12088
Lam TT, Wang J, Shen Y, Zhou B, Duan L, Cheung CL et al (2013) The genesis and source of the H7N9 influenza viruses causing human infections in China. Nature 502:241–244
Lindsley WG, Blachere FM, Thewlis RE, Vishnu A, Davis KA, Cao G et al. (2010) Measurements of airborne influenza virus in aerosol particles from human coughs. PLoS One 5, e15100
Linster M, van Boheemen S, De Graaf M, Schrauwen EJ, Lexmond P, Manz B et al (2014) Identification, characterization, and natural selection of mutations driving airborne transmission of A/H5N1 virus. Cell. 157(2):329–339
Liu D, Shi W, Shi Y, Wang D, Xiao H, Li W et al (2013) Origin and diversity of novel avian influenza A H7N9 viruses causing human infection: phylogenetic, structural, and coalescent analyses. Lancet 381:1926–1932
Lopez-Martinez I, Balish A, Barrera-Badillo G, Jones J, Nunez-Garcia TE, Jang Y et al (2013) Highly pathogenic avian influenza A(H7N3) virus in poultry workers, Mexico, 2012. Emerg Infect Dis 19:1531–1534
Lowen AC, Mubareka S, Tumpey TM, Garcia-Sastre A, Palese P (2006) The guinea pig as a transmission model for human influenza viruses. Proc Natl Acad Sci USA 103:9988–9992
Lu X, Shi Y, Zhang W, Zhang Y, Qi J, Gao GF (2013) Structure and receptor-binding properties of an airborne transmissible avian influenza A virus hemagglutinin H5 (VN1203mut). Protein Cell 4:502–511

Maines TR, Chen LM, Matsuoka Y, Chen H, Rowe T, Ortin J et al (2006) Lack of transmission of H5N1 avian-human reassortant influenza viruses in a ferret model. Proc Natl Acad Sci USA 103:12121–12126

Maines TR, Chen LM, Van Hoeven N, Tumpey TM, Blixt O, Belser JA et al (2011) Effect of receptor binding domain mutations on receptor binding and transmissibility of avian influenza H5N1 viruses. Virology 413:139–147

Matrosovich M, Tuzikov A, Bovin N, Gambaryan A, Klimov A, Castrucci MR et al (2000) Early alterations of the receptor-binding properties of H1, H2, and H3 avian influenza virus hemagglutinins after their introduction into mammals. J Virol 74:8502–8512

Matrosovich MN, Krauss S, Webster RG (2001) H9N2 influenza A viruses from poultry in Asia have human virus-like receptor specificity. Virology 281:156–162

Mehle A, Doudna JA (2009) Adaptive strategies of the influenza virus polymerase for replication in humans. Proc Natl Acad Sci USA 106:21312–21316

Mehle A, Doudna JA (2008) An inhibitory activity in human cells restricts the function of an avian-like influenza virus polymerase. Cell Host Microbe 4:111–122

Munster VJ, Baas C, Lexmond P, Waldenstrom J, Wallensten A, Fransson T et al (2007) Spatial, temporal, and species variation in prevalence of influenza A viruses in wild migratory birds. PLoS Pathog 3:e61

Munster VJ, de Wit E, van den Brand JM, Herfst S, Schrauwen EJ, Bestebroer TM et al (2009) Pathogenesis and transmission of swine-origin 2009 A(H1N1) influenza virus in ferrets. Science 325:481–483

Octaviani CP, Ozawa M, Yamada S, Goto H, Kawaoka Y (2010) High level of genetic compatibility between swine-origin H1N1 and highly pathogenic avian H5N1 influenza viruses. J Virol 84:10918–10922

Pan American Health Organization (2004) Emerging and reemerging infectious diseases, region of the Americas. http://new.paho.org/hq/dmdocuments/2010/Avian_Influenza_Egypt_070503.pdf

Pappas C, Viswanathan K, Chandrasekaran A, Raman R, Katz JM, Sasisekharan R et al (2010) Receptor specificity and transmission of H2N2 subtype viruses isolated from the pandemic of 1957. PLoS One 5:e11158

Peiris M, Yuen KY, Leung CW, Chan KH, Ip PL, Lai RW et al (1999) Human infection with influenza H9N2. Lancet 354:916–917

ProMED-mail (2013) http://www.promedmail.org/?archiveid=20131217.2121718

Puzelli S, Di Trani L, Fabiani C, Campitelli L, De Marco MA, Capua I et al (2005) Serological analysis of serum samples from humans exposed to avian H7 influenza viruses in Italy between 1999 and 2003. J Infect Dis 192:1318–1322

Qi X, Qian YH, Bao CJ, Guo XL, Cui LB, Tang FY et al (2013) Probable person to person transmission of novel avian influenza A (H7N9) virus in Eastern China, 2013: epidemiological investigation. BMJ 347:f4752

Richard M, Schrauwen EJ, de Graaf M, Bestebroer TM, Spronken MI, van Boheemen S et al (2013) Limited airborne transmission of H7N9 influenza A virus between ferrets. Nature 501:560–563

Roberts KL, Shelton H, Scull M, Pickles R, Barclay WS (2011) Lack of transmission of a human influenza virus with avian receptor specificity between ferrets is not due to decreased virus shedding but rather a lower infectivity in vivo. J Gen Virol 92:1822–1831

Rogers GN, Paulson JC (1983) Receptor determinants of human and animal influenza virus isolates: differences in receptor specificity of the H3 hemagglutinin based on species of origin. Virology 127:361–373

Russell RJ, Stevens DJ, Haire LF, Gamblin SJ, Skehel JJ (2006) Avian and human receptor binding by hemagglutinins of influenza A viruses. Glycoconj J 23:85–92

Schrauwen EJ, Bestebroer TM, Rimmelzwaan GF, Osterhaus AD, Fouchier RA, Herfst S (2013) Reassortment between Avian H5N1 and human influenza viruses is mainly restricted to the matrix and neuraminidase gene segments. PLoS One 8:e59889

Shaw ML, Palese P (2013) Orthomyxoviruses. In: Knipe DM, Howley PM (ed.) Fields virology, vol. 1, pp 1151–1185. Wolters Kluwer Health/Lippincott Williams & Wilkins

Shi Y, Zhang W, Wang F, Qi J, Wu Y, Song H et al (2013) Structures and receptor binding of hemagglutinins from human-infecting H7N9 influenza viruses. Science 342:243–247
Shinya K, Ebina M, Yamada S, Ono M, Kasai N, Kawaoka Y (2006) Avian flu: influenza virus receptors in the human airway. Nature 440:435–436
Sims LD, Ellis TM, Liu KK, Dyrting K, Wong H, Peiris M et al (2003) Avian influenza in Hong Kong 1997-2002. Avian Dis 47:832–838
Smith W, Andrewes CH, Laidlaw PP (1933) A virus obtained from influenza patients. Lancet 222:66–68
Sorrell EM, Schrauwen EJ, Linster M, De Graaf M, Herfst S, Fouchier RA (2011) Predicting 'airborne' influenza viruses: (trans-) mission impossible? Curr Opin Virol 1:635–642
Sorrell EM, Wan H, Araya Y, Song H, Perez DR (2009) Minimal molecular constraints for respiratory droplet transmission of an avian-human H9N2 influenza A virus. Proc Natl Acad Sci USA 106:7565–7570
Srinivasan K, Raman R, Jayaraman A, Viswanathan K, Sasisekharan R (2013) Quantitative description of glycan-receptor binding of influenza A virus H7 hemagglutinin. PLoS One 8:e49597
Steel J, Lowen AC, Mubareka S, Palese P (2009) Transmission of influenza virus in a mammalian host is increased by PB2 amino acids 627K or 627E/701N. PLoS Pathog 5:e1000252
Stevens J, Blixt O, Glaser L, Taubenberger JK, Palese P, Paulson JC et al (2006) Glycan microarray analysis of the hemagglutinins from modern and pandemic influenza viruses reveals different receptor specificities. J Mol Biol 355:1143–1155
Subbarao EK, London W, Murphy BR (1993) A single amino acid in the PB2 gene of influenza A virus is a determinant of host range. J Virol 67:1761–1764
Tharakaraman K, Jayaraman A, Raman R, Viswanathan K, Stebbins NW, Johnson D et al (2013) Glycan receptor binding of the influenza A virus H7N9 hemagglutinin. Cell 153:1486–1493
Tong S, Li Y, Rivailler P, Conrardy C, Castillo DA, Chen LM et al (2012) A distinct lineage of influenza A virus from bats. Proc Natl Acad Sci USA 109:4269–4274
Tong S, Zhu X, Li Y, Shi M, Zhang J, Bourgeois M et al (2013) New world bats harbor diverse influenza A viruses. PLOS pathogens 9:e1003657
Tumpey TM, Maines TR, Van Hoeven N, Glaser L, Solorzano A, Pappas C et al (2007) A two-amino acid change in the hemagglutinin of the 1918 influenza virus abolishes transmission. Science 315:655–659
Tweed SA, Skowronski DM, David ST, Larder A, Petric M, Lees W et al (2004) Human illness from avian influenza H7N3, British Columbia. Emerg Infect Dis 10:2196–2199
Ungchusak K, Auewarakul P, Dowell SF, Kitphati R, Auwanit W, Puthavathana P et al (2005) Probable person-to-person transmission of avian influenza A (H5N1). N Engl J Med 352:333–340
Van Hoeven N, Pappas C, Belser JA, Maines TR, Zeng H, Garcia-Sastre A et al (2009) Human HA and polymerase subunit PB2 proteins confer transmission of an avian influenza virus through the air. Proc Natl Acad Sci USA 106:3366–3371
van Riel D, Leijten LM, de Graaf M, Siegers JY, Short KR, Spronken MI et al (2013) Novel avian-origin influenza A (H7N9) virus attaches to epithelium in both upper and lower respiratory tract of humans. Am J Pathol 183:1137–1143
van Riel D, Munster VJ, de Wit E, Rimmelzwaan GF, Fouchier RA, Osterhaus AD et al (2006) H5N1 Virus Attachment to Lower Respiratory Tract. Science 312:399
Wang H, Feng Z, Shu Y, Yu H, Zhou L, Zu R et al (2008) Probable limited person-to-person transmission of highly pathogenic avian influenza A (H5N1) virus in China. Lancet 371:1427–1434
Watanabe T, Kiso M, Fukuyama S, Nakajima N, Imai M, Yamada S et al (2013) Characterization of H7N9 influenza A viruses isolated from humans. Nature 501:551–555
Webster RG, Bean WJ, Gorman OT, Chambers TM, Kawaoka Y (1992) Evolution and ecology of influenza A viruses. Microbiol Rev 56:152–179
WHO (2014) Influenza at the human-animal interface. http://www.who.int/influenza/human_animal_interface/EN_GIP_20140124CumulativeNumberH5N1cases.pdf?ua=1

Xiong X, Coombs PJ, Martin SR, Liu J, Xiao H, McCauley JW et al (2013a) Receptor binding by a ferret-transmissible H5 avian influenza virus. Nature 497:392–396
Xiong X, Martin SR, Haire LF, Wharton SA, Daniels RS, Bennett MS et al (2013b) Receptor binding by an H7N9 influenza virus from humans. Nature 499:496–499
Yamada S, Hatta M, Staker BL, Watanabe S, Imai M, Shinya K et al (2010) Biological and structural characterization of a host-adapting amino acid in influenza virus. PLoS Pathog 6:e1001034
Yang H, Carney PJ, Chang JC, Villanueva JM, Stevens J (2013) Structural Analysis of the Hemagglutinin from the Recent 2013 H7N9 Influenza Virus. J Virol 87:12433–12446
Yang H, Chen LM, Carney PJ, Donis RO, Stevens J (2010) Structures of receptor complexes of a North American H7N2 influenza hemagglutinin with a loop deletion in the receptor binding site. PLoS Pathog 6:e1001081
Yen HL, Lipatov AS, Ilyushina NA, Govorkova EA, Franks J, Yilmaz N et al (2007) Inefficient transmission of H5N1 influenza viruses in a ferret contact model. J Virol 81:6890–6898
Yuan J, Zhang L, Kan X, Jiang L, Yang J, Guo Z et al (2013) Origin and molecular characteristics of a novel 2013 avian influenza A H6N1 virus causing human infection in Taiwan. Clin Infect Dis 57:1167–1169
Zhang Q, Shi J, Deng G, Guo J, Zeng X, He X et al (2013a) H7N9 influenza viruses are transmissible in ferrets by respiratory droplet. Science 341:410–414
Zhang W, Shi Y, Lu X, Shu Y, Qi J, Gao GF (2013b) An airborne transmissible avian influenza H5 hemagglutinin seen at the atomic level. Science 340:1463–1467
Zhang Y, Zhang Q, Kong H, Jiang Y, Gao Y, Deng G et al (2013c) H5N1 hybrid viruses bearing 2009/H1N1 virus genes transmit in guinea pigs by respiratory droplet. Science 340:1459–1463
Zhu H, Wang D, Kelvin DJ, Li L, Zheng Z, Yoon SW et al (2013) Infectivity, transmission, and pathology of human-isolated H7N9 influenza virus in ferrets and pigs. Science 341:183–186

Transmission in the Guinea Pig Model

Anice C. Lowen, Nicole M. Bouvier and John Steel

Abstract The ability of an influenza virus to transmit efficiently from human-to-human is a major factor in determining the epidemiological impact of that strain. The use of a relevant animal model to identify viral determinants of transmission, as well as host and environmental factors affecting transmission efficiency, is therefore critical for public health. The characterization of newly emerging influenza viruses in terms of their potential to transmit in a mammalian host is furthermore an important part of pandemic risk assessment. For these reasons, a guinea pig model of influenza virus transmission was developed in 2006. The guinea pig provides an important alternative to preexisting models for influenza. Most influenza viruses do not readily transmit among mice. Ferrets, while highly relevant, are expensive and can be difficult to obtain in high numbers. Moreover, it is generally accepted that efforts to accurately model human disease are strengthened by the use of multiple animal species. Herein, we provide an overview of influenza virus infectivity, growth, and transmission in the guinea pig and highlight knowledge gained on the topic of influenza virus transmission using the guinea pig model.

Abbreviations

Pan/99	A/Panama/2007/1999 (H3N2)
PR8	A/Puerto Rico/8/1934 (H1N1)
SC/09	A/Sichuan/1/2009 (H1N1)
Bris/59	A/Brisbane/59/2007 (H1N1)
DK/35	A/duck/Guanxi/35/2001 (H5N1)

A. C. Lowen · J. Steel (✉)
Department of Microbiology and Immunology, Emory University School of Medicine, Atlanta, GA, USA
e-mail: john.steel@emory.edu

N. M. Bouvier
Department of Microbiology, Icahn School of Medicine at Mount Sinai, New York, NY, USA

Current Topics in Microbiology and Immunology (2014) 385: 157–183
DOI: 10.1007/82_2014_390

Published Online: 8 July 2014

NY/1253	A/New York/08-1253/2008 (H1N1)
NY/1326	A/New York/08-1326/2008 (H1N1)
AH/1	A/Anhui/1/2013 (H7N9)
SH/1	A/Shanghai/1/2013 (H7N9)
Cal/09	A/California/04/2009 (H1N1pdm09)
sH1N1	Seasonal influenza A(H1N1)
pH1N1	2009 pandemic influenza A(H1N1)
PFU	Plaque-forming unit
ID_{50}	50 % infectious dose
EID50	50 % egg infectious dose
TCID50	50 % tissue culture infectious dose
MUNANA	2'-(4-methylumbelliferyl)-α-D-N-acetylneuraminic acid
MAA II	Maakia amurensis agglutinin II
SNA	Sambucus nigra agglutinin

Contents

1 Introduction

The guinea pig (*Cavia porcellus*) has been used as a model to study influenza virus infection and immune responses since the 1970s (Wetherbee 1973; Phair et al. 1979; Azoulay-Dupuis et al. 1984). The utility of this species for the study of influenza virus transmission was, however, first described in 2006 (Lowen et al. 2006). The development of the guinea pig transmission model offered an important alternative to the better-established ferret and mouse models. Unlike mice, guinea pigs support efficient transmission of influenza viruses adapted to human hosts and are highly susceptible to infection with a broad range of influenza A and B viruses—without prior adaptation of the viruses through serial passage. Compared to ferrets, guinea pigs are relatively small and inexpensive, and seronegative guinea pigs are easily obtained, facilitating the performance of experiments that are adequately powered to gain statistically significant results. The utility of the guinea pig transmission model is evidenced by the expansion of its use since 2006. Research performed in the model has contributed important insights into influenza virus transmission, including the effects of viral, host, and environmental factors. This chapter describes influenza virus infection, growth, and transmission in guinea pigs; highlights how these properties differ among influenza viruses adapted to human, swine, and avian hosts; and provides an overview of knowledge gained through the study of influenza virus transmission in the guinea pig model.

2 Influenza Virus Infection in the Guinea Pig

2.1 Susceptibility to Infection

Guinea pigs are highly susceptible to infection by the intranasal route with influenza viruses derived from human, avian, and swine hosts. Determinations of 50 % infectious dose (ID50) for a handful of human and avian isolates have yielded values ranging from 3 PFU to 66 PFU (Lowen et al. 2006; Bouvier et al. 2008; Steel et al. 2009; Gabbard et al. 2013). While precise ID50 values have not been reported for swine isolates, the success of low dose inoculations with isolates of both North American and Eurasian swine influenza virus lineages indicates that the ID50 is approximately 100 PFU or less in each case (J. Steel, unpublished data). These results for ID50 in guinea pigs are similar to what has been reported for humans and ferrets (Alford et al. 1966), indicating that guinea pigs and ferrets have comparable natural susceptibilities to infection with a broad range of influenza viruses (Gustin et al. 2011; Lakdawala et al. 2011). Indeed, a sero-survey of domestic guinea pigs in Ecuador, obtained either from farms or live animal markets, revealed antibodies to influenza A and B viruses in the majority of samples tested (Leyva-Grado et al. 2012). These results suggest that guinea pigs in close contact with humans naturally acquire influenza virus infection.

Nevertheless, among the guinea pigs we have obtained from laboratory animal vendors since 2006, none has been found to be seropositive in hemagglutination inhibition assays (JS, AL and NB, unpublished data); we consider this to be an important advantage of the guinea pig over the ferret model.

2.2 Viral Growth in the Guinea Pig Respiratory Tract

Following intranasal inoculation, influenza virus replication in the guinea pig is largely confined to the upper respiratory tract (Lowen et al. 2006; Gabbard et al. 2013; Seibert et al. 2013). While nasal lavage fluid and homogenates of nasal turbinate yield high virus titers, infectious virus is typically present at lower levels or not detected in the lung (Lowen et al. 2006; Gabbard et al. 2013; Seibert et al. 2013). These findings are supported by histological staining for influenza virus antigen: on days 2 and 4 post-infection, viral protein is easily detected in the nasal tissues of guinea pigs infected with the human seasonal A/Panama/2007/1999 (H3N2) [Pan/99] virus, but not found in the lungs of the same animals (Gabbard et al. 2013). In humans, seasonal influenza virus infection is mainly confined to the upper respiratory tract, while growth in the lung is more rare and associated with severe disease (Treanor 2010). Also in ferrets, seasonal influenza viruses target mainly the upper respiratory tract (van der Laan et al. 2008; Zeng et al. 2013). Thus, in broad terms, the tropism of seasonal human influenza viruses in the guinea pig is similar to that typically observed in human hosts, as well as that of ferrets.

Interesting exceptions are seen with influenza A/Anhui/1/2013 (H7N9) virus, an early isolate from the 2013 outbreak in China, and with the highly pathogenic A/duck/Guangxi/35/2001 (H5N1) [DK/35] virus. When inoculated at high dose intranasally (10^6 PFU or EID50), these viruses initiated productive infection in the lung of guinea pigs (Gao et al. 2009; Gabbard et al. 2013). Under the same conditions, Pan/99 and A/rhea/North Carolina/39482/1993 (H7N1) viruses did not become established in the lung, despite successful delivery of the intranasal inoculum to this tissue (Gabbard et al. 2013). Interestingly, disruption of α2,3 receptor binding in the DK/35 virus background also abrogated the growth of this virus in the guinea pig lung, suggesting that receptor usage is a major determinant of tropism for the guinea pig lower respiratory tract (Gao et al. 2009). Similarly, while human adapted strains remain localized in the upper respiratory tract of macaques, infections with highly pathogenic H5N1 and 2013 H7N9 viruses spread to the lung in these animals (Shinya et al. 2012; Watanabe et al. 2013). In humans, highly pathogenic H5N1 and A/Anhui/1/2013-like H7N9 viruses have caused lower respiratory tract complications such as pneumonia and acute respiratory distress syndrome (ARDS) in a high proportion of the identified cases (Hien et al. 2004; Gao et al. 2013). Thus, tropism of the avian-like H5N1 and H7N9 viruses is broadened not only in guinea pigs, but also in humans and nonhuman primates.

Collection of nasal lavage samples at multiple time points after infection is normally used to track influenza viral replication in guinea pigs. Compared to tissue

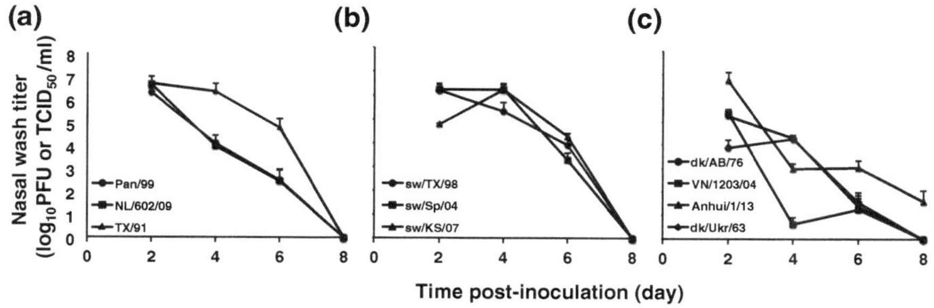

Fig. 1 Shedding patterns of influenza viruses adapted to human, swine and avian hosts in the guinea pig model. Average viral titers detected in nasal washings collected on days 2, 4, 6, and 8 post-infection are plotted. Groups of four guinea pigs were inoculated with the indicated virus strains at a dose of 10^3 (human and avian isolates) or 10^4 (swine isolates) PFU. *Error bars* indicate standard deviation. Results for A/Panama/2007/1999 (H3N2), A/Netherlands/602/2009 (H1N1) and A/Texas/36/1991 (H1N1) are shown in (**a**); results for A/swine/Texas/4199-2/1998 (H3N2), A/swine/Spain/53207/2004 (H1N1) and A/swine/Kansas/77778/2007 (H1N1) are shown in (**b**); and results for A/duck/Alberta/35/1976 (H1N1), A/duck/Ukraine/1963 (H3N8), A/Anhui/1/2013 (H7N9) and A/Viet Nam/1203/2004 (H5N1) are shown in (**c**). All titers are in units of PFU/ml, except for those of A/Anhui/1/2013, which are in TCID50/ml

collection typically used in mice, this method has the advantages of allowing serial sampling from the same animal and reducing animal numbers. Depending on the viral strain and inoculum dose, viral growth in the guinea pig upper respiratory tract peaks between days 2 and 4 after infection, and the infection is usually cleared by day 8 (Fig. 1). At the peak of shedding, titers obtained in guinea pig nasal washings are in the range of 10^6–10^7 PFU/ml for human seasonal viruses (Lowen et al. 2006; Mubareka et al. 2009; Steel et al. 2011; Pica et al. 2012) and approximately 10^5–10^7 PFU/ml for swine influenza viruses (Steel et al. 2010, 2011; Sun et al. 2010). Avian strains tested to date grew to maximum titers of approximately 10^4–10^5 PFU/ml, with the notable exception of A/Anhui/1/2013 (H7N9) virus, which yielded 10^6–10^7 $TCID_{50}$/ml on day 2 post-infection (Steel et al. 2009; Gabbard et al. 2013; Zhang et al. 2013b). These trends in kinetics and peak titers are broadly comparable to those seen in ferrets (Maines et al. 2005, 2006; Belser et al. 2011; Barman et al. 2012) and the growth of human seasonal strains in guinea pigs is similar to that seen in experimentally infected humans (Carrat et al. 2008).

2.3 *Receptor Distribution in the Guinea Pig Respiratory Tract*

Influenza viruses attach to their host cells via surface exposed glycoproteins or glycolipids in which sialic acids are linked by an α2,6 or α2,3 bond to galactose. In general, influenza viruses adapted to avian species bind preferentially to α2,3 linked glycans, and strains adapted to humans bind mainly α2,6 linked receptors.

The distribution of α2,6 and α2,3 sialylated glycans within host tissues is thought to be a major determinant of host susceptibility and viral tropism, and is therefore an important feature of animal models for influenza. Receptor distribution in the guinea pig respiratory tract has been examined by two methods: (i) staining of fixed tissue sections with lectins specific for α2,6 or α2,3 sialylated glycans (Gao et al. 2009; Sun et al. 2010) and (ii) binding to and detection of whole influenza virus on fixed tissue sections (termed 'virus histochemistry') (Gabbard et al. 2013; Siegers et al. In Press).

Staining with the lectins MAA II for α2,3 and SNA for α2,6 sialylated glycans revealed a mixture of both sugar types in the guinea pig nasal respiratory epithelia and trachea, while the lung was decorated mainly with α2,3 sialylated glycans (Gao et al. 2009; Sun et al. 2010). By comparison, similar staining of tissues derived from the human respiratory tract indicated that human upper airway epithelia showed mainly α2,6 linked sialic acid with α2,3 forms present at low levels, while trachea and bronchi showed predominantly α2,6 sialic acids and both α2,6 and α2,3 sialylated glycans were prevalent in the lower lung (Shinya et al. 2006; Nicholls et al. 2007).

By virus histochemistry, the H3N2 subtype human influenza viruses studied (Pan/99 and A/Netherlands/213/03) attached mainly to the guinea pig upper respiratory tract and the trachea, with little to no binding detected on bronchiolar and alveolar epithelia. A 2009 pandemic isolate, A/Netherlands/602/2009 (H1N1), showed high levels of binding to nasal concha as well as low and moderate positivity in the bronchioles and alveoli, respectively. The low pathogenic avian strains examined, including two 2013 H7N9 isolates, bound throughout the respiratory tract of the guinea pig. In contrast, two highly pathogenic avian-like viruses, A/VN/1194/04 (H5N1) and A/NL/219/03 (H7N7), did not show attachment in the upper respiratory tract but bound efficiently to tracheal and lung epithelia. Thus, overall, influenza virus receptors in the guinea pig appear to support attachment of human-adapted viruses mainly to the upper respiratory tract, and avian-adapted viruses either throughout the respiratory tract or to mainly lower respiratory tract tissues (Gabbard et al. 2013; Siegers et al. In Press). There is appreciable overlap between these attachment patterns and those seen in human samples. When human tissues are used as the substrate, avian influenza viruses tend to bind mainly to the lower respiratory tract, while human influenza viruses bind to both upper and lower tracts (van Riel et al. 2006, 2007, 2010, 2013; Siegers et al. In Press).

2.4 Pathology in the Guinea Pig Respiratory Tract Following Influenza Virus Infection

Overt signs of disease are not readily apparent in influenza virus infected guinea pigs (Steel et al. 2009; Van Hoeven et al. 2009a). Nevertheless, examination of respiratory tissues from infected guinea pigs has revealed significant histopathological changes. Inspection of the upper airways of infected animals, the main site of viral replication, revealed rhinitis characterized by heavy nasal mucus secretion, mild to severe intraepithelial and lamina proprial inflammation, and mild to severe vacuolation of epithelia (Tang and Chong 2009; Gabbard et al. 2013). When high doses ($\sim 10^6$ PFU or TCID50) of virus were used, marked histopathology in the lower respiratory tract of infected guinea pigs is also observed. Mild to severe broncho-interstitial pneumonia characterized by the infiltration of immune cells and stripping of ciliated epithelia have been reported with a number of different influenza strains (Azoulay-Dupuis et al. 1984; Kwon et al. 2009; Tang and Chong 2009; Van Hoeven et al. 2009a). As expected based on results from other animal models, the severity of lung pathology was greater for the 1918 pandemic strain and a highly pathogenic H5N1 subtype virus than either a low pathogenic avian isolate or a seasonal human influenza virus (Van Hoeven et al. 2009a).

2.5 *Outbred Versus Inbred Guinea Pigs*

Almost exclusively, the animals used for influenza virus research since 2006 have been outbred Hartley strain guinea pigs, which are readily available from laboratory animal vendors. Viral growth and signs of disease following infection with seasonal influenza viruses have also been evaluated in inbred strain 2 and strain 13 guinea pigs, but no consistent differences to the Hartley strain were noted (Mubareka et al. 2009 and A. Lowen and P. Palese unpublished data).

3 Influenza Virus Transmission in the Guinea Pig

To study the transmissibility of influenza viruses in animal models, two different models are used: contact and respiratory droplet. In contact models, an inoculated animal is placed in the same cage with a naïve recipient animal. In this setting transmission can proceed through direct or indirect contact with respiratory secretions, or by the short-range spread of respiratory droplets. In respiratory droplet models, inoculated and naïve animals are placed in separate cages and exposure is achieved by placing the cages in proximity to one another, such that air exchange can occur between them. Thus, in a respiratory droplet model, influenza viruses can transmit only via droplets traveling through the air. Due to the practicalities of performing transmission experiments, most study designs do not distinguish between the spread of small droplet aerosols and larger droplets that settle out of the air more rapidly.

The set up of both contact and respiratory droplet models varies among laboratories. For example, the ratio between inoculated and contact animals used is often 1:1, but may be 1:2 or higher. An important variable for respiratory droplet experiments is the rate and directionality of airflow between cages. While the precise relationship between air exchange rates and transmission levels has not been defined, anecdotal evidence suggests this parameter may account for differing results with a particular strain of influenza virus. Cage size and permeability, ambient humidity, and temperature (discussed further below) and inoculum dose also frequently differ among research groups and may impact transmission outcomes.

3.1 Transmission of Human Influenza Viruses in the Guinea Pig

As indicated in Table 1, influenza viruses adapted to humans generally transmit well among guinea pigs. Although there is strain-to-strain variability, the following generalizations can be made. Seasonal H3N2 viruses and 2009 H1N1 pandemic strains show similar and high efficiency of transmission in both contact and respiratory droplet models (Lowen et al. 2006; Steel et al. 2009, 2010, 2011). Influenza B viruses show high levels of transmission in a contact model and intermediate transmissibility in a respiratory droplet model (Pica et al. 2012). Viruses derived from the seasonal H1N1 lineage that circulated prior to 2009 tend to show lower transmissibility in guinea pigs (Mubareka et al. 2009; Bouvier et al. 2012).

For a given strain of influenza virus, transmission in a contact model occurs more rapidly than transmission in a respiratory droplet model (Lowen et al. 2006; Steel et al. 2010; Pica et al. 2012). In addition, the rate and efficiency of transmission among guinea pigs placed in separate cages have been found to decline as the cages are moved further apart (Lowen et al. 2006; Mubareka et al. 2009). These observations most likely reflect the dilution of infectious bio-aerosols with distance from the shedding host.

3.2 Transmission of Avian and Swine Influenza Viruses in the Guinea Pig

Most low pathogenic avian influenza viruses tested to date have not transmitted in the guinea pig model (Table 1). Exceptions include the human isolates, A/Anhui/1/2013 and A/Shanghai/1/2013 (H7N9), which are representatives of the 2013 outbreak in China (Gabbard et al. 2013; Hai et al. 2013) and two H9N2 subtype strains isolated from chickens in Shandong (Lv et al. 2012). In addition, certain highly pathogenic avian influenza viruses of the H5N1 subtype have shown efficient transmission among co-caged guinea pigs (Gao et al. 2009; Steel et al. 2009).

Results obtained in the guinea pig model with influenza viruses adapted to swine hosts vary among strains tested. A diverse set of Chinese isolates, representing classical, European avian-like, North American triple reassortant, and human-like swine lineages showed no transmission in a contact model (Sun et al.

Table 1 Transmission efficiency of influenza A and B viruses in the guinea pig model

Virus strain	Proportion of contacts infected[a]	Reference(s)
A/Panama/2007/1999 (H3N2)	29/29 (C), 29/32 (RD)	(Lowen et al. 2006, 2007, 2008, 2009; Bouvier et al. 2008; Mubareka et al. 2009; Steel et al. 2009, 2010)
A/Jianxi/262/2005 (H3N2)	2/3 (C)	(Sun et al. 2010)
A/Wisconsin/67/2005 (H3N2)	5/8 (C)	(Lowen et al. 2009)
A/Puerto Rico/8/1934 (H1N1)	0/13 (C)	(Ince et al. 2013; Seladi-Schulman et al. 2013)
A/Texas/36/1991 (H1N1)	1/4 (RD)	(Mubareka et al. 2009)
A/New York/1253/2008 (H1N1)	12/16 (C), 2/8 (RD)	(Bouvier et al. 2012)
A/New York/1326/2008 (H1N1)	4/4 (C), 3/8 (RD)	(Bouvier et al. 2012)
A/Guandong/41/2006 (H1N1)	1/3 (C)	(Sun et al. 2010)
A/Beijing/317/2009 (H1N1)	3/3 (C)	(Sun et al. 2010)
A/California/04/2009 (H1N1)	11/11 (C), 20/20 (RD)	(Seibert et al. 2010; Steel et al. 2010; Ince et al. 2013)
A/Hansa Hamburg/01/2009 (H1N1)	8/8 (C)	(Seibert et al. 2010; Kaminski et al. 2013)
A/Korea/1/2009 (H1N1)	4/6 (C)	(Kim et al. 2013)
A/Netherlands/602/2009 (H1N1)	4/4 (C), 8/8 (RD)	(Chutinimitkul et al. 2010; Steel et al. 2010; Seladi-Schulman et al. 2013)
A/Sichuan/1/2009 (H1N1)	8/8 (RD)	(Zhang et al. 2012, 2013b)
B/Victoria/2/1987	4/4 (C), 2/4 (RD)	(Pica et al. 2012)
B/Florida/4/2006	2/4 (RD)	(Pica et al. 2012)
A/Anhui/1/2013 (H7N9)	4/4 (C)	(Gabbard et al. 2013)
A/Shanghai/1/2013 (H7N9)	1/4 (RD)	(Hai et al. 2013)
A/Viet Nam/1203/2004 (H5N1)	3/4 (C), 0/4 (RD)	(Steel et al. 2009, JS & AL unpublished)
A/duck/Fujian/17/2001 (H5N1)	0/3 (C)	(Gao et al. 2009)
A/duck/Guangxi/22/2001 (H5N1)	0/3 (C)	(Gao et al. 2009)
A/duck/Guangxi/35/2001 (H5N1)	3/3 (C)	(Gao et al. 2009)
A/duck/Shanghai/13/2001 (H5N1)	0/3 (C)	(Gao et al. 2009)
A/duck/Guangdong/22/2002 (H5N1)	0/3 (C)	(Gao et al. 2009)

(continued)

Table 1 (continued)

Virus strain	Proportion of contacts infected[a]	Reference(s)
A/bar-headed goose/Qinghai/3/2005 (H5N1)	3/3 (C)	(Gao et al. 2009)
A/quail/Hong Kong/G1/1997 (H9N2)	0/3 (C)	(Sun et al. 2010)
A/chicken/Shandong/ZB/2007 (H9N2)	0/3 (C)	(Sun et al. 2010)
A/chicken/Hebei/LC/2008 (H9N2)	0/3 (C)	(Sun et al. 2010)
A/chicken/Shandong/A/2009 (H9N2)	1/3 (C), 0/3 (RD)	(Lv et al. 2012)
A/chicken/Shandong/M/2009 (H9N2)	3/3 (C), 0/3 (RD)	(Lv et al. 2012)
A/duck/Ukraine/1/1963 (H3N8)	0/4 (C)	(Gabbard et al. 2013)
A/duck/Alberta/35/1976 (H1N1)	0/4 (C)	(Gabbard et al. 2013)
A/rhea/North Carolina/39482/1994 (H7N1)	0/4 (C)	(Gabbard et al. 2013)
A/swine/Texas/4199-2/1998 (H3N2)	1/4 (RD)	(Steel et al. 2010)
A/swine/Guangdong/7/2006 (H3N2)	0/3 (C)	(Sun et al. 2010)
A/swine/Guangdong/211/2006 (H3N2)	0/3 (C)	(Sun et al. 2010)
A/swine/Guangdong/811/2006 (H3N2)	0/3 (C)	(Sun et al. 2010)
A/swine/Guangdong/968/2006 (H3N2)	0/3 (C)	(Sun et al. 2010)
A/swine/Guangdong/1222/2006 (H1N2)	0/3 (C)	(Sun et al. 2010)
A/swine/Guangdong/33/2006 (H1N1)	0/3 (C)	(Sun et al. 2010)
A/swine/Fujian/204/2007 (H1N1)	0/3 (C)	(Sun et al. 2010)

Data include those obtained under regulated conditions of 20 °C and 20 % relative humidity, and under standard animal room conditions. RD = respiratory droplet. C = contact

[a] As determined by virus detection in nasal washings

2010). We have observed inefficient respiratory droplet transmission of the TRIG lineage A/swine/Texas/4199-2/1998 (H3N2) virus (Steel et al. 2010) and limited transmission by a contact route of the TRIG A/swine/Kansas/77778/2007 (H1N1) virus and the European avian-like A/swine/Spain/53207/2004 (H1N1) virus (J. Steel, unpublished data).

4 Dependence of Influenza Virus Transmission on Environmental Factors

The guinea pig model was used to give much needed experimental insight into the underlying causes of influenza seasonality. Namely, we used the model to test the effects of two environmental conditions that vary with the seasons, relative humidity, and temperature, on influenza virus transmission. Infected and exposed guinea pigs were housed in environmental test chambers for the duration of the exposure period. Transmission efficiency was determined at temperatures of 5, 20, and 30 °C and relative humidities of 20, 35, 50, 65, and 80 %. Using several diverse influenza viruses, including Pan/99, A/Netherlands/602/2009 (H1N1), A/New York/08-1253/2008 (H1N1), B/Victoria/2/1987 and B/Florida/4/2006, transmission was found to be markedly more efficient at 5 °C than at 20 °C (Lowen et al. 2007; Steel et al. 2011; Bouvier et al. 2012; Pica et al. 2012). Continuing this trend, transmission by respiratory droplet was highly inefficient when temperatures were increased to 30 °C (Lowen et al. 2007; Steel et al. 2011). Both high (80 %) and intermediate (50 %) humidities were found to reduce transmission efficiency in a respiratory droplet model (Lowen et al. 2007; Steel et al. 2011), in line with earlier data on the stability of influenza viruses in an aerosol (Schaffer et al. 1976). Although some effect of humidity and temperature have been observed when using a contact transmission model, the effects are dampened compared to those on respiratory droplet transmission, most likely due to the higher overall efficiency of transmission among co-caged animals (Lowen et al. 2008; Pica et al. 2012). Overall, examinations of influenza virus transmission under varying conditions of humidity and temperature strongly suggest that the seasonality of influenza is caused by improved transmission under cold, dry conditions found in the winter-time in temperature regions of the world.

5 Identification of Viral Determinants of Transmission in the Guinea Pig Model

When coupled with reverse genetics systems that enable the targeted mutagenesis of influenza virus genomes, the guinea pig model allows the identification of viral traits important for transmission. The fact that diverse human-adapted influenza viruses transmit well among guinea pigs, while most avian and swine adapted strains do not, supports the likelihood that viral factors found to be important for guinea pig-to-guinea pig transmission will also be important for human-to-human transmission.

5.1 Determinants in the PB2 Protein

Certain amino acids located in the surface-exposed "627 domain" of PB2 have been found to be highly important for optimal replication in mammalian species (Subbarao et al. 1993; Gabriel et al. 2005). In particular, the amino acid changes E627K, D701N and G590S/Q591R, have each been implicated in the adaptation to humans of influenza viruses derived from avian and swine reservoirs (de Jong et al. 2006; Mehle and Doudna 2009; de Wit et al. 2010). The guinea pig model was used to test the roles of 627K and 701N in supporting transmission among mammals (Gao et al. 2009; Steel et al. 2009). K627E and K627E / D701N mutant viruses were generated in the backgrounds of the highly pathogenic A/Viet Nam/1203/2004 (H5N1) virus and the human seasonal Pan/99 (H3N2) virus (Steel et al. 2009). The H5N1 mutants were then compared to the wild-type virus in a contact transmission model, while the H3N2 strains were tested in a respiratory droplet model. In both viral backgrounds, the presence of either human-like adaptation (PB2 627K or 701N) led to greater transmission efficiency over the avian-like amino acid sequence. Similarly, in the context of the highly pathogenic A/duck/Guangxi/35/2001 (H5N1) virus, which is transmissible among co-housed guinea pigs, the presence of the wild-type asparagine at PB2 701 was required for transmission (Gao et al. 2009). This work identified PB2 627K and 701N as determinants of, not only growth and pathogenicity, but also transmission in mammals. Analogous results were also obtained in the ferret model, using the 1918 H1N1 pandemic strain (Van Hoeven et al. 2009b). The observation that these residues contribute to the transmission of H3N2, H1N1, and H5N1 influenza viruses suggests that adaptations in the PB2 627 domain may be a prerequisite for mammalian transmission common to all influenza A subtypes.

Outside of the 627 domain, the position PB2 271 was also shown to affect transmission. Specifically, the mutation A271T in the background of a 2009 pandemic virus abolished transmission by a respiratory droplet route. Both the A and T polymorphisms are prevalent in swine influenza viruses, but 271A is found mainly in the triple reassortant swine lineage and is rare in the classical and Eurasian avian-like swine viruses (Zhang et al. 2012).

5.2 Determinants Within the M Segment

Evidence that the influenza virus M segment encodes determinants of transmission was first obtained through the study of the swine-origin 2009 pandemic strain.

Although human infection with swine influenza viruses occurs occasionally when individuals are in close contact with pigs, these swine-adapted strains do not typically transmit onward from human-to-human. In contrast, the 2009 pandemic strain originated in the swine reservoir, crossed the species barrier to humans and spread rapidly through the human population, affecting an estimated 1–2 billion people world-wide (Van Kerkhove et al. 2013). These events raised the question, what viral traits differentiating the 2009 pandemic strain from other swine influenza viruses allowed its transmission among humans? One hypothesis related to the highly unusual genotype of this virus: six segments were derived from the North American triple reassortant swine lineage and two segments, NA and M, from the Eurasian swine lineage (Smith et al. 2009).

Studies in the guinea pig model have revealed an important role for the pandemic M segment in supporting the transmissibility of the 2009 virus (Chou et al. 2011). When the M segment from the pandemic isolate, A/California/4/2009 (H1N1) [Cal/09], was included in the background of the laboratory adapted and non-transmissible A/Puerto Rico/8/1934 (H1N1) [PR8] virus, the resultant 7 + 1 reassortant virus transmitted to five of eight respiratory droplet contacts. Similarly, in the background of a North American triple reassortant swine influenza virus, A/swine/Texas/1998 (H3N2), inclusion of the HA, NA, and M segments of Cal/09 resulted in six of eight contacts becoming infected. Inclusion of the Cal/09 HA and NA alone in the A/swine/Texas/1998 background resulted in transmission to only two of eight contact animals. The importance of the Eurasian origin M segment to the highly transmissible phenotype of the 2009 pandemic virus is also supported by data obtained in ferret (Lakdawala et al. 2011) and pig (Ma et al. 2012) models.

A prominent role for the M segment in determining viral fitness and transmissibility in guinea pigs is further supported by two recent efforts to adapt PR8 virus to guinea pigs through serial passage (Ince et al. 2013; Seladi-Schulman et al. 2013). Three independent lineages of guinea pig adapted PR8 virus, generated in two different laboratories, carried coding changes in the M1 open reading frame. All of the viruses passaged in guinea pigs exhibited improved growth and transmission among co-caged animals. M1 mutations identified in the first study, at positions 62 and 166, were confirmed to contribute to these adapted phenotypes (Ince et al. 2013), but individual M1 changes identified in the second study were not sufficient to support improved growth or transmission (Seladi-Schulman et al. 2013).

5.3 2009 Pandemic PA and NS Segments Confer Transmissibility to a Highly Pathogenic H5N1 Virus

To gauge the risk of human-transmissible H5 subtype viruses arising through reassortment between a highly pathogenic avian H5N1 influenza virus and strains of the 2009 pandemic lineage, Zhang et al. tested the transmissibility of a broad set of 21 reassortants in the guinea pig model (Zhang et al. 2013b). The viruses were generated by reverse genetics using the isolates A/duck/Guangxi/35/2001 (H5N1) [DK/35] and A/Sichuan/1/2009 (H1N1)[SC/09] and all gene combinations tested included the H5 HA. The wild-type DK/35 H5N1 virus transmitted efficiently among guinea pigs placed in the same cage, but not by a respiratory droplet route. Like other 2009 pandemic strains, SC/09 transmitted with high efficiency by respiratory droplet. The report by Zhang et al. is rich in data, but the most striking results were the highly efficient transmission of DK/35-based reassortant viruses carrying only the PA or NS gene segments of SC/09. It should be noted that the DK/35 virus H5 HA protein shows partial binding to 2,6 linked sialic acids and that this ability to bind human-type receptors was required for its transmission among co-caged guinea pigs (Gao et al. 2009). Thus, when combined with a semi-adapted HA protein in a highly pathogenic H5N1 virus background, the PA or NS gene segments of the 2009 pandemic virus are sufficient to support transmission among guinea pigs in the absence of direct or indirect contact (Zhang et al. 2013b).

5.4 Determinants in the HA Protein

The receptor binding specificity of the HA protein has been identified as an important determinant of transmission efficiency in a number of contexts. In guinea pigs, robust binding to alpha 2,6 linked sialic acids was shown to be required for the respiratory droplet transmission of the pandemic strain A/Sichaun/1/2009 (H1N1) (Zhang et al. 2012) and the contact transmission of A/duck/Guangxi/35/2001 (H5N1) (Gao et al. 2009).

Transmission and overall fitness were also affected by a natural polymorphism found at position 147 of human-adapted, H1 subtype hemagglutinins (Kim et al. 2013). K147 was shown to stabilize the interaction of the 2009 H1 hemagglutinin with sialic acid (Xu et al. 2012). The function of this site within the HA was found to be particularly important to maintaining fitness upon the introduction of glycosylation sites that were acquired by H1N1 viruses circulating in humans from 1918 to 2009, but have not yet been acquired by viruses of the 2009 pandemic lineage. Thus, the presence of K147 may allow the evolution of increased glycosylation as a means of immune escape, without the acquisition of fitness defects (Kim et al. 2013).

6 Impact of Antiviral Drug Treatment and Resistance on Influenza Virus Transmission

Although timely vaccination is clearly the best way to confer relatively long-lasting protection against influenza-associated morbidity and mortality on an individual or population level, subtype-specific vaccines are generally not immediately available in a pandemic that follows an antigenic shift. In this case, antiviral drugs, used either prophylactically or therapeutically, would be of great value if they could prevent or reduce transmission among humans while a specific vaccine is in preparation.

Very little research has been performed to date on the efficacy of antiviral drugs to prevent influenza virus infection in or transmission among guinea pigs. Because influenza virus-infected guinea pigs do not display obvious clinical signs such as fever or cough, demonstrating a decrease in symptoms or signs is not a practicable endpoint in this model. Antiviral efficacy in this species can, however, be assessed by measuring reductions in nasopharyngeal viral replication in or transmission among susceptible guinea pigs.

6.1 Treatment with Oseltamivir

In an unpublished experiment presented in Fig. 2, we investigated the impact of an NA inhibitor, oseltamivir, on the respiratory droplet transmission of influenza virus among guinea pigs. Two groups of guinea pigs were inoculated with Pan/99 virus (10^3 PFU per animal) and treated with 10 doses of oral oseltamivir in 1.5 % sucrose-water (75 mg/kg/day in two divided doses, administered 12 h apart) or a sucrose-water placebo, starting 4 h before inoculation and continuing for 5 days. The placebo-treated guinea pigs demonstrated normal nasopharyngeal virus shedding kinetics, and 3 of 4 animals transmitted Pan/99 by respiratory droplets to a naive partner animal (Fig. 2). However, among the Pan/99-inoculated, oseltamivir-treated guinea pigs, only one animal displayed a typical virus shedding pattern, while two guinea pigs had no detectable virus in any nasal washes, and five guinea pigs demonstrated protracted viral replication kinetics, with low virus titers ($<10^3$ PFU/ml) on day 2 post-inoculation and peak shedding delayed by 2–6 days. Importantly, none of the eight naïve guinea pigs exposed to these Pan/99-inoculated, oseltamivir-treated guinea pigs became infected by respiratory droplet transmission (N.M. Bouvier and M. Michta, unpublished data). These experiments demonstrate that NA enzymatic function is required for maximally efficient influenza virus transmission among guinea pigs and suggest that treatment of influenza patients with NA inhibitors may prevent or reduce transmission among unvaccinated humans, though clinical trials would be required to confirm this hypothesis.

6.2 Treatment with Type I Interferon

Daily intranasal treatment with recombinant human type I interferon (Horisberger and de Staritzky 1987) was also highly effective in limiting the growth of influenza virus in inoculated, treated guinea pigs (Van Hoeven et al. 2009a; Steel et al. 2010)

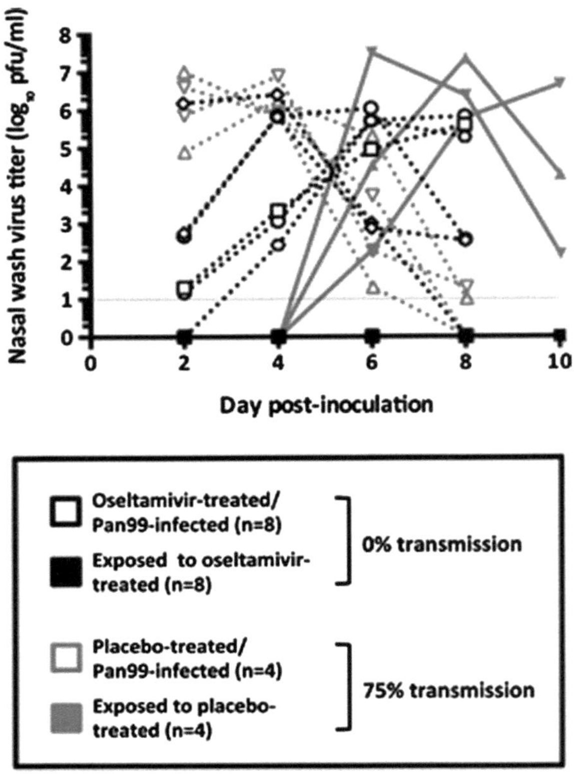

Fig. 2 Oseltamivir alters the kinetics of influenza virus shedding and prevents respiratory droplet transmission in guinea pigs. *Lines* represent the Pan/99 virus titer in each individual guinea pig nasal wash, plotted as a function of day post-inoculation. Eight Pan/99-inoculated, oseltamivir-treated guinea pigs failed to transmit virus to a naïve partner animal, while 3 of the 4 naïve guinea pigs paired with Pan/99-inoculated, placebo-treated animals became infected by respiratory droplet transmission. *Black lines* and *symbols* represent the oseltamivir treatment group, and *grey lines* and *symbols* represent the placebo treatment group. *Dotted lines* and *open symbols* represent intranasally inoculated guinea pigs, and *solid lines* and *closed symbols* represent naïve guinea pigs exposed, starting 24 h post-inoculation, to the respiratory droplets exhaled by inoculated animals. (Unpublished data, N.M. Bouvier and M. Michta)

and in preventing transmission from these animals to untreated contacts in the same cage (Steel et al. 2010). Furthermore, treatment of naïve contact guinea pigs was effective in preventing their infection via transmission from untreated, infected cage mates (Steel et al. 2010). Similarly reduced viral titers were observed in interferon-treated ferrets challenged with seasonal influenza viruses (Kugel et al. 2009). Thus, activation of the innate immune response through interferon treatment is an effective means of limiting influenza virus transmission, both from and to interferon treated animals.

6.3 Transmission Potential of Oseltamivir Resistant Influenza Viruses

Early in the preclinical development of the NA inhibitor class of antiviral medications, it was observed that mutations in the viral NA could confer relative resistance to the inhibitory action of oseltamivir and, to a lesser degree, zanamivir. Type and subtype-specific point mutations—most commonly, in N2 numbering, H274Y or N294S in the N1 subtype of influenza A viruses; E119V or R292K in the N2 subtype; and R152K or D198N in influenza B viruses (Govorkova 2012)—were observed in cell culture with viruses passaged in the presence of sub-inhibitory NA inhibitor concentrations (Gubareva et al. 1996, 1997; McKimm-Breschkin et al. 1998; Tai et al. 1998; Barnett et al. 1999; Baz et al. 2007), and in the clinic, in influenza patients treated with oseltamivir (Gubareva et al. 1998, 2001; Kiso et al. 2004; de Jong et al. 2005).

Between 2002 and 2005, some research was performed in the ferret contact transmission model to elucidate the impact of oseltamivir resistance mutations on the fitness of several influenza A viruses isolated in the prior decade. The H274Y NA mutation in an A/New Caledonia/20/1999 (H1N1)-like virus was observed to compromise its infectivity and ferret transmissibility, relative to its wild-type parental isolate (Herlocher et al. 2004). Among A/(H3N2) viruses, both A/Wuhan/359/1995-like and A/Sydney/05/1997-like viruses with the NA-R292K mutation transmitted very poorly between donor and recipient ferrets (Herlocher et al. 2002; Yen et al. 2005); however, Wuhan/95-like viruses encoding the NA-E119V substitution transmitted as efficiently between ferrets as the oseltamivir-sensitive parental virus (Herlocher et al. 2004; Yen et al. 2005).

However, the NA-E119V oseltamivir-resistance mutation is rarely seen in human isolates, except those obtained from oseltamivir-treated patients. Thus, it was hypothesized that this mutation may confer a slight fitness deficit sufficient to prevent widespread human-to-human transmission of viruses encoding it, but subtle enough to have been imperceptible in the ferret contact transmission experiments performed previously. Reverse genetics-derived Pan/99 viruses with and without the NA-E119V mutation were found to transmit with equal efficiency in the guinea pig contact transmission model, as had been seen with similar isolates in the ferret model (Herlocher et al. 2004; Yen et al. 2005). However, in the guinea pig respiratory droplet transmission model, wild-type Pan/99 again transmitted efficiently, to 7 of 8 exposed guinea pigs, while Pan/99 encoding the NA-E119V oseltamivir-resistance mutation transmitted to only 2 of 8 exposed guinea pigs ($p < 0.05$, by Fisher's exact test) (Bouvier et al. 2008). Thus, the results previously observed in the ferret contact transmission model were successfully replicated in the guinea pig model; however, the guinea pig respiratory droplet transmission experiments suggested that a subtle fitness penalty was indeed associated with acquisition of the E119V oseltamivir-resistance mutation.

Prior to 2008, the vast majority of in vitro and in vivo data suggested that oseltamivir resistance mutations in the influenza virus NA came at some cost to viral fitness (Carr et al. 2002; Herlocher et al. 2002, 2004; Ives et al. 2002; Yen et al. 2005; Zurcher et al. 2006; Bouvier et al. 2008), leading many to conclude that most oseltamivir-resistant viruses were "of limited clinical significance" (Tai et al. 1998) or "unlikely to be of clinical consequence in man" (Ives et al. 2002). However, coincident with the circulation of a new antigenic drift variant characterized by A/Brisbane/59/2007 (H1N1) (Bris/59), in 2008 a sudden increase in the prevalence of oseltamivir resistance among seasonal influenza A(H1N1) (sH1N1) virus isolates was noted. Within 5 months, 25 % of European sH1N1 isolates encoded the NA-H274Y oseltamivir-resistance mutation; by 2009, it was found in 96 % of sH1N1 isolates worldwide (World Health Organization 2009).

In the ferret contact transmission model, oseltamivir-resistant (NA-H274Y) and -sensitive (NA-H274) viruses were shown to replicate and transmit equally well (Abed et al. 2011). However, enhanced transmission efficiency among humans was one of only a few hypotheses that could account for the unprecedented, exponential increase in prevalence of the NA-H274Y mutation in sH1N1 viruses (Chao et al. 2012). With paired oseltamivir-sensitive and -resistant Bris/59-like clinical isolates from the New York State Department of Health, we demonstrated that the oseltamivir-resistant isolate transmitted more efficiently among guinea pigs than its oseltamivir-sensitive counterpart. Using reverse genetics, we rescued reassortants between the oseltamivir-sensitive (A/New York/08-1253/2008, "NY/1253") and -resistant (A/New York/08-1326/2008; "NY/1326") isolates and found that expression of oseltamivir-resistant NY/1326 NA, in a virus backbone comprising the other seven segments from the oseltamivir-sensitive isolate NY/1253, was sufficient to significantly enhance transmission efficiency among guinea pigs ($p = 0.009$ by the logrank test). Because NY/1253 encodes two non-consensus residues (S336N and M430L) in its NA gene, we also reassorted it with the Bris/59 NA to create a virus that differs only by only two amino acids (NA-H274Y and D354G) from the 7:1 reassortant encoding the NY/1326 NA. This reassortant also transmitted less efficiently than the one encoding the 1326 NA, though it was narrowly nonsignificant ($p = 0.078$) (Fig. 3) (Bouvier et al. 2012). These data support the hypothesis that the exponential increase in oseltamivir-resistant Bris/59-like sH1N1 influenza viruses resulted from enhanced human-to-human transmissibility conferred primarily by the oseltamivir-resistant NA. These experiments also highlight an advantage of the guinea pig model; namely, lower cost relative to the ferret model, which enables the use of adequately powered experimental groups. A subtle but statistically significant enhancement in transmission efficiency could be demonstrated with experimental groups containing eight guinea pig pairs, whereas prior experiments using groups of four ferret pairs were insufficiently powered to reveal a significant difference in transmissibility between oseltamivir-sensitive and -resistant Bris/59-like viruses (Abed et al. 2011).

The guinea pig model has recently been employed to investigate the mammalian transmissibility of avian-origin influenza A(H7N9) viruses (Gabbard et al. 2013; Hai et al. 2013), which were first isolated from humans in the spring of

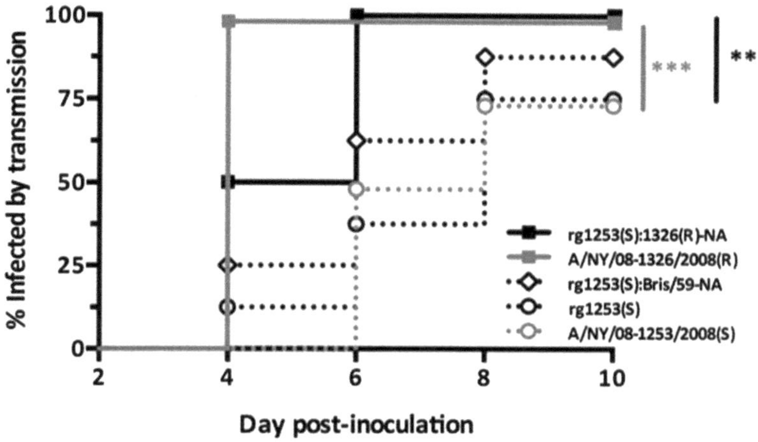

Fig. 3 Guinea pig transmission of A/Brisbane/59/2007-like seasonal influenza A(H1N1) viruses is enhanced by expression of an oseltamivir-resistant NA. A time-to-event (Kaplan-Meier) analysis of data reported previously (Bouvier et al. 2012) demonstrates that a 7:1 reassortant encoding the oseltamivir-resistant NA typical of Brisbane/59-like sH1N1 viruses, expressed in the context of the remaining seven segments from an oseltamivir-sensitive isolate, transmits significantly more rapidly than does the oseltamivir-sensitive isolate itself, similar to the transmission kinetics of the wild-type clinical isolates themselves. *Black lines* represent reverse genetics-derived viruses, and *grey lines* represent wild-type clinical isolates. *Solid lines* represent viruses with an oseltamivir-resistant NA, and *dotted lines* represent viruses with an oseltamivir-sensitive NA. **, $p < 0.01$; ***, $p < 0.001$

2013. These novel viruses have been shown to transmit efficiently by contact in both ferrets and guinea pigs (Belser et al. 2013; Gabbard et al. 2013; Zhu et al. 2013) but generally less efficiently by respiratory droplets in both species (Belser et al. 2013; Gabbard et al. 2013; Hai et al. 2013; Richard et al. 2013; Watanabe et al. 2013; Xu et al. 2013; Zhang et al. 2013a; Zhu et al. 2013). In contrast to previous ferret transmission experiments demonstrating inefficient transmission of H3N2 viruses encoding the NA-R292K oseltamivir-resistance mutation (Herlocher et al. 2002; Yen et al. 2005), a reverse genetics-derived clone of the oseltamivir-resistant clinical isolate A/Shanghai/1/2013 (SH/1), encoding the NA-R292K mutation, transmitted no less efficiently than a 7:1 reassortant that expressed the oseltamivir-sensitive NA-R292 from A/Anhui/1/2013 (AH/1) in a SH/1 backbone. Because PR8 reassortants encoding the NAs from both SH/1 and AH/1 demonstrated roughly equivalent decrements in NA enzyme velocity and substrate affinity and in hemagglutination activity, it was hypothesized that reduced HA expression accompanying the oseltamivir-resistant NA may offset the decrements in NA enzymatic function conferred by the NA-R292K mutation (Hai et al. 2013). These experiments underscore the multigenic nature of fitness in influenza A viruses.

7 Impact of Vaccination on Influenza Virus Transmission

7.1 *Potential for Vaccination to Limit Transmission*

Blocking transmission through vaccination has the potential to control influenza in all age groups, thereby limiting the burden of disease. Nevertheless, vaccines are not typically evaluated in terms of the efficiency of transmission to or from vaccinated individuals. To demonstrate the value of this approach, we tested immunization through natural infection, intramuscular administration of killed influenza virus, or intranasal infection with a live attenuated influenza virus (Lowen et al. 2009). Immunized guinea pigs were challenged either through intranasal inoculation or through exposure to acutely infected guinea pigs. In addition, naïve animals were housed with vaccinated and challenged animals to assess the reduction in transmission efficiency achieved through vaccination. Immunity acquired through natural infection was found to fully protect from challenge with homologous and heterologous strains and, therefore, block any onward transmission. The live attenuated vaccine tested, encoding a truncated NS1 protein, also fully blocked transmission from vaccinated animals, but allowed partial transmission of the heterologous challenge virus to vaccinated guinea pigs. The killed, whole virus vaccine did not provide sterilizing protection, even against homologous challenge, by either intranasal or contact exposure routes. The killed vaccine, however, reduced but did not block onward transmission from vaccinated guinea pigs to naïve cagemates (Lowen et al. 2009).

A similar study design in the ferret model recently generated comparable results (Houser et al. 2013). Vaccination of ferrets with the 2010/2011 trivalent inactivated vaccine reduced transmission only marginally. The homologous seasonal H3N2 virus and a heterologous H3N2v swine-like virus transmitted from vaccinated and challenged ferrets to the majority of naïve contacts. In contrast, immunization through infection with H3N2 seasonal strains prevented transmission of the heterologous H3N2v challenge virus to all but one of nine contact ferrets (Houser et al. 2013).

Thus, the data obtained in both the guinea pig and ferret models suggest that the public health impact of influenza vaccination could be improved through the optimization of vaccines to more efficiently block transmission.

7.2 *Role of Secreted IgA in Mediating Protection from Transmission*

Expanding upon the work of Lowen and colleagues, (Lowen et al. 2009), Seibert and colleagues (Seibert et al. 2013) investigated the impact of immunoglobulin isotypes on transmission efficiency. With 30D1, a neutralizing mouse monoclonal IgG_{2b} antibody directed at the globular head of the HA of the 2009 pandemic

H1N1 (pH1N1) virus A/California/04/2009 (Cal/09), they observed that passive immunization by intramuscular administration of this antibody (at 10 mg/kg) did not protect naïve guinea pigs from infection by transmission of Cal/09 from inoculated partner animals, despite achieving high serum antibody titers after immunization. In contrast, a single intranasal (IN) administration of 30D1 (900 ng per animal) in naïve guinea pigs, which were subsequently exposed to Cal/09-inoculated partners, was sufficient to completely abrogate infection by transmission, suggesting that prevention of infection by respiratory droplets is mediated by the presence of neutralizing antibody at the respiratory mucosa, not in serum. To confirm this finding with a physiologically relevant isotype, the variable region of 30D1 was cloned into the murine IgA heavy chain gene, and expressed with murine κ and J chains. Guinea pigs immunized intramuscularly with this 30D1 IgA construct at 1 mg/kg were not protected from infection with Cal/09 by transmission, but 7 of 8 guinea pigs immunized with 5 mg/kg of 30D1 IgA were protected. The 30D1 IgA antibody was detectible by ELISA in the nasal washes of guinea pigs immunized with 5 mg/kg of 30D1 IgA, but not in those given the lower dose. Collectively, these experiments indicate that mucosal immunity, particularly the expression of sufficient quantities of neutralizing antibodies at the mucosal surfaces of the respiratory tract, is more important than serum IgG in preventing transmission of influenza viruses by respiratory droplets. These results, together with those of Lowen, Steel et al. (Lowen et al. 2009) and Houser et al. (Houser et al. 2013), suggest that inactivated influenza vaccines, which stimulate a primarily IgG antibody response, may not optimally protect against transmission of influenza viruses, and thus the clinical efficacy of influenza vaccines intended to block human-to-human transmission may be more accurately assessed with a correlate of protection other than serum IgG titers.

8 Conclusions and Perspectives

Research carried out in the guinea pig model over the past eight years has demonstrated the utility of this system for the study of influenza virus transmission. The robust growth and efficient transmission in guinea pigs of influenza viruses adapted to human hosts, compared to the poor growth and lack of transmission of most avian adapted strains, strongly supports the relevance of the model to transmission in the human population. While the absence of measurable signs of disease limits the use of the guinea pig for pathogenesis studies, analysis of tissues using histological methods allows the virulence of an infection to be gauged. Further characterization of the pathology induced by influenza viruses in the guinea pig is warranted. As more research groups adopt the guinea pig as a model system for influenza research, a need has arisen for more data on the differences and similarities between ferret and guinea pig models. A direct comparison, in which influenza virus growth, transmission, and disease are evaluated under standardized conditions in both species, would be highly valuable to the ultimate goal of understanding influenza in humans. Finally, despite significant recent progress, much remains to be learned about the factors driving influenza virus transmission. The reasons for the acute dependence of transmission on humidity and temperature are not yet clear. Similarly, mechanisms underlying the contributions of the M1 and/or M2 proteins and polymerase components to transmission phenotypes remain largely unknown. Ultimately, a comprehensive model of the viral traits required to support influenza virus transmission in avian versus mammalian species is needed. The complex relationships among viral components and between viral and host factors have made, and will continue to make, this aim difficult to attain. If achieved, however, an in-depth understanding of the requirements for sustained influenza virus transmission in both avian and mammalian reservoirs would be invaluable to public health efforts aimed at controlling influenza.

Acknowledgments Research in the authors' laboratories is supported by the NIH under R01 AI099000 (to AL) and the Center for Excellence in Influenza Research and Surveillance (CEIRS) contract number HHSN272201400004C (to JS and AL).

References

Abed Y, Pizzorno A, Bouhy X, Boivin G (2011) Role of permissive neuraminidase mutations in influenza A/Brisbane/59/2007-like (H1N1) viruses. PLoS Pathog 7:e1002431. doi:10.1371/journal.ppat.1002431

Alford RH, Kasel JA, Gerone PJ, Knight V (1966) Human influenza resulting from aerosol inhalation. Proc Soc Exp Biol Med 122:800–804

Azoulay-Dupuis E, Lambre CR, Soler P, Moreau J, Thibon M (1984) Lung alterations in guinea-pigs infected with influenza virus. J Comp Pathol 94:273–283. doi:10.1016/0021-9975(84)90046-X

Barman S, Krylov PS, Fabrizio TP et al (2012) Pathogenicity and transmissibility of North American triple reassortant swine influenza A viruses in ferrets. PLoS Pathog 8:e1002791. doi:10.1371/journal.ppat.1002791, PPATHOGENS-D-12-00034

Barnett JM, Cadman A, Burrell FM, Madar SH, Lewis AP, Tisdale M, Bethell R (1999) In vitro selection and characterisation of influenza B/Beijing/1/87 isolates with altered susceptibility to zanamivir. Virology 265:286–295. doi:10.1006/viro.1999.0058

Baz M, Abed Y, Boivin G (2007) Characterization of drug-resistant recombinant influenza A/H1N1 viruses selected in vitro with peramivir and zanamivir. Antiviral Res 74:159–162. doi:10.1016/j.antiviral.2006.10.012

Belser JA, Gustin KM, Maines TR, Blau DM, Zaki SR, Katz JM, Tumpey TM (2011) Pathogenesis and transmission of triple-reassortant swine H1N1 influenza viruses isolated before the 2009 H1N1 pandemic. J Virol 85:1563–1572. doi:10.1128/JVI.02231-10

Belser JA, Gustin KM, Pearce MB et al (2013) Pathogenesis and transmission of avian influenza A (H7N9) virus in ferrets and mice. Nature 501:556–559. doi:10.1038/nature12391

Bouvier NM, Lowen AC, Palese P (2008) Oseltamivir-resistant influenza A viruses are transmitted efficiently among guinea pigs by direct contact but not by aerosol. J Virol. 82:10052–10058. doi:10.1128/JVI.01226-08

Bouvier NM, Rahmat S, Pica N (2012) Enhanced mammalian transmissibility of seasonal influenza A/H1N1 viruses encoding an oseltamivir-resistant neuraminidase. J Virol 86:7268–7279. doi:10.1128/JVI.07242-12

Carr J, Ives J, Kelly L et al (2002) Influenza virus carrying neuraminidase with reduced sensitivity to oseltamivir carboxylate has altered properties in vitro and is compromised for infectivity and replicative ability in vivo. Antiviral Res 54:79–88
Carrat F, Vergu E, Ferguson NM, Lemaitre M, Cauchemez S, Leach S, Valleron AJ (2008) Time lines of infection and disease in human influenza: a review of volunteer challenge studies. Am J Epidemiol 167:775–785. doi:10.1093/aje/kwm375
Chao DL, Bloom JD, Kochin BF, Antia R, Longini IM Jr (2012) The global spread of drug-resistant influenza. J R Soc Interface R Soc 9:648–656. doi:10.1098/rsif.2011.0427
Chou YY, Albrecht RA, Pica N et al (2011) The M segment of the 2009 new pandemic H1N1 influenza virus is critical for its high transmission efficiency in the guinea pig model. J Virol 85:11235–11241. doi:10.1128/JVI.05794-11
Chutinimitkul S, Herfst S, Steel J et al (2010) Virulence-associated substitution D222G in the hemagglutinin of 2009 pandemic influenza A(H1N1) virus affects receptor binding. J Virol 84:11802–11813. doi:10.1128/JVI.01136-10
de Jong MD, Simmons CP, Thanh TT et al (2006) Fatal outcome of human influenza A (H5N1) is associated with high viral load and hypercytokinemia. Nat Med 12:1203–1207. doi:10.1038/nm1477
de Jong MD, Tran TT, Truong HK et al (2005) Oseltamivir resistance during treatment of influenza A (H5N1) infection. N Engl j Med 353:2667–2672. doi:10.1056/NEJMoa054512
de Wit E, Munster VJ, van Riel D et al (2010) Molecular determinants of adaptation of highly pathogenic avian influenza H7N7 viruses to efficient replication in the human host. J Virol 84:1597–1606. doi:10.1128/JVI.01783-09
Gabbard JD, Dlugolenski D, Van Riel D et al (2013) Novel H7N9 influenza virus shows low infectious dose, high growth and efficient contact transmission in the guinea pig model. J Virol. doi:10.1128/JVI.02959-13
Gabriel G, Dauber B, Wolff T, Planz O, Klenk HD, Stech J (2005) The viral polymerase mediates adaptation of an avian influenza virus to a mammalian host. Proc Natl Acad Sci U S A 102:18590–18595. doi:10.1073/pnas.0507415102
Gao HN, Lu HZ, Cao B et al (2013) Clinical findings in 111 cases of influenza A (H7N9) virus infection. N Engl J Med 368:2277–2285. doi:10.1056/NEJMoa1305584
Gao Y, Zhang Y, Shinya K et al (2009) Identification of amino acids in HA and PB2 critical for the transmission of H5N1 avian influenza viruses in a mammalian host. PLoS Pathog 5:e1000709. doi:10.1371/journal.ppat.1000709
Govorkova EA (2012) Consequences of resistance: in vitro fitness, in vivo infectivity, and transmissibility of oseltamivir-resistant influenza A viruses. Influenza Other Respir Viruses 7:50–57
Gubareva LV, Bethell R, Hart GJ, Murti KG, Penn CR, Webster RG (1996) Characterization of mutants of influenza A virus selected with the neuraminidase inhibitor 4-guanidino-Neu5Ac2en. J Virol 70:1818–1827
Gubareva LV, Kaiser L, Matrosovich MN, Soo-Hoo Y, Hayden FG (2001) Selection of influenza virus mutants in experimentally infected volunteers treated with oseltamivir. J Infect Dis 183:523–531. doi:10.1086/318537
Gubareva LV, Matrosovich MN, Brenner MK, Bethell RC, Webster RG (1998) Evidence for zanamivir resistance in an immunocompromised child infected with influenza B virus. J Infect Dis 178:1257–1262
Gubareva LV, Robinson MJ, Bethell RC, Webster RG (1997) Catalytic and framework mutations in the neuraminidase active site of influenza viruses that are resistant to 4-guanidino-Neu5Ac2en. J Virol 71:3385–3390
Gustin KM, Belser JA, Wadford DA, Pearce MB, Katz JM, Tumpey TM, Maines TR (2011) Influenza virus aerosol exposure and analytical system for ferrets. Proc Natl Acad Sci U S A 108:8432–8437. doi:10.1073/pnas.1100768108
Hai R, Schmolke M, Leyva-Grado VH et al (2013) Influenza A(H7N9) virus gains neuraminidase inhibitor resistance without loss of in vivo virulence or transmissibility. Nature Commun 4:2854. doi:10.1038/ncomms3854

Herlocher ML, Carr J, Ives J, Elias S, Truscon R, Roberts N, Monto AS (2002) Influenza virus carrying an R292K mutation in the neuraminidase gene is not transmitted in ferrets. Antiviral Res 54:99–111

Herlocher ML, Truscon R, Elias S, Yen HL, Roberts NA, Ohmit SE, Monto AS (2004) Influenza viruses resistant to the antiviral drug oseltamivir: transmission studies in ferrets. J Infect Dis 190:1627–1630. doi:10.1086/424572

Hien TT, de Jong M, Farrar J (2004) Avian influenza–a challenge to global health care structures. N Engl J Med 351:2363–2365. doi:10.1056/NEJMp048267

Horisberger MA, de Staritzky K (1987) A recombinant human interferon-alpha B/D hybrid with a broad host-range. J Gen Virol 68(Pt 3):945–948

Houser KV, Pearce MB, Katz JM, Tumpey TM (2013) Impact of prior seasonal H3N2 influenza vaccination or infection on protection and transmission of emerging variants of influenza A(H3N2)v virus in ferrets. J Virol 87:13480–13489. doi:10.1128/JVI.02434-13

Ince WL, Gueye-Mbaye A, Bennink JR, Yewdell JW (2013) Reassortment complements spontaneous mutation in influenza A virus NP and M1 genes to accelerate adaptation to a new host. J Virol 87:4330–4338. doi:10.1128/JVI.02749-12

Ives JA, Carr JA, Mendel DB et al (2002) The H274Y mutation in the influenza A/H1N1 neuraminidase active site following oseltamivir phosphate treatment leave virus severely compromised both in vitro and in vivo. Antiviral Res 55:307–317

Kaminski MM, Ohnemus A, Staeheli P, Rubbenstroth D (2013) Pandemic 2009 H1N1 influenza A virus carrying a Q136K mutation in the neuraminidase gene is resistant to zanamivir but exhibits reduced fitness in the guinea pig transmission model. J Virol 87:1912–1915. doi:10.1128/JVI.02507-12

Kim JI, Lee I, Park S et al (2013) Genetic requirement for hemagglutinin glycosylation and its implications for influenza A H1N1 virus evolution. J Virol 87:7539–7549. doi:10.1128/JVI.00373-13

Kiso M, Mitamura K, Sakai-Tagawa Y et al (2004) Resistant influenza A viruses in children treated with oseltamivir: descriptive study. Lancet 364:759–765. doi:10.1016/S0140-6736(04)16934-1

Kugel D, Kochs G, Obojes K et al (2009) Intranasal administration of alpha interferon reduces seasonal influenza A virus morbidity in ferrets. J Virol 83:3843–3851. doi:10.1128/JVI.02453-08

Kwon YK, Lipatov AS, Swayne DE (2009) Bronchointerstitial pneumonia in guinea pigs following inoculation with H5N1 high pathogenicity avian influenza virus. Vet Pathol 46:138–141. doi:10.1354/vp.46-1-138

Lakdawala SS, Lamirande EW, Jr Suguitan AL et al (2011) Eurasian-origin gene segments contribute to the transmissibility, aerosol release, and morphology of the 2009 pandemic H1N1 influenza virus. PLoS Pathog 7:e1002443. doi: 10.1371/journal.ppat.1002443, PPATHOGENS-D-11-01769

Leyva-Grado VH, Mubareka S, Krammer F, Cardenas WB, Palese P (2012) Influenza virus infection in guinea pigs raised as livestock, Ecuador. Emerg Infect Dis 18:1135–1138. doi:10.3201/eid1807.111930

Lowen AC, Mubareka S, Steel J, Palese P (2007) Influenza virus transmission is dependent on relative humidity and temperature. PLoS Pathog 3:1470–1476. doi:07-PLPA-RA-0426, doi:10.1371/journal.ppat.0030151

Lowen AC, Mubareka S, Tumpey TM, Garcia-Sastre A, Palese P (2006) The guinea pig as a transmission model for human influenza viruses. Proc Natl Acad Sci U S A 103:9988–9992. doi:10.1073/pnas.0604157103

Lowen AC, Steel J, Mubareka S, Carnero E, Garcia-Sastre A, Palese P (2009) Blocking interhost transmission of influenza virus by vaccination in the guinea pig model. J Virol 83:2803–2818. doi:10.1128/JVI.02424-08

Lowen AC, Steel J, Mubareka S, Palese P (2008) High temperature (30 degrees C) blocks aerosol but not contact transmission of influenza virus. J Virol 82:5650–5652. doi:10.1128/JVI.00325-08

Lv J, Wei B, Yang Y et al (2012) Experimental transmission in guinea pigs of H9N2 avian influenza viruses from indoor air of chicken houses. Virus Res 170:102–108. doi:10.1016/j.virusres.2012.09.003, S0168-1702(12)00321-8

Ma W, Liu Q, Bawa B et al (2012) The neuraminidase and matrix genes of the 2009 pandemic influenza H1N1 virus cooperate functionally to facilitate efficient replication and transmissibility in pigs. J Gen Virol 93:1261–1268. doi:10.1099/vir.0.040535-0

Maines TR, Chen LM, Matsuoka Y et al (2006) Lack of transmission of H5N1 avian-human reassortant influenza viruses in a ferret model. Proc Natl Acad Sci U S A 103:12121–12126. doi:10.1073/pnas.0605134103

Maines TR, Lu XH, Erb SM et al (2005) Avian influenza (H5N1) viruses isolated from humans in Asia in 2004 exhibit increased virulence in mammals. J Virol 79:11788–11800. doi:10.1128/JVI.79.18.11788-11800.2005

McKimm-Breschkin JL, Sahasrabudhe A, Blick TJ et al (1998) Mutations in a conserved residue in the influenza virus neuraminidase active site decreases sensitivity to Neu5Ac2en-derived inhibitors. J Virol 72:2456–2462

Mehle A, Doudna JA (2009) Adaptive strategies of the influenza virus polymerase for replication in humans. Proc Natl Acad Sci U S A 106:21312–21316. doi:10.1073/pnas.0911915106

Mubareka S, Lowen AC, Steel J, Coates AL, Garcia-Sastre A, Palese P (2009) Transmission of influenza virus via aerosols and fomites in the guinea pig model. J Infect Dis 199:858–865

Nicholls JM, Bourne AJ, Chen H, Guan Y, Peiris JS (2007) Sialic acid receptor detection in the human respiratory tract: evidence for widespread distribution of potential binding sites for human and avian influenza viruses. Respir Res 8:73. doi:10.1186/1465-9921-8-73

Phair JP, Kauffman CA, Jennings R, Potter CW (1979) Influenza virus infection of the guinea pig: immune response and resistance. Med Microbiol Immunol 165:241–254

Pica N, Chou YY, Bouvier NM, Palese P (2012) Transmission of influenza B viruses in the guinea pig. J Virol 86:4279–4287. doi:10.1128/JVI.06645-11

Richard M, Schrauwen EJ, de Graaf M et al (2013) Limited airborne transmission of H7N9 influenza A virus between ferrets. Nature 501:560–563. doi:10.1038/nature12476

Schaffer FL, Soergel ME, Straube DC (1976) Survival of airborne influenza virus: effects of propagating host, relative humidity, and composition of spray fluids. Arch Virol 51:263–273

Seibert CW, Kaminski M, Philipp J et al (2010) Oseltamivir-resistant variants of the 2009 pandemic H1N1 influenza A virus are not attenuated in the guinea pig and ferret transmission models. J Virol 84:11219–11226. doi:10.1128/JVI.01424-10

Seibert CW, Rahmat S, Krause JC et al (2013) Recombinant IgA is sufficient to prevent influenza virus transmission in guinea pigs. J Virol 87:7793–7804. doi:10.1128/JVI.00979-13

Seladi-Schulman J, Steel J, Lowen AC (2013) Spherical influenza viruses have a fitness advantage in embryonated eggs, while filament-producing strains are selected in vivo. J Virol 87:13343–13353. doi:10.1128/JVI.02004-13

Shinya K, Ebina M, Yamada S, Ono M, Kasai N, Kawaoka Y (2006) Avian flu: influenza virus receptors in the human airway. Nature 440:435–436. doi:10.1038/440435a

Shinya K, Gao Y, Cilloniz C et al (2012) Integrated clinical, pathologic, virologic, and transcriptomic analysis of H5N1 influenza virus-induced viral pneumonia in the rhesus macaque. J Virol 86:6055–6066. doi:10.1128/JVI.00365-12

Siegers JY, Short KR, Leijten LME et al (In Press) Novel avian-origin influenza A (H7N9) virus attachment to the mammalian respiratory tract. J Virol

Smith GJ, Vijaykrishna D, Bahl J et al (2009) Origins and evolutionary genomics of the 2009 swine-origin H1N1 influenza A epidemic. Nature 459:1122–1125. doi:10.1038/nature08182

Steel J, Lowen AC, Mubareka S, Palese P (2009) Transmission of influenza virus in a mammalian host is increased by PB2 amino acids 627K or 627E/701N. PLoS Pathog 5:e1000252. doi:10.1371/journal.ppat.1000252

Steel J, Palese P, Lowen AC (2011) Transmission of a 2009 pandemic influenza virus shows a sensitivity to temperature and humidity similar to that of an H3N2 seasonal strain. J Virol 85:1400–1402. doi:10.1128/JVI.02186-10

Steel J, Staeheli P, Mubareka S, Garcia-Sastre A, Palese P, Lowen AC (2010) Transmission of pandemic H1N1 influenza virus and impact of prior exposure to seasonal strains or interferon treatment. J Virol 84:21–26. doi:10.1128/JVI.01732-09
Subbarao EK, London W, Murphy BR (1993) A single amino acid in the PB2 gene of influenza A virus is a determinant of host range. J Virol 67:1761–1764
Sun Y, Bi Y, Pu J et al (2010) Guinea pig model for evaluating the potential public health risk of swine and avian influenza viruses. PLoS One 5:e15537. doi:10.1371/journal.pone.0015537
Tai CY, Escarpe PA, Sidwell RW et al (1998) Characterization of human influenza virus variants selected in vitro in the presence of the neuraminidase inhibitor GS 4071. Antimicrobial agents and chemotherapy 42:3234–3241
Tang X, Chong KT (2009) Histopathology and growth kinetics of influenza viruses (H1N1 and H3N2) in the upper and lower airways of guinea pigs. J Gen Virol 90:386–391. doi:10.1099/vir.0.007054-0
Treanor JJ (2010) Influenza viruses, including avian influenza and swine influenza. In: Mandell GL, Bennett JE, Dolin R (eds) Mandell, Douglas, and Bennett's principles and practices of infectious diseases, 7th edn. Elsevier, Philadelphia
van der Laan JW, Herberts C, Lambkin-Williams R, Boyers A, Mann AJ, Oxford J (2008) Animal models in influenza vaccine testing. Expert Rev Vaccines 7:783–793. doi:10.1586/14760584.7.6.783
Van Hoeven N, Belser JA, Szretter KJ et al (2009a) Pathogenesis of 1918 pandemic and H5N1 influenza virus infections in a guinea pig model: antiviral potential of exogenous alpha interferon to reduce virus shedding. J Virol 83:2851–2861. doi:10.1128/JVI.02174-08
Van Hoeven N, Pappas C, Belser JA et al (2009b) Human HA and polymerase subunit PB2 proteins confer transmission of an avian influenza virus through the air. Proc Natl Acad Sci U S A 106:3366–3371. doi:10.1073/pnas.0813172106
Van Kerkhove MD, Hirve S, Koukounari A, Mounts AW (2013) Estimating age-specific cumulative incidence for the 2009 influenza pandemic: a meta-analysis of A(H1N1)pdm09 serological studies from 19 countries. Influenza Other Respir Viruses 7:872–886. doi:10.1111/irv.12074
van Riel D, den Bakker MA, Leijten LM et al (2010) Seasonal and pandemic human influenza viruses attach better to human upper respiratory tract epithelium than avian influenza viruses. Am J Pathol 176:1614–1618. doi: S0002-9440(10)60476-2, doi:10.2353/ajpath.2010.090949
van Riel D, Leijten LM, de Graaf M et al (2013) Novel avian-origin influenza A (H7N9) virus attaches to epithelium in both upper and lower respiratory tract of humans. Am J Pathol 183:1137–1143. doi: S0002-9440(13)00457-4, doi:10.1016/j.ajpath.2013.06.011
van Riel D, Munster VJ, de Wit E, Rimmelzwaan GF, Fouchier RA, Osterhaus AD, Kuiken T (2006) H5N1 virus attachment to lower respiratory tract. Science 312:399. doi:10.1126/science.1125548
van Riel D, Munster VJ, de Wit E, Rimmelzwaan GF, Fouchier RA, Osterhaus AD, Kuiken T (2007) Human and avian influenza viruses target different cells in the lower respiratory tract of humans and other mammals. Am J Pathol 171:1215–1223. doi: S0002-9440(10)62385-1, doi:10.2353/ajpath.2007.070248
Watanabe T, Kiso M, Fukuyama S et al (2013) Characterization of H7N9 influenza A viruses isolated from humans. Nature 501:551–555. doi:10.1038/nature12392
Wetherbee RE (1973) Induction of systemic delayed hypersensitivity during experimental viral infection of the respiratory tract with a myxovirus or paramyxovirus. J Immunol 111:157–163
World Health Organization (2009) Table 2: seasonal influenza A(H1N1) virus resistant to oseltamivir, last quarter 2008 to 31 March 2009. World Health Organization, Geneva, Switzerland
Xu L, Bao L, Deng W et al (2013) Novel avian-origin human influenza A(H7N9) can be transmitted between ferrets via respiratory droplets. J Infect Dis. doi:10.1093/infdis/jit474
Xu R, McBride R, Nycholat CM, Paulson JC, Wilson IA (2012) Structural characterization of the hemagglutinin receptor specificity from the 2009 H1N1 influenza pandemic. J Virol 86:982–990. doi:10.1128/JVI.06322-11

Yen HL, Herlocher LM, Hoffmann E, Matrosovich MN, Monto AS, Webster RG, Govorkova EA (2005) Neuraminidase inhibitor-resistant influenza viruses may differ substantially in fitness and transmissibility. Antimicrob Agents Chemother 49:4075–4084. doi:10.1128/AAC.49.10.4075-4084.2005

Zeng H, Goldsmith CS, Maines TR et al (2013) Tropism and infectivity of influenza virus, including highly pathogenic avian H5N1 virus, in ferret tracheal differentiated primary epithelial cell cultures. J Virol 87:2597–2607. doi:10.1128/JVI.02885-12

Zhang Q, Shi J, Deng G et al (2013a) H7N9 influenza viruses are transmissible in ferrets by respiratory droplet. Science 341:410–414. doi:10.1126/science.1240532

Zhang Y, Zhang Q, Gao Y et al (2012) Key molecular factors in hemagglutinin and PB2 contribute to efficient transmission of the 2009 H1N1 pandemic influenza virus. J Virol 86:9666–9674. doi:10.1128/JVI.00958-12

Zhang Y, Zhang Q, Kong H et al (2013b) H5N1 hybrid viruses bearing 2009/H1N1 virus genes transmit in guinea pigs by respiratory droplet. Science 340:1459–1463. doi:10.1126/science.1229455

Zhu H, Wang D, Kelvin DJ et al (2013) Infectivity, transmission, and pathology of human-isolated H7N9 influenza virus in ferrets and pigs. Science 341:183–186. doi:10.1126/science.1239844

Zurcher T, Yates PJ, Daly J et al (2006) Mutations conferring zanamivir resistance in human influenza virus N2 neuraminidases compromise virus fitness and are not stably maintained in vitro. J Antimicrob Chemother 58:723–732. doi:10.1093/jac/dkl321

Enhancement of Influenza Virus Transmission by Gene Reassortment

Chengjun Li and Hualan Chen

Abstract Influenza A virus is characterized by a genome composed of eight single-stranded, negative sense RNA segments, which allow for reassortment between different strains when they co-infect the same host cell. Reassortment is an important driving force for the evolution of influenza viruses. The ability of reassortment allows influenza virus to endlessly reinvent itself and pose a constant threat to the health of humans and other animals. Of the four human influenza pandemics since the beginning of the last century, three of them were caused by reassortant viruses bearing genes of avian, human or swine influenza virus origin. In the past decade, great efforts have been made to understand the transmissibility of influenza viruses. The use of reverse genetics technology has made it substantially easier to generate reassortant viruses and evaluate the contribution of individual virus gene on virus transmissibility in animal models such as ferrets and guinea pigs. H5, H7, and H9 avian influenza viruses represent the top three subtypes that are candidates to cause the next human influenza pandemic. Many studies have been conducted to determine whether the transmission of these avian influenza viruses could be enhanced by acquisition of gene segments from human influenza viruses. Moreover, the 2009 pdmH1N1 viruses and the triple reassortant swine influenza viruses were extensively studied to identify the gene segments that contribute to their transmissibility. These studies have greatly deepened our understanding of the transmissibility of reassortant influenza viruses, which, in turn, has improved our ability to be prepared for reassortant influenza virus with enhanced transmissibility and pandemic potential.

C. Li · H. Chen (✉)
State Key Laboratory of Veterinary Biotechnology, Harbin Veterinary Research Institute, Chinese Academy of Agricultural Sciences, 427 Maduan Street, Harbin 150001, Nangang, China
e-mail: chenhualan@caas.cn

C. Li
e-mail: lichengjun@caas.cn

Current Topics in Microbiology and Immunology (2014) 385: 185–204
DOI: 10.1007/82_2014_389

Published Online: 22 July 2014

Contents

1 Introduction

Influenza A virus (hereafter influenza virus) infects a wide range of avian and mammalian hosts. The genome of influenza virus is composed of eight single-stranded, negative-sense RNA segments. This unique genome arrangement allows gene exchange between two different influenza viruses when they infect the same cell, resulting in the generation of new reassortant viruses. From the last century to the present day, four human influenza pandemics have occurred. Among them, the 1918 H1N1 pandemic is believed to have been caused by a progenitor avian influenza virus that adapted in a mammalian host (Taubenberger et al. 1997). In contrast, the other three human influenza pandemics were all caused by reassortant viruses. A progenitor avian virus was the donor of the HA and PB1 genes, with or without the NA gene, for the 1957 H2N2 and the 1968 H3N2 pandemic virus. The rest of the genes of these two pandemic viruses were derived from a contemporary circulating human virus (Kawaoka et al. 1989; Lindstrom et al. 2004). The emergence of the 2009 pandemic H1N1 virus (hereafter 2009 pdmH1N1 virus) involved more complicated reassortment events with PB2 and PA genes of North American avian virus origin, a PB1 gene of human H3N2 virus origin, HA, NP, and NS genes of classical swine virus origin, and NA and M genes of Eurasian avian-like swine virus origin (Garten et al. 2009; Neumann et al. 2009). Compared with genetic adaptation of a progenitor avian virus in a mammalian host, reassortment is likely to be a more economical and efficient way to generate pandemic viruses.

The most prominent property of the pandemic and epidemic influenza viruses is their efficient and sustained transmission among humans. The 2009 pdmH1N1 viruses spread to over 215 countries and territories from April 2009 to August 2010 and caused more than 18,000 deaths worldwide (Lakdawala et al. 2011). There are three modes of influenza virus transmission: contact transmission, droplet transmission, and aerosol transmission. Droplet transmission and aerosol transmission are caused by large and small respiratory particles, respectively. These two modes of transmission are generally collectively called respiratory droplet transmission (Lakdawala et al. 2011). It is generally believed that the ability to transmit efficiently via respiratory droplet is an essential property for pandemic and epidemic influenza viruses.

The transmissibility of influenza viruses is determined by multiple virus genes. The transmission of influenza virus has been studied extensively in the ferret and guinea pig models. However, many unknowns still exist regarding the precise mechanism of influenza transmission. Therefore, one cannot rule out the contribution of any of the eight gene segments to virus transmissibility. The reassortment process has the potential to bring together a subset of eight gene segments that are compatible and create a transmissible virus. At least four elements must be met for an influenza virus to become transmissible in humans: attachment to the preferential α-2,6 sialic acid receptor, efficient replication in the human upper respiratory tract, efficient virus release into air, and stable maintenance in the environment.

Influenza pandemics have occurred sporadically. Although influenza virologists are not able to predict which subtype or strain of influenza virus will cause the next pandemic, they all agree that another pandemic will occur sometime in the future. Currently, the H5, H7, and H9 avian influenza viruses are thought to be the most likely to cause the next pandemic (Belser et al. 2008; Palese 2004). Some of the potential candidates arose and have been circulating in birds for a long period, whereas others have only recently emerged; for example, the H7N9 influenza virus in 2013 (Gao et al. 2013; Shi et al. 2013). Fortunately, these avian influenza viruses still lack the ability to transmit efficiently among humans. To determine whether they possess the potential to acquire this trait and cause a new pandemic is a pressing research question, the answer to which would allow us to be better prepared for a future pandemic and to take necessary countermeasures. On the other hand, the importance of swine to the emergence of pandemic influenza viruses is well documented, as demonstrated by the emergence of the 2009 pdmH1N1 pandemic virus. The 2009 pdmH1N1 viruses are still actively reassorting with other swine influenza viruses (Ducatez et al. 2011; Howard et al. 2011; Kitikoon et al. 2012; Liu et al. 2012; Moreno et al. 2011; Zhu et al. 2011); therefore, it is important to identify the gene segments that are responsible for the superior transmissibility of this virus. In addition, the triple reassortant swine influenza viruses are widely circulating in the world (Choi et al. 2002, 2013; Pascua et al. 2008; Richt et al. 2003; Webby et al. 2004). In this review, we summarize our understanding of how gene reassortment between different influenza viruses can result in enhanced transmission in animal models. This knowledge will deepen our understanding of how influenza pandemics arise and will support the development of effective prevention and control strategies.

2 Avian Influenza Viruses

2.1 H5N1 Avian Influenza Virus

The H5N1 avian influenza virus has spread to more than 60 countries since it was first isolated in 1996 (Chen et al. 2004; Li et al. 2014; Swayne 2012). In addition to being the primary cause of enzootic diseases in poultry and wild birds worldwide, the H5N1 virus also sporadically infects humans. To date, 664 human cases of H5N1 virus infection have been confirmed; of these, 391 cases were fatal [see the World Health Organization (WHO) website, http://www.who.int]. Thus, the severity of the disease caused by the H5N1 virus infection, coupled with the lack of immunity among humans to H5N1 viruses, has incited great fear that the H5N1 viruses will cause the next pandemic if they acquire efficient and sustained transmission among humans. Currently, the H5N1 viruses are widely circulating over a large geographical region. Several studies have demonstrated that H5N1 viruses are substantially compatible with human H3N2 and 2009 pdmH1N1 viruses in the generation of reassortant viruses (Chen et al. 2008; Li et al. 2010; Zhang et al. 2013b). Therefore, the opportunity for H5N1 viruses to reassort with a human or swine influenza virus is huge, which could result in the emergence of reassortant H5 viruses with enhanced transmissibility and pandemic potential. To be better prepared for such a scenario, the transmissibility of reassortant H5 viruses between H5N1 viruses and human or swine viruses must be carefully investigated.

The first attempt to evaluate the transmissibility of reassortant H5 viruses was reported in 2006 (Maines et al. 2006). The authors generated two reassortant H5N1 viruses with the four RNP genes (PB2, PB1, PA and NP) or six internal genes from a human H3N2 virus and the remaining four or two genes from a human H5N1 virus isolate, as well as a reassortant H3N2 virus with the HA and NA genes from the human H3N2 virus and the remaining six genes from the human H5N1 virus. They found that neither of the two reassortant H5N1 viruses could transmit to contact ferrets via respiratory droplets. After passaging the reassortant H5N1 viruses containing the four RNP genes of the human H3N2 virus five times in ferrets, transmission was still not detected in the contact ferrets. In addition, the authors found that the reassortant H3N2 virus with the six internal genes from the H5N1 virus was unable to transmit as efficiently as the wild-type H3N2 virus because no virus shedding was detected in the nasal washes of the contact ferrets, even though one of the contact ferrets seroconverted. In another study by Jackson et al., the authors co-infected ferrets to generate reassortant viruses between an H5N1 virus and a human H3N2 virus (Jackson et al. 2009). The authors chose five reassortant H5 viruses and examined their transmissibility in ferrets in a direct contact transmission model. They found that none of the reassortant H5 viruses transmitted to the contact ferrets. These studies show that the reassortant H5 viruses between H5N1 and H3N2 viruses still lack certain properties to render them transmissible in this animal model.

The pdmH1N1 virus caused the first human influenza pandemic of this century in 2009 (Dawood et al. 2009). Schrauwen et al. generated reassortant viruses between H5N1 and 2009 pdmH1N1 viruses by co-infecting the two parental viruses in MDCK cells (Schrauwen et al. 2013). They conducted their transmission study with the mixture of reassortant viruses produced in MDCK cells and found that none of the reassortant viruses in the mixture was transmitted to the exposed ferrets via the respiratory droplet route. Similarly, a study by Cline et al. created the reassortant virus CA/09-483HA, containing the HA gene of A/Hong Kong/483/97 (H5N1) and the other seven genes from A/California/4/09 (2009 pdmH1N1) virus (Cline et al. 2011). Although this reassortant H5 virus replicated as well as the parental 2009 pdmH1N1 virus, it did not transmit to contact ferrets.

Several studies have shown that certain wild-type H5N1 influenza viruses could be transmitted from inoculated animals to naïve animals via direct contact (Gao et al. 2009; Steel et al. 2009; Yen et al. 2007), and revealed that mutations, including HA T160A, PB2 E627K and PB2 D701N, are critical for the direct contact transmission of H5N1 viruses in guinea pigs (Gao et al. 2009; Steel et al. 2009). However, the first report that a reassortant H5 virus could undergo respiratory droplet transmission came in 2012. Chen et al. showed that a reassortant H5 virus with the Q196R/Q226L/G228S mutations in HA, an N2 NA from a human H3N2 virus, and the six internal genes from an H5N1 virus, could transmit among ferrets via the respiratory droplet route, as shown by detection of virus shedding in the nasal secretions of one of two exposed ferrets (Chen et al. 2012). Also in 2012, Imai et al. published a study showing that a reassortant H5 virus containing four mutations in HA (N158D/N224K/Q226L/T318I) in the background of a 2009 pdmH1N1 virus (A/California/04/2009) could transmit efficiently among ferrets by respiratory droplets (Imai et al. 2012). In these two studies, the introduction of mutations into the HA gene conferred α-2,6 sialic acid receptor binding specificity to the resultant viruses, which is believed to be a prerequisite for efficient transmission of influenza viruses in humans (Shinya et al. 2006; van Riel et al. 2006). In addition, the virus genes that originated from the human H3N2 or 2009 pdmH1N1 virus most probably played a critical role in enhancing the transmissibility of these reassortant H5 viruses.

More recently, Zhang et al. conducted a systemic study to determine whether reassortment between an authentic avian H5N1 virus and a 2009 pdmH1N1 virus could create reassortant H5 viruses that were highly transmissible via the respiratory droplet route in a guinea pig model (Zhang et al. 2013b). The authors successfully rescued all of the 127 reassortant viruses containing the HA gene from an avian H5N1 virus, A/duck/Guangxi/35/2001 (DK/35), with the other seven genes of different combinations between DK/35 and a 2009 pdmH1N1 virus, A/Sichuan/1/2009 (SC/09). They tested a large number of reassortant H5 viruses for their transmissibility in guinea pigs and found that many of the reassortant H5 viruses were transmissible via respiratory droplets (Table 1). Their findings revealed that both the PA and NS genes of SC/09 virus could make the reassortant H5 viruses highly transmissible, and that the NA and M genes of SC/09 virus also contributed to the transmission of the reassortant viruses. This study represents the most comprehensive research to date examining the transmissibility of reassortant H5 viruses, providing abundant evidence that transmissible H5 viruses can be generated by reassortment between an authentic H5N1 avian influenza virus and the highly transmissible 2009 pdmH1N1 virus.

Table 1 Replication and respiratory droplet transmission of DK/35 (H5N1) and its reassortants with SC/09 (2009 pdmH1N1) virus in inoculated or exposed guinea pigs

Virus name	Viral replication positive/total		Seroconversion positive/total (HI titers)		Respiratory droplet transmission
	Inoculated	Exposed	Inoculated	Exposed	
DK/35 (H5N1)	3/3	0/3	3/3 (40–80)	0/3	None
SC/09 (H1N1)	3/3	3/3	3/3 (320–1280)	3/3 (640)	Highly efficient
r1	3/3	0/3	3/3 (20–80)	0/3	None
r12	3/3	0/3	3/3 (20–40)	0/3	None
r123*	6/6	0/6	6/6 (40–80)	0/6	None
r125	3/3	0/3	3/3 (40–80)	0/3	None
r13*	6/6	0/6	6/6 (40–80)	0/6	None
r135*	6/6	0/6	6/6 (20–80)	0/6	None
r15	3/3	0/3	3/3 (20–40)	0/3	None
r2	3/3	0/3	3/3 (40–80)	0/3	None
r23*	6/6	0/6	6/6 (20–80)	0/6	None
r235*	6/6	0/6	6/6 (40–80)	0/6	None
r25	3/3	0/3	3/3 (20–40)	0/3	None
r3*	6/6	6/6	6/6 (80)	6/6 (20–40)	Highly efficient
r35*	6/6	6/6	6/6 (40–80)	6/6 (20–40)	Highly efficient
r5	3/3	0/3	3/3 (40–80)	0/3	None
r1235*	6/6	0/6	6/6 (20–80)	0/6	None
r1235678	3/3	2/3	3/3 (40–80)	2/3 (10–20)	Efficient
r678	3/3	3/3	3/3 (40–80)	3/3 (10–80)	Highly efficient
r6	3/3	1/3	3/3 (80)	1/3 (20)	Less efficient
r7	3/3	1/3	3/3 (40–80)	1/3 (40)	Less efficient
r8*	6/6	6/6	6/6 (20–160)	6/6 (20–80)	Highly efficient
r3678	3/3	3/3	3/3 (40–80)	3/3 (40–80)	Highly efficient

In the virus name, *r* denotes reassortant. The numbers in the virus name indicate segments derived from the SC/09(H1N1) virus as follows: *1* PB2, *2* PB1, *3* PA, *5* NP, *6* NA, *7* M and *8* NS. The virus segments derived from the DK/35(H5N1) virus were not assigned numbers. "*" in the virus name indicates that the transmission experiment was conducted twice for this virus and the combined data from both experiments are presented in the table. Three guinea pigs were inoculated intranasally (i.n.) with 10^6 EID_{50} of the test virus and housed in separate cages within an isolator. Twenty-four hours later, three naïve guinea pigs were placed in adjacent cages. Nasal washes were collected on days 2, 4, and 6 post-inoculation (p.i.) from the inoculated animals or on days 1, 3, 5, 7, and 9 post-exposure (p.e.) from the exposed animals and titrated in eggs to test for respiratory droplet transmission. Sera were collected from all animals on day 21 p.i. for hemagglutinin inhibition antibody detection. Respiratory droplet transmission was confirmed when virus was detected in the nasal washes and by seroconversion of the naïve exposed animals at the end of the 3-week observation period. Adapted from Zhang et al. (Zhang et al. 2013b)

2.2 H9N2 Avian Influenza Virus

Since its isolation was first reported in turkeys in the United States in 1966 (Homme and Easterday 1970), the H9N2 avian influenza virus has been circulating widely in the world (Alexander 2000; Li et al. 2005), and has greatly expanded its host species, including pigs (Peiris et al. 2001; Xu et al. 2004; Yu et al. 2008, 2011). H9N2 avian influenza virus also sporadically infects humans (Butt et al. 2005; Peiris et al. 1999). The infection of H9N2 virus in humans causes only mild influenza-like illness. As a result, H9N2 influenza virus could spread silently and human H9N2 infection might go unrecognized and unreported. Serological studies conducted in many countries have demonstrated that a proportion of the human population (particularly poultry farmers) is seropositive for H9N2 virus (Blair et al. 2013; Coman et al. 2013; Gray et al. 2011; Okoye et al. 2013; Uyeki et al. 2012; Wang et al. 2014; Yu et al. 2013). During circulation, many H9N2 isolates have gained the 226L mutation in their HA protein (Matrosovich et al. 2001), resulting in enhanced binding to α-2,6 sialic acid receptors, a typical feature of human influenza viruses, and increased replication in human airway epithelial cells and in ferrets (Wan and Perez 2007; Wan et al. 2008). These findings have led to the concern that H9N2 virus may have the potential to cause a pandemic by reassorting with human influenza viruses, because most of the human population remains immunologically naïve to the H9N2 influenza virus. To ascertain the potential risk, several studies have been conducted to investigate the transmissibility of reassortant H9 viruses between avian and human influenza viruses.

Wan et al. examined the transmissibility of H9N2 avian influenza viruses in ferrets and found that none of them could transmit via the aerosol route, although two were transmitted to exposed ferrets via direct contact (Wan et al. 2008). These authors also created a reassortant virus with the HA and NA genes from an H9N2 virus and the remaining six genes from a human H3N2 virus. They found that the replication and direct contact transmission of this reassortant virus was enhanced in ferrets compared with the parental H9N2 virus. However, this reassortant virus was still unable to transmit via the aerosol route. This finding demonstrated that reassortment can promote virus transmission, but that certain properties are still missing for reassortant H9N2 virus to become aerosol transmissible. Based on these findings, Sorrell et al. further adapted this reassortant H9N2 virus in ferrets (Sorrell et al. 2009). After ten serial passages, these authors obtained an avian-human H9N2 reassortant capable of transmitting via respiratory droplet. Sequence analysis revealed that two mutations, T189A in HA1 and G192R in HA2, had occurred in the HA, which was critical for the respiratory droplet transmission of the reassortant H9N2 virus. The authors concluded that minimal molecular changes are needed for the respiratory droplet transmission of an H9N2 avian-human influenza virus.

In another study, Kimble et al. created four reassortant H9 viruses between avian H9N2 and 2009 pdmH1N1 viruses (Kimble et al. 2011). All four reassortant viruses contained the six internal genes of the 2009 pdmH1N1 virus, an H9 HA gene with either the wild-type sequence or with two mutations (T189A in HA1 and G192R in

HA2), and an NA gene from either the H9N2 (with or without a I28V mutation) or the 2009 pdmH1N1 virus. The authors found that three out of the four reassortant viruses could transmit among ferrets via respiratory droplets. Interestingly, the reassortant virus with the wild-type HA and NA genes from the H9N2 virus in the background of the 2009 pdmH1N1 virus efficiently transmitted among ferrets via respiratory droplets without the need of prior adaptation in ferrets. This result contrasts sharply with that of the reassortant H9N2 virus containing the six internal genes of a human H3N2 virus, which required ten serial passages in ferrets to achieve efficient respiratory droplet transmission (Sorrell et al. 2009). Such data highlight the superiority of the 2009 pdmH1N1 virus with respect to its transmissibility and ability to enhance the transmission of avian-human reassortant viruses. As the authors suggested, their study also emphasized the importance of the balance between HA and NA for the generation of transmissible reassortant viruses, because the reassortant viruses with wild-type H9 HA and the 2009 pdmH1N1 NA did not transmit via respiratory droplets. However, when the two HA mutations (T189A in HA1 and G192R in HA2) were present in a reassortant virus with the same gene constellation, the balance between the mutant HA and 2009 pdmH1N1 NA was restored and this virus could efficiently transmit among ferrets via respiratory droplets.

2.3 H7N9 Avian Influenza Virus

On February, 2013, a new human influenza outbreak occurred in China that was caused by a previously undescribed H7N9 influenza virus (Gao et al. 2013). As of February 28, 2014, 375 cases of human infection with H7N9 virus have been reported with a lethality rate of approximately 30 % (http://www.who.int). Several studies have shown that the H7N9 viruses bind to the human-like α-2,6 sialic acid receptor, although some isolates still retain avian-like α-2,3 sialic acid receptor binding activity (Belser et al. 2013; Watanabe et al. 2013; Xiong et al. 2013; Zhang et al. 2013a; Zhou et al. 2013). Importantly, transmission studies have shown that the H7N9 viruses have acquired the ability to transmit among ferrets via respiratory droplets (Belser et al. 2013; Richard et al. 2013; Watanabe et al. 2013; Zhang et al. 2013a; Zhu et al. 2013).

The H7N9 virus is a triple reassortant, whose HA gene shares the highest homology with those of H7N3 viruses isolated in ducks in the Zhejiang province of China. Its NA gene was most likely derived from H4N9 and H11N9 viruses isolated from ducks or from H2N9 and H7N9 viruses from wild birds, and its six internal genes are closely related to those of the H9N2 avian influenza viruses in chickens and bramblings (Gao et al. 2013; Shi et al. 2013). The H7N9 influenza viruses acquired respiratory droplet transmissibility among ferrets soon after they emerged, although it is not efficient for most isolates. This is in clear contrast with other avian influenza viruses, such as H5N1 viruses, which still lack respiratory droplet transmissibility despite circulating for nearly two decades in poultry and wild birds. Although there is no direct evidence, the gene constellation of the H7N9 viruses may have contributed to its enhanced transmissibility compared with other avian influenza viruses. The H7N9 virus is difficult to eradicate because it is nonpathogenic to avian species (Zhang et al. 2013a), thus it can be maintained and spread silently in poultry. The biggest concern regarding the H7N9 virus is that it may evolve into a highly transmissible virus among humans by acquiring new mutations or by reassorting with other human influenza viruses, such as the 2009 pdmH1N1 virus. To protect the public's health, it is essential to further evaluate the pandemic potential of the H7N9 virus by performing transmission studies to identify mutations or gene constellations that could confer enhanced transmissibility in animal models.

3 Triple Reassortant Swine Influenza Virus

The H1N1 classical swine influenza virus was dominant in the swine population in the United States prior to 1998 (Vincent et al. 2008). In 1998, triple reassortant events occurred among the H1N1 classical swine influenza virus, the human H3N2 virus, and avian influenza viruses, resulting in the emergence of triple reassortants in the swine farms (Lekcharoensuk et al. 2006; Ma et al. 2006; Webby et al. 2000, 2004; Zhou et al. 1999). All of the triple reassortant viruses possessed a similar constellation of internal genes, with NP, M, and NS from classical swine virus, PB1 from the human virus, and PB2 and PA from the avian viruses. This constellation of internal genes has been referred to as the triple reassortant internal gene (TRIG) cassette (Vincent et al. 2008). The triple reassortant swine influenza viruses have been maintained as the dominant viruses in the swine population in North America since 1998 (Choi et al. 2002; Richt et al. 2003; Webby et al. 2004), and they have since spread to pigs of countries in Asia (Choi et al. 2013; Pascua et al. 2008). These reassortant viruses have caused sporadic human infection (Bastien et al. 2010; Cox et al. 2011; Olsen et al. 2006; Pearce et al. 2012; Shinde et al. 2009), suggesting that they may further reassort with other human or avian influenza viruses, leading to the emergence of a new reassortant virus with pandemic potential. Moreover, they share six genes with the 2009 pdmH1N1 virus, including PB2, PB1, PA, HA, NP and NS (Smith et al. 2009). Therefore, it is important to understand the transmissibility of the triple reassortant swine influenza viruses.

Several studies have evaluated the transmissibility of the triple reassortant swine influenza viruses in ferrets. Belser et al. reported that two triple reassortant swine viruses that were isolated from human cases transmitted inefficiently among ferrets by respiratory droplets since seroconversion was detected in only one or two of three contact ferrets (Belser et al. 2011). Barman et al. found that the transmissibility of the triple reassortant swine viruses via respiratory droplets was dependent on the origins of their HA and NA genes (Barman et al. 2012). The triple reassortant viruses with human-like HA and NA genes could transmit efficiently among ferrets, whereas viruses with swine-like HA and NA genes were transmitted inefficiently or not transmitted among ferrets; viruses with swine-like HA and human-like NA were intermediate in transmission. In a study by Pascua et al., two H3N2 triple reassortant viruses could not transmit or transmitted minimally among ferrets via respiratory droplets, while an H1N2 triple reassortant virus with swine-like HA and human-like NA transmitted efficiently after acquiring two mutations, HA 225G and NA 315N, during its replication in ferrets (Pascua et al. 2012). When these two mutations were incorporated into another nontransmissible H1N2 triple reassortant swine virus, the resultant virus could not transmit among ferrets (Pascua et al. 2013). However, this nontransmissible H1N2 virus could be rendered transmissible when the PB2, PB1 and PA genes of a transmissible H1N2 virus were introduced into the genome of the nontransmissible H1N2 virus. These studies suggested that subtle differences exist in the genome of the triple reassortant swine viruses, which may account for their different transmissibility characteristics.

The co-circulation of multiple lineages of swine influenza viruses worldwide has facilitated further reassortment between different virus lineages. Among them, the reassortment between 2009 pdmH1N1 virus and H3N2 triple reassortant swine virus has resulted in the generation of new reassortant H3N2 viruses in the United States, designated as rH3N2p, which contain one to five gene segments from the 2009 pdmH1N1 virus (Kitikoon et al. 2012). The common feature of the rH3N2p viruses is that they all acquired their M gene from the 2009 pdmH1N1 virus (Ducatez et al. 2011; Liu et al. 2012). Since July 2011, the rH3N2p virus has frequently caused human infections in the United States and the virus isolated from humans has been designated as A(H3N2)v (Kitikoon et al. 2012). To date, more than 330 human cases of A(H3N2)v virus infection have been reported (Houser et al. 2013). Pearce et al. examined the transmissibility of four human isolates of swine H3N2 influenza viruses, including three isolates (KS/09, MN/10 and PA/10) with the same gene constellation as the H3N2-TRIG virus (H3N2 triple reassortant swine virus) and one isolate (IN11) containing the M gene of the 2009 pdmH1N1 virus and the remaining genes from the H3N2-TRIG virus (Pearce et al. 2012). The authors found that all four viruses transmitted efficiently to direct contact ferrets. With the exception of KS/09 virus, which spread to only two of three respiratory droplet contact ferrets and showed a delayed transmission profile, the other three viruses, MN/10, PA/10 and IN/11, efficiently transmitted among ferrets via respiratory droplets. Although several studies have shown the role of the M gene in enhancing the respiratory droplet transmission of the 2009 pdmH1N1 virus (Chou et al. 2011; Lakdawala et al. 2011), the IN/11 virus exhibited a similar respiratory droplet transmission pattern among ferrets to that seen with the MN/10 and PA/10 viruses. Therefore, it is unclear whether the 2009 pdmH1N1 M gene in the IN/11 virus and other A(H3N2)v viruses can enhance virus spread to humans.

4 2009 pdmH1N1 Influenza Virus

The 2009 pdmH1N1 pandemic taught us that an influenza pandemic can also be caused by a reassortant virus with HA and NA genes of the same subtype as a previous pandemic or epidemic as long as the human population lacks immunity to the new virus. The 2009 pdmH1N1 virus spread to over 215 countries and territories between April 2009 and August 2010 (Lakdawala et al. 2011). Thus, it probably showed the most efficient human-to-human transmission of the four influenza pandemics to date. The 2009 pdmH1N1 virus frequently jumps back into the swine population and is continuously reassorting with other swine influenza viruses (Ducatez et al. 2011; Howard et al. 2011; Kitikoon et al. 2012; Liu et al. 2012; Moreno et al. 2011; Zhu et al. 2011). Due to its superior transmissibility, the 2009 pdmH1N1 virus could be the gene donor for the generation of a new pandemic influenza virus. Therefore, studies of the contributions of its gene segments to its superior transmissibility will provide invaluable information.

During surveillance studies in pigs in southern China, Zhu et al. isolated a reassortant virus possessing the Eurasian avian-like swine H1N1 surface genes and the six internal genes of the 2009 pdmH1N1 virus (Zhu et al. 2011). This virus was efficiently transmitted among pigs via both direct and aerosol contact. Moreover, it also was transmitted from infected pigs to ferrets via the aerosol route. This study highlights the importance of the 2009 pdmH1N1 virus in the generation of new reassortant viruses with enhanced transmissibility and pandemic potential. The role of the 2009 pdmH1N1 virus in the generation of new transmissible viruses was also confirmed by an experimental study in ferrets. Angel et al. generated two reassortant viruses, one possessing the HA and NA genes of the human H3N2 virus and the remaining genes of the 2009 pdmH1N1 virus, and the other possessing the HA and NA genes of the 2009 pdmH1N1 virus and the other six genes of a human H3N2 virus (Angel et al. 2013). Both reassortant viruses transmitted efficiently to ferrets via direct contact and respiratory droplets. The authors also co-infected ferrets with these two viruses to further select reassortant viruses after serial passages. One reassortant H1N2 virus, containing the PB1 and NA genes of the human H3N2 virus and the remaining genes from the 2009 pdmH1N1 virus, was obtained. This reassortant virus could transmit among ferrets via both direct contact and the aerosol route.

The length of the NA gene also affects the transmissibility of reassortant viruses. Blumenkrantz et al. generated two reassortant viruses with NA genes that had either a short stalk or an elongated stalk from an H5N1 virus and the remaining seven genes from the 2009 pdmH1N1 virus (Blumenkrantz et al. 2013). They found that the reassortant with a short stalk replicated less efficiently than one with an elongated stalk in its NA protein. Although both viruses could transmit among ferrets by direct contact, only the reassortant virus with the elongated stalk could transmit via respiratory droplets. The enhanced transmission of the reassortant with the elongated stalk correlated with its higher neuraminidase activity, which could help the virus penetrate the mucus layer in the respiratory tract, thereby facilitating infection and spread.

Several studies have attempted to evaluate the specific role of the viral genes of the 2009 pdmH1N1 virus in its superior transmissibility (Table 2). The NA and M segments of the 2009 pdmH1N1 virus were derived from the Eurasian avian-like swine influenza viruses. Studies have shown that the NA and M segments, or M alone, are critical for the efficient transmission of the 2009 pdmH1N1 virus in humans. Chou et al. demonstrated that the PR8 virus does not transmit via aerosol in guinea pigs, but a single reassortant PR8 virus containing the M gene of the 2009 pdmH1N1 virus transmitted to 5 of 8 guinea pigs via aerosol (Chou et al. 2011). These authors also tested the transmission-enhancing activity of the M gene in the background of an H3N2 triple reassortant virus, A/Swine/Texas/4199-2/1998 (sw/TX/98). When the HA and NA genes of sw/TX/98 were replaced with those of the 2009 pdmH1N1 virus, a transmission rate of only 25 % was observed in guinea pigs via aerosol, which is the same as that of the wild-type sw/TX/98 virus. However, when the M gene of the 2009 pdmH1N1 virus was included together with the HA and NA genes, the transmission rate for the resultant reassortant virus was 75 % in guinea pigs via aerosol. This study demonstrated that the M gene of the 2009 pdmH1N1 virus is critical for its efficient transmission. Lakdawala et al. examined the contribution of the Eurasian origin NA and M genes on the transmission of the 2009 pdmH1N1 virus in the ferret model (Lakdawala et al. 2011). They found that the 2009 pdmH1N1 virus, A/California/07/2009, transmitted to 100 % of ferrets via respiratory droplets, whereas a 6:2 reassortant virus with NA and M from a triple reassortant swine virus (A/Ohio/02/2007) and the other six genes from A/California/07/2009 virus only transmitted to 50 % of the exposed ferrets. Compared with the progenitor triple reassortant swine virus (A/Ohio/02/2007) and the Eurasian swine virus (A/Thailand/271/2005), the NA gene of the 2009 pdmH1N1 virus possessed higher neuraminidase activity. Moreover, its NA and M genes conferred the filamentous morphology to a significantly larger proportion of the 2009 pdmH1N1 virus particles and were responsible for the efficient release of large amounts of viral-RNA containing influenza particles into the air. In a similar study by Campbell et al., the contribution of the three surface genes of the 2009 pdmH1N1 virus on direct contact transmission among guinea pigs was determined (Campbell et al. 2014). These authors found that the HA, NA, and M genes of the 2009 pdmH1N1 virus were important for the enhancement of contact transmission of a reassortant PR8 virus, with the M gene playing the most important role. Interestingly, the authors found that replacement of the PR8 M gene with that of the 2009 pdmH1N1 virus increased the neuraminidase activity, and that the introduction of the M or NA gene alone or in combination could bestow the reassortant PR8 virus with filamentous morphology. Increased neuraminidase activity and a filamentous morphology are correlated with enhanced contact transmissibility of reassortant PR8 virus. Bialas et al. found that three residues in the M1 protein are critical for the efficient formation of virus-like particles and the maintenance of the spherical morphology of the 2009 pdmH1N1 virus, which is believed by the authors to be important for the efficient transmission of the 2009

Table 2 The role of viral genes from 2009 pdmH1N1 virus in its superior transmissibility

Reassortant or mutant viruses with genes of 2009 pdmH1N1 virus	Animal model	Transmissibility of the reassortant or mutant viruses	Refs
Multiple reassortants between 2009 pdmH1N1 virus (A/California/04/09, Cal/09) and PR8 virus or H3N2 triple reassortant swine virus (A/swine/Texas/1998, sw/Tx/98)	Guinea pig	PR8:Cal/09 M could transmit to 5 of 8 guinea pigs by aerosol; sw/Tx/98:Cal/09HANA only transmitted to 2 of 8 aerosol contact guinea pigs, whereas sw/Tx/98:Cal/09HANAM could transmit to 6 of 8 aerosol contact guinea pigs	Chou et al. (2011)
A 6:2 reassortant virus with NA and M from a triple reassortant swine virus (A/Ohio/02/2007) and the other six genes from a 2009 pdmH1N1 virus (A/California/07/2009)	Ferret	A/California/07/2009 transmitted to 4/4 of ferrets by respiratory droplets, whereas the 6:2 reassortant only transmitted to 2/4 of the exposed ferrets	Lakdawala et al. (2011)
Five reassortants between a 2009 pdmH1N1 virus (A/Netherlands/602/2009, NL602) and PR8 virus	Guinea pig	When guinea pigs were inoculated with 1000 PFU of reassortant viruses, PR8 NL602 M + NA + HA transmitted to all 8 direct contact guinea pigs, PR8 NL602 M and PR8 NL602 NA + M transmitted to 75 % (12/16 and 6/8, respectively) of the guinea pigs, whereas PR8 NL602 NA + HA only transmitted to 25 % (2/8) of the direct contact guinea pigs	Campbell et al. (2014)
A reassortant virus with the NA gene of a 2009 pdmH1N1 virus, A/HK/415742/09 (HK415742) and the rest of its genes from a triple reassortant swine virus, A/swine/Hong Kong/915/04 (sw915)	Ferret	Sw915 virus transmitted to 1/3 ferrets by respiratory droplets, whereas the reassortant virus with the NA of HK415742 in the background of sw915 transmitted to 3/3 ferrets	Yen et al. (2011)
Ten reassortant viruses with the genes of a 2009 pdmH1N1 virus, A/Beijing/7/2009 (BJ09), and the rest of their genes from a reassortant virus, rH1N1, containing the NA and M genes of an Eurasian avian-like swine H1N1 virus, A/swine/Fujian/204/2007 and the other 6 genes of a triple-reassortant H1N2 swine influenza virus, A/swine/Guangdong/1222/2006	Guinea pig	rH1N1 could not transmit to the direct contact guinea pigs, whereas multiple reassortant viruses with HA and/or NS genes of BJ/09 virus in the background of rH1N1 virus could transmit to the exposed guinea pigs by direct contact	Zhao et al. (2011)
A 2009 pdmH1N1 virus, A/Sichuan/1/2009 (SC/09), and three mutant SC/09 viruses with HA Q226R, PB2 A271T or both mutations	Guinea pig, ferret	SC/09 virus transmitted to 5/5 guinea pigs and 3/3 ferrets by respiratory droplets, SC/09 HA/226R and SC/09 PB2/271T transmitted to 0/5 guinea pigs and 3/3 ferrets, whereas SC/09 HA/226R + PB2/271T transmitted to 0/3 of the exposed ferrets	Zhang et al. (2012)

pdmH1N1 virus (Bialas et al. 2012). However, the spherical morphology of the 2009 pdmH1N1 virus observed in this study is in clear contrast with other studies in which the virus particles are predominantly filamentous (Campbell et al. 2014; Lakdawala et al. 2011). What underlies the discrepancy between these studies is unclear; however, they all agreed on the importance of the M gene for the transmission of the 2009 pdmH1N1 virus.

Yen et al. demonstrated that the balance between the HA and NA activity of the 2009 pdmH1N1 virus is critical for efficient virus transmission in ferrets (Yen et al. 2011). They found that a swine H1N2 influenza virus, sw915, possessing the same gene constellation as the 2009 pdmH1N1 virus except for the N2 NA gene, transmitted inefficiently via respiratory droplet. When they introduced the NA gene of the 2009 pdmH1N1 virus into the sw915 virus, the authors found that the resultant reassortant virus transmitted efficiently among ferrets.

Zhao et al. generated a reassortant virus with the same gene constellation as the 2009 pdmH1N1 virus, containing the NA and M genes of a Eurasian avian-like swine virus and the remaining six genes from a triple reassortant swine H1N2 virus (Zhao et al. 2011). In contrast with the 2009 pdmH1N1 virus, which efficiently transmitted among guinea pigs, the artificial H1N1 virus did not transmit among guinea pigs by direct contact. When the transmissibility of artificial H1N1 viruses with single gene replacements from the 2009 pdmH1N1 virus was assessed, the authors found that the HA and NS genes of the 2009 pdmH1N1 virus were important for its transmissibility in guinea pigs. They also found that replacement of both the HA and NA genes of the artificial H1N1 virus with those of the 2009 pdmH1N1 virus could largely recapture the transmissibility to that of the 2009 pdmH1N1 virus in guinea pigs.

Moreover, Zhang et al.'s study established that the HA and PB2 genes of the 2009 pdmH1N1 virus are critical for virus transmission in guinea pig and ferret models (Zhang et al. 2012). The authors found that the HA Q226R mutation of A/Sichuan/1/2009 virus, which switched the receptor-binding preference from human α-2,6 to avian α-2,3 sialic acid, resulted in a virus incapable of respiratory droplet transmission in guinea pigs. The PB2 A271T mutation also abolished the virus respiratory droplet transmission in guinea pigs. Moreover, the HA Q226R mutation and PB2 A271T mutation, when combined together, abolished the virus respiratory droplet transmission in ferrets.

5 Conclusions

Unlike viruses with non-segmented genomes, influenza viruses have the ability to exchange gene segments among different strains when they co-infect the same host cell. The continual reassortment of influenza viruses presents a constant threat to public health especially when a reassortant virus with enhanced transmissibility among humans emerges. It is well established that all of the influenza viruses that caused human pandemics were originally derived from avian influenza viruses.

Currently, the H5, H7, and H9 avian influenza viruses are considered to be the most likely to cause the next human pandemic. During their circulation, these viruses have greatly expanded their host range and geographical distribution, significantly increased their genetic diversity, and some have acquired the human-type receptor-binding specificity. Moreover, we also have to deal with the threat posed by newly emerging influenza viruses, such as the current H7N9 influenza virus. Given that most humans lack immunity to these subtypes of avian influenza viruses, public health will be at risk if these viruses acquire enhanced transmissibility by reassorting with contemporary human or swine influenza viruses and cause a new pandemic.

Over the last decade, great strides have been made in understanding the transmissibility of influenza viruses. With the aid of reverse genetics technology, reassortant influenza viruses can be easily generated and their transmissibility can be easily evaluated in animal models such as ferrets and guinea pigs. These studies have greatly deepened our understanding of the transmissibility and pandemic potential of different reassortant influenza viruses. Influenza virus has the ability to keep surprising us by giving rise to new pandemic and epidemic strains. For our part, we must strengthen our surveillance efforts to detect reassortant viruses as soon as they emerge, determine the roles of their individual genes in transmissibility, and determine whether their transmissibility could be further enhanced by reassortment with other human, swine, or avian influenza viruses.

Acknowledgments We thank S. Watson for editing the manuscript. This work was supported by the Ministry of Agriculture (CARS-42-G08) and by the Ministry of Science and Technology (2012ZX10004214).

References

Alexander DJ (2000) A review of avian influenza in different bird species. Vet Microbiol 74:3–13

Angel M, Kimble JB, Pena L, Wan H, Perez DR (2013) In vivo selection of H1N2 influenza virus reassortants in the ferret model. J Virol 87:3277–3283

Barman S, Krylov PS, Fabrizio TP, Franks J, Turner JC, Seiler P, Wang D, Rehg JE, Erickson GA, Gramer M, Webster RG, Webby RJ (2012) Pathogenicity and transmissibility of North American triple reassortant swine influenza A viruses in ferrets. PLoS Pathog 8:e1002791

Bastien N, Antonishyn NA, Brandt K, Wong CE, Chokani K, Vegh N, Horsman GB, Tyler S, Graham MR, Plummer FA, Levett PN, Li Y (2010) Human infection with a triple-reassortant swine influenza A(H1N1) virus containing the hemagglutinin and neuraminidase genes of seasonal influenza virus. J Infect Dis 201:1178–1182

Belser JA, Blixt O, Chen LM, Pappas C, Maines TR, Van Hoeven N, Donis R, Busch J, McBride R, Paulson JC, Katz JM, Tumpey TM (2008) Contemporary North American influenza H7 viruses possess human receptor specificity: implications for virus transmissibility. Proc Natl Acad Sci U S A 105:7558–7563

Belser JA, Gustin KM, Maines TR, Blau DM, Zaki SR, Katz JM, Tumpey TM (2011) Pathogenesis and transmission of triple-reassortant swine H1N1 influenza viruses isolated before the 2009 H1N1 pandemic. J Virol 85:1563–1572

Belser JA, Gustin KM, Pearce MB, Maines TR, Zeng H, Pappas C, Sun X, Carney PJ, Villanueva JM, Stevens J, Katz JM, Tumpey TM (2013) Pathogenesis and transmission of avian influenza A (H7N9) virus in ferrets and mice. Nature 501:556–559

Bialas KM, Desmet EA, Takimoto T (2012) Specific residues in the 2009 H1N1 swine-origin influenza matrix protein influence virion morphology and efficiency of viral spread in vitro. PLoS ONE 7:e50595

Blair PJ, Putnam SD, Krueger WS, Chum C, Wierzba TF, Heil GL, Yasuda CY, Williams M, Kasper MR, Friary JA, Capuano AW, Saphonn V, Peiris M, Shao H, Perez DR, Gray GC (2013) Evidence for avian H9N2 influenza virus infections among rural villagers in Cambodia. J Infect Public Health 6:69–79

Blumenkrantz D, Roberts KL, Shelton H, Lycett S, Barclay WS (2013) The short stalk length of highly pathogenic avian influenza H5N1 virus neuraminidase limits transmission of pandemic H1N1 virus in ferrets. J Virol 87:10539–10551

Butt KM, Smith GJ, Chen H, Zhang LJ, Leung YH, Xu KM, Lim W, Webster RG, Yuen KY, Peiris JS, Guan Y (2005) Human infection with an avian H9N2 influenza A virus in Hong Kong in 2003. J Clin Microbiol 43:5760–5767

Campbell PJ, Danzy S, Kyriakis CS, Deymier MJ, Lowen AC, Steel J (2014) The M segment of the 2009 pandemic influenza virus confers increased neuraminidase activity, filamentous morphology, and efficient contact transmissibility to A/Puerto Rico/8/1934-based reassortant viruses. J Virol 88:3802–3814

Chen H, Deng G, Li Z, Tian G, Li Y, Jiao P, Zhang L, Liu Z, Webster RG, Yu K (2004) The evolution of H5N1 influenza viruses in ducks in southern China. Proc Natl Acad Sci U S A 101:10452–10457

Chen LM, Blixt O, Stevens J, Lipatov AS, Davis CT, Collins BE, Cox NJ, Paulson JC, Donis RO (2012) In vitro evolution of H5N1 avian influenza virus toward human-type receptor specificity. Virology 422:105–113

Chen LM, Davis CT, Zhou H, Cox NJ, Donis RO (2008) Genetic compatibility and virulence of reassortants derived from contemporary avian H5N1 and human H3N2 influenza A viruses. PLoS Pathog 4:e1000072

Choi YK, Goyal SM, Joo HS (2002) Prevalence of swine influenza virus subtypes on swine farms in the United States. Arch Virol 147:1209–1220

Choi YK, Pascua PN, Song MS (2013) Swine influenza viruses: an Asian perspective. Curr Top Microbiol Immunol 370:147–172

Chou YY, Albrecht RA, Pica N, Lowen AC, Richt JA, Garcia-Sastre A, Palese P, Hai R (2011) The M segment of the 2009 new pandemic H1N1 influenza virus is critical for its high transmission efficiency in the guinea pig model. J Virol 85:11235–11241

Cline TD, Karlsson EA, Freiden P, Seufzer BJ, Rehg JE, Webby RJ, Schultz-Cherry S (2011) Increased pathogenicity of a reassortant 2009 pandemic H1N1 influenza virus containing an H5N1 hemagglutinin. J Virol 85:12262–12270

Coman A, Maftei DN, Krueger WS, Heil GL, Friary JA, Chereches RM, Sirlincan E, Bria P, Dragnea C, Kasler I, Gray GC (2013) Serological evidence for avian H9N2 influenza virus infections among Romanian agriculture workers. J Infect Public Health 6:438–447

Cox CM, Neises D, Garten RJ, Bryant B, Hesse RA, Anderson GA, Trevino-Garrison I, Shu B, Lindstrom S, Klimov AI, Finelli L (2011) Swine influenza virus A (H3N2) infection in human, Kansas, USA, 2009. Emerg Infect Dis 17:1143–1144

Dawood FS, Jain S, Finelli L, Shaw MW, Lindstrom S, Garten RJ, Gubareva LV, Xu X, Bridges CB, Uyeki TM (2009) Emergence of a novel swine-origin influenza A (H1N1) virus in humans. N Engl J Med 360:2605–2615

Ducatez MF, Hause B, Stigger-Rosser E, Darnell D, Corzo C, Juleen K, Simonson R, Brockwell-Staats C, Rubrum A, Wang D, Webb A, Crumpton JC, Lowe J, Gramer M, Webby RJ (2011) Multiple reassortment between pandemic (H1N1) 2009 and endemic influenza viruses in pigs, United States. Emerg Infect Dis 17:1624–1629

Gao R, Cao B, Hu Y, Feng Z, Wang D, Hu W, Chen J, Jie Z, Qiu H, Xu K, Xu X, Lu H, Zhu W, Gao Z, Xiang N, Shen Y, He Z, Gu Y, Zhang Z, Yang Y, Zhao X, Zhou L, Li X, Zou S,

Zhang Y, Yang L, Guo J, Dong J, Li Q, Dong L, Zhu Y, Bai T, Wang S, Hao P, Yang W, Han J, Yu H, Li D, Gao GF, Wu G, Wang Y, Yuan Z, Shu Y (2013) Human infection with a novel avian-origin influenza A (H7N9) virus. N Engl J Med 368:1888–1897

Gao Y, Zhang Y, Shinya K, Deng G, Jiang Y, Li Z, Guan Y, Tian G, Li Y, Shi J, Liu L, Zeng X, Bu Z, Xia X, Kawaoka Y, Chen H (2009) Identification of amino acids in HA and PB2 critical for the transmission of H5N1 avian influenza viruses in a mammalian host. PLoS Pathog 5: e1000709

Garten RJ, Davis CT, Russell CA, Shu B, Lindstrom S, Balish A, Sessions WM, Xu X, Skepner E, Deyde V, Okomo-Adhiambo M, Gubareva L, Barnes J, Smith CB, Emery SL, Hillman MJ, Rivailler P, Smagala J, de Graaf M, Burke DF, Fouchier RA, Pappas C, Alpuche-Aranda CM, Lopez-Gatell H, Olivera H, Lopez I, Myers CA, Faix D, Blair PJ, Yu C, Keene KM, Dotson PD Jr, Boxrud D, Sambol AR, Abid SH, St George K, Bannerman T, Moore AL, Stringer DJ, Blevins P, Demmler-Harrison GJ, Ginsberg M, Kriner P, Waterman S, Smole S, Guevara HF, Belongia EA, Clark PA, Beatrice ST, Donis R, Katz J, Finelli L, Bridges CB, Shaw M, Jernigan DB, Uyeki TM, Smith DJ, Klimov AI, Cox NJ (2009) Antigenic and genetic characteristics of swine-origin 2009 A(H1N1) influenza viruses circulating in humans. Science 325:197–201

Gray GC, Ferguson DD, Lowther PE, Heil GL, Friary JA (2011) A national study of US bird banders for evidence of avian influenza virus infections. J Clin Virol 51:132–135

Homme PJ, Easterday BC (1970) Avian influenza virus infections. I. Characteristics of influenza A-turkey-wisconsin-1966 virus. Avian Dis 14:66–74

Houser KV, Pearce MB, Katz JM, Tumpey TM (2013) Impact of prior seasonal H3N2 influenza vaccination or infection on protection and transmission of emerging variants of influenza A(H3N2)v virus in ferrets. J Virol 87:13480–13489

Howard WA, Essen SC, Strugnell BW, Russell C, Barass L, Reid SM, Brown IH (2011) Reassortant pandemic (H1N1) 2009 virus in pigs, United Kingdom. Emerg Infect Dis 17:1049–1052

Imai M, Watanabe T, Hatta M, Das SC, Ozawa M, Shinya K, Zhong G, Hanson A, Katsura H, Watanabe S, Li C, Kawakami E, Yamada S, Kiso M, Suzuki Y, Maher EA, Neumann G, Kawaoka Y (2012) Experimental adaptation of an influenza H5 HA confers respiratory droplet transmission to a reassortant H5 HA/H1N1 virus in ferrets. Nature 486:420–428

Jackson S, Van Hoeven N, Chen LM, Maines TR, Cox NJ, Katz JM, Donis RO (2009) Reassortment between avian H5N1 and human H3N2 influenza viruses in ferrets: a public health risk assessment. J Virol 83:8131–8140

Kawaoka Y, Krauss S, Webster RG (1989) Avian-to-human transmission of the PB1 gene of influenza a viruses in the 1957 and 1968 pandemics. J Virol 63:4603–4608

Kimble JB, Sorrell E, Shao H, Martin PL, Perez DR (2011) Compatibility of H9N2 avian influenza surface genes and 2009 pandemic H1N1 internal genes for transmission in the ferret model. Proc Natl Acad Sci U S A 108:12084–12088

Kitikoon P, Vincent AL, Gauger PC, Schlink SN, Bayles DO, Gramer MR, Darnell D, Webby RJ, Lager KM, Swenson SL, Klimov A (2012) Pathogenicity and transmission in pigs of the novel A(H3N2)V influenza virus isolated from humans and characterization of swine H3N2 viruses isolated in 2010–2011. J Virol 86:6804–6814

Lakdawala SS, Lamirande EW, Suguitan AL Jr, Wang W, Santos CP, Vogel L, Matsuoka Y, Lindsley WG, Jin H, Subbarao K (2011) Eurasian-origin gene segments contribute to the transmissibility, aerosol release, and morphology of the 2009 pandemic H1N1 influenza virus. PLoS Pathog 7:e1002443

Lekcharoensuk P, Lager KM, Vemulapalli R, Woodruff M, Vincent AL, Richt JA (2006) Novel swine influenza virus subtype H3N1, United States. Emerg Infect Dis 12:787–794

Li C, Bu Z, Chen H (2014) Avian influenza vaccines against H5N1 'bird flu'. Trends Biotechnol 32:147–156

Li C, Hatta M, Nidom CA, Muramoto Y, Watanabe S, Neumann G, Kawaoka Y (2010) Reassortment between avian H5N1 and human H3N2 influenza viruses creates hybrid viruses with substantial virulence. Proc Natl Acad Sci U S A 107:4687–4692

Li C, Yu K, Tian G, Yu D, Liu L, Jing B, Ping J, Chen H (2005) Evolution of H9N2 influenza viruses from domestic poultry in Mainland China. Virology 340:70–83
Lindstrom SE, Cox NJ, Klimov A (2004) Genetic analysis of human H2N2 and early H3N2 influenza viruses, 1957–1972: evidence for genetic divergence and multiple reassortment events. Virology 328:101–119
Liu Q, Ma J, Liu H, Qi W, Anderson J, Henry SC, Hesse RA, Richt JA, Ma W (2012) Emergence of novel reassortant H3N2 swine influenza viruses with the 2009 pandemic H1N1 genes in the United States. Arch Virol 157:555–562
Ma W, Gramer M, Rossow K, Yoon KJ (2006) Isolation and genetic characterization of new reassortant H3N1 swine influenza virus from pigs in the midwestern United States. J Virol 80:5092–5096
Maines TR, Chen LM, Matsuoka Y, Chen H, Rowe T, Ortin J, Falcon A, Nguyen TH, le Mai Q, Sedyaningsih ER, Harun S, Tumpey TM, Donis RO, Cox NJ, Subbarao K, Katz JM (2006) Lack of transmission of H5N1 avian-human reassortant influenza viruses in a ferret model. Proc Natl Acad Sci U S A 103:12121–12126
Matrosovich MN, Krauss S, Webster RG (2001) H9N2 influenza A viruses from poultry in Asia have human virus-like receptor specificity. Virology 281:156–162
Moreno A, Di Trani L, Faccini S, Vaccari G, Nigrelli D, Boniotti MB, Falcone E, Boni A, Chiapponi C, Sozzi E, Cordioli P (2011) Novel H1N2 swine influenza reassortant strain in pigs derived from the pandemic H1N1/2009 virus. Vet Microbiol 149:472–477
Neumann G, Noda T, Kawaoka Y (2009) Emergence and pandemic potential of swine-origin H1N1 influenza virus. Nature 459:931–939
Okoye J, Eze D, Krueger WS, Heil GL, Friary JA, Gray GC (2013) Serologic evidence of avian influenza virus infections among Nigerian agricultural workers. J Med Virol 85:670–676
Olsen CW, Karasin AI, Carman S, Li Y, Bastien N, Ojkic D, Alves D, Charbonneau G, Henning BM, Low DE, Burton L, Broukhanski G (2006) Triple reassortant H3N2 influenza A viruses, Canada, 2005. Emerg Infect Dis 12:1132–1135
Palese P (2004) Influenza: old and new threats. Nat Med 10:S82–S87
Pascua PN, Song MS, Kwon HI, Lim GJ, Kim EH, Park SJ, Lee OJ, Kim CJ, Webby RJ, Webster RG, Choi YK (2013) The homologous tripartite viral RNA polymerase of A/swine/Korea/CT1204/2009(H1N2) influenza virus synergistically drives efficient replication and promotes respiratory droplet transmission in ferrets. J Virol 87:10552–10562
Pascua PN, Song MS, Lee JH, Baek YH, Kwon HI, Park SJ, Choi EH, Lim GJ, Lee OJ, Kim SW, Kim CJ, Sung MH, Kim MH, Yoon SW, Govorkova EA, Webby RJ, Webster RG, Choi YK (2012) Virulence and transmissibility of H1N2 influenza virus in ferrets imply the continuing threat of triple-reassortant swine viruses. Proc Natl Acad Sci U S A 109:15900–15905
Pascua PN, Song MS, Lee JH, Choi HW, Han JH, Kim JH, Yoo GJ, Kim CJ, Choi YK (2008) Seroprevalence and genetic evolutions of swine influenza viruses under vaccination pressure in Korean swine herds. Virus Res 138:43–49
Pearce MB, Jayaraman A, Pappas C, Belser JA, Zeng H, Gustin KM, Maines TR, Sun X, Raman R, Cox NJ, Sasisekharan R, Katz JM, Tumpey TM (2012) Pathogenesis and transmission of swine origin A(H3N2)V influenza viruses in ferrets. Proc Natl Acad Sci U S A 109:3944–3949
Peiris JS, Guan Y, Markwell D, Ghose P, Webster RG, Shortridge KF (2001) Cocirculation of avian H9N2 and contemporary "human" H3N2 influenza A viruses in pigs in southeastern China: potential for genetic reassortment? J Virol 75:9679–9686
Peiris M, Yuen KY, Leung CW, Chan KH, Ip PL, Lai RW, Orr WK, Shortridge KF (1999) Human infection with influenza H9N2. Lancet 354:916–917
Richard M, Schrauwen EJ, de Graaf M, Bestebroer TM, Spronken MI, van Boheemen S, de Meulder D, Lexmond P, Linster M, Herfst S, Smith DJ, van den Brand JM, Burke DF, Kuiken T, Rimmelzwaan GF, Osterhaus AD, Fouchier RA (2013) Limited airborne transmission of H7N9 influenza A virus between ferrets. Nature 501:560–563
Richt JA, Lager KM, Janke BH, Woods RD, Webster RG, Webby RJ (2003) Pathogenic and antigenic properties of phylogenetically distinct reassortant H3N2 swine influenza viruses cocirculating in the United States. J Clin Microbiol 41:3198–3205

Schrauwen EJ, Bestebroer TM, Rimmelzwaan GF, Osterhaus AD, Fouchier RA, Herfst S (2013) Reassortment between avian H5N1 and human influenza viruses is mainly restricted to the matrix and neuraminidase gene segments. PLoS ONE 8:e59889

Shi J, Deng G, Liu P, Zhou J, Guan L, Li W, Li X, Guo J, Wang G, Fan J, Wang J, Li Y, Jiang Y, Liu L, Tian G, Li C, Chen H (2013) Isolation and characterization of H7N9 viruses from live poultry markets-implication of the source of current H7N9 infection in humans. Chin Sci Bull 58:1857–1863

Shinde V, Bridges CB, Uyeki TM, Shu B, Balish A, Xu X, Lindstrom S, Gubareva LV, Deyde V, Garten RJ, Harris M, Gerber S, Vagasky S, Smith F, Pascoe N, Martin K, Dufficy D, Ritger K, Conover C, Quinlisk P, Klimov A, Bresee JS, Finelli L (2009) Triple-reassortant swine influenza A (H1) in humans in the United States, 2005–2009. N Engl J Med 360:2616–2625

Shinya K, Ebina M, Yamada S, Ono M, Kasai N, Kawaoka Y (2006) Avian flu: Influenza virus receptors in the human airway. Nature 440:435–436

Smith GJ, Vijaykrishna D, Bahl J, Lycett SJ, Worobey M, Pybus OG, Ma SK, Cheung CL, Raghwani J, Bhatt S, Peiris JS, Guan Y, Rambaut A (2009) Origins and evolutionary genomics of the 2009 swine-origin H1N1 influenza A epidemic. Nature 459:1122–1125

Sorrell EM, Wan H, Araya Y, Song H, Perez DR (2009) Minimal molecular constraints for respiratory droplet transmission of an avian-human H9N2 influenza A virus. Proc Natl Acad Sci U S A 106:7565–7570

Steel J, Lowen AC, Mubareka S, Palese P (2009) Transmission of influenza virus in a mammalian host is increased by PB2 amino acids 627K or 627E/701N. PLoS Pathog 5:e1000252

Swayne DE (2012) Impact of vaccines and vaccination on global control of avian influenza. Avian Dis 56:818–828

Taubenberger JK, Reid AH, Krafft AE, Bijwaard KE, Fanning TG (1997) Initial genetic characterization of the 1918 "Spanish" influenza virus. Science 275:1793–1796

Uyeki TM, Nguyen DC, Rowe T, Lu X, Hu-Primmer J, Huynh LP, Hang NL, Katz JM (2012) Seroprevalence of antibodies to avian influenza A (H5) and A (H9) viruses among market poultry workers, Hanoi, Vietnam, 2001. PLoS ONE 7:e43948

van Riel D, Munster VJ, de Wit E, Rimmelzwaan GF, Fouchier RA, Osterhaus AD, Kuiken T (2006) H5N1 virus attachment to lower respiratory tract. Science 312:399

Vincent AL, Ma W, Lager KM, Janke BH, Richt JA (2008) Swine influenza viruses a North American perspective. Adv Virus Res 72:127–154

Wan H, Perez DR (2007) Amino acid 226 in the hemagglutinin of H9N2 influenza viruses determines cell tropism and replication in human airway epithelial cells. J Virol 81:5181–5191

Wan H, Sorrell EM, Song H, Hossain MJ, Ramirez-Nieto G, Monne I, Stevens J, Cattoli G, Capua I, Chen LM, Donis RO, Busch J, Paulson JC, Brockwell C, Webby R, Blanco J, Al-Natour MQ, Perez DR (2008) Replication and transmission of H9N2 influenza viruses in ferrets: evaluation of pandemic potential. PLoS ONE 3:e2923

Wang Q, Ju L, Liu P, Zhou J, Lv X, Li L, Shen H, Su H, Jiang L, Jiang Q (2014) Serological and virological surveillance of avian influenza A virus H9N2 subtype in humans and poultry in Shanghai, China, between 2008 and 2010. Zoonoses Public Health. doi: 10.1111/zph.12133

Watanabe T, Kiso M, Fukuyama S, Nakajima N, Imai M, Yamada S, Murakami S, Yamayoshi S, Iwatsuki-Horimoto K, Sakoda Y, Takashita E, McBride R, Noda T, Hatta M, Imai H, Zhao D, Kishida N, Shirakura M, de Vries RP, Shichinohe S, Okamatsu M, Tamura T, Tomita Y, Fujimoto N, Goto K, Katsura H, Kawakami E, Ishikawa I, Watanabe S, Ito M, Sakai-Tagawa Y, Sugita Y, Uraki R, Yamaji R, Eisfeld AJ, Zhong G, Fan S, Ping J, Maher EA, Hanson A, Uchida Y, Saito T, Ozawa M, Neumann G, Kida H, Odagiri T, Paulson JC, Hasegawa H, Tashiro M, Kawaoka Y (2013) Characterization of H7N9 influenza A viruses isolated from humans. Nature 501:551–555

Webby RJ, Rossow K, Erickson G, Sims Y, Webster R (2004) Multiple lineages of antigenically and genetically diverse influenza A virus co-circulate in the United States swine population. Virus Res 103:67–73

Webby RJ, Swenson SL, Krauss SL, Gerrish PJ, Goyal SM, Webster RG (2000) Evolution of swine H3N2 influenza viruses in the United States. J Virol 74:8243–8251

Xiong X, Martin SR, Haire LF, Wharton SA, Daniels RS, Bennett MS, McCauley JW, Collins PJ, Walker PA, Skehel JJ, Gamblin SJ (2013) Receptor binding by an H7N9 influenza virus from humans. Nature 499:496–499
Xu C, Fan W, Wei R, Zhao H (2004) Isolation and identification of swine influenza recombinant A/Swine/Shandong/1/2003(H9N2) virus. Microbes Infect 6:919–925
Yen HL, Liang CH, Wu CY, Forrest HL, Ferguson A, Choy KT, Jones J, Wong DD, Cheung PP, Hsu CH, Li OT, Yuen KM, Chan RW, Poon LL, Chan MC, Nicholls JM, Krauss S, Wong CH, Guan Y, Webster RG, Webby RJ, Peiris M (2011) Hemagglutinin-neuraminidase balance confers respiratory-droplet transmissibility of the pandemic H1N1 influenza virus in ferrets. Proc Natl Acad Sci U S A 108:14264–14269
Yen HL, Lipatov AS, Ilyushina NA, Govorkova EA, Franks J, Yilmaz N, Douglas A, Hay A, Krauss S, Rehg JE, Hoffmann E, Webster RG (2007) Inefficient transmission of H5N1 influenza viruses in a ferret contact model. J Virol 81:6890–6898
Yu H, Hua RH, Wei TC, Zhou YJ, Tian ZJ, Li GX, Liu TQ, Tong GZ (2008) Isolation and genetic characterization of avian origin H9N2 influenza viruses from pigs in China. Vet Microbiol 131:82–92
Yu H, Zhou YJ, Li GX, Ma JH, Yan LP, Wang B, Yang FR, Huang M, Tong GZ (2011) Genetic diversity of H9N2 influenza viruses from pigs in China: a potential threat to human health? Vet Microbiol 149:254–261
Yu Q, Liu L, Pu J, Zhao J, Sun Y, Shen G, Wei H, Zhu J, Zheng R, Xiong D, Liu X, Liu J (2013) Risk perceptions for avian influenza virus infection among poultry workers, China. Emerg Infect Dis 19:313–316
Zhang Q, Shi J, Deng G, Guo J, Zeng X, He X, Kong H, Gu C, Li X, Liu J, Wang G, Chen Y, Liu L, Liang L, Li Y, Fan J, Wang J, Li W, Guan L, Li Q, Yang H, Chen P, Jiang L, Guan Y, Xin X, Jiang Y, Tian G, Wang X, Qiao C, Li C, Bu Z, Chen H (2013a) H7N9 influenza viruses are transmissible in ferrets by respiratory droplet. Science 341:410–414
Zhang Y, Zhang Q, Gao Y, He X, Kong H, Jiang Y, Guan Y, Xia X, Shu Y, Kawaoka Y, Bu Z, Chen H (2012) Key molecular factors in hemagglutinin and PB2 contribute to efficient transmission of the 2009 H1N1 pandemic influenza virus. J Virol 86:9666–9674
Zhang Y, Zhang Q, Kong H, Jiang Y, Gao Y, Deng G, Shi J, Tian G, Liu L, Liu J, Guan Y, Bu Z, Chen H (2013b) H5N1 hybrid viruses bearing 2009/H1N1 virus genes transmit in guinea pigs by respiratory droplet. Science 340:1459–1463
Zhao X, Sun Y, Pu J, Fan L, Shi W, Hu Y, Yang J, Xu Q, Wang J, Hou D, Ma G, Liu J (2011) Characterization of an artificial swine-origin influenza virus with the same gene combination as H1N1/2009 virus: A genesis clue of pandemic strain. PLoS ONE 6:e22091
Zhou J, Wang D, Gao R, Zhao B, Song J, Qi X, Zhang Y, Shi Y, Yang L, Zhu W, Bai T, Qin K, Lan Y, Zou S, Guo J, Dong J, Dong L, Wei H, Li X, Lu J, Liu L, Zhao X, Huang W, Wen L, Bo H, Xin L, Chen Y, Xu C, Pei Y, Yang Y, Zhang X, Wang S, Feng Z, Han J, Yang W, Gao GF, Wu G, Li D, Wang Y, Shu Y (2013) Biological features of novel avian influenza A (H7N9) virus. Nature 499:500–503
Zhou NN, Senne DA, Landgraf JS, Swenson SL, Erickson G, Rossow K, Liu L, Yoon K, Krauss S, Webster RG (1999) Genetic reassortment of avian, swine, and human influenza A viruses in American pigs. J Virol 73:8851–8856
Zhu H, Wang D, Kelvin DJ, Li L, Zheng Z, Yoon SW, Wong SS, Farooqui A, Wang J, Banner D, Chen R, Zheng R, Zhou J, Zhang Y, Hong W, Dong W, Cai Q, Roehrl MH, Huang SS, Kelvin AA, Yao T, Zhou B, Chen X, Leung GM, Poon LL, Webster RG, Webby RJ, Peiris JS, Guan Y, Shu Y (2013) Infectivity, transmission, and pathology of human-isolated H7N9 influenza virus in ferrets and pigs. Science 341:183–186
Zhu H, Zhou B, Fan X, Lam TT, Wang J, Chen A, Chen X, Chen H, Webster RG, Webby R, Peiris JS, Smith DK, Guan Y (2011) Novel reassortment of Eurasian avian-like and pandemic/2009 influenza viruses in swine: Infectious potential for humans. J Virol 85:10432–10439

Swine and Influenza: A Challenge to One Health Research

Robert E. Kahn, Wenjun Ma and Jürgen A. Richt

Abstract The challenge of increasing swine production and a rising number of novel and known swine influenza viruses has prompted a considerable boost in research into how and why pigs have become such significant hosts for influenza viruses. The ecology of influenza A viruses is rather complicated, involving multiple host species and a segmented genome. Wild aquatic birds are the reservoir for the majority of influenza A viruses, but novel influenza viruses were recently identified in bats. Occasionally, influenza A viruses can be transmitted to mammals from avian species and this event could lead to the generation of human pandemic strains. Swine are thought to be "mixing vessels" because they are susceptible to infection with both avian and mammalian influenza viruses; and novel influenza viruses can be generated in pigs by reassortment. At present, it is difficult to predict which viruses might cause a human pandemic. Therefore, both human and veterinary research needs to give more attention to the potential cross-species transmission capacity of influenza A viruses.

Contents

R.E. Kahn · W. Ma · J.A. Richt (✉)
College of Veterinary Medicine, Diagnostic Medicine/Pathobiology,
Kansas State University, K-224B Mosier Hall, Manhattan,
KS 66506-5601, USA
e-mail: jricht@vet.k-state.edu

Current Topics in Microbiology and Immunology (2014) 385: 205–218
DOI: 10.1007/82_2014_392

Published Online: 10 July 2014

1 Introduction

The extensive expansion of pig, poultry, and cattle/milk production since 1980 has rightly been called "the second livestock revolution", following the first revolution in the nineteenth century as Europe became urbanized (Coker et al. 2011; Delgado et al. 1999). Since the last century, in which three major influenza outbreaks occurred, both large-scale intensive production and small-scale backyard farming practices for pigs and poultry are now continually subject to the threat of influenza. Therefore, understanding influenza pathogenesis and transmission in numerous species (including humans) and implementing effective control measures has become essential for global health and food security; yet achieving the prevention and/or mitigation of influenza outbreaks and human pandemics is still difficult.

The ecology of influenza A viruses is complicated, involving multiple host species and a segmented viral genome with eight segments. Influenza viruses use two mechanisms for molecular evolution: antigenic shift and antigenic drift. Antigenic shift is the process by which two or more different influenza viruses infect one cell or one host and their segmented genomes mix. This mechanism is also called reassortment, supports rapid evolution, and can create new influenza subtypes. Antigenic drift is the natural mutation process that occurs over time due to the lack of viral RNA polymerase proofreading activity, and may allow evasion of preexisting host immunity, or vaccine mismatch. The majority of influenza A viruses likely originate from avian hosts (Krauss et al. 2007); however, nucleic acids obtained recently from bat samples indicate that bats may be a reservoir of novel bat influenza viruses (Tong et al. 2012, 2013). The influenza taxonomy has been based on the antigenic properties of the two surface glycoproteins, hemagglutinin (HA), and neuraminidase (NA). There are 18 HA and 11 NA subtypes (Wu et al. 2014) that have been found naturally paired in many combinations: H1N1, H3N2, H5N1, H7N9, etc. Occasionally, influenza A viruses are transmitted from birds to mammals, an event which can lead to the development of human pandemic strains by direct or indirect transmission to humans (Schrauwen and Fouchier 2014).

2 The Interaction Between Influenza and Its Receptors: The Mixing Vessel Hypothesis

Attachment of influenza A virus to a host cell is mediated through a receptor binding site (RBS) on the HA and a specific molecular species of sialic acid (SA) linked to galactose on the host cell's surface (Suzuki 2005). Two major linkages between SA and the penultimate galactose residues of carbohydrate side chains are found in nature: *N*-acetylneuraminic acid-α2,3-galactose (NeuAc α(2,3)-Gal), and NeuAc α(2,6)-Gal. Different HAs have different recognition specificities for these linkages and the distribution of specific SAs expressed on cell surfaces varies among animal species. Most influenza A viruses isolated from humans preferentially recognize NeuAcα(2,6)Gal, whereas most avian isolates preferentially recognize NeuAcα(2,3)Gal (Rogers and Paulson 1983; Rogers and D'Souza 1989). Swine influenza viruses (SIVs) bind equally well to both NeuAcα(2,6)Gal and NeuAcα(2,3)Gal or with a slight predominance toward NeuAcα(2,6)Gal (Ito et al. 1997; Suzuki et al. 1997). The predominant receptors are NeuAcα(2,6)Gal on epithelial cells in the human respiratory tract (Baum and Paulson 1990) and NeuAcα(2,3)Gal on intestinal epithelial cells of ducks (Suzuki 2005). These results led to the hypothesis that the type of linkage of the SA molecules to galactose contributes to the host range of influenza viruses. Surprisingly, both "human" (α2,6) and "avian" (α2,3) receptors have been detected in the pig trachea (Ito et al. 1998; Suzuki et al. 2000), in the respiratory tract of humans (Shinya et al. 2006), in the trachea and intestine of quail (Wan and Perez 2006) and other minor poultry species (Kimble et al. 2010), indicating that both, avian- and mammalian-like influenza viruses can infect these hosts.

In 1985, Scholtissek and colleagues proposed that human influenza viruses might not be able to cross the species barrier to birds directly, and avian influenza viruses might not infect humans directly without prior reassortment in pigs. This established the notion that swine might be a "mixing vessel" for the generation of pandemic influenza viruses (Scholtissek et al. 1985).

The "mixing vessel" theory explains how simultaneous infection of swine with avian-like and mammalian-like viruses can result in the creation of novel reassortant viruses (Fig. 1a; Ito et al. 1998; Brown 2000, 2013; Ma et al. 2009). Such human–avian virus reassortants were responsible for the 1957 and 1968 pandemics. Furthermore, it has been experimentally shown that swine can be infected with avian influenza viruses of various HA subtypes, from H4 to H13, and reassortant viruses can be generated in swine co-infected with both avian- and mammalian-like viruses (Kida et al. 1994). It is likely that other animal species (including humans) expressing both receptor types are able to support replication of both avian and mammalian influenza viruses and generate reassortant viruses. A critical component, however, for efficient replication of avian influenza viruses in various mammalian species is their adaptation to the mammalian receptor repertoire. It was shown that with continued replication, some avian-like swine viruses can acquire the ability to recognize human virus receptors (Fig. 1b; Ito et al. 1998). This molecular adaptation helps explain the emergence of pandemic influenza viruses. An example of reassortment between avian and mammalian influenza viruses and adaptation to the mammalian receptor repertoire occurred in 2006 when H2N3 influenza viruses were isolated from two pig farms in the state of Missouri, USA (Ma et al. 2007). These H2N3 viruses were of public health concern because of the known pandemic potential of H2 viruses as shown during the 1957 pandemic. Sequencing of the two viruses showed that these H2N3 viruses were reassortants comprising avian virus-like HA, NA, and PA genes with the remaining viral segments derived from the endemic triple reassortant SIVs; this is consistent with the mixing vessel theory. In addition, the HA of the H2N3 viruses was already partially adapted to the mammalian receptor repertoire, specifically by

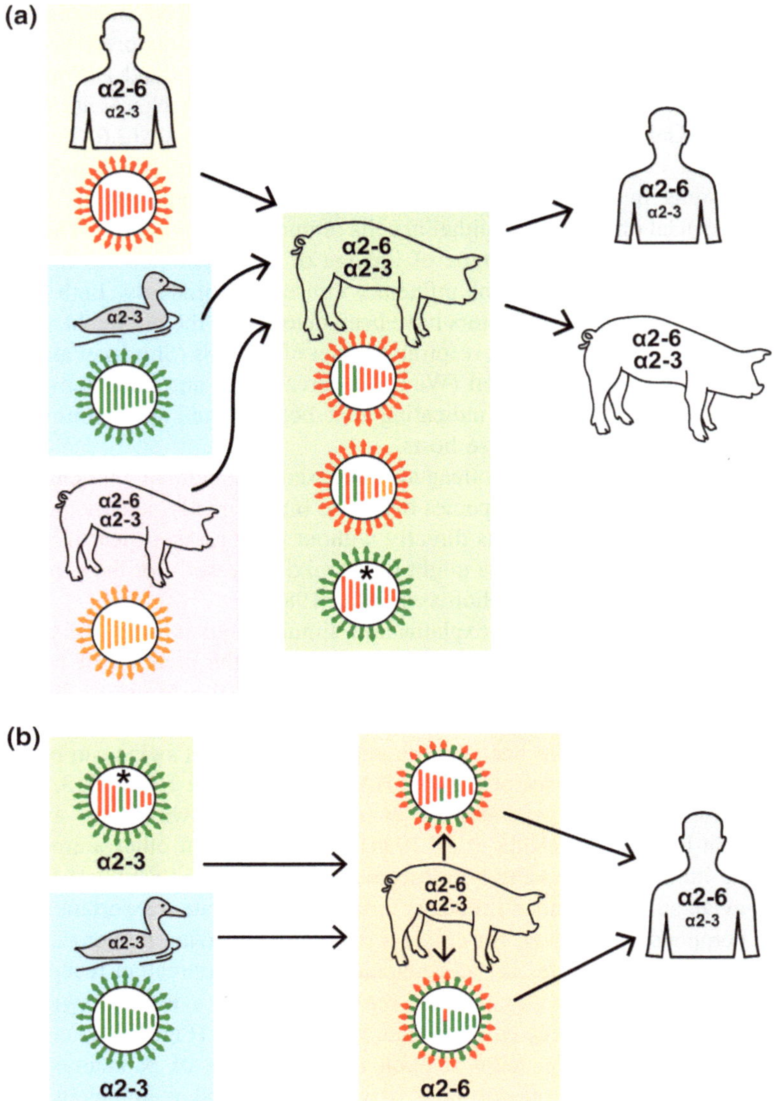

Fig. 1 Models for the generation of mammalian-like influenza viruses in pigs. **a** Reassortment model: in a classical genetic reassortment model, avian, swine, and human viruses can replicate in pigs because pigs express both, α2,3 and α2,6 receptor types. This sets the stage for the emergence of reassortant viruses that can infect other mammalian species including humans. However, some of these reassortant viruses (*) might need an adaptation step before they can efficiently infect humans. The surface proteins on each particle recognize α2,3 (*green*) and α2,6 (*red*, *orange*) receptor types and the segments in the center the viral genome. Adapted from Ito et al. (1998). **b** Adaptation model: avian influenza viruses and reassortant viruses with avian-like receptor binding properties (*) acquire the ability to replicate efficiently in humans during adaptation to the α2,6 receptor in pigs. This change is mediated mainly by mutations in the HA gene. Adapted from Ito et al. (1998)

having a leucine at codon position 226 (Ma et al. 2007). It should be noted that this mutation is linked to better binding to the α2,6 mammalian receptor. The latter finding supports the concept that influenza viruses reassorted in swine can undergo adaptation and can acquire the ability to recognize the human/mammalian virus receptor (Fig. 1b).

3 Are Pigs Maintenance Hosts for Avian Influenza Viruses?

Since swine can support reassortment of influenza A viruses from different hosts, and replication in pigs can promote adaptation of viruses to the mammalian receptor conformation (Fig. 1), there is great interest in the ability of swine to act as maintenance hosts for avian influenza viruses, which have been sporadically isolated from humans. Such viruses include the H5N1, H7N9, H7N7, and H9N2 viruses. The data available on the replication of highly pathogenic H5N1 viruses in swine are not conclusive. The isolation of H5N1 viruses from swine has been reported in Indonesia and China, both countries that are endemic for H5N1 in poultry (Liu et al. 2011). To date, there is no evidence for sustained pig-to-pig transmission of H5N1 viruses, but regular infection potential via environmental contamination should not be ignored. Experimental infection of swine with H5N1 suggests that this virus is most likely not maintained in the swine population (Choi et al. 2005; Lipatov et al. 2008). These experimental data suggest that swine are poorly susceptible to H5N1 infection and do not appear to be major contributors to the ecology of H5N1 viruses.

H9N2 subtype avian influenza viruses are endemic in land-based poultry in Asia and have been documented to infect humans and pigs sporadically (Peiris et al. 1999, 2001). Since both H9N2 and pandemic H1N1 viruses (pH1N1) are able to infect swine, reassortant H9 viruses in the background of the pH1N1 virus were evaluated for pathogenicity and transmission in pigs by replacing either both the HA and NA genes or only the HA gene with the respective genes from the H9N2 virus. The parental H9N2 replicated poorly and was not transmitted in pigs, whereas both reassortant viruses replicated and were transmitted efficiently in pigs (Qiao et al. 2012). These results demonstrated that reassortant H9 viruses containing pH1N1 genes show enhanced replication and transmissibility in pigs compared with the parental H9N2 virus, indicating that they may pose a threat to humans if such reassortants arise in swine.

The highly pathogenic H7N7 virus responsible for a poultry epidemic in the Netherlands in 2003 was transmitted cross-species to humans and to pigs (de Jong et al. 2009). However, no infectious virus could be isolated from pigs, and only serological evidence for H7 swine infection on farms with H7N7-infected poultry was evident, suggesting infection was via environmental contaminants. Recently, a novel low pathogenic avian H7N9 virus has caused more than 300 human

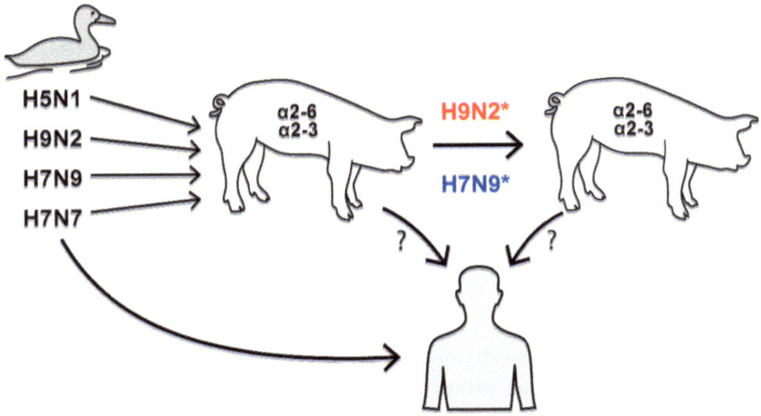

Fig. 2 Transmission of certain avian influenza viruses to pigs. Avian influenza viruses are transmitted to humans directly or indirectly via pigs. Certain molecular signatures—like mutations within HA or reassortment (*)—are needed for the H9N2 and H7N9 viruses to transmit between pigs (see text)

infections in China. The H7N9 virus is a reassortant virus containing avian lineage genomic RNA segments of viruses that circulated in waterfowl and terrestrial birds (e.g., chicken and brambling) (Kageyama et al. 2013; Liu et al. 2013). Many of the H7N9 viruses circulating in avian reservoirs contain putative mammalian signatures in the HA and PB2 genes. Studies with reverse genetically derived H7N9 viruses revealed that critical mutations in the HA gene (Q226L) are enabling the H7N9 viruses to transmit between pigs (Liu et al. 2014). This suggests that the myriads of H7N9 genotypes circulating in avian species in China and closely related strains (e.g., H7N7) have the potential for further adaptation to humans or other mammalian hosts (e.g., pigs) resulting in strains capable of sustained human-to-human transmission.

In addition, avian-like H6N6 viruses were recently isolated from domestic pigs in China, most likely originating from domestic ducks (Zhang et al. 2011; Zhao et al. 2013). One of the H6N6 isolate replicated in mice without prior adaptation, indicating that some H6N6 viruses already may be adapted to the mammalian host potentially posing a threat to animal and human health. In summary, it seems that wholly avian influenza viruses under field conditions do not replicate and transmit very efficiently in domestic pigs even if they contain various mammalian-like signatures. Under certain circumstances (reassortment, RBS adaptation) these viruses might be transmitted from pig-to-pig, and could be maintained in the pig population and/or cross the species barrier into humans (Fig. 2).

4 Transmission of Swine Influenza Viruses to Humans

The first serological evidence of SIV infecting humans was reported in Czechoslovakia in 1958 in a laboratory worker, with subsequent human-to-human transmission of the H1N1 virus among family members (Kluska et al. 1961). The first reported death of a human due to SIV was described in 1974 in Minnesota, where a 16-year old male pig farmer with Hodgkin's disease died of H1N1-induced pneumonia (Smith et al. 1976). The first large-scale occurrence of human-to-human transmission of SIV occurred on a military base in Fort Dix, New Jersey, in 1976 (Gaydos et al. 1977a, b; Hodder et al. 1977; Top and Russell 1977; Gaydos et al. 2006). Thirteen soldiers were serologically diagnosed with the classical H1N1 SIV infection, which was confirmed in some cases by virus isolation; one soldier died. Subsequent serological investigation at Fort Dix identified 230 additional suspected cases. A nationwide vaccine campaign was initiated in the United States with 45 million people vaccinated (Neustadt and Fineberg 1978/2005; Silverstein 1981). However, the campaign was halted because of lack of evidence of SIV transmission outside the military base and a significant number of cases of Guillain-Barre syndrome apparently associated with the vaccine (Marks and Halpin 1980; Sencer and Millar 2006).

Until 1998, all human infections in the United States with SIV were caused by the classical H1N1 virus. In the late 1990s, a dramatic change in SIV epidemiology occurred with the advent of the triple reassortant H3N2 SIVs. These viruses represented reassortment events between avian, swine, and human influenza viruses (Zhou et al. 1999) and are now endemic in US swine and US turkey populations (Yassine et al. 2007). The triple reassortant H3N2 SIVs subsequently reassorted with endemic H1N1 viruses and novel triple reassortant H1N1, H1N2, and H3N1 viruses emerged (Vincent et al. 2008). Active human infections with the triple reassortant H3N2 and H1N1 SIVs were reported over the next 10 years (Krueger and Gray 2013). Rather surprisingly to influenza epidemiologists, in April 2009, a novel quadruple reassortant H1N1 SIV was detected in two children in California. This virus contained genetic material from classical swine H1N1 (HA, NP, NS genes), North America triple reassortant H3N2 (avian PA and PB2 genes, and human PB1 gene), and avian-like Eurasian swine H1N1 (NA, M genes) viruses. This novel H1N1 influenza virus ultimately spread worldwide causing the first influenza pandemic in the twenty-first century. Interestingly, the H1N1 flu pandemic claimed roughly the same number of lives in 2009 as seasonal flu did on average in each of the 4 years preceding the pandemic. Researchers suggest between 123,000 and 203,000 people died worldwide from pandemic flu in the final 9 months of 2009. A striking regional heterogeneity was observed, with almost 20-fold higher mortality in some countries in the Americas than in Europe. In contrast to seasonal flu, the majority of victims were younger than 65 years of age (Simonsen et al. 2013). The 2009 human pandemic H1N1 (pH1N1) virus further complicated our understanding of SIV ecology and cross-transmissibility. It reminded us also that transmission of influenza viruses occurs not only from pigs to humans, but also from humans to pigs (Keenliside 2013), so-called reverse zoonosis. The naming of the virus as "swine flu" led to a drop in the demand for pork in many regions of the world and a subsequent drop in the price of pork. Losses to pork producers in North America are estimated at hundreds of millions of dollars, since several countries closed their borders to imports of pork from North America. The pH1N1 virus has been maintained since 2009 by swine-to-swine transmission; and next generation reassortant viruses with triple reassortant H1N1 and H3N2 viruses have been subsequently observed in swine in the United States (Ducatez et al. 2011; Liu et al. 2012; Vincent et al. 2014). Importantly, such next-generation H3N2 and H1N2 viruses were isolated from humans in the US throughout 2011 and 2012 (Epperson et al. 2013; Komadina et al. 2014). These variant H3N2 or H1N2 (H3N2v; H1N2v) viruses contain the matrix gene from the pH1N1 within the genetic background of triple reassortant H3N2 and H1N2 SIVs. Most of these transmission events happened at agricultural fairs and swine shows, which have been recognized as playing an important role in viral transmission between pigs and humans (Bowman et al. 2014). These recent events ensure the intense ongoing interest of both public and veterinary health sectors in swine influenza in North America.

In contrast to the US, Europe documented only a dozen human cases with SIV infection and the virus isolates have been H1N1 avian-like or H3N2 human–avian reassortant SIVs (Krueger and Gray 2013). While only a few human infections with SIVs have been reported in Asia, including H3N2, H1N1, H1N2 subtypes (Krueger and Gray 2013), the potential for further infections of humans in Asia is very high (Zhu et al. 2013).

In the past 20 years, human cases of SIV infection have been increasing. This seems to be in concert with the development of modern pig farming and the emergence of triple reassortant SIVs in US swine in 1998. In addition, surveillance of humans for SIV infections in general has improved (Milinovich et al. 2014). Although most SIV infections in humans are generally mild or subclinical, they can be quite serious in patients with underlying medical conditions. Excluding cases of the 2009 pH1N1 virus, about 10 % of human SIV cases have resulted in death. The reader may refer to two excellent reviews on the transmission of SIVs to humans for more detailed information (Myers et al. 2007; Krueger and Gray 2013).

5 Pig Production and Its Global Environmental Context

The environmental impact of pork production tends to be less than that of beef or lamb, but higher than for poultry and almost all non-meat sources of protein (Kleanthous 2009). The drive for intensive agricultural production of cheap meat began during World War II was supported by extensive agricultural subsidies in Europe and the US, and has been highly successful ever since. However, national and international statistics document that farm animals now consume a third of the world's cereal crops, 90 % of its soy meal and nearly 30 % of fish caught globally

(Lymbery and Oakeshott 2014). Not only are these food resources diverted from human consumption, but also half of the antibiotics used in the world are now being given to farm animals. Additionally, industrial farming of animals in small spaces and back-yard farms, also popular now in industrial nations, which allow close contact between animals and humans, especially children, have led to new viruses being isolated. While the world population is estimated to increase from 7.2 billion in 2013 to 9.6 billion in 2050—an increase of 33 %—the United Nations Food and Agriculture Organization has estimated that world livestock production will almost double over the same time period; this is likely to increase the number of animals slaughtered from 70 billion a year to 120 billion a year (Worldometers 2014; Lymbery and Oakeshott 2014). The question of how many and how humanely animals are to be raised is closely linked to the increasing demand for meat in human diets, as well as the public health implications arising from the sheer scale of animal farming, both intensive agriculture and small-scale farming.

The considerable global increase in both farm animals and newly emerging viruses is instructive. Because there is an animal reservoir for influenza viruses in food animals and wildlife, eradication of these viruses is impossible; therefore, it is important to understand the ecology of influenza A viruses and its mechanisms of intra and interspecies transmission. There is increasing evidence that a critical source for novel influenza viruses is Southeast Asia. For example, in 2013 over half of the world's pig population was raised in China, with 716 million pigs slaughtered. In addition, China produces one-third of all meat and two-fifths of all eggs in the world (Weis 2013; National Bureau of Statistics of China 2014). As a consequence, novel avian H9N2, H6N6 and H5N1 viruses have been detected in Chinese pigs; some of them have shown partial adaptation to mammalian hosts. Importantly, as the numbers of pigs and novel SIVs have increased globally, so has the amount of research being done on the causes of and mitigation strategies for SIVs (Richt and Webby 2013).

6 One Health for Influenza

If a One Health perspective is adopted for influenza in which human, animal, and environmental factors are considered, there can be little doubt that the greatly increased demand for meat in the human diet as well as the inherent characteristics of intensive pig farming and back-yard farms are highly significant. Although pigs are being raised in very different conditions in the tightly confined spaces of intensive industrial farming and with considerable interaction with other species (including humans) in back-yard farms, in both these situations new viruses are emerging that might mix with endemic viruses, which are prevalent in the herd. Current research is therefore focused especially on molecular surveillance of currently circulating SIVs (Morens and Taubenberger 2014).

The current field and laboratory research is essential to control both old and new influenza viruses. Of equal importance is a holistic investigation into the conditions in which swine are being raised and careful study of how co-infections potentially leading to a higher susceptibility to influenza viruses, in both people and animals, can be prevented or mitigated through improved environmental awareness. Both the recent emergence of the 2009 pH1N1 virus and the increasing number of SIV infections in humans, especially in the US, suggest that novel SIVs will continue to cause sporadic infections in humans whose severity will be unpredictable (Richt et al. 2013).

7 Conclusion

Since no single intervention can prevent influenza viruses from reassorting and rapidly mutating, efforts should be made to better monitor and prepare for their emergence. Improved surveillance of swine and human populations for novel influenza viruses, along with wider seasonal vaccination coverage among agricultural workers, health care workers, and the general public are critical mitigation strategies in the battle against emerging zoonotic SIVs. The realization that humans and other mammals can frequently be infected with H5N1 highly pathogenic avian influenza virus following exposure to infected birds has complicated the role of pigs as the mixing vessel in the transmission of avian viruses to humans and the epidemiology of influenza A viruses as a whole. Why certain viruses can directly infect humans and others seem to require an intermediate host is not fully understood. Certainly, the infection dose may be a critical factor. The control of influenza in humans and animals is challenging due to rapid virus evolution. If there is limited genetic change in the circulating viruses, then it is relatively easy to produce efficacious inactivated virus vaccines. However, if the rate of antigenic drift and shift accelerates, then it becomes difficult for vaccine production to keep up with circulating viruses (Osterholm et al. 2012). In the case of swine, there is clear evidence that swine can generate novel influenza A viruses that have the potential to infect humans and other animal species.

At present, it is impossible to predict which virus might cause the next human pandemic. History would suggest that the likelihood of such an event is high, although its timing and severity is certainly unclear. Therefore, it seems important to minimize the risk of transmission of SIVs to people as well as the risk of transmission of avian viruses to swine. Surveillance, control, and prevention strategies for swine influenza are necessary, not only for reducing economic losses for pork producers, but also for public health purposes. For a proactive approach to control of zoonotic influenza infections to be effective, farmers and veterinarians need to share data. In order to be successful, future surveillance and reporting policies must include provisions to protect the livelihoods of farmers (e.g., indemnity payments) and farm workers. A One Health perspective should be adopted for influenza in which human, animal, and environmental factors are considered.

Acknowledgments This work was partially supported by the U.S. Department of Homeland Security under Grant Award Number 2010-ST-AG0001 to CEEZAD, the Kansas Bioscience Authority and NIH grant HHSN266200700005C. We also would like to thank Mal Hoover for her excellent work on the figures and Jessica Green for proofreading.

References

Baum LG, Paulson JC (1990) Sialyloligosaccharides of the respiratory epithelium in the selection of human influenza virus receptor specificity. Acta Histochem Suppl 40:35–38

Bowman AS, Workman JD, Nolting JM, Nelson SW, Siemons RD (2014) Exploration of risk factors contributing to the presence of influenza virus in swine at agricultural fairs. Emerg Microbes Infect 3:e5

Brown IH (2000) The epidemiology and evolution of influenza viruses in pigs. Vet Microbiol 74:29–46

Brown IH (2013) History and epidemiology of swine influenza in Europe. Cur Top Microbiol Immunol 370:133–146

Choi YK, Nguyen TD, Ozaki H, Webby RJ, Puthavathana P, Buranathal C, Chaisingh A, Auewarakul P, Hanh NT, Ma SK, Hui PY, Guan Y, Peiris JS, Webster RG (2005) Studies of H5N1 influenza virus infection of pigs by using viruses isolated in Vietnam and Thailand in 2004. J Virol 79:10821–10825

Coker R, Rushton J, Mounier-Jack S, Karimuribo E, Lutumba P, Kambarage D, Pfeiffer DU, Stark K, Rweyemamu M (2011) Towards a conceptual framework to support one-health research for policy on emerging zoonoses. Lancet Infect Dis 11:326–331

de Jong MC, Stegeman A, van der Goot J, Koch G (2009) Intra- and interspecies transmission of H7N7 highly pathogenic avian influenza virus during the avian influenza epidemic in the Netherlands in 2003. Rev Sci Tech 28:333–340

Delgado C, Rosegrant M, Steinfeld H, Ehui S, Courbois C (1999) Livestock to 2020. The next food revolution. Food, agriculture and the environment. Discussion paper 28. International Food Policy Research Institute, Washington DC

Ducatez MF, Hause B, Stigger-Rosser E, Darnell D, Corzo C, Juleen K, Simonson R, Brockwell-Staats C, Rubrum A, Wang D, Webb A, Crumpton JC, Lowe J, Gramer M, Webby RJ (2011) Multiple reassortment between pandemic (H1N1) 2009 and endemic influenza viruses in pigs, United States. Emerg Infect Dis 17:1624–1629

Epperson S, Jhung M, Richards S, Quinlisk P, Ball L, Moll M, Boulton R, Haddy L, Biggerstaff M, Brammer L, Trock S, Burns E, Gomez T, Wong KK, Katz J, Lindstrom S, Klimov A, Bresee JS, Jernigan DB, Cox N, Finelli L (2013) Human infections with influenza A(H3N2) variant virus in the United States, 2011–2012. Clin Infect Dis 57(Suppl 1):S4–S11

Gaydos JC, Hodder RA, Top FH Jr, Allen RG, Soden VJ, Nowosiwsky T, Russell PK (1977a) Swine influenza A at Fort Dix, New Jersey (January–February 1976). II. Transmission and morbidity in units with cases. J Infect Dis 136(Suppl):S363–S368

Gaydos JC, Hodder RA, Top FH Jr, Soden VJ, Allen RG, Bartley JD, Zabkar JH, Nowosiwsky T, Russell PK (1977b) Swine influenza A at Fort Dix, New Jersey (January–February 1976). I. Case finding and clinical study of cases. J Infect Dis 136(Suppl):S356–S362

Gaydos JC, Top FH Jr, Hodder RA, Russell PK (2006) Swine influenza A outbreak, Fort Dix, New Jersey, 1976. Emerg Infect Dis 12:23–28

Hodder RA, Gaydos JC, Allen RG, Top FH Jr, Nowosiwsky T, Russell PK (1977) Swine influenza A at Fort Dix, New Jersey (January–February 1976). III. Extent of spread and duration of the outbreak. J Infect Dis 136(Suppl):S369–S375

Ito T, Suzuki Y, Mitnaul L, Vines A, Kida H, Kawaoka Y (1997) Receptor specificity of influenza A viruses correlates with the agglutination of erythrocytes from different animal species. Virology 227:493–499

Ito T, Couceiro JN, Kelm S, Baum LG, Krauss S, Castrucci MR, Donatelli I, Kida H, Paulson JC, Webster RG, Kawaoka Y (1998) Molecular basis for the generation in pigs of influenza A viruses with pandemic potential. J Virol 72:7367–7373

Kageyama T, Fujisaki S, Takashita E, Xu H, Yamada S, Uchida Y, Neumann G, Saito T, Kawaoka Y, Tashiro M (2013) Genetic analysis of novel avian A(H7N9) influenza viruses isolated from patients in China, February to April 2013. Euro Surveill 18:20453

Keenliside J (2013) Pandemic influenza A H1N1 in Swine and other animals. Curr Top Microbiol Immunol 370:259–271

Kida H, Ito T, Yasuda J, Shimizu Y, Itakura C, Shortridge KF, Kawaoka Y, Webster RG (1994) Potential for transmission of avian influenza viruses to pigs. J Gen Virol 75:2183–2188

Kimble B, Nieto GR, Perez DR (2010) Characterization of influenza virus sialic acid receptors in minor poultry species. Virol J 7:365

Kleanthous A (2009) Pigs and the environment. Here Tomorrow/BPEX, London

Kluska V, Macku M, Mensik J (1961) Demonstration of antibodies against swine influenza viruses inman. Cesk Pediatr 16:408–414

Komadina N, McVernon J, Hall R, Leder K (2014) A historical perspective of influenza A(H1N2) virus. Emerg Infect Dis 20:6–12

Krauss S, Obert CA, Franks J, Walker D, Jones K, Seiler P, Niles L, Pryor SP, Obenauer JC, Naeve CW, Widjaja L, Webby RJ, Webster RG (2007) Influenza in migratory birds and evidence of limited intercontinental virus exchange. PLoS Pathog 3:e167

Krueger WS, Gray GC (2013) Swine influenza virus infections in man. Curr Top Microbiol Immunol 370:201–225

Lipatov AS, Kwon YK, Sarmento LV, Lager KM, Spackman E, Suarez DL, Swayne DE (2008) Domestic pigs have low susceptibility to H5N1 highly pathogenic avian influenza viruses. PLoS Pathog 4:e1000102

Liu W, Wei M-T, Tong Y, Tang F, Zhang L, Fang L, Yang H, Cao W-C (2011) Seroprevalence and genetic characteristics of five subtypes of influenza A viruses in the Chinese pig population: a pooled data analysis. Vet J 187:200–206

Liu Q, Ma J, Liu H, Qi W, Anderson J, Henry SC, Hesse RA, Richt JA, Ma W (2012) Emergence of novel reassortant H3N2 swine influenza viruses with the 2009 pandemic H1N1 genes in the United States. Arch Virol 157:555–562

Liu D, Shi W, Shi Y, Wang D, Xiao H, Li W, Bi Y, Wu Y, Li X, Yan J, Liu W, Zhao G, Yang W, Wang Y, Ma J, Shu Y, Lei F, Gao GF (2013) Origin and diversity of novel avian influenza A H7N9 viruses causing human infection: phylogenetic, structural, and coalescent analyses. Lancet 381:1926–1932

Liu Q, Bin Z, Ma W, Bawa B, Ma J, Wang W, Lang Y, Lyoo Y, Halpin RA, Lin X, Stockwell TB, Webby RJ, Wentworth D, Richt JA (2014) Analysis of recombinant H7N9 wild type and mutant viruses in pigs shows Q226L in HA is important for transmission. J Virol 88(14):8153–8165

Lymbery P, Oakeshott I (2014) Farmageddon: the true cost of cheap meat. Bloomsbury, London

Ma W, Vincent AL, Gramer MR, Brockwell CB, Lager KM, Janke BH, Gauger PC, Patnayak DP, Webby RJ, Richt JA (2007) Identification of H2N3 influenza A viruses from swine in the United States. Proc Natl Acad Sci USA 104:20949–20954

Ma W, Kahn RE, Richt JA (2009) The pig as a mixing vessel for influenza viruses: human and veterinary implications. J Mol Genet Med 3:158–166

Marks JS, Halpin TJ (1980) Guillain-Barre syndrome in recipients of A/New Jersey influenza vaccine. JAMA 243:2490–2494

Milinovich GJ, Williams GM, Clements ACA, Hu W (2014) Internet-based surveillance systems for monitoring emerging infectious diseases. Lancet Infect Dis 14:160–168

Morens DM, Taubenberger JK (2014) A possible outbreak of swine influenza, 1892. Lancet Infect Dis 14:169–172

Myers KP, Olsen CW, Gray GC (2007) Cases of swine influenza in humans: a review of the literature. Clin Infect Dis 44:1084–1088

National Bureau of Statistics of China (2014) China's economy showed good momentum of steady growth in the year of 2013. www.stats.gov.cn/english/pressrelease/201401/tp140120_502079.html

Neustadt RE, Fineberg HV (1978/2005) The swine flu affair: decision-making on a slippery disease. University Press of the Pacific, Honolulu

Osterholm MT, Kelley NS, Sommer A, Belongia EA (2012) Efficacy and effectiveness of influenza vaccines: a systematic review and meta-analysis. Lancet Infect Dis 12:36–44

Peiris M, Yuen KY, Leung CW, Chan KH, Ip PL, Lai RW, Orr WK, Shortridge KF (1999) Human infection with influenza H9N2. Lancet 354:916–917

Peiris JS, Guan Y, Markwell D, Ghose P, Webster RG, Shortridge KF (2001) Cocirculation of avian H9N2 and contemporary "human" H3N2 influenza A viruses in pigs in southeastern China: potential for genetic reassortment? J Virol 75:9679–9686

Qiao C, Liu Q, Bawa B, Shen H, Qi W, Chen Y, Mok CK, Garcia-Sastre A, Richt JA, Ma W (2012) Pathogenicity and transmissibility of reassortant H9 influenza viruses with genes from pandemic H1N1 virus. J Gen Virol 93:2337–2345

Richt JA, Webby R (eds) (2013) Swine influenza. Curr Top Microbiol Immunol 370:1–300

Richt JA, Webby R, Kahn RE (2013) The pandemic H1N1 influenza experience. In: Mackenzie JS, Jeggo M, Daszak PS, Richt JA (eds) One health: the human–animal-environment interfaces in emerging infectious disease: the concept and examples of a one health approach. Current topics in microbiology and immunology, vol 365, pp 269–280

Rogers GN, D'Souza BL (1989) Receptor binding properties of human and animal H1 influenza virus isolates. Virology 173:317–322

Rogers GN, Paulson JC (1983) Receptor determinants of human and animal influenza virus isolates: differences in receptor specificity of the H3 hemagglutinin based on species of origin. Virology 127:361–373

Scholtissek C, Burger H, Kistner O, Shortridge KF (1985) The nucleoprotein as a possible major factor in determining host specificity of influenza H3N2 viruses. Virology 147:287–294

Schrauwen EJA, Fouchier RAM (2014) Host adaptation and transmission of influenza A viruses in mammals. Emerg Microbes Infect 3:e9

Sencer DJ, Millar JD (2006) Reflections on the 1976 swine flu vaccination program. Emerg Infect Dis 12:29–33

Shinya K, Ebina M, Yamada S, Ono M, Kasai N, Kawaoka Y (2006) Avian flu: influenza virus receptors in the human airway. Nature 440:435–436

Silverstein AM (1981) Pure politics and impure science: the swine flu affair. Johns Hopkins University Press, Baltimore

Simonsen L, Spreeuwenberg P, Lustig R, Taylor RJ, Fleming DM, Kroneman M, Van Kerkhove MD, Mounts AW, Paget WJ (2013) Global mortality estimates for the 2009 influenza pandemic from the GLaMOR project: a modeling study. PLoS Med 10:e1001558

Smith TF, Burgert EO Jr, Dowdle WR, Noble GR, Campbell RJ, Van Scoy RE (1976) Isolation of swine influenza virus from autopsy lung tissue of man. N Engl J Med 294:708–710

Suzuki Y (2005) Sialobiology of influenza: molecular mechanism of host range variation of influenza viruses. Biol Pharm Bull 28:399–408

Suzuki T, Horiike G, Yamazaki Y, Kawabe K, Masuda H, Miyamoto D, Matsuda M, Nishimura SI, Yamagata T, Ito T, Kida H, Kawaoka Y, Suzuki Y (1997) Swine influenza virus strains recognize sialylsugar chains containing the molecular species of sialic acid predominantly present in the swine tracheal epithelium. FEBS Lett 404:192–196

Suzuki Y, Ito T, Suzuki T, Holland RE Jr, Chambers TM, Kiso M, Ishida H, Kawaoka Y (2000) Sialic acid species as a determinant of the host range of influenza A viruses. J Virol 74:11825–11831

Tong S, Li Y, Rivailler P, Conrardy C, Castillo DA, Chen LM, Recuenco S, Ellison JA, Davis CT, York IA, Turmelle AS, Moran D, Rogers S, Shi M, Tao Y, Weil MR, Tang K, Rowe LA, Sammons S, Xu X, Frace M, Lindblade KA, Cox NJ, Anderson LJ, Rupprecht CE, Donis RO (2012) A distinct lineage of influenza A virus from bats. Proc Natl Acad Sci USA 109:4269–4274

Tong S, Zhu X, Li Y, Shi M, Zhang J, Bourgeois M, Yang H, Chen X, Recuenco S, Gomez J, Chen LM, Johnson A, Tao Y, Dreyfus C, Yu W, McBride R, Carney PJ, Gilbert AT, Chang J, Guo Z, Davis CT, Paulson JC, Stevens J, Rupprecht CE, Holmes EC, Wilson IA, Donis RO (2013) New world bats harbor diverse influenza A viruses. PLoS Pathog 9:e1003657

Top FH Jr, Russell PK (1977) Swine influenza A at Fort Dix, New Jersey (January–February 1976). IV. Summary and speculation. J Infect Dis 136(Suppl):S376–S380

Vincent AL, Ma W, Lager KM, Janke BH, Richt JA (2008) Swine influenza viruses a North American perspective. Adv Virus Res 72:127–154

Vincent A, Awada L, Brown I, Chen H, Claes F, Dauphin G, Donis R, Culhane M, Hamilton K, Lewis N, Mumford E, Nguyen T, Parchariyanon S, Pasick J, Pavade G, Pereda A, Peiris F (2014) Review of influenza A virus in swine worldwide: a call for increased surveillance and research. Zoonoses Public Health 61:4–17

Wan H, Perez DR (2006) Quail carry sialic acid receptors compatible with binding of avian and human influenza viruses. Virology 346:278–286

Weis T (2013) The ecological hoofprint: the global burden of industrial livestock. Zed Books, London

Worldometers (2014) World population. www.worldometers.info/world-population/

Wu Y, Tefsen B, Shi Y, Gao GF (2014) Bat-derived influenza-like viruses H17N10 and H18N11. Trends Microbiol 22:183–191

Yassine HM, Al-Natour MQ, Lee CW, Saif YM (2007) Interspecies and intraspecies transmission of triple reassortant H3N2 influenza A viruses. Virol J 4:129

Zhang G, Kong W, Qi W, Long LP, Cao Z, Huang L, Qi H, Cao N, Wang W, Zhao F, Ning Z, Liao M, Wan XF (2011) Identification of an H6N6 swine influenza virus in southern China. Infect Genet Evol 11:1174–1177

Zhao G, Chen C, Huang J, Wang Y, Peng D, Liu X (2013) Characterisation of one H6N6 influenza virus isolated from swine in China. Res Vet Sci 95:434–436

Zhou NN, Senne DA, Landgraf JS, Swenson SL, Erickson G, Rossow K, Liu L, Yoon K, Krauss S, Webster RG (1999) Genetic reassortment of avian, swine, and human influenza A viruses in American pigs. J Virol 73:8851–8856

Zhu H, Webby R, Lam TT, Smith DK, Peiris JS, Guan Y (2013) History of swine influenza viruses in Asia. Curr Top Microbiol Imunol 370:57–68

Part IV
Animal Models and Pathogenesis

Influenza Pathobiology and Pathogenesis in Avian Species

Monique S. França and Justin D. Brown

Abstract Wild birds in the orders Anseriformes and Charadriiformes are the natural and asymptomatic reservoirs of influenza A viruses representing all of the avian hemagglutinin (HA) and neuraminidase (NA) subtypes. Transmission of avian influenza (AI) viruses from wild birds to gallinaceous poultry species occurs regularly and outcomes vary, ranging from asymptomatic infections to mortality. Circulation of H5 and H7 low pathogenic AI (LPAI) viruses in gallinaceous poultry may result in mutations in the HA protein cleavage site and the emergence of highly pathogenic AI (HPAI) viruses, which in poultry can cause severe disease with high economic losses. Since 2002, various wild bird species also have succumbed to infection with the Eurasian H5N1 HPAI viruses. The pathogenesis of AI is complex and the ability of these viruses to produce disease and death in avian species is dependent on various host, viral and environmental factors, which are not completely understood.

Abbreviations

AI	Avian influenza
HA	Hemagglutinin
HPAI	Highly pathogenic avian influenza
LPAI	Low pathogenic avian influenza
NA	Neuraminidase

M.S. França (✉)
Poultry Diagnostic and Research Center, The University of Georgia, 953 College Station Road, Athens, GA 30602, USA
e-mail: mfranca@uga.edu

J.D. Brown
Pennsylvania Game Commission, Orchard Road, University Park, PA 16802, USA
e-mail: judbrow@pa.gov

Current Topics in Microbiology and Immunology (2014) 385: 221–242
DOI: 10.1007/82_2014_385

Published Online: 12 July 2014

Contents

1 Introduction

Avian influenza (AI) viruses have been detected in more than 105 bird species from different taxa; however, species in the orders Anseriformes (i.e. ducks, geese, swans) and Charadriiformes (i.e., gulls, terns, and shorebirds) are the natural reservoirs for most subtypes of type A influenza viruses, including hemagglutinin (HA) 1–16 and neuraminidase (NA) 1–9 (Stallknecht 1998). With rare exceptions, AI virus infection in wild birds is silent and not associated with any overt signs of disease or lesions.

Waterfowl may directly or indirectly transmit AI viruses to aberrant hosts, including gallinaceous poultry. Wild bird-origin low pathogenic AI (LPAI) viruses are generally poorly adapted for gallinaceous poultry and the outcomes of infections with these viruses range from infection without clinical signs to severe disease with high mortality (Easterday et al. 1997). Circulation of LPAI viruses in domestic poultry species may occasionally result in adaptation and efficient replication in these birds. AI viruses are classified into low pathogenicity or high pathogenicity viruses based on the sequence of the cleavage site of the HA molecule and the severity of the disease they cause in chickens (*Gallus gallus domesticus*) (OIE 2009). Adaptation of LPAI viruses in gallinaceous poultry may result in the acquisition of multiple basic amino acids in the HA cleavage site of H5 and H7 LPAI viruses and the emergence of highly pathogenic AI (HPAI) viruses.

LPAI and HPAI viruses rarely spill back from domestic poultry to wild birds. The first report of AI in wild birds occurred in 1961 during a mortality event of common terns (*Sterna hirundo*) caused by a H5N3 HPAI virus in South Africa (Becker 1966). This is, until H5N1 HPAI, the only sustained HPAI virus outbreak in wild birds without apparent exposure to infected poultry. Since 2002, various aquatic and terrestrial wild bird species have succumbed to infection with the Eurasian H5N1 HPAI viruses. Although some of these infections have been associated with exposure to infected domestic poultry (including waterfowl), there are reports of wild bird infection without an obvious link to domestic birds. While it is clear that some strains of H5N1 HPAI virus are able to cause mortality in select wild avian species, it is still unclear if this virus is able to be maintained in wild bird populations.

Susceptibility to infection and pathogenesis of AI viruses in avian species are determined by various host and viral factors, including host species, age at infection, host immunity, secondary infections, and virus strain. The focus of this chapter is to summarize the existing knowledge on AI pathobiology and pathogenesis in birds. Although globally there are over 10,000 taxonomically-diverse species of birds, AI infection and pathogenesis have only been thoroughly studied in select species in the Orders Anseriformes (e.g., ducks) and Galliformes (e.g., chickens and turkeys). Consequently, the following sections will largely concentrate on these two avian groups, but will incorporate information on other avian taxa when possible.

2 Influenza Pathogenesis in Avian Species

2.1 *Galliformes*

2.1.1 Low Pathogenicity Avian Influenza

The HA protein of LPAI viruses requires trypsin-like proteases to be cleaved and these enzymes are available in mucosal epithelial cells of the respiratory and intestinal tracts; this limits the tissue distribution of these viruses (Klenk et al. 1975; Suarez 2008). The predominant site of initial viral replication in gallinaceous poultry is the epithelium of the nasal cavity, followed by infection and replication in other sites of the respiratory and intestinal tracts (Swayne 2007; Swayne and Pantin-Jackwood 2008). LPAI viruses have the ability to replicate in renal tubules and pancreatic acinar cells in these birds after intravenous inoculation, as these cells also contain trypsin-like enzymes (Swayne et al. 1994; Swayne and Slemons 1995). LPAI viruses can also replicate in oviduct epithelium of turkey (*Meleagris gallopavo*) breeders and can be transmitted in eggs (Pillai et al. 2009, 2010).

Infection with LPAI viruses in gallinaceous poultry may result in subclinical infections or clinical disease. Clinical disease in these birds is usually characterized by high morbidity (>50 %) and low mortality (<5 %), although higher mortality ranging from 5 to 97 % can occur in birds with secondary bacterial infections, young birds, severely stressed birds, and in laying chickens (Johnson and Maxfield 1976; Brugh and Beard 1986; Capua and Marangon 2000; Swayne and Pantin-Jackwood 2008; Swayne et al. 2013). Domestic poultry can have nonspecific clinical signs (huddling, depression, decreased feed and water consumption, weight loss, ruffled feathers), respiratory signs (sneezing, coughing, snicking, rales and ocular discharge), drop in egg production, decrease in egg quality, decrease in hatchability and occasionally greenish or yellowish diarrhea (Capua and Terregino 2009; Swayne et al. 2013). Outbreaks of LPAI have been more frequently reported in turkeys, less frequently in laying chickens, and rarely

Fig. 1 Turkey naturally infected with H9N2 LPAI virus exhibiting swelling of the infraorbital sinuses. Courtesy of Dr. H. L. Shivaprasad

in other domesticated poultry species (Swayne et al. 2013). Turkeys are more susceptible to clinical disease caused by LPAI than chickens (Capua and Terregino 2009).

Gross and microscopic lesions caused by LPAI viruses in gallinaceous poultry vary depending on host and viral factors such as bird species, age at infection, immune status of the host, virus isolate, route of exposure, and presence of secondary bacterial infections (Swayne et al. 2013). Previously published research papers, book chapters and reviews have reported or summarized the pathobiology of LPAI in gallinaceous poultry (Slemons and Swayne 1990; Shalaby et al. 1994; Swayne et al. 1994, 2013; Mo et al. 1997; Ziegler et al. 1999; Capua and Marangon 2000; Capua and Mutinelli 2001a; Hooper and Selleck 2003; Swayne and Pantin-Jackwood 2008; Pantin-Jackwood and Swayne 2009; Kuiken et al. 2010). Gross lesions seen in the upper respiratory tract include catarrhal to fibrinous rhinitis, sinusitis, laryngitis, and tracheitis. Turkeys can have swollen infraorbital sinuses (Fig. 1) with mucoid to caseous exudate. The trachea can be edematous, congested and with mild hemorrhages. Occasionally, caseous exudate occludes the lumen of the trachea, resulting in death due to suffocation. The lungs may be congested and edematous, but fibrinous bronchopneumonia (Fig. 2) and airsacculitis (Fig. 3) may occur in birds with secondary bacterial infections such as *Mycoplasma gallisepticum*, *Escherichia coli* and *Pasteurella multocida* (Stipkovits et al. 2012; Swayne et al. 2013). Birds in egg production can have egg-yolk peritonitis (Fig. 4) catarrhal to fibrinous salpingitis, ovarian regression, and hemorrhage of ovarian follicles. Laid eggs may be misshapen and with loss of pigmentation. Other gross lesions include enteritis, typhlitis, and pancreatic lesions including hemorrhages and pale areas of necrosis, which are usually seen in turkeys (Capua and Mutinelli 2001a; Swayne and Pantin-Jackwood 2008). Swollen kidneys, urolithiasis, and visceral urate deposition have also been reported in naturally infected laying hens and in intravenously inoculated chickens (Swayne et al. 1994; Kinde et al. 2003). The bursa of Fabricius and thymus may be atrophied.

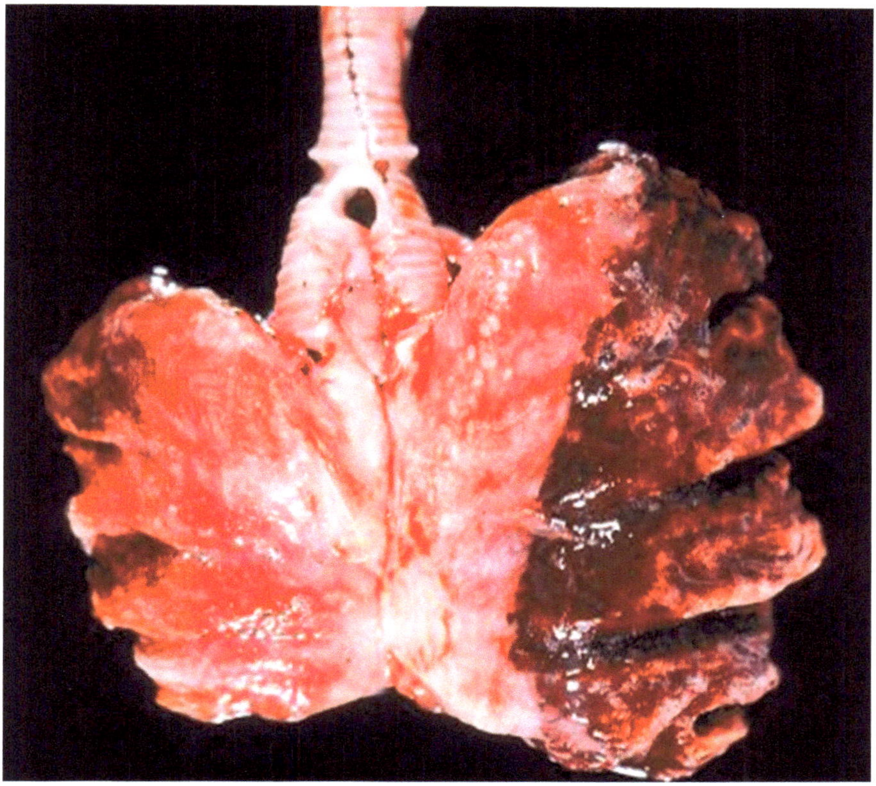

Fig. 2 Lung with pneumonia from a turkey naturally infected with H9N2 LPAI virus. *Source* Avian Influenza slide study set by Dr. N. L. Tablanta and Dr. H. L Shivaprasad, 2011. Reprinted with permission from American Association of Avian Pathologists (AAAP)

Microscopically, gallinaceous poultry infected with LPAI viruses usually have heterophilic to lymphoplasmacytic conjunctivitis, rhinitis, sinusitis, tracheitis, and bronchitis. The lungs may have pneumonia, which is usually in the ventromedial portion. Nephrosis, renal tubular necrosis, and interstitial nephritis have been occasionally reported in field cases and in experimentally infected birds, especially in birds inoculated via the intravenous route (Swayne et al. 1994; Kinde et al. 2003; Swayne and Pantin-Jackwood 2008). Pancreatic acinar cell necrosis and pancreatitis have been reported in field cases and in experimentally infected birds, and occur more commonly in turkeys than chickens (Capua and Mutinelli 2001a). The bursa of Fabricius, thymus, spleen, and other lymphoid tissues can have lymphocyte depletion, necrosis, and apoptosis. Inflammation and necrosis of the oviduct epithelium can occur (Pillai et al. 2009). AI virus antigen has been demonstrated in epithelial cells of the respiratory tract, intestine, kidney, pancreas, and oviduct (Swayne and Slemons 1994; Mutinelli et al. 2003; Pillai et al. 2009; Spackman et al. 2010).

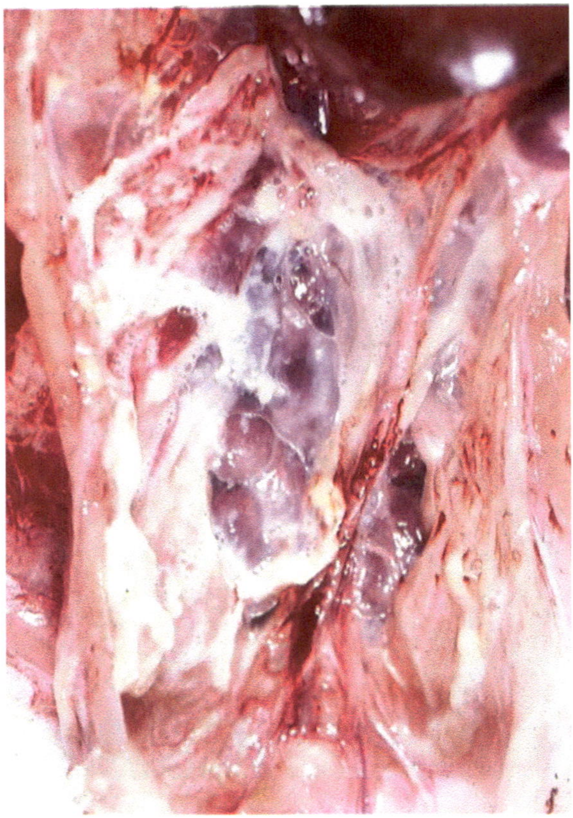

Fig. 3 Air sac with caseous and foamy exudate in a turkey naturally infected with H9N2 LPAI virus. *Source* Avian Influenza slide study set by Dr. N. L. Tablanta and Dr. H. L Shivaprasad, 2011. Reprinted with permission from American Association of Avian Pathologists (AAAP)

2.1.2 High Pathogenicity Avian Influenza

HPAI viruses have multiple basic amino acids (arginine and lysine) at the cleavage site of the HA protein. This allows the HA of HPAI viruses able to be cleaved by furin-like proteases present in many organs, which results in systemic disease with high mortality in gallinaceous poultry. Death results from multiple organ failure and damage to cardiovascular and nervous systems (Swayne and Halvorson 2008). HPAI viruses cause tissue damage through necrosis and apoptosis of infected cells (Perkins and Swayne 2001). Necrosis is usually associated with high viral titers and detection of abundant viral antigen in tissues (Swayne and Halvorson 2008). Infection of vascular endothelium causes altered vascular permeability and endothelial cell injury leading to edema, hemorrhage, disseminated vascular thrombosis, and ischemia in tissues. Cellular mediators such as cytokines also produce indirect tissue damage in HPAI virus infections (Swayne and Pantin-Jackwood 2008).

Virus strain, species, age at infection, and immune status of the host influence the pathogenesis of HPAI viruses in birds. Initial replication of HPAI viruses in

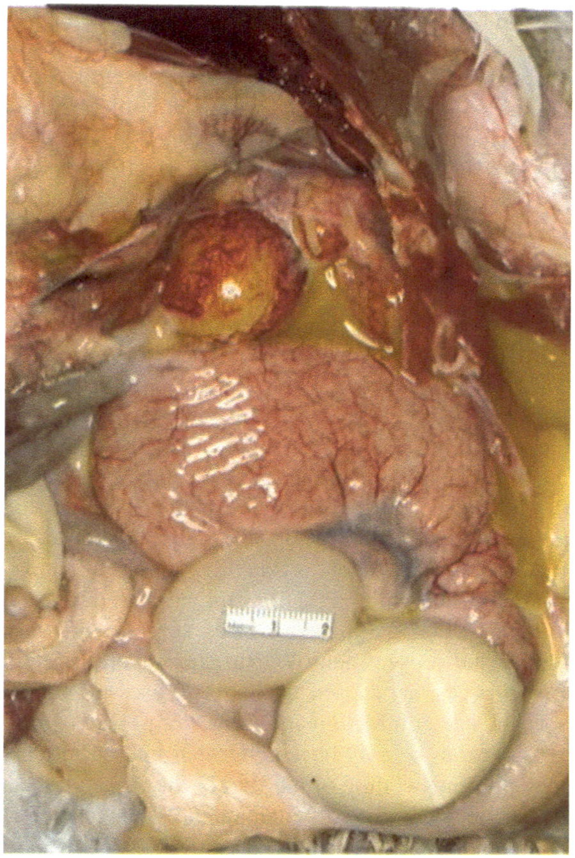

Fig. 4 Egg-yolk peritonitis, oviduct edema and shell less eggs in a White Leghorn chicken naturally infected with H6N3 LPAI virus. *Source* Avian Influenza slide study set by Dr. N. L. Tablanta and Dr. H. L Shivaprasad, 2011. Reprinted with permission from American Association of Avian Pathologists (AAAP)

gallinaceous poultry occur in epithelial cells of the upper respiratory tract followed by infection of capillary endothelium in the submucosa and systemic spread (Swayne 2007). HPAI viruses infect and replicate in heterophils, macrophages, and endothelial cells and are disseminated via the lymphatic and vascular systems to multiple tissues (Swayne and Pantin-Jackwood 2008). Different HPAI viruses have different tissue tropism and these viruses may be epitheliotropic, endotheliotropic, neurotropic or pantropic (Pantin-Jackwood and Swayne 2009). Lesions in the brain are associated with infection of the vascular endothelium followed by dissemination of the virus into the neuroparenchyma (Kobayashi et al. 1996a; Swayne 2007). Viremia observed post infection with some HPAI viruses is associated with minimal or absence of replication in endothelial cells and these viruses are usually associated with prolonged survival; however, extensive tissue damage still occurs due to higher viral replication in multiple organs, such as, the central and autonomic nervous systems, heart, pancreas, and adrenal glands.

Several previously published research papers, book chapters, and reviews have reported or summarized the pathobiology of HPAI in gallinaceous poultry (Acland et al. 1984; Kobayashi et al. 1996b; Mo et al. 1997; Capua and Marangon 2000; Capua et al. 2000b; Capua and Mutinelli 2001a; Swayne and Pantin-Jackwood 2008; Pantin-Jackwood and Swayne 2009; Kuiken et al. 2010; Swayne et al. 2013). HPAI viruses, including H5N1 HPAI viruses, cause systemic disease and high mortality that can reach up to 100 % in chickens and turkeys. Clinical signs vary with the organs affected, duration and severity of the disease, bird species, and virus strain. The disease spreads rapidly and peak mortality usually occurs in 3–5 days in floor-reared poultry flocks. The disease spreads more slowly in cage-reared poultry with peak mortality usually occurring after 10–15 days. Clinical signs may not be apparent in domestic turkeys and chickens until sudden death. Turkeys and chickens with peracute to acute disease may be markedly depressed, lethargic, with difficulty in standing and in a comatose state. Feed and water consumption are usually markedly decreased and poultry houses may be very quiet due to reduction of normal bird vocalization. Birds that survive the acute phase of the infection may present neurological signs including incoordination, head and neck tremors, torticollis, opisthotonus, circling movements, paresis, and paralysis. Respiratory signs such as coughing, sneezing, and rales may be seen, but are less common than with LPAI. Layers and breeders have a drastic decline in egg production. Other gallinaceous birds exhibit similar clinical signs, but may survive longer and present neurological signs (Perkins and Swayne 2001).

Gross lesions caused by HPAI viruses in gallinaceous poultry depend on host species, age at infection and virus strain. Birds with peracute disease may not have gross lesions or may have congested internal organs. Chickens with acute disease may present swelling of the skin of the face, comb, snood, wattles, upper neck, and leg shanks (Fig. 5) with edema and hemorrhages, which are more easily visible in nonfeathered skin. The comb, wattles, (Fig. 6) and snood may also be cyanotic and have areas of necrosis. The conjunctiva and trachea may be congested, edematous and hemorrhagic. Pulmonary congestion, edema, and hemorrhage may also occur. Hemorrhages may be seen in serosal and mucosal surfaces of the gastrointestinal tract and are more prominent in the mucosa of the proventriculus and ventriculus. Hemorrhages are less frequently seen in cecal tonsils, Peyer's patches, and Meckel's diverticulum. Pale foci of necrosis may be present in visceral organs and are more frequently seen in the pancreas, heart, spleen, and less frequently in the liver and kidneys. The heart may have hemorrhages in the epicardium and coronary fat. Skeletal muscles can also develop hemorrhages. The bursa of Fabricius and thymus may be atrophied. Microscopically, gallinaceous poultry have multi-organ necrosis and/or inflammation with lesions more frequently seen in the brain, heart, lung, pancreas, adrenal glands, skin, and lymphoid organs (Swayne and Pantin-Jackwood 2008). Birds with peracute disease may not present microscopic lesions or may have mild necrotizing and inflammatory lesions in visceral organs. Birds with acute disease have predominantly necrosis, hemorrhage, edema, and inflammatory cells in multiple visceral organs and brain. Lesions seen in the heart include cardiomyocyte degeneration, necrosis (Fig. 7) and lymphohistiocytic

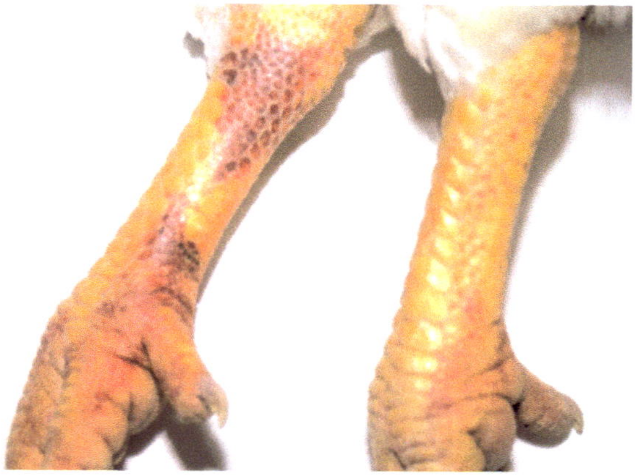

Fig. 5 Broiler breeder naturally infected with HPAI virus of the H7N1 subtype, showing hemorrhages on the shank. *Source* Capua and Terregino 2009. Reprinted with permission from Springer-Verlag

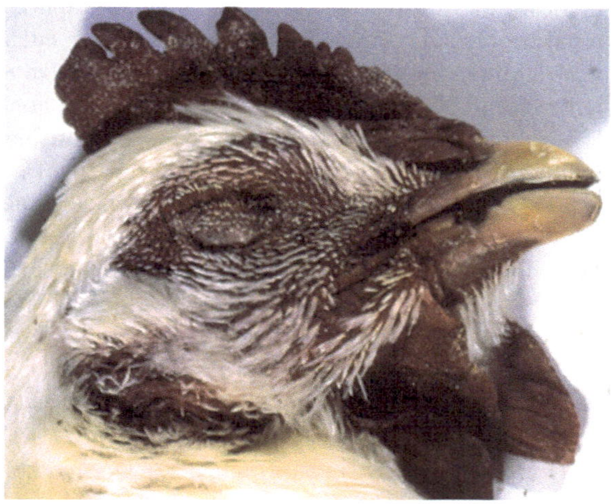

Fig. 6 Broiler breeder naturally infected with HPAI virus of the H7N1 subtype, displaying congestion and cyanosis of the comb and wattles. *Source* Capua and Terregino 2009. Reprinted with permission from Springer-Verlag

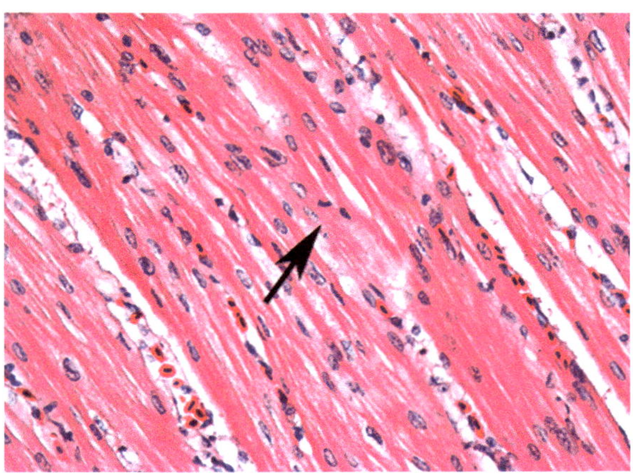

Fig. 7 Photomicrograph of the heart with cardiomyocyte necrosis from a chicken experimentally infected with H5N1 HPAI virus. Courtesy of Dr. Mary J. Pantin-Jackwood

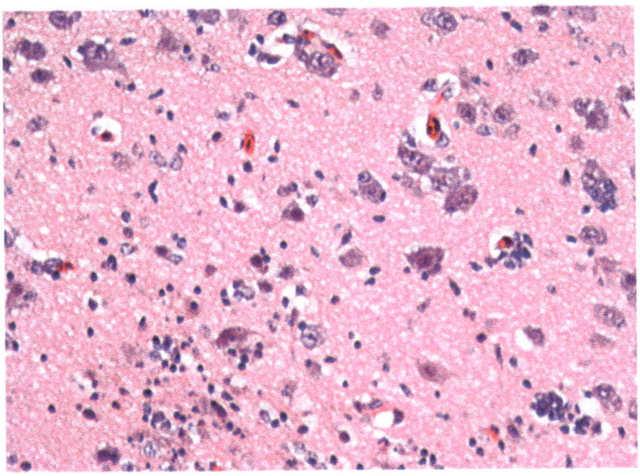

Fig. 8 Photomicrograph of the cerebrum with lymphocytic encephalitis, neuronal satellitosis and neuronophagia from a chicken experimentally infected with H5N1 HPAI virus. Courtesy of Dr. Mary J. Pantin-Jackwood

myocarditis. Lesions in the brain include lymphocytic perivascular cuffing, neuronal necrosis, neuronophagia (Fig. 8), gliosis, edema of the neuropil, meningitis, ventriculitis, choroid plexitis, and periventricular necrosis. Edema, hemorrhages with vasculitis and microthrombosis are seen in nonfeathered skin. Other common lesions include pancreatic acinar degeneration and necrosis, pancreatitis, renal tubular necrosis, interstitial nephritis, necrosis of corticotrophic cells in the adrenals, skeletal muscle degeneration, and vascular necrosis. Rhinitis and tracheitis with ulceration and hemorrhages may also occur. The lungs may have interstitial pneumonia, necrosis of air capillaries and hemorrhages. Immunohistochemistry reveals AI viral antigen in vascular endothelium, necrotic cells of various visceral organs (Fig. 9), brain (Fig. 10), autonomic neurons and skin. Lymphoid organs including bursa of Fabricius, spleen, thymus, and cecal tonsils may have lymphoid depletion, necrosis, and apoptosis. The cecal tonsils may be hemorrhagic in some cases. In lymphoid organs, AI virus antigen is rarely seen in lymphocytes and is more frequently seen in macrophages and endothelial cells (Swayne and Pantin-Jackwood 2008). The ovary and oviduct may have inflammation and hemorrhage associated with AI viral antigen (Silva et al. 2013).

Gallinaceous poultry species other than chickens and turkeys may survive longer and present more severe necrotizing and inflammatory lesions in visceral organs and brain (Perkins and Swayne 2001).

2.2 *Anseriformes*

2.2.1 Low Pathogenicity Avian Influenza

The primary site of viral replication in mallards (*Anas platyrhynchos*) is the enterocytes lining the intestinal tract and, in young ducks, the bursa of Fabricius (Slemons and Easterday 1978; Webster et al. 1978; Kida et al. 1980; Franca et al.

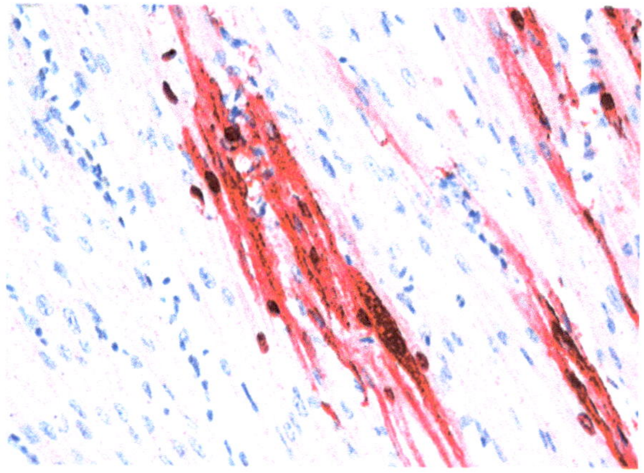

Fig. 9 Photomicrograph of the heart with AI viral antigen in cardiomyocytes from a chicken experimentally infected with H5N1 HPAI virus. Courtesy of Dr. Mary J. Pantin-Jackwood

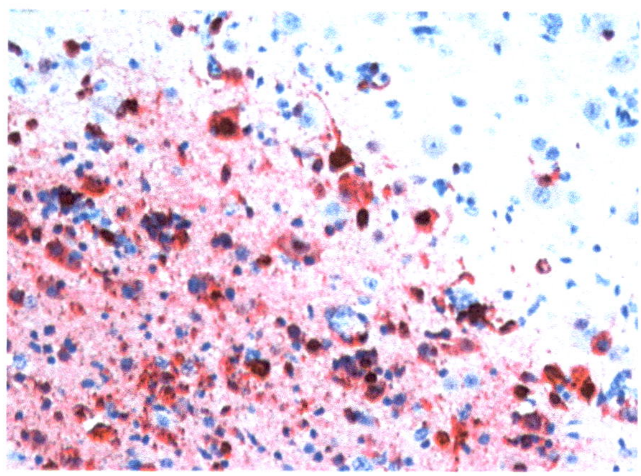

Fig. 10 Photomicrograph of the cerebrum with AI viral antigen in neurons and neuroparenchyma from a chicken experimentally infected with H5N1 HPAI virus. Courtesy of Dr. Mary J. Pantin-Jackwood

2012). Viral replication is associated with high concentrations of viral shedding in the feces, which can exceed 10^7 EID_{50}/g of feces (Webster et al. 1978; Brown et al. 2012b). Viral replication, and associated shedding, also occur in the epithelium of the upper respiratory tract; however, this is greatly reduced and only occurs during the first few days of infection (Slemons and Easterday 1978; Webster et al. 1978; Kida et al. 1980; Spackman et al. 2009; Franca et al. 2012). Susceptibility and extent of viral replication/shedding vary between AI virus strains, and is dependent on the degree of viral host adaptation; waterfowl-origin viruses replicate more efficiently in ducks than viruses isolated from chickens, gulls, or mammals (Webster et al. 1978; Kida et al. 1980; Hinshaw et al. 1982; Alexander et al. 1986; Swayne 2007; Mundt et al. 2009; Costa et al. 2011).

Despite high levels of viral replication and shedding, clinical signs or lesions are rarely observed with natural or experimental infections (Daoust et al. 2011; Franca et al. 2012; Kuiken 2013). However, there is a growing body of evidence to suggest that infection with AI viruses is not inconsequential to the host and that subclinical effects may occur, including changes in weight gain (Latorre-Margalef et al. 2009), behavior (van Gils et al. 2007), immune function (Laudert et al. 1993), and reproduction (Laudert et al. 1993). Whether these effects are truly associated with viral infection or reflect study design (i.e. unnatural routes of inoculation or low sample size) is not known, nor is the significance of these potential effects on wild avian populations.

When disease has been observed in domestic ducks and geese, there often are confounding factors associated with these reports, including bacterial co-infections, stress, unnatural routes of inoculation, or poor husbandry, making it difficult to determine the association between infection with LPAI viruses and clinical signs (Swayne and Halvorson 2008). When clinical signs have been observed, they typically are nonspecific (depression, ruffled feathers, reluctance to move) or reflective of respiratory disease (Alexander et al. 1978; Swayne and Halvorson 2008).

Although there are only a limited number of studies examining the pathogenesis of AI in waterfowl, particularly natural infections, gross or microscopic lesions associated with viral replication are often absent or mild. Reported lesions associated with LPAI in ducks and geese have been respiratory in nature, including tracheitis, laryngitis, sinusitis, conjunctivitis, and pneumonia (Cooley et al. 1989; Swayne and Pantin-Jackwood 2008; Franca et al. 2012).

2.2.2 High Pathogenicity Avian Influenza

With the exception of some lineages of Eurasian H5N1 HPAI viruses, which will be addressed separately below, natural and experimental infections with HPAI viruses are less common in domestic or wild waterfowl than gallinaceous poultry. Although domestic geese (*Anser anser domesticus*) and ducks have been infected during outbreaks of HPAI, clinical disease or lesions have been rarely reported (Alexander et al. 2008; Swayne 2008). Consistent with field observations, waterfowl have generally been more resistant to experimental challenge with HPAI virus strains than gallinaceous poultry, as evidenced by resistance to infection, no or mild clinical disease and lesions, and reduced viral shedding (Alexander et al. 1978, 1986; Wood et al. 1985, 1995; Swayne and Pantin-Jackwood 2008). The reason for the increased resistance of waterfowl to HPAI viruses is not completely known, but may be associated with host-adaptation of HPAI viruses to gallinaceous poultry or innate immunity and expression of RIG-I, an antiviral mechanism that mediates interferon response, which is found in cells of ducks, but apparently is not present in chicken cells (Barber et al. 2010). However, as with LPAI viruses, there are exceptions to these general trends, including: domestic geese challenged with H7N7 HPAI viruses excreted virus via the respiratory and intestinal tracts for several days without clinical signs (Rohm et al. 1996); ducks challenged with two HPAI virus strains remained asymptomatic despite the fact that both strains replicated systemically and were isolated from multiple visceral tissues (Wood et al. 1995); ducks challenged with HPAI virus

Fig. 11 A female wood duck with severe neurologic clinical signs of disease after intranasal inoculation with an Asian strain of highly pathogenic avian influenza H5N1 virus. *Source* Brown et al. (2006). Susceptibility of North American ducks and gulls to H5N1 highly pathogenic avian influenza viruses. Emerg Infect Dis, Nov publication (cited 2014 February 14). Available from http://wwwnc.cdc.gov/eid/article/12/11/06-0652_article.htm

strains had pneumonia and airsacculitis in the absence of clinical disease (Cooley et al. 1989); ducks challenged with a H7N1 HPAI virus exhibited high morbidity (depression, paralysis, swollen eyes) and mortality (Alexander et al. 1978); and during an outbreak of H7N1 HPAI virus in Italy during 1999–2000, muscovy ducks (*Cairina moschata*) and domestic geese were reportedly found dead or exhibiting severe neurologic clinical signs with viral replication and associated lesions in the central nervous system and pancreas (Capua and Mutinelli 2001b). These examples of clinical disease, systemic viral replication, extensive shedding, or lesions are uncommon for HPAI viruses in waterfowl; however, they highlight the variability between virus strains, as well as unpredictability of AI viruses in various hosts.

H5N1 HPAI Viruses. Since 2002, some Eurasian lineages of H5N1 HPAI viruses have emerged that are virulent for domestic and wild waterfowl. The pathogenicity of these viruses varies dramatically between virus strains and waterfowl hosts, ranging from asymptomatic infections with low levels of viral shedding and no lesions to acute and fatal infections with extensive replication and lesions in multiple visceral organs and the brain (Perkins and Swayne 2002a; Brown et al. 2006, 2008; Pantin-Jackwood and Swayne 2007; Pasick et al. 2007; Swayne 2007; Keawcharoen et al. 2008). Field and experimental data suggest that species of swans, geese, and diving ducks are highly susceptible to infection and are most likely to experience morbidity and mortality. Premonitory disease in these species, even during experimental studies, is brief and often the birds are found dead. When clinical signs are detected, they typically are neurologic (Fig. 11), including ataxia, severe depression, listlessness, and seizures. Consistent with the clinical presentation, H5N1 HPAI virus replication and associated lesions in these susceptible species are extensive in the central and peripheral nervous systems. Similarly to gallinaceous poultry, viral replication and lesions are also most frequently detected in the pancreas, adrenal gland, brain, and heart. Lesions and AI antigen are less often seen in the air sacs, lungs, spleen, and kidneys (Brown et al. 2006, 2008; Pantin-Jackwood and Swayne 2007; Pasick et al. 2007; Keawcharoen et al. 2008).

2.3 Other Avian Taxa

Aside from waterfowl, AI virus infection is most consistently detected in species in the Order Charadriiformes, principally gulls, terns, and shorebirds; however, the epidemiologic patterns of infection in gulls and terns and in shorebirds are different and distinct from those in waterfowl (Stallknecht et al. 2007). With gulls, AI virus can be detected throughout the year, but highest prevalence is reportedly in juvenile birds on breeding colonies in the spring and summer (Olsen et al. 2006; Stallknecht et al. 2007; Velarde et al. 2010). A wide-diversity of HA subtypes have been reportedly isolated from gulls, but H13 and H16 viruses are the most common and are believed to be strongly host-adapted to gulls (Fouchier et al. 2007). The only site globally where AI virus was consistently detected from shorebirds at a high prevalence is at Delaware Bay on the East coast of the US during spring migratory stopover (Kawaoka et al. 1988; Krauss et al. 2004; Hanson et al. 2008). Virus has been detected from a wide diversity of other wild and domestic avian species (Olsen et al. 2006), but at a low prevalence which was not consistent between years. With the exception of H5N1 HPAI virus, there have been rare reports of HPAI virus from avian species outside the orders Anseriformes and Galliformes.

2.3.1 Low Pathogenicity Avian Influenza

Wild birds. Natural LPAI virus infections in wild birds, aside from waterfowl, have not been associated with clinical signs or microscopic lesions (Olsen et al. 2006; Stallknecht et al. 2007; Swayne 2007; Stallknecht and Brown 2008; Brown et al. 2011). Although very few LPAI virus challenge studies have been conducted in wild birds, aside from waterfowl, the experimental data are consistent with field observations. Inoculation of gulls and shorebirds with multiple subtypes of LPAI virus have resulted in asymptomatic infections (Bahl and Pomeroy 1977; Costa et al. 2011; Brown et al. 2012a; Curran et al. 2013; Hall et al. 2013). In these studies, as opposed to waterfowl, LPAI viral shedding was reportedly equivalent to or greater via the respiratory tract, relative to the feces. Few of these studies characterized tissue tropism or examined tissues for lesions; however, respiratory shedding in juvenile ring-billed gulls (*Larus delawarensis*) infected with a H13N9 LPAI virus, was associated with low levels of viral replication in tracheal epithelium and mild inflammation (Brown et al. 2012a).

No morbidity or mortality has been reported with natural or experimental LPAI virus infection in wild birds outside the orders Anseriformes and Charadriiformes. Experimentally, very few studies have been conducted on wild avian species other than gulls, shorebirds, or waterfowl.

Ratites. Ostriches infected with AI virus may have respiratory signs, ocular discharge, reluctance to eat, prostration, and green diarrhea (Allwright et al. 1993; Manvell et al. 1998, 2003; Mutinelli et al. 2003; Olivier 2006). These clinical signs are usually more evident in young birds and may not be seen in adult birds. Severity of the disease increases with environmental conditions and secondary bacterial infections, and mortality may be as high as 30 % in ostriches (*Struthio camelus*) (Capua and Terregino 2009). Rheas, emus and ostriches infected with LPAI viruses may have conjunctivitis, fibrinous sinusitis, tracheitis, bronchitis, perihepatitis, pericarditis, and pneumonia (Woolcock et al. 2000; Swayne and Pantin-Jackwood 2008; Capua and Terregino 2009). Ostriches can also have catarrhal enteritis, necrohemorrhagic lesions in the pancreas and hepatic necrosis (Capua and Terregino 2009).

2.3.2 High Pathogenicity Avian Influenza

Wild birds. Aside from infections with the Eurasian H5N1 HPAI viruses, the only significant mortality associated with HPAI in wild birds was during an outbreak of H5N3 HPAI in common terns in South Africa (Becker 1966) in which approximately 1,300 common terns died. Reported premonitory clinical signs exhibited by terns were nonspecific, consisting of ruffled feathers and a reluctance to move. Multiple HPAI challenge studies have been conducted in gulls and terns, which have yield varying results depending on virus strains and host species. The majority of ring-billed gulls challenged with a H5N2 HPAI virus seroconverted and shed low levels of virus (Wood et al. 1985). A single gull ($n = 8$) in this trial died and the virus was re-isolated from multiple tissues (intestines, lung, and spleen); however, it was not definitively determined that the mortality was caused by HPAI virus infection. The remaining seven inoculated gulls in this study remained asymptomatic and excreted low concentrations of virus via the respiratory tract. Juvenile laughing gulls (*Leucophaeus atricilla*) (Perkins and Swayne 2002b) and adult common terns (Becker 1967) were challenged with the tern-origin H5N3 HPAI virus described above. The HPAI virus replicated systemically in common terns resulting in high morbidity and mortality. Nonspecific clinical signs were observed, including ruffled feathers, drooped wings, and depression, which rapidly progressed to death. Viral replication was detected with immunohistochemistry in the lungs, heart, skeletal muscle, brain, and less frequently in the spleen and kidneys in terns that died. Laughing gulls challenged with the H5N3 HPAI virus remained asymptomatic, but excreted low levels of virus in oropharyngeal and cloacal swabs up to 10 and 7 DPI, respectively. Although no clinical signs were detected in the gulls, viral replication was identified in the pancreas and liver and mild to moderate airsacculitis, interstitial pneumonia, and, less frequently, pancreatitis and hepatitis were detected. Collectively, the experimental data on gulls indicate that they are permissive hosts for a wide-diversity of AI viruses, including LPAI and HPAI strains, but the susceptibility and pathobiology varies dramatically with viral strain, host species, and likely a variety of other factors related to host, environment (including study design), and virus.

Aside from infections with the Eurasian H5N1 HPAI viruses, isolations of HPAI from other wild avian species, outside the orders Anseriformes and Charadriiformes, have been uncommon. Sporadic mortality involving individual or low numbers of birds has been reported from peridomestic, terrestrial species of the orders Passeriformes or Columbiformes during outbreaks of HPAI in gallinaceous poultry (Morgan and Kelly 1990; Alexander 2000). Often, little pathology or diagnostic information is available from these cases. Experimentally, European starlings (*Sternus vulgaris*) and house sparrows (*Passer domesticus*) were inoculated with two strains of an H7N7 HPAI virus (Nestorowicz et al. 1987). Both strains were highly virulent for the starlings as evidenced by systemic replication and 100 % mortality in inoculated and contact-exposed birds. The H7N7 HPAI virus strains were less virulent for house sparrows, but mortality was observed in a subset of the challenged birds. In the sparrows that died, virus replicated systemically and exhibited a similar tropism to the starlings.

Ratites. Ostriches infected with HPAI viruses may be asymptomatic or may be depressed and may have reduced feed consumption, respiratory signs, neurological signs, and swelling of the neck due to edema (Manvell et al. 1998; Capua et al. 2000a; Mutinelli et al. 2003; Olivier 2006). Juvenile ostriches are reported to be usually more susceptible to morbidity and mortality, although some viruses may cause asymptomatic infections in young birds (Manvell et al. 1998). Ostriches naturally infected with H7N1 HPAI virus in Italy presented with severe hemorrhagic enteritis, pancreatic hemorrhages, urate deposits in kidney, tracheal and pulmonary congestion and petechiae in the epicardium (Manvell et al. 1998; Mutinelli et al. 2003). Gross lesions in juvenile ostriches infected with an H5N2 HPAI virus in South Africa included generalized congestion, hepatosplenomegaly, and airsacculitis (Howerth et al. 2012). Microscopically, ostriches infected with HPAI viruses can have necrosis in liver, kidney, spleen, pancreas, intestine, and brain (Capua et al. 2000a; Mutinelli et al. 2003; Howerth et al. 2012).

H5N1 HPAI Viruses. H5N1 HPAI virus infection in other avian taxa has resulted in lesions similar to that described above for waterfowl (i.e. acute neurologic disease). Birds in the orders Ciconiiformes, Columbiformes, Passeriformes, Accipitriformes, Falconiformes, Gruiformes, Pelecaniformes, Phoenicopteriformes, Podicipediformes, and Strigiformes have succumbed to natural infections with the Eurasian H5N1 HPAI viruses since 2002 (Ellis et al. 2004; Liu et al. 2005; Stallknecht and Brown 2008; Swayne and Pantin-Jackwood 2008; Marinova-Petkova et al. 2012). Experimental infections with H5N1 HPAI viruses in terrestrial birds have shown marked variations in virulence between birds in different taxonomic groups as well as between species belonging to the same order. In one experiment, infections with H5N1 HPAI virus did not result in mortality in European starlings and house sparrows, while zebra finches (*Taeniopygia guttata*), house finches (*Carpodacus mexicanus*), and budgerigars (*Melopsittacus undulatus*) had significant morbidity and mortality (Perkins and Swayne 2003). Zebra finches were the most severely affected species with lesions and viral antigen in multiple organs (Perkins and Swayne 2003). Other studies showed high levels of viral shedding prior to development of clinical signs in house sparrows, suggesting that these birds could potentially play a role in an outbreak in domestic poultry (Boon et al. 2007; Brown et al. 2009). Within the order Columbiformes, rock pigeons (*Columba livia*) have been reported to be resistant to experimental infection with LPAI and HPAI viruses, but some H5N1 HPAI viruses have caused mortality and severe brain lesions in these birds (Klopfleisch et al. 2006; Boon et al. 2007; Jia et al. 2008; Brown et al. 2009). Neurotropism is a common feature in terrestrial bird species that succumb to H5N1 HPAI virus infection and these birds usually have non-suppurative encephalitis with viral antigen in neurons.

3 Conclusions

The susceptibility to infection and the severity of the clinical disease caused by AI viruses in avian species are dependent on various factors including species, age at infection, immunity, viral strain, route of exposure, environmental factors, and co-infections. In general, infections with LPAI viruses in gallinaceous poultry cause high morbidity and low mortality with respiratory disease and drops in egg production, while HPAI viruses cause high mortality with systemic disease in these birds. By contrast, infections with LPAI viruses in wild birds are usually asymptomatic. Clinical outcomes of HPAI virus infections in wild birds have been highly variable and unpredictable; although asymptomatic infections are most common, high mortality with neurological signs and systemic disease have been reported in some species infected with certain HPAI virus strains. The specific factors involved with natural resistance to HPAI in some bird species, such as ducks and pigeons, as well as the tissue tropism and transmission routes of AI viruses in different wild bird species are largely unknown. Various experimental and surveillance studies in the past 15 years have improved our knowledge of AI in avian species; however, our understanding on the pathobiology and epidemiology of LPAI and HPAI viruses in wild birds is still scarce. Additional field and experimental studies are therefore necessary in order to untangle the complexity of host, viral, and environmental factors that determine differences in susceptibility to infection, pathogenesis, and transmission of AI viruses in avian species.

Acknowledgments The authors thank Dr. H.L. Shivaprasad for providing Fig. 1 and Dr. Mary J. Pantin-Jackwood for providing Figs. 7, 8, 9 and 10. The authors also thank Dr. H.L. Shivaprasad and Dr. David E. Stallknecht for reviewing this chapter.

References

Acland HM, Silverman Bachin LA, Eckroade RJ (1984) Lesions in broiler and layer chickens in an outbreak of highly pathogenic avian influenza virus infection. Vet Pathol 21:564–569

Alexander DJ (2000) A review of avian influenza in different bird species. Vet Microbiol 74:3–13

Alexander DJ, Allan WH, Parsons DG et al (1978) The pathogenicity of four avian influenza viruses for fowls, turkeys and ducks. Res Vet Sci 24:242–247

Alexander DJ, Capua I, Koch G (2008) Highly pathogenic avian influenza oubreaks in Europe, Asia, and Africa since 1959, excluding the Asian H5N1 virus outbreaks. In: Swayne DE (ed) avian influenza, 1st edn. Blackwell Publishing, Ames, pp 217–237
Alexander DJ, Parsons G, Manvell RJ (1986) Experimental assessment of the pathogenicity of eight avian influenza A viruses of H5 subtype for chickens, turkeys, ducks and quail. Avian Pathol 15:647–662
Allwright DM, Burger WP, Geyer A et al (1993) Isolation of an influenza A virus from ostriches (*Struthio camelus*). Avian Pathol 22:59–65
Bahl AK, Pomeroy BS (1977) Experimental exposure of Franklins' gulls (*Larvus pipixcan*) and mallards (*Anas platyrhynchos*) to a turkey influenza A virus A/Turkey/Minn/BF/71 (Hav6Neq2). J Wildl Dis 13:420–426
Barber MR, Aldridge JR Jr, Webster RG et al (2010) Association of RIG-I with innate immunity of ducks to influenza. Proc Natl Acad Sci U S A 107:5913–5918
Becker WB (1966) The isolation and classification of tern virus: influenza virus A/Tern/South Africa/1961. J Hyg 64:309–320
Becker WB (1967) Experimental infection of common terns with tern virus: influenza virus A/Tern/South Africa/1961. J Hyg (Lond) 65:61–65
Boon AC, Sandbulte MR, Seiler P et al (2007) Role of terrestrial wild birds in ecology of influenza A virus (H5N1). Emerg Infect Dis 13:1720–1724
Brown J, Poulson R, Carter D et al (2012a) Susceptibility of avian species to North American H13 low pathogenic avian influenza viruses. Avian Dis 56:969–975
Brown J, Swayne DE, Costa T et al (2011) Experimental infection studies of avian influenza in wild birds as a complement to surveillance. In: Majumdar SK, Brenner FJ, Huffman JE, McLean RG, Panah AI, Pietrobon PJ, Keeler SP, Shive S (eds) Pandemic influenza viruses: science, surveillance, and public health, 1st edn. The Pennsylvania Academy of Sciences, Easton, pp 176–200
Brown JD, Berghaus RD, Costa TP et al (2012b) Intestinal excretion of a wild bird-origin H3N8 low pathogenic avian influenza virus in mallards (*Anas platyrhynchos*). J Wildl Dis 48:991–998
Brown JD, Stallknecht DE, Beck JR et al (2006) Susceptibility of North American ducks and gulls to H5N1 highly pathogenic avian influenza viruses. Emerg Infect Dis 12:1663–1670
Brown JD, Stallknecht DE, Berghaus RD et al (2009) Infectious and lethal doses of H5N1 highly pathogenic avian influenza virus for house sparrows (*Passer domesticus*) and rock pigeons (*Columbia livia*). J Vet Diagn Invest 21:437–445
Brown JD, Stallknecht DE, Swayne DE (2008) Experimental infection of swans and geese with highly pathogenic avian influenza virus (H5N1) of Asian lineage. Emerg Infect Dis 14:136–142
Brugh M, Beard CW (1986) Influence of dietary calcium stress on lethality of avian influenza viruses for laying chickens. Avian Dis 30:672–678
Capua I, Marangon S (2000) The avian influenza epidemic in Italy, 1999–2000: a review. Avian Pathol 29:289–294
Capua I, Mutinelli F (2001a) Avian influenza. Color atlas and text. PAPI, Bologna
Capua I, Mutinelli F (2001b) Mortality in Muscovy ducks (*Cairina moschata*) and domestic geese (*Anser anser var. domestica*) associated with natural infection with a highly pathogenic avian influenza virus of H7N1 subtype. Avian Pathol 30:179–183
Capua I, Mutinelli F, Bozza MA et al (2000a) Highly pathogenic avian influenza (H7N1) in ostriches (*Struthio camelus*). Avian Pathol 29:643–646
Capua I, Mutinelli F, Marangon S et al (2000b) H7N1 avian influenza in Italy (1999 to 2000) in intensively reared chickens and turkeys. Avian Pathol 29:537–543
Capua I, Terregino C (2009) Clinical traits and pathology of avian influenza infections, guidelines for farm visit and differential diagnosis. In: Capua I, Alexander, DJ (eds) Avian influenza and newcastle disease—a field and laboratory manual, 3rd edn. Springer, Milan, pp 45–71
Cooley AJ, Van Campen H, Philpott MS et al (1989) Pathological lesions in the lungs of ducks infected with influenza A viruses. Vet Pathol 26:1–5

Costa TP, Brown JD, Howerth EW et al (2011) Variation in viral shedding patterns between different wild bird species infected experimentally with low-pathogenicity avian influenza viruses that originated from wild birds. Avian Pathol 40:119–124

Curran JM, Robertson ID, Ellis TM et al (2013) Variation in the responses of wild species of duck, gull, and wader to inoculation with a wild-bird-origin H6N2 low pathogenicity avian influenza virus. Avian Dis 57:581–586

Daoust PY, Kibenge FS, Fouchier RA et al (2011) Replication of low pathogenic avian influenza virus in naturally infected Mallard ducks (*Anas platyrhynchos*) causes no morphologic lesions. J Wildl Dis 47:401–409

Easterday BC, Hinshaw VS, Halvorson DA (1997) Influenza. In: Calnek BW, Barnes HJ, Beard CW, McDougald LR, Saif YM (eds) Diseases of poultry, 10th edn. Iowa State University Press, Ames, pp 583–605

Ellis TM, Bousfield RB, Bissett LA et al (2004) Investigation of outbreaks of highly pathogenic H5N1 avian influenza in waterfowl and wild birds in Hong Kong in late 2002. Avian Pathol 33:492–505

Fouchier RAM, Munster VJ, Keawcharoen J et al (2007) Virology of avian influenza in relation to wild birds. J Wildl Dis 43:S7–S14

Franca M, Stallknecht DE, Poulson R et al (2012) The pathogenesis of low pathogenic avian influenza in mallards. Avian Dis 56:976–980

Hall JS, Krauss S, Franson JC et al (2013) Avian influenza in shorebirds: experimental infection of ruddy turnstones (*Arenaria interpres*) with avian influenza virus. Influenza Other Respir Viruses 7:85–92

Hanson BA, Luttrell MP, Goekjian VH et al (2008) Is the occurrence of avian influenza virus in Charadriiformes species and location dependent? J Wildl Dis 44:351–361

Hinshaw VS, Air GM, Gibbs AJ et al (1982) Antigenic and genetic characterization of a novel hemagglutinin subtype of influenza A viruses from gulls. J Virol 42:865–872

Hooper P, Selleck P (2003) Pathology of low and high virulent influenza virus infections. Avian Dis 47:134–141

Howerth EW, Olivier A, Franca M et al (2012) Pathobiology of highly pathogenic avian influenza virus H5N2 infection in juvenile ostriches from South Africa. Avian Dis 56:966–968

Jia B, Shi J, Li Y et al (2008) Pathogenicity of Chinese H5N1 highly pathogenic avian influenza viruses in pigeons. Arch Virol 153:1821–1826

Johnson DC, Maxfield BG (1976) An occurrence of avian influenza virus infection in laying chickens. Avian Dis 20:422–424

Kawaoka Y, Chambers TM, Sladen WL et al (1988) Is the gene pool of influenza viruses in shorebirds and gulls different from that in wild ducks? Virology 163:247–250

Keawcharoen J, van Riel D, van Amerongen G et al (2008) Wild ducks as long-distance vectors of highly pathogenic avian influenza virus (H5N1). Emerg Infect Dis 14:600–607

Kida H, Yanagawa R, Matsuoka Y (1980) Duck influenza lacking evidence of disease signs and immune response. Infect Immun 30:547–553

Kinde H, Read DH, Daft BM et al (2003) The occurrence of avian influenza A subtype H6N2 in commercial layer flocks in Southern California (2000-02): clinicopathologic findings. Avian Dis 47:1214–1218

Klenk HD, Rott R, Orlich M et al (1975) Activation of influenza A viruses by trypsin treatment. Virology 68:426–439

Klopfleisch R, Werner O, Mundt E et al (2006) Neurotropism of highly pathogenic avian influenza virus A/chicken/Indonesia/2003 (H5N1) in experimentally infected pigeons (*Columbia livia f. domestica*). Vet Pathol 43:463–470

Kobayashi Y, Horimoto T, Kawaoka Y et al (1996a) Neuropathological studies of chickens infected with highly pathogenic avian influenza viruses. J Comp Pathol 114:131–147

Kobayashi Y, Horimoto T, Kawaoka Y et al (1996b) Pathological studies of chickens experimentally infected with two highly pathogenic avian influenza viruses. Avian Pathol 25:285–304

Krauss S, Walker D, Pryor SP et al (2004) Influenza A viruses of migrating wild aquatic birds in North America. Vector Borne Zoonotic Dis 4:177–189

Kuiken T (2013) Is low pathogenic avian influenza virus virulent for wild waterbirds? Proc Biol Sci 280:20130990

Kuiken T, van den Brand J, van Riel D et al (2010) Comparative pathology of select agent influenza a virus infections. Vet Pathol 47:893–914

Latorre-Margalef N, Gunnarsson G, Munster VJ et al (2009) Effects of influenza A virus infection on migrating mallard ducks. Proc Biol Sci 276:1029–1036

Laudert EA, Sivanandan V, Halvorson DA (1993) Effect of intravenous inoculation of avian influenza virus on reproduction and growth in mallard ducks. J Wildl Dis 29:523–526

Liu J, Xiao H, Lei F et al (2005) Highly pathogenic H5N1 influenza virus infection in migratory birds. Science 309:1206

Manvell RJ, English C, Jorgensen PH et al (2003) Pathogenesis of H7 influenza A viruses isolated from ostriches in the homologous host infected experimentally. Avian Dis 47:1150–1153

Manvell RJ, Jorgensen PH, Nielsen OL et al (1998) Experimental assessment of the pathogenicity of two avian influenza A H5 viruses in ostrich chicks (Struthio camelus) and chickens. Avian Pathol 27:400–404

Marinova-Petkova A, Georgiev G, Seiler P et al (2012) Spread of influenza virus A (H5N1) clade 2.3.2.1 to Bulgaria in common buzzards. Emerg Infect Dis 18:1596–1602

Mo IP, Brugh M, Fletcher OJ et al (1997) Comparative pathology of chickens experimentally inoculated with avian influenza viruses of low and high pathogenicity. Avian Dis 41:125–136

Morgan IR, Kelly AP (1990) Epidemiology of an avian influenza outbreak in Victoria in 1985. Aust Vet J 67:125–128

Mundt E, Gay L, Jones L et al (2009) Replication and pathogenesis associated with H5N1, H5N2, and H5N3 low-pathogenic avian influenza virus infection in chickens and ducks. Arch Virol 154:1241–1248

Mutinelli F, Capua I, Terregino C et al (2003) Clinical, gross, and microscopic findings in different avian species naturally infected during the H7N1 low- and high-pathogenicity avian influenza epidemics in Italy during 1999 and 2000. Avian Dis 47:844–848

Nestorowicz A, Kawaoka Y, Bean WJ et al (1987) Molecular analysis of the hemagglutinin genes of Australian H7N7 influenza viruses: role of passerine birds in maintenance or transmission? Virology 160:411–418

OIE (2009) Avian influenza. Terrestrial animal health code. World Organisation for Animal Health, Paris, pp 1–20

Olivier AJ (2006) Ecology and epidemiology of avian influenza in ostriches. Dev Biol (Basel) 124:51–57

Olsen B, Munster VJ, Wallensten A et al (2006) Global patterns of influenza a virus in wild birds. Science 312:384–388

Pantin-Jackwood MJ, Swayne DE (2007) Pathobiology of Asian highly pathogenic avian influenza H5N1 virus infections in ducks. Avian Dis 51:250–259

Pantin-Jackwood MJ, Swayne DE (2009) Pathogenesis and pathobiology of avian influenza virus infection in birds. Rev Sci Tech 28:113–136

Pasick J, Berhane Y, Embury-Hyatt C et al (2007) Susceptibility of Canada geese (*Branta canadensis*) to highly pathogenic avian influenza virus (H5N1). Emerg Infect Dis 13:1821–1827

Perkins LE, Swayne DE (2001) Pathobiology of A/chicken/Hong Kong/220/97 (H5N1) avian influenza virus in seven gallinaceous species. Vet Pathol 38:149–164

Perkins LE, Swayne DE (2002a) Pathogenicity of a Hong Kong-origin H5N1 highly pathogenic avian influenza virus for emus, geese, ducks, and pigeons. Avian Dis 46:53–63

Perkins LE, Swayne DE (2002b) Susceptibility of laughing gulls (*Larus atricilla*) to H5N1 and H5N3 highly pathogenic avian influenza viruses. Avian Dis 46:877–885

Perkins LE, Swayne DE (2003) Varied pathogenicity of a Hong Kong-origin H5N1 avian influenza virus in four passerine species and budgerigars. Vet Pathol 40:14–24

Pillai SP, Pantin-Jackwood M, Jadhao SJ et al (2009) Pathobiology of triple reassortant H3N2 influenza viruses in breeder turkeys and its potential implication for vaccine studies in turkeys. Vaccine 27:819–824

Pillai SP, Saif YM, Lee CW (2010) Detection of influenza A viruses in eggs laid by infected turkeys. Avian Dis 54:830–833

Rohm C, Suss J, Pohle V et al (1996) Different hemagglutinin cleavage site variants of H7N7 in an influenza outbreak in chickens in Leipzig, Germany. Virology 218:253–257

Shalaby AA, Slemons RD, Swayne DE (1994) Pathological studies of A/chicken/Alabama/7395/75 (H4N8) influenza virus in specific-pathogen-free laying hens. Avian Dis 38:22–32

Silva MS, Rissi DR, Pantin-Jackwood M et al (2013) High-pathogenicity avian influenza virus in the reproductive tract of chickens. Vet Pathol 50:956–960

Slemons RD, Easterday BC (1978) Virus replication in the digestive tract of ducks exposed by aerosol to type-A influenza. Avian Dis 22:367–377

Slemons RD, Swayne DE (1990) Replication of a waterfowl-origin influenza virus in the kidney and intestine of chickens. Avian Dis 34:277–284

Spackman E, Gelb J Jr, Preskenis LA et al (2010) The pathogenesis of low pathogenicity H7 avian influenza viruses in chickens, ducks and turkeys. Virol J 7:331

Spackman E, Pantin-Jackwood MJ, Swayne DE et al (2009) An evaluation of avian influenza diagnostic methods with domestic duck specimens. Avian Dis 53:276–280

Stallknecht DE (1998) Ecology and epidemiology of avian influenza viruses in wild bird populations: waterfowl, shorebirds, pelicans, cormorants, etc. In: Swayne DE, Slemons RD (eds) Proceedings of the 4th international symposium on avian influenza, Athens, GA, 1997

Stallknecht DE, Brown JD (2008) Ecology of avian influenza in wild birds. In: Swayne DE (ed) Avian influenza, 1st edn. Blackwell Publishing, Ames, pp 43–58

Stallknecht DE, Nagy E, Hunter DB et al (2007) Avian influenza. In: Thomas NJ, Hunter DB, Atkinson CT (eds) Infectious diseases of wild birds, 1st edn. Blackwell Publishing, Ames, pp 108–130

Stipkovits L, Glavits R, Palfi V et al (2012) Pathologic lesions caused by coinfection of *Mycoplasma gallisepticum* and H3N8 low pathogenic avian influenza virus in chickens. Vet Pathol 49:273–283

Suarez DL (2008) Influenza A virus. In: Swayne DE (ed) Avian influenza, 1st edn. Blackwell Publishing, Ames, pp 3–22

Swayne DE (2007) Understanding the complex pathobiology of high pathogenicity avian influenza viruses in birds. Avian Dis 51:242–249

Swayne DE (2008) High pathogenicity avian influenza in the Americas. In: Swayne DE (ed) Avian influenza, 1st edn. Blackwell Publishing, Ames, pp 191–216

Swayne DE, Halvorson DA (2008) Influenza. In: Saif YM, Glisson JR, Fadly AM, McDougald LR, Nolan L (eds) Diseases of poultry, 12th edn. Blackwell Publishing, Ames, pp 153–184

Swayne DE, Pantin-Jackwood M (2008) Pathobiology of avian influenza virus infections in birds and mammals. In: Swayne DE (ed) Avian influenza, 1st edn. Blackwell Publishing, Ames, pp 87–122

Swayne DE, Radin MJ, Hoepf TM et al (1994) Acute renal failure as the cause of death in chickens following intravenous inoculation with avian influenza virus A/Chicken/Alabama/7395/75 (H4N8). Avian Dis 38:151–157

Swayne DE, Slemons RD (1994) Comparative pathology of a chicken-origin and two duck-origin influenza virus isolates in chickens: the effect of route of inoculation. Vet Pathol 31:237–245

Swayne DE, Slemons RD (1995) Comparative pathology of intravenously inoculated wild duck- and turkey-origin type A influenza viruses in chickens. Avian Dis 39:74–84

Swayne DE, Suarez DL, Sims LD (2013) Influenza. In: Swayne DE, Glisson JR, McDougald LR, Nolan LK, Suarez DL, Nair V (eds) Diseases of poultry, 13th edn. Wiley, Ames, pp 191–218

van Gils JA, Munster VJ, Radersma R et al (2007) Hampered foraging and migratory performance in swans infected with low-pathogenic avian influenza A virus. PLoS ONE 2:e184

Velarde R, Calvin SE, Ojkic D et al (2010) Avian influenza virus H13 circulating in ring-billed gulls (*Larus delawarensis*) in Southern Ontario, Canada. Avian Dis 54:411–419

Webster RG, Yakhno M, Hinshaw VS et al (1978) Intestinal influenza: replication and characterization of influenza viruses in ducks. Virology 84:268–278

Wood GW, Parsons G, Alexander DJ (1995) Replication of influenza A viruses of high and low pathogenicity for chickens at different sites in chickens and ducks following intranasal inoculation. Avian Pathol 24:545–551

Wood JM, Webster RG, Nettles VF (1985) Host range of A/Chicken/Pennsylvania/83 (H5N2) influenza virus. Avian Dis 29:198–207

Woolcock PR, Shivaprasad HL, De Rosa M (2000) Isolation of avian influenza virus (H10N7) from an emu (*Dromaius novaehollandiae*) with conjunctivitis and respiratory disease. Avian Dis 44:737–744

Ziegler AF, Davison S, Acland H et al (1999) Characteristics of H7N2 (nonpathogenic) avian influenza virus infections in commercial layers, in Pennsylvania, 1997-98. Avian Dis 43:142–149

Molecular Determinants of Influenza Virus Pathogenesis in Mice

Ram P. Kamal, Jaqueline M. Katz and Ian A. York

Abstract Mice are widely used for studying influenza virus pathogenesis and immunology because of their low cost, the wide availability of mouse-specific reagents, and the large number of mouse strains available, including knockout and transgenic strains. However, mice do not fully recapitulate the signs of influenza infection of humans: transmission of influenza between mice is much less efficient than in humans, and influenza viruses often require adaptation before they are able to efficiently replicate in mice. In the process of mouse adaptation, influenza viruses acquire mutations that enhance their ability to attach to mouse cells, replicate within the cells, and suppress immunity, among other functions. Many such mouse-adaptive mutations have been identified, covering all 8 genomic segments of the virus. Identification and analysis of these mutations have provided insight into the molecular determinants of influenza virulence and pathogenesis, not only in mice but also in humans and other species. In particular, several mouse-adaptive mutations of avian influenza viruses have proved to be general mammalian-adaptive changes that are potential markers of pre-pandemic viruses. As well as evaluating influenza pathogenesis, mice have also been used as models for evaluation of novel vaccines and anti-viral therapies. Mice can be a useful animal model for studying influenza biology as long as differences between human and mice infections are taken into account.

R.P. Kamal (✉) · J.M. Katz · I.A. York
Influenza Division, Centers for Disease Control and Prevention, Atlanta, GA, USA
e-mail: RKamal@cdc.gov

J.M. Katz
e-mail: JKatz@cdc.gov

I.A. York
e-mail: IYork@cdc.gov

R.P. Kamal
Battelle Memorial Institute, Atlanta, GA, USA

Current Topics in Microbiology and Immunology (2014) 385: 243–274
DOI: 10.1007/82_2014_388

Published Online: 20 July 2014

Contents

1 Introduction

Waterfowl are the natural hosts for most influenza viruses, harboring 16 of the 18 known hemagglutinin (HA) and 9 of the 11 known neuraminidase (NA) subtypes of influenza A viruses. A limited number of influenza A virus subtypes have become established in humans and other mammalian hosts and have been responsible for tens of millions of human deaths following their emergence in humans, during the initial pandemic phase and during subsequent sustained seasonal transmission in the endemic phase. Besides influenza viruses that are endemic human pathogens, several avian- or swine-origin subtypes are capable of zoonotic infection, with the potential of causing another pandemic. Therefore, the mechanisms of influenza pathogenicity are therefore the subject of intense research, especially the molecular mechanisms that allow non-human influenza viruses to become adapted to efficient human infection and transmission. Mice are a popular animal model for influenza research. We focus this review on molecular determinants that confer increased pathogenicity to influenza viruses in mice.

As with any disease model, the mouse has both advantages and disadvantages for the study of influenza pathogenesis. The relatively low cost of mice, the ease of handling, the wide range of mouse-specific reagents, and the availability of many inbred, transgenic, and knockout mouse strains make them an attractive model. On the other hand, some aspects of influenza infection and disease differ between mice and humans, complicating the application of mouse-based research to human health.

Initial attempts to characterize influenza viruses from infected humans involved transmission experiments between humans and many different species, including mice; however, despite an early claim of mouse infection that was not reproduced (Gibson and Connor 1918), "all such attempts were entirely unsuccessful until the ferret was used" (Smith et al. 1933). However, shortly after transmission to ferrets was demonstrated, several groups demonstrated that outbred mice could be infected with both swine and human influenza viruses (Andrewes et al. 1934; Francis 1934; Shope 1935). These early investigators observed signs that included weight loss, lethargy, increased respiratory effort, and loss of appetite, as well as lung pathology with extensive consolidation and fluid accumulation. Significant mortality was seen in mice following infection with high doses of the viral filtrates. Notably, there was evidence of adaptation of some, but not all, influenza viruses, with some isolates showing a marked increase in mice virulence after repeated passages in mice (Andrewes et al. 1934; Francis 1934; Francis and Magill 1935; Shope 1935; Stuart-Harris 1939) while other isolates were virulent even without mouse adaptation (Shope 1935). All these findings still hold true today.

2 Influenza Infection of Mice

While some human seasonal influenza A and B viruses can replicate in the murine respiratory tract, mice are not a natural host for influenza viruses, and adaptation through repeated passage is usually required to increase the virulence of the human-origin viruses for this species (Andrewes et al. 1934; Beaudette et al. 1957; Francis 1934; Francis and Magill 1935; Hirst 1947; Shope 1935). Importantly, even non-mouse-adapted influenza viruses that cause minimal or no disease in mice are often capable of some replication in the respiratory tract (Hirst 1947), since otherwise mouse passage would not allow adaptation to occur. The adaptation process increases the ability to replicate and may also increase virulence, either directly as a result of increased replication or through other factors.

A number of mouse-adapted strains of virus are commonly used, with A/Puerto Rico/8/1934(H1N1) (PR8) (Francis 1934; Francis and Magill 1935) and A/WSN/1933(H1N1) (WSN) (Stuart-Harris 1939) being the most common in early studies. The WSN virus was adapted through repeated mouse brain passage, resulting in a strain that in mice is neurotropic and highly virulent (Stuart-Harris 1939). PR8 is also widely used in reassortment as a recipient for HA and NA genes from other influenza viruses. A reassortant virus (X-31) containing HA and NA from an H3N2 pandemic virus, A/Aichi/2/1968(H3N2), and remaining gene segments from PR8 (Kilbourne et al. 1971) is often used as a prototype H3N2 virus in mouse studies.

The virulence of non-mouse-adapted strains of the influenza virus to mice varies widely (reviewed in Bouvier and Lowen 2010). Contemporary seasonal H3N2 and former (pre-2009 pandemic) seasonal H1N1 viruses generally replicate poorly in mice, even after extensive passage. However, pandemic H1N1 viruses including the 1918 H1N1 pandemic virus (Tumpey et al. 2005) and the 2009 A(H1N1)pdm09 pandemic strain (H1N1pdm09) (Belser et al. 2010; Maines et al. 2009) replicate efficiently in mouse respiratory tissues and cause disease in mice without prior adaptation, although some adaptation was required to derive mouse-lethal H1N1pdm09 viruses. Some swine viruses are also pathogenic to mice without adaptation (Belser et al. 2010), as are a number of avian influenza (AI) viruses (AIV). AI viruses are divided into "low-pathogenicity AIV (LPAIV)" and highly pathogenic AIV (HPAIV) based on their pathogenicity in chickens, and some members of both groups are able to infect and cause disease in mice without adaptation (Driskell et al. 2010). In general, LPAIV are less likely to be highly virulent in mice than are HPAIV; the latter, especially those of the H5N1 HPAI lineage, often cause very severe disease in mice (Gao et al. 1999; Gubareva et al. 1998; Lu et al. 1999). However, even some LPAI viruses, such as H7N9, can cause severe disease in mice, albeit at higher doses than required for many HPAI viruses (Xu et al. 2013).

Mice are typically inoculated with influenza virus by intranasal droplets; aerosol infection may also be used, and may be associated with increased morbidity and mortality (reviewed in Gustin et al. 2012). Depending on the virus, volume of inoculum, and level of sedation, after intranasal inoculation the virus initiates replication in the nasal turbinates and may then descend to the lower respiratory tract, which is the main site of replication. Unlike humans and ferrets, mice infected with influenza virus do not exhibit sneezing, coughing, or marked fever. When disease is present, the most common manifestations are weight loss, ruffled fur, shivering, hunched posture, hypothermia, and reduced activity. Weight loss, measured daily over the course of infection, is a readily quantifiable parameter used to evaluate morbidity. Pathologic and histopathologic evidence of lung edema, epithelial cell damage, and inflammation are often apparent. Viral lung titers are commonly used as an indicator of viral replication, especially when evaluating immunity or antiviral treatment. Depending on the challenge dose, mortality (including euthanasia based on humane endpoints) may be used as a measure of disease severity (Lu et al. 1999; Matsuoka et al. 2009a; Mount and Belz 2010; Narasaraju et al. 2009; O'Donnell and Subbarao 2011; Raut et al. 1975; Tripp and Tompkins 2009; van der Laan et al. 2008; Ward 1997; Xu et al. 2013). Mice have also been used to model the comparative ability of influenza viruses of different subtypes to use the eye as a portal of entry and initiation of infection (Belser et al. 2009).

Although most viruses infectious for mice replicate primarily in the respiratory tract, some viruses are capable of efficient extrapulmonary dissemination. One factor that permits systemic spread is the presence of a polybasic cleavage site in HA (discussed further below), which allows viral replication outside the respiratory tract. H5N1 HPAI viruses, which usually possess this polybasic cleavage site, frequently show systemic spread after a few days of infection, including to the brain (Lu et al. 1999; Spesock et al. 2011). Similarly, the neurotropic H1N1 strain WSN is capable of spread outside the respiratory tract (Ward 1996). In these cases, neurological signs such as ataxia and hind-limb paralysis may be seen, and the disease is generally fatal.

Even in susceptible mice infected with mouse-adapted strains of influenza, the virus is rarely spontaneously transmitted between animals, even by direct contact (Lowen et al. 2006). In the occasional cases where influenza transmission between mice has been reported (Andrewes et al. 1934; Eaton 1940; Edenborough et al.

2012; Price et al. 2014; Schulman 1967, 1968; Schulman and Kilbourne 1962; 1963a, b), the transmission is much more efficient through direct contact than via aerosol spread (Price et al. 2014), and transmission only occurs among a minority of animals.

3 Host Genetics Influencing Susceptibility to Influenza

Mouse strains differ markedly in their resistance to influenza infection. Most laboratory mice are much more susceptible to influenza viral infection than wild mice (Haller 1981; Lindenmann 1964), since laboratory strains carry a large deletion in the myxovirus resistance 1 (MX1) gene (Staeheli et al. 1988), an interferon (IFN)-induced gene that confers protection against many viruses (Haller et al. 2007). However, even strains of mice that lack MX1 have very different susceptibilities, with A/J mice being more susceptible than the widely-used C57BL/6 and BALB/c strains (Srivastava et al. 2009) and DBA mice being still more susceptible (Alberts et al. 2010; Pica et al. 2011; Srivastava et al. 2009; Trammell et al. 2012).

Several recent reviews (Horby et al. 2012, 2013; Korth et al. 2013; Trammell and Toth 2008) have discussed genetic variations of mice associated with influenza susceptibility. With the exception of MX1 (Ferris et al. 2013), the function and relative importance of most of these genetic factors are not well understood, although some are potentially associated with inflammation, such as *Tnfrsf21* (Boivin et al. 2012), *IL16*, and Nox4 (Ferris et al. 2013). Importantly, crosses between resistant and susceptible mice display a gradient of resistance, suggesting that resistance is polygenic rather than involving a limited number of genes (Boon et al. 2011; Bottomly et al. 2012; Ferris et al. 2013; Horby et al. 2013).

A major factor influencing the severity of influenza disease in mice is the inflammatory response induced by the virus, with increased inflammation being associated with increased disease severity (Askovich et al. 2013; Brandes et al. 2013; Cilloniz et al. 2010; Song et al. 2013a; Wyde and Cate 1978; Wyde et al. 1978). Many genes associated with inflammation are strongly upregulated following influenza infection of mice, especially following highly virulent viruses (Boon et al. 2011; Josset et al. 2012; Kash et al. 2006; Korth et al. 2013; Trammell and Toth 2008). Many of these genes are differentially regulated in different mouse strains (Alberts et al. 2010; Ding et al. 2008), with susceptible strains showing a more robust inflammatory response to influenza infection than do the more resistant strains (Alberts et al. 2010; Boon et al. 2011; Srivastava et al. 2009; Trammell et al. 2012), suggesting that at least one reason for differential influenza susceptibility may be variations in immune responses. However, the genetic loci involved are as yet unclear, and the increased inflammation may be secondary to increased viral load, which may be controlled by other host factors (Boon et al. 2011; Ferris et al. 2013).

One advantage of the mouse as a model system is the relative ease of generating mutant mice with alterations in specific genes. Knockout and transgenic mice have demonstrated the importance of many genes in influenza pathogenesis. Examples include transgenic mice in which MX1 expression was restored, conferring greatly increased influenza resistance, and knockout mice lacking RIG-I (Kato et al. 2006) or the IFN-inducible gene IFITM3 (Everitt et al. 2012), which are highly susceptible to influenza infection. Conversely, mouse knockouts lacking IL-17RA, IL-15, CC-chemokine receptor 2 (CCR2), or a combination of tumour necrosis factor (TNF) and IL-1 receptors, were relatively resistant to mortality due to influenza virus infection (reviewed in Medina and Garcia-Sastre 2011), further suggesting that at least in some cases influenza mortality in mice may be due to an overactive inflammatory response. It is important to note that while complete knockout of a gene can demonstrate the role of that gene in the response to influenza, it does not tell us about differential responses associated with allelic variation.

4 Viral Determinants of Influenza Virulence in Mice

Many components of influenza viruses contribute to their ability to infect mice. Identification of the viral molecular determinants involved in mouse virulence can be approached in a number of ways. By analyzing strains of influenza with differential virulence in mice (e.g., viruses before and after mouse adaptation), genetic changes associated with mouse virulence can be identified. However, specific genetic changes in these viruses are not necessarily associated with mouse virulence; some changes may have been selected as a result of other laboratory properties, such as growth in eggs, or may simply represent genetic drift. For example, the PR8 lineage of the virus, which has been grown in the laboratory for many decades, has diverged into multiple strains with different genome sequences, all of which are still capable of causing disease in mice to varying extents (Blazejewska et al. 2011; Grimm et al. 2007; Liedmann et al. 2013).

Another approach to identifying mouse virulence-associated variants is to introduce specific mutations into viruses, and then test the resulting strains for their ability to infect mice. This reverse genetics approach may also be used to confirm the functional importance of the variants found in mouse-adapted viral strains. A disadvantage with this approach is that it becomes difficult to test multiple mutations that act together to alter virulence.

In spite of the technical difficulties involved, a number of mouse adaptation determinants have been identified, spanning all eight segments of the influenza genome (summarized in Table 1). Some of these variants may be specific to mice, while others represent adaptations to mammals versus birds, or even broader virulence determinants that are important for infecting cells in general. Many of the general mammalian determinants have been reviewed recently (Basler and Aguilar 2008; Pepin et al. 2010; Reperant et al. 2012).

Table 1 Summary of influenza virus variants and mutations that affect replication and virulence in mice

Protein	Virus	Mouse	Mutation	Effect	Reference(s)
HA	A/Vietnam/1203/2004(H5N1)	BALB/c	Deletion of polybasic cleavage site	Reduced respiratory and systemic replication. Reduced proinflammatory cytokines in lungs	Suguitan et al. (2012)
	A/Hong Kong/156/97(H5N1)	ddY	P211T	Decreased virulence	Hiromoto et al. (2000)
	Reassortant WSN-HA + A/Hong Kong/1/68(H1N1xH3N2)	BALB/c	T160N, P162S	Increased virulence	Keleta et al. (2008)
	A/Aichi/2/68(H3N2)	BALB/c	G218E	Increased virulence	Narasaraju et al. (2009)
	A/Mallard duck/Pennsylvania/10218/84(H5N2)		S203F, E273G, L320P	Adaptation and virulence	Smirnov et al. (2000)
	A/California/04/09(H1N1pdm09), A/Tennessee/1-560/09(H1N1pdm09)	BALB/c	D222G	Mouse adaptation and virulence	Ilyushina et al. (2010)
	A/Brisbane/59/2007(H1N1)	BALB/c	T89I, N125T, D221G	Mouse adaptation and virulence	Xu et al. (2011a)
	A/Hamburg/04/09(H1N1pdm09)	BALB/c	D222G K163E	Mouse adaptation and pathogenicity	Seyer et al. (2012)
	A/Hong Kong/1/68(H3N2)	CD-1	G218W	Mouse adaptation and virulence	Ping et al. (2010)
	A/Hong Kong/1/68(H3N2)	CD-1	T156N	Mouse adaptation and virulence	Ping et al. (2010)
	Reassortant WSN-HA + A/Hong Kong/1/68(H1N1xH3N2)	BALB/c	N154S	Mouse adaptation	Keleta et al. (2008)
	A/Vietnam/1203/2004(H5N1) NS1-deleted	Outbred	K58I	Increased immunogenicity	Krenn et al. (2011)
	A/black duck/New Jersey/1580/78(H2N3)		E216D, K307R, T318I	Mice adaptation and virulence	Govorkova et al. (2000)
	A/JapanxBellamy/57(H2N1)		K25T, S203F		

(continued)

Table 1 (continued)

Protein	Virus	Mouse	Mutation	Effect	Reference(s)
M1	A/black duck/New Jersey/1580/78(H2N3)		N30D, Q214H	Mice adaptation and virulence	Govorkova et al. (2000)
	A/JapanxBellamy/57(H2N1)		M179K		
	A/Wisconsin/33		A41V + T139A	Mice adaptation neurovirulence	Ward (1995)
	A/FM/1/47(H1N1)		T139A	Mice adaptation and virulence	Brown and Bailly (1999)
	A/duck/Guangxi/53/2002(H5N1)	BALB/c	N30D + T215A	Increased virulence	Fan et al. (2009)
M2	A/black duck/New Jersey/1580/78(H2N3), A/JapanxBellamy/57(H2N1)		N93S	Mice adaptation and virulence	Govorkova et al. (2000)
NS1	Reassortant A/Hong Kong/1/68(H3N2) x PR8	BALB/c	F103L + M106I	Increased infectivity and virulence	Dankar et al. (2011)
	A/Puerto Rico/8/34(H1N1)	PKR KO, IFNAR/IL28R KO	R184G	Increased replication in mice lungs	Steidle et al. (2010)
	A/Aichi/2/68(H3N2)	BALB/c	D125G	Increased virulence	Narasaraju et al. (2009)
	A/Hong Kong/156/97(H5N1)	ddY	D101N	Increased virulence	Hiromoto et al. (2000)
	A/Duck/Guangxi/12/03(H5N1)	BALB/c	P42S	Increased virulence	Jiao et al. (2008)
	A/Duck/Guangxi/27/03(H5N1)	BALB/c	S42P	Attenuation	
	A/Duck/Shandong/093/2004(H5N1)	BALB/c	E92D	Decreased virulence	Long et al. (2008)
PA	A/Hong Kong/156/97(H5N1)	ddY	G631S	Increased virulence	Hiromoto et al. (2000)
	A/Hamburg/04/09(H1N1pdm09)	BALB/c	F35L	Mouse adaptation and pathogenicity	Seyer et al. (2012)

(continued)

Table 1 (continued)

Protein	Virus	Mouse	Mutation	Effect	Reference(s)
PB1	A/Hong Kong/156/97(H5N1)	ddY	H456Y, S712P	Increased virulence	Hiromoto et al. (2000)
	A/Cambodia/P0322095/2005(H5N1)	BALB/c	473L + 598L	Decrease in mouse infectivity	Xu et al. (2012)
	A/Wisconsin/33(H1N1) PB2K627E	BALB/c	L473V + L598P	Restore mouse infection/replication	
	Reassortant A/California/07/2009(H1N1pdm09) + WSN	BALB/c	473L	Decrease in mouse infectivity	
PB1-F2	A/Viet Nam/1203/2004(H5N1)	C57BL/6/A	N66S	Increased virulence	Schmolke et al. (2011)
				No effect	
	Reassortant A/WSN/33(H1N1) + A/HK/156/97(H5N1) PB1, A/Brevig Mission/1918(H1N1)	C57BL/6	N66S	Increased virulence	Conenello et al. (2007)
	Reassortant A/Puerto Rico/8/34(H1N1) + A/Wuhan/359/95(H3N2) PB1	DBA/2	P62L + H75R + S82L	Increased virulence	Alymova et al. (2011)
	A/Puerto Rico/8/34 (H1N1)	BALB/c	I68T + L69Q + V70G	Reduced primary viral infectivity and secondary bacterial pneumonia	Alymova et al. (2014)
PB2	A/Hong Kong/1/68(H3N2)	CD-1	K482R, D701N, D740N	Increased mouse virulence	Ping et al. (2011)
	A/Hong Kong/1/68 (H3N2)	CD-1	D701N	Mouse adaptation and virulence	Ping et al. (2010)
	A/Hong Kong/156/97(H5N1)	ddY	N701D	Increased virulence	Hiromoto et al. (2000)
	A/Seal/Massachussetts/1/80(H7N7)	BALB/c	D701N	Increased virulence	Gabriel et al. (2005)
	A/duck/Guangxi/22/2001(H5N1)	BALB/c	D701N	Adaptation and increased virulence	Li et al. (2005)
	A/New York/1682/2009(H1N1pdm09)	BALB/c	D701N	Increased virulence	Zhou et al. (2013)
	A/chicken/Netherlands/621557/03(H7N7)	BALB/c	E627K	Mouse adaptation and increased virulence	de Jong et al. (2013)

(continued)

Table 1 (continued)

Protein	Virus	Mouse	Mutation	Effect	Reference(s)
	A/chicken/Guangdong/Ts/2004(H9N2), A/chicken/Guangdong/V/2008(H9N2)	BALB/c	E627K	Reduced host antiviral response	Tian et al. (2012)
	A/chicken/Shandong/16/05(H9N2)	BALB/c	M147L + E627K	Increased virulence	Wang et al. (2012)
	A/Hong Kong/483/97(H5N1) A/Hong Kong/486/97(H5N1)	BALB/c	627K	Increased virulence	Chen et al. (2007), Hatta et al. (2001)
	A/Netherlands/33/03(H7N7) A/Netherlands/219/03(H7N7)	BALB/c	627K	Increased virulence	Munster et al. (2007)
	A/Shanghai/2/2013(H7N9)	BALB/c	E627K	Increased virulence	Mok et al. (2014)
	A/Anhui/1/2013(H7N9)	BALB/c	E627K	Increased replication and virulence	Zhang et al. (2014)
	Reassortant 1918 H1N1 x avian H1N1	BALB/c	E627K	Increased mouse virulence	Qi et al. (2012)
	A/equine/London/1416/73(H7N7)	BALB/c	E627K	Adaptation and increased neuronal invasion	Shinya et al. (2007)
NP	A/Hong Kong/1/68(H3N2)	CD-1	D34N, D290N, D480N	Increased mouse virulence Adaptation	Ping et al. (2010)
	A/Hong Kong/156/97(H5N1)	ddY	E127K	Increased virulence	Hiromoto et al. (2000)
	A/Seal/Massachussetts/1/80(H7N7)	BALB/c	N319K	Mouse adaptation	Gabriel et al. (2005, 2008)
NA	r A/Lyon/969/2009(H1N1pdm09) +/− NA from A/Lyon/1337/2007(H1N1)	Balb/cByJ	H275Y	Increased infectivity	Ferraris et al. (2012)
	A/California/04/09(H1N1pdm09)	Balb/c	H275Y	Increased virulence	Song et al. (2013b)

4.1 Hemagglutinin

Influenza viruses enter cells after binding to cell-surface receptors consisting of sialic acid (SA) moieties. Sialic acid is a complex structure that can be attached to galactose in a number of different linkages (α2,3; α2,6; α2,8). Influenza viruses of different host origins preferentially bind to different SA configurations, with human viruses preferring SA with an α2,6 linkage to galactose while avian viruses preferentially bind SA with α2,3 linkages (Ge and Wang 2011; Imai and Kawaoka 2012; Shinya et al. 2006). The distribution of SA in the host is therefore an important factor determining influenza pathogenesis. In humans, α2,6-linked SA dominate in the upper airways (Shinya et al. 2006), and human influenza viruses mainly bind to α2,6-linked SA. In mice, α2,3-linked SA are found in both the upper and lower respiratory tract (Ibricevic et al. 2006), so that many human wild-type viruses are poorly or not infectious for mice.

As the protein required for receptor binding, HA is a critical component of influenza infectivity. Multiple studies have linked variants in the HA segment of the influenza genome with enhanced mouse infectivity and virulence (Brown 1990; Kobasa et al. 2004; Pappas et al. 2008; Rudneva et al. 1986; Uraki et al. 2013; Watanabe et al. 2013). In some cases, mouse adaptation of human influenza viruses has been associated with a shift in receptor preference from to α2,6- to α2,3-linked SA (Ilyushina et al. 2010; Keleta et al. 2008). However, several lines of evidence suggest that this is not the sole reason for mouse adaptation. As noted above, even viruses that are avirulent in mice may replicate in the lungs (Hirst 1947). Some viruses that preferentially bind α2,6-linked SA, such as the 1918 pandemic H1N1, replicate efficiently and cause significant disease in mice (Qi et al. 2009; Tumpey et al. 2005), while many LPAIV that preferentially bind α2,3-linked SA cause only minimal morbidity in this species (Driskell et al. 2010).

HPAIV generally have a polybasic cleavage site in their HA (Bosch et al. 1981; Webster and Rott 1987), while HA from mammalian viruses and LPAIV usually have only a single basic residue at their cleavage site (Garten and Klenk 1999; Klenk and Garten 1994). The presence of the polybasic residues allows the HA to be cleaved by enzymes that are expressed in a range of tissues, whereas the standard cleavage site can only be cleaved by enzymes in the respiratory tract. Since HA cleavage is required for viral infectivity, the polybasic cleavage site broadens the tissue range of the virus and is strongly associated with the systemic spread and increased infectivity and virulence in birds and mammals, including mice (Stienke-Grober et al. 1992; Suguitan et al. 2012; Tanaka et al. 2003; Zhang et al. 2012). In H7 viruses, a similar phenotype has arisen on several occasions through non-homologous recombination between HA and other viral or host genes (reviewed in Belser and Tumpey 2014). Similarly, sequence variation at the cleavage site of the WSN strain that confers a broader range of enzyme cleavage is associated with increased neurovirulence (Sun et al. 2010). However, not all HPAIV with a polybasic cleavage site are highly virulent in mice (Gao et al. 1999; Hiromoto et al. 2000; Hu et al. 2012; Katz et al. 2000; Lu et al. 1999).

Mutations in residues in three regions that surround the receptor-binding site (the 190-helix, the 220-loop, and the 130-loop; reviewed in Imai and Kawaoka 2012) have been associated with mouse adaptation (Govorkova et al. 2000; Hiromoto et al. 2000; Ilyushina et al. 2010; Keleta et al. 2008; Koerner et al. 2012; Narasaraju et al. 2009; Ping et al. 2010, 2011; Smirnov et al. 2000; Song et al. 2013b). This is probably due to changes in receptor binding preference (Imai and Kawaoka 2012; Pekosz et al. 2009).

One mouse-adaptive change of particular interest is at position 222 (in H1 numbering; 225 in H3 numbering). A D222G mutation has repeatedly appeared during mouse adaptation of H1N1 viruses (Ilyushina et al. 2010; Seyer et al. 2012; Smee et al. 2007; Song et al. 2013b; Xu et al. 2011a) and, at least in some cases, increased mouse virulence (Abed et al. 2011; Zheng et al. 2010). This variant is associated with increased pathogenicity of H1N1pdm09 in humans (Baldanti et al. 2011; Kilander et al. 2010) and increases α2,3-linked SA binding of HA (Abed et al. 2011; Belser et al. 2011; Chutinimitkul et al. 2010).

HA is a glycoprotein, with multiple sites of carbohydrate attachment. Loss of glycosylation sites from HA is a common observation during mouse adaptation (Chen et al. 2007; Reading et al. 2007; Shilov and Sinitsyn 1994; Smee et al. 2007; Ward 1996, 1997). Studies with natural variants and reverse genetics viruses have shown that virulence in mice decreases as the number of glycans increases (Kaverin et al. 2002; Medina et al. 2013; Reading et al. 2009; Rudneva et al. 2005; Sun et al. 2013a; Vigerust et al. 2007; Yen et al. 2009). In one exception to the rule, adding glycosylation sites to the HA of H1N1pdm09 viruses increased viral virulence in mice, perhaps related to increased inflammation due to enhanced innate immune recognition (Zhang et al. 2013). Some of this effect may be due to changing receptor specificity (Das et al. 2011; Ohuchi et al. 1997; Sun et al. 2013a; Wang et al. 2009; Yen et al. 2009; Zhang et al. 2013) and thus may be specific to mouse infection. However, increasing glycosylation also renders influenza virus more susceptible to innate immune control (Kaverin et al. 2002; Reading et al. 2007, 2009; Shilov and Sinitsyn 1994; Tate et al. 2011; Vigerust et al. 2007), so that loss of glycosylation is probably an adaptive factor for mammalian infections in general. The effect of glycosylation is complicated by the ability of glycans to shield antigenic sites from antibody recognition, so that increasing glycosylation may confer some protection against population immunity while reducing viral infectivity (Das et al. 2011; Job et al. 2013; Medina et al. 2013; Rudneva et al. 2005); however, since mice infected with influenza are generally immunologically naïve to the virus, this is less likely to affect mouse adaptation.

Other changes in HA associated with mouse virulence, including 78, 162, 321, and 354 in HA1 (Keleta et al. 2008; Koerner et al. 2012; Ping et al. 2011; Seyer et al. 2012; Xu et al. 2011b) and 58, 154, and 156 in HA2 (Keleta et al. 2008; Krenn et al. 2011; Ping et al. 2010), may alter HA stability and fusion efficiency (Smeenk et al. 1996; Ward 1997; Zaraket et al. 2013), which in turn may alter cell tropism. Other changes in HA may be associated with altered inflammatory response to the virus (Brown and Bailly 1999; Watanabe et al. 2013; Xu et al. 2011a).

4.2 Neuraminidase

While HA binds to SA, NA cleaves SA from glycoprotein carbohydrate chains, allowing newly-formed virions to escape from the cell surface. NA activity must balance HA activity for optimal viral entry and exit, and it is therefore not surprising that the NA segment affects mouse virulence (Gen et al. 2013; Pappas et al. 2008; Tumpey et al. 2004; Zhang et al. 2011). In some cases, enzymatic activity of NA has been linked to mouse virulence. For example, NA contains a stalk region that varies in length between and within influenza subtypes and influences its enzymatic function (Castrucci and Kawaoka 1993; Matsuoka et al. 2009b; Zhou et al. 2009). NA segments from either H5N1 (Matsuoka et al. 2009b; Zhou et al. 2009) or H9N2 viruses (Sun et al. 2013b) containing a short stalk confer increased mouse pathogenicity to reassortant influenza viruses, compared to NA with full-length stalks. Loss of a glycosylation site near the catalytic site of NA was found in two mouse-adapted viruses (Brown and Bailly 1999; Li et al. 1993) and may alter the enzymatic activity of NA (Brown and Bailly 1999) or even change HA functionality (Goto and Kawaoka 1998).

NA mutations, such as H275Y (H274Y in N2 numbering), associated with resistance to the NA inhibitor antiviral drug oseltamivir, are occasionally observed in viruses isolated from humans, At least on the H1N1pdm09 background, the H275Y variant is associated with increased virulence and infectivity in mice even in the absence of antiviral treatment (Ferraris et al. 2012; Song et al. 2013b), especially when compensatory mutations (such as V234M and R222Q in NA (N2 numbering) and K153E in HA) are present (Bloom et al. 2010; Song et al. 2013b).

4.3 Ribonucleoprotein Complex

Within an influenza A virion, each genomic segment is associated with a set of proteins comprised of NP and the RNA-dependent RNA polymerase complex (PB2, PB1, and PA). This complex is collectively termed the ribonucleoprotein (RNP) complex (reviewed in Noda and Kawaoka 2010), and acts as a single unit throughout much of the viral life cycle. Variants in RNP components are frequently linked to mouse adaptation, and in several cases the same functional adaptation (such as improved nuclear import) may be achieved through mutations in any of several components of the RNP.

PB2 is a major virulence determinant of influenza viruses, and mouse-adapted viruses frequently contain mutations in this gene. One of the most common changes as avian influenza viruses are adapted to mouse infection is an E627K change in PB2 (de Jong et al. 2013; Kim et al. 2010; Ping et al. 2010; Song et al. 2011; Tian et al. 2012; Wang et al. 2012; Wu et al. 2009; Zhang et al. 2011), and this change is strongly linked to increased mouse virulence in H5N1 (Chen et al. 2007; Hatta et al. 2001; Li et al. 2009; Maines et al. 2005) and H7N7 (Munster et al. 2007) HPAI viruses, H7N9 LPAI viruses (Mok et al. 2014; Zhang et al. 2014), and the 1918

H1N1 (Qi et al. 2012). Many H5N1 HPAIV and H7N9 LPAIV isolated from zoonotic human infections possess this substitution, whereas viruses isolated from avian species generally do not (Fonville et al. 2013; Manz et al. 2013). The same effect was seen in an equine H7N7 virus as it became mouse adapted (Shinya et al. 2004, 2007). Less commonly found during mouse adaptation, but also strongly associated with increased mouse virulence, is a D701N variant in PB2 (Gabriel et al. 2005; Li et al. 2005; Ping et al. 2010, 2011; Zhou et al. 2013). Both of these changes (E627K and D701N) are considered to be general markers of mammalian, not just mouse, virulence (Gabriel et al. 2007, 2013; Steel et al. 2009; Subbarao et al. 1993). Polymerase complexes which possess the E627K mutation are more active in mammalian cells (Subbarao et al. 1993), at least in part due to their interaction with importin-α isoforms (Gabriel et al. 2008, 2011; Hudjetz and Gabriel 2012). Importins are components of the nuclear import pathway (Whittaker et al. 2000) that are required for influenza virus nuclear entry and therefore replication (Hutchinson and Fodor 2012). Species-specific differences in importins affect influenza RNP nuclear import, and mutations in the RNP that enhance nuclear import in a new host appear to be strongly adaptive. As well as PB2, changes in NP (N319K) alter interactions with importins and therefore affect mouse adaptation and host range in general (Gabriel et al. 2005, 2008, 2011). These and other compensatory changes (e.g. PB2 Q591K/L) (Gabriel et al. 2013; Yamada et al. 2010) mean that the E627K and D701N variants are not absolutely required for mouse virulence (Li et al. 2009; Manz et al. 2012; Xu et al. 2012; Yamada et al. 2010).

As well as interactions with importins, adaptive changes in PB2, including E627K, may play other roles in polymerase functions, such as enhancing the ability of the polymerase to act at the lower temperatures associated with mammalian respiratory tracts vs. avian gastrointestinal tracts (reviewed in Manz et al. 2013).

Mouse-adaptive substitutions in NP seem to be rare, but include D34N and D480N of H3N2 (Ping et al. 2011). Mouse-adaptive changes in the PB1 and PA genes include several that compensate for the lack of E627K in an avian-origin RNP complex, such as L472V and L598P in PB1 (Xu et al. 2012). Multiple other substitutions in these genes that have been associated with changes in mouse virulence have been described (Hiromoto et al. 2000; Ilyushina et al. 2010; Liedmann et al. 2013; Manz et al. 2013; Ping et al. 2011; Song et al. 2013b; Zhang et al. 2011). The mechanisms by which these substitutions affect mouse virulence are not known, although many of the mutations are at sites involved in the protein-protein interactions involved in polymerase complex assembly (Manz et al. 2013; Ping et al. 2011). The PB1 gene also encodes a small protein known as PB1-F2, expressed from an alternate open reading frame. The significance of this protein in influenza virus mouse adaptation and virulence will be addressed below.

4.4 M1 and M2

M1 and M2 are produced by differential splicing from the M gene segment. M1 is an internal structural protein that helps mediate virion assembly, interacting with HA and NA as well as the RNP. M2 is an ion channel that is required for viral entry into cells and subsequent replication (reviewed in Rossman and Lamb 2011). Early investigations into mouse adaptive mutations identified changes in M1 (A41V and T139A) as important virulence determinants for the mouse-adapted WSN strain (Ward 1995, 1996, 1997). The T139A mutation has also been identified in the mouse-adapted A/FM/1/47-MA(H1N1) strain (Brown and Bailly 1999; Smeenk and Brown 1994; Smeenk et al. 1996). Changes in M1 and in M2 have also been observed during mouse adaptation of H2 (Govorkova et al. 2000), H3N2 (Brown et al. 2001; Ping et al. 2010), and H9N2 (Zhang et al. 2011) viruses, although the functional importance of these changes is unclear. Using reverse genetics, N30D and T215A in the M1 protein were shown to increase the virulence of H5N1 virus in mice (Fan et al. 2009). A V15I/T substitution in M1 protein was a common amino acid substitution found in H5N1 viruses exhibiting high virulence in mice (Katz et al. 2000). Functional domains of M1 and M2 are not well understood as yet, and the mechanism(s) by which these substitutions in M1 and M2 may alter mouse pathogenicity are not yet clear.

4.5 NS1 and NEP

NS1 is an important non-structural protein expressed from a transcript from the NS gene segment. Alternate splicing of this transcript produces a second protein, nuclear export protein (NEP; also known as NS2) that has several roles in viral replication (Paterson and Fodor 2012). While NS1 is non-essential for viral replication in cultured cells (Garcia-Sastre et al. 1998), it is critical for natural infection because of its ability to inhibit the host innate immune response (Donelan et al. 2003) (reviewed in Hale et al. 2008) through species-specific interactions with multiple host proteins (Rajsbaum et al. 2012). Therefore, it is not surprising that mutations in NS1 are frequently observed during mouse adaptation of both human (Brown et al. 2001; Forbes et al. 2012; Ping et al. 2011) and avian (Dankar et al. 2011; Zhang et al. 2011) influenza viruses. For example, adaptation of the human H3N2 virus to mice resulted in 12 different NS1 mutations, most of which enhanced the ability of the virus to block IFN induction in mice (Forbes et al. 2012). However, NS1 mutations can apparently alter mouse virulence without affecting the IFN response (Forbes et al. 2012; Steidle et al. 2010).

Reassortant and reverse genetics experiments have provided more detailed information on the amino acid residues in NS1 associated in mouse pathogenesis, identifying a number of functional motifs that interact with host proteins and affect NS1 activity. One such motif is the four-residue C-terminal PDZ domain ligand.

This region interacts with multiple members of the PDZ domain family of proteins, which are often involved in localized intracellular signaling. The PDZ domain ligand sequence has different canonical sequences in human (typically RSKV) and avian (ESEV) NS1 proteins, and the avian sequence binds to more human PDZ domain proteins than does the human sequence, potentially disrupting multiple protein-protein interactions (Obenauer et al. 2006). In mice, the avian sequence confers increased virulence to low-virulence viruses (Fan et al. 2013; Jackson et al. 2008; Zielecki et al. 2010). However, the impact of this sequence is context-dependent (Hale et al. 2010; Zielecki et al. 2010), so that in some cases effects are only seen when certain other changes are present (Fan et al. 2013).

Another functional motif in NS1 includes residues F103 and M106, which are involved in binding to the cleavage and polyadenylation specificity factor (CPSF30), which is required for 3' end processing of cellular pre-mRNAs and is involved in IFN induction following influenza virus infection. H5N1 HPAI viruses containing NS1 without this motif are significantly less virulent in mice, showing greatly reduced systemic spread (Dankar et al. 2011, 2013; Spesock et al. 2011). Similarly, an R184G mutation in NS1, which also affects CPSF30 binding, reduces the mouse virulence of PR8; however, this mutation alters mouse virulence without apparently altering the IFN response (Steidle et al. 2010).

Other NS1 variants that have been linked with increased mouse virulence include P42S, which increases the ability of H5N1 NS1 to block IFN responses in mice (Jiao et al. 2008), and a deletion of residues 80–84 combined with a D92E change in H5N1 NS1, which increased virulence by an unknown mechanism (Long et al. 2008).

NEP, which plays several roles in viral replication, may be an important factor in the adaptation of AIV to human cells (Manz et al. 2012; Paterson and Fodor 2012). NEP mutations have been observed during mouse adaption, but the functional importance of these changes is not known (Forbes et al. 2012; Manz et al. 2013).

4.6 PB1-F2

PB1-F2 is a small gene that is expressed from an alternative open reading frame of segment 2, which also encodes the PB1 protein. Although multiple functions have been proposed for PB1-F2, several groups have found that this gene can antagonize IFN expression (Conenello et al. 2011; Dudek et al. 2011; Le Goffic et al. 2010; Varga et al. 2011, 2012). PB1-F2 is non-essential for viral replication and is frequently truncated in influenza viruses of mammals, but is usually full length in avian influenza viruses, suggesting that it may play an important role in the viral pathogenesis in birds versus mammals (Pasricha et al. 2013; Zell et al. 2007). In mice, the expression of full-length PB1-F2 (compared to a truncated protein) increases the virulence of several influenza viruses, including H5N1 and reassortant viruses expressing PB1-F2 from 1918 H1N1 or H5N1 viruses in the context of PR8 and WSN (Belser et al. 2010; Dudek et al. 2011; Leymarie et al. 2013; McAuley et al. 2007, 2013; Zamarin et al. 2006) (reviewed in Basler and Aguilar 2008; Conenello and Palese 2007; Krumbholz et al. 2011). The presence of full-length PB1-F2 in the PR8 virus is associated with increased susceptibility to secondary bacterial pneumonia (Alymova et al. 2014; Huber 2012; Iverson et al. 2011; McAuley et al. 2007). For reasons that are as yet unclear, the influence of PB1-F2 on pathogenicity is virus-dependent (McAuley et al. 2010) (reviewed in Krumbholz et al. 2011). For example, restoring full-length PB1-F2 expression to H1N1pdm09 viruses (which contain a stop codon 11 aa after the protein start site) does not alter virulence in mice (Meunier and von Messling 2012; Ozawa et al. 2011; Wang et al. 2010).

Analysis of reassortant and reverse-genetic viruses has shown that 66S in PB1-F2 is associated with high mouse virulence in H5N1 HPAIV and in 1918 H1N1, while 66N is associated with low-virulence viruses (Conenello et al. 2007, 2011; Schmolke et al. 2011), probably by altering IFN responses (Conenello et al. 2011; Varga et al. 2011). Several other residues located between 62 and 82 of PB1-F2 (L62, I68, L69, V70, R75, R79, and L82) have also been linked to increased inflammation, secondary bacterial infection, and lung pathology in mice (Alymova et al. 2011, 2014).

5 Conclusions

Are mice a good model for human influenza infection and disease? First, it is important to note that mice are used for many aspects of the influenza virus research other than the pathogenesis studies we discuss here. For example, mice are widely used to evaluate antibody and cell-mediated immune responses to influenza and are often a first-line preclinical model used to evaluate novel vaccines (Bodewes et al. 2010; Tripp and Tompkins 2009; van der Laan et al. 2008), and antiviral therapies including therapeutic antibodies (Mancini et al. 2011), and antiviral drugs (Barnard 2009; Gubareva et al. 1998; Ison et al. 2006; Sidwell and Smee 2000). It is also worth noting that mouse experiments are not performed in vacuum; potentially interesting host or viral genetic determinants can be further analyzed in other models, such as ferrets and non-human primates.

If mice are a useful model for human infection with influenza virus adaptation and pathogenesis, then the changes associated with mouse adaptation should also be adaptive for humans. In many cases, mouse adaptive changes seem to be similar to those selected as avian viruses adapt to human infection (Brown et al. 2001; Pepin et al. 2010; Reperant et al. 2012). In general, the changes associated with mouse virulence can be divided into three groups. First are changes in the way the virus interacts with its receptors, which includes many of the HA and NA adaptive changes. These are most likely to be mouse-specific, since SA distribution in mice is different from humans (Ibricevic et al. 2006; Shinya et al. 2006). Nevertheless, some of these changes are informative for human disease. For example, the D222G

(D225G in H3 numbering) substitution in HA is a mutation associated with mouse adaptation (Ilyushina et al. 2010; Seyer et al. 2012; Smee et al. 2007; Song et al. 2013b; Xu et al. 2011a), and also has been associated with severe disease in humans (Baldanti et al. 2011; Kilander et al. 2010). Similarly, some mouse-adaptive changes in the HA from H3N2 viruses mirror the changes associated with human adaptation of avian H3N2 (Keleta et al. 2008).

A second group of molecular changes associated with mouse adaptation and virulence is of those found in viral proteins required for replication. Examples of such changes include PB2 E627K and D701N, and related compensatory mutations. These are often associated with mouse adaptation, and are also repeatedly seen in H5N1 and H7N9 viruses isolated from humans, while the same viral subtypes isolated from birds are less likely to have these mutations. These mutations therefore do seem to be broadly relevant to mammalian infection, including humans (Gabriel et al. 2007, 2013; Manz et al. 2013; Steel et al. 2009).

A third group includes changes in non-essential proteins such as PB1-F2 and NS1. Such mutations are commonly seen during mouse adaptation and have major effects on virulence in mice. These proteins tend to modulate the host response to infection. Again, in many cases, these interactions seem to be broadly similar for mammalian infection. However, some interactions seem to be different for mice and humans, since mouse adaptation of human viruses leads to multiple mutations in NS1 (Brown et al. 2001; Forbes et al. 2012; Ping et al. 2011). This is not surprising, since genes associated with immune function evolve more rapidly than those involved with basic cellular functions (Hughes et al. 2005), so that such targets are more likely to differ between different mammalian species. For these genes, the overall pattern of change may be significant while the precise mutations involved may not always extrapolate to virus adaptation to humans.

Thus, studies of influenza infection in mice can in many ways parallel or predict virus adaptation to and pathogenesis in humans (Pepin et al. 2010; Reperant et al. 2012). Overall, mice are a convenient and cost-effective animal model that, as long as observations are interpreted in the light of physiology and evolution, can enhance our understanding of influenza virus virulence determinants for mammals, including humans.

References

Abed Y, Pizzorno A, Hamelin ME, Leung A, Joubert P, Couture C, Kobasa D, Boivin G (2011) The 2009 pandemic H1N1 D222G hemagglutinin mutation alters receptor specificity and increases virulence in mice but not in ferrets. J Infect Dis 204:1008–1016. doi:10.1093/infdis/jir483

Alberts R, Srivastava B, Wu H, Viegas N, Geffers R, Klawonn F, Novoselova N, do Valle TZ, Panthier JJ, Schughart K (2010) Gene expression changes in the host response between resistant and susceptible inbred mouse strains after influenza A infection. Microbes Infect 12:309–318. doi:10.1016/j.micinf.2010.01.008

Alymova IV, Green AM, van de Velde N, McAuley JL, Boyd KL, Ghoneim HE, McCullers JA (2011) Immunopathogenic and antibacterial effects of H3N2 influenza A virus PB1-F2 map to amino acid residues 62, 75, 79, and 82. J Virol 85:12324–12333. doi:10.1128/JVI.05872-11

Alymova IV, Samarasinghe A, Vogel P, Green AM, Weinlich R, McCullers JA (2014) A novel cytotoxic sequence contributes to influenza A viral protein PB1-F2 pathogenicity and predisposition to secondary bacterial infection. J Virol 88:503–515. doi:10.1128/JVI.01373-13

Andrewes CH, Laidlaw PP, Smith W (1934) The susceptibility of mice to the viruses of human and swine influenza. The Lancet 224:859–862

Askovich PS, Sanders CJ, Rosenberger CM, Diercks AH, Dash P, Navarro G, Vogel P, Doherty PC, Thomas PG, Aderem A (2013) Differential host response, rather than early viral replication efficiency, correlates with pathogenicity caused by influenza viruses. PLoS One 8:e74863. doi:10.1371/journal.pone.0074863

Baldanti F, Campanini G, Piralla A, Rovida F, Braschi A, Mojoli F, Iotti G, Belliato M, Conaldi PG, Arcadipane A, Pariani E, Zanetti A, Minoli L, Emmi V (2011) Severe outcome of influenza A/H1N1/09v infection associated with 222G/N polymorphisms in the haemagglutinin: a multicentre study. Clin Microbiol Infect 17:1166–1169. doi:10.1111/j.1469-0691.2010.03403.x

Barnard DL (2009) Animal models for the study of influenza pathogenesis and therapy. Antiviral Res 82:A110–A122. doi:10.1016/j.antiviral.2008.12.014

Basler CF, Aguilar PV (2008) Progress in identifying virulence determinants of the 1918 H1N1 and the Southeast Asian H5N1 influenza A viruses. Antiviral Res 79:166–178. doi:10.1016/j.antiviral.2008.04.006

Beaudette FR, Byrne RJ, Kaschula VR (1957) The adaptation of equine influenza virus to infant mice by the intracerebral route. Cornell Vet 47:137–143

Belser JA, Jayaraman A, Raman R, Pappas C, Zeng H, Cox NJ, Katz JM, Sasisekharan R, Tumpey TM (2011) Effect of D222G mutation in the hemagglutinin protein on receptor binding, pathogenesis and transmissibility of the 2009 pandemic H1N1 influenza virus. PLoS One 6:e25091. doi:10.1371/journal.pone.0025091

Belser JA, Tumpey TM (2014) Mammalian models for the study of H7 virus pathogenesis and transmission. Current Topics in Microbiology and Immunology (in press)

Belser JA, Wadford DA, Pappas C, Gustin KM, Maines TR, Pearce MB, Zeng H, Swayne DE, Pantin-Jackwood M, Katz JM, Tumpey TM (2010) Pathogenesis of pandemic influenza A (H1N1) and triple-reassortant swine influenza A (H1) viruses in mice. J Virol 84:4194–4203. doi:10.1128/JVI.02742-09

Belser JA, Wadford DA, Xu J, Katz JM, Tumpey TM (2009) Ocular infection of mice with influenza A (H7) viruses: a site of primary replication and spread to the respiratory tract. J Virol 83:7075–7084. doi:10.1128/JVI.00535-09

Blazejewska P, Koscinski L, Viegas N, Anhlan D, Ludwig S, Schughart K (2011) Pathogenicity of different PR8 influenza A virus variants in mice is determined by both viral and host factors. Virology 412:36–45. doi:10.1016/j.virol.2010.12.047

Bloom JD, Gong LI, Baltimore D (2010) Permissive secondary mutations enable the evolution of influenza oseltamivir resistance. Science 328:1272–1275. doi:10.1126/science.1187816

Bodewes R, Rimmelzwaan GF, Osterhaus AD (2010) Animal models for the preclinical evaluation of candidate influenza vaccines. Expert Rev Vaccines 9:59–72. doi:10.1586/erv.09.148

Boivin GA, Pothlichet J, Skamene E, Brown EG, Loredo-Osti JC, Sladek R, Vidal SM (2012) Mapping of clinical and expression quantitative trait loci in a sex-dependent effect of host susceptibility to mouse-adapted influenza H3N2/HK/1/68. J Immunol 188:3949–3960. doi:10.4049/jimmunol.1103320

Boon AC, Finkelstein D, Zheng M, Liao G, Allard J, Klumpp K, Webster R, Peltz G, Webby RJ (2011) H5N1 influenza virus pathogenesis in genetically diverse mice is mediated at the level of viral load. MBio 2. doi:10.1128/mBio.00171, 10.1128/mBio.00171

Bosch FX, Garten W, Klenk HD, Rott R (1981) Proteolytic cleavage of influenza virus hemagglutinins: primary structure of the connecting peptide between HA1 and HA2 determines proteolytic cleavability and pathogenicity of avian influenza viruses. Virology 113:725–735

Bottomly D, Ferris MT, Aicher LD, Rosenzweig E, Whitmore A, Aylor DL, Haagmans BL, Gralinski LE, Bradel-Tretheway BG, Bryan JT, Threadgill DW, de Villena FP, Baric RS, Katze MG, Heise M, McWeeney SK (2012) Expression quantitative trait Loci for extreme host response to influenza a in pre-collaborative cross mice. G3 (Bethesda) 2: 213–221. doi: 10.1534/g3.111.001800

Bouvier NM, Lowen AC (2010) Animal models for influenza virus pathogenesis and transmission. Viruses 2:1530–1563. doi:10.3390/v20801530

Brandes M, Klauschen F, Kuchen S, Germain RN (2013) A systems analysis identifies a feed forward inflammatory circuit leading to lethal influenza infection. Cell 154:197–212. doi:10. 1016/j.cell.2013.06.013

Brown EG (1990) Increased virulence of a mouse-adapted variant of influenza A/FM/1/47 virus is controlled by mutations in genome segments 4, 5, 7, and 8. J Virol 64:4523–4533

Brown EG, Bailly JE (1999) Genetic analysis of mouse-adapted influenza A virus identifies roles for the NA, PB1, and PB2 genes in virulence. Virus Res 61:63–76

Brown EG, Liu H, Kit LC, Baird S, Nesrallah M (2001) Pattern of mutation in the genome of influenza A virus on adaptation to increased virulence in the mouse lung: identification of functional themes. Proc Natl Acad Sci U S A 98:6883–6888. doi:10.1073/pnas.111165798

Castrucci MR, Kawaoka Y (1993) Biologic importance of neuraminidase stalk length in influenza A virus. J Virol 67:759–764

Chen H, Bright RA, Subbarao K, Smith C, Cox NJ, Katz JM, Matsuoka Y (2007) Polygenic virulence factors involved in pathogenesis of 1997 Hong Kong H5N1 influenza viruses in mice. Virus Res 128:159–163. doi:10.1016/j.virusres.2007.04.017

Chutinimitkul S, Herfst S, Steel J, Lowen AC, Ye J, van Riel D, Schrauwen EJ, Bestebroer TM, Koel B, Burke DF, Sutherland-Cash KH, Whittleston CS, Russell CA, Wales DJ, Smith DJ, Jonges M, Meijer A, Koopmans M, Rimmelzwaan GF, Kuiken T, Osterhaus AD, Garcia-Sastre A, Perez DR, Fouchier RA (2010) Virulence-associated substitution D222G in hemagglutinin of 2009 pandemic influenza a (H1N1) virus affects receptor binding. J Virol. doi:10.1128/JVI.01136-10

Cilloniz C, Pantin-Jackwood MJ, Ni C, Goodman AG, Peng X, Proll SC, Carter VS, Rosenzweig ER, Szretter KJ, Katz JM, Korth MJ, Swayne DE, Tumpey TM, Katze MG (2010) Lethal dissemination of H5N1 influenza virus is associated with dysregulation of inflammation and lipoxin signaling in a mouse model of infection. J Virol 84:7613–7624. doi:10.1128/JVI. 00553-10

Conenello GM, Palese P (2007) Influenza A virus PB1-F2: a small protein with a big punch. Cell Host Microbe 2:207–209. doi:10.1016/j.chom.2007.09.010

Conenello GM, Tisoncik JR, Rosenzweig E, Varga ZT, Palese P, Katze MG (2011) A single N66S mutation in the PB1-F2 protein of influenza A virus increases virulence by inhibiting the early interferon response in vivo. J Virol 85:652–662. doi:10.1128/JVI.01987-10

Conenello GM, Zamarin D, Perrone LA, Tumpey T, Palese P (2007) A single mutation in the PB1-F2 of H5N1 (HK/97) and 1918 influenza A viruses contributes to increased virulence. PLoS Pathog 3:1414–1421. doi:10.1371/journal.ppat.0030141

Dankar SK, Miranda E, Forbes NE, Pelchat M, Tavassoli A, Selman M, Ping J, Jia J, Brown EG (2013) Influenza A/Hong Kong/156/1997(H5N1) virus NS1 gene mutations F103L and M106I both increase IFN antagonism, virulence and cytoplasmic localization but differ in binding to RIG-I and CPSF30. Virol J 10:243. doi:10.1186/1743-422X-10-243

Dankar SK, Wang S, Ping J, Forbes NE, Keleta L, Li Y, Brown EG (2011) Influenza A virus NS1 gene mutations F103L and M106I increase replication and virulence. Virol J 8:13. doi:10. 1186/1743-422X-8-13

Das SR, Hensley SE, David A, Schmidt L, Gibbs JS, Puigbo P, Ince WL, Bennink JR, Yewdell JW (2011) Fitness costs limit influenza A virus hemagglutinin glycosylation as an immune evasion strategy. Proc Natl Acad Sci U S A 108:E1417–E1422. doi:10.1073/pnas.1108754108
de Jong RM, Stockhofe-Zurwieden N, Verheij ES, de Boer-Luijtze EA, Ruiter SJ, de Leeuw OS, Cornelissen LA (2013) Rapid emergence of a virulent PB2 E627K variant during adaptation of highly pathogenic avian influenza H7N7 virus to mice. Virol J 10:276. doi:10.1186/1743-422X-10-276
Ding M, Lu L, Toth LA (2008) Gene expression in lung and basal forebrain during influenza infection in mice. Genes Brain Behav 7:173–183. doi:10.1111/j.1601-183X.2007.00335.x
Donelan N, Basler C, Garcia-Sastre A (2003) A recombinant influenza A virus expressing an RNA-binding-defective NS1 protein induces high levels of beta interferon and is attenuated in mice. J Virol 77:13257–13266
Driskell EA, Jones CA, Stallknecht DE, Howerth EW, Tompkins SM (2010) Avian influenza virus isolates from wild birds replicate and cause disease in a mouse model of infection. Virology 399:280–289. doi:10.1016/j.virol.2010.01.005
Dudek SE, Wixler L, Nordhoff C, Nordmann A, Anhlan D, Wixler V, Ludwig S (2011) The influenza virus PB1-F2 protein has interferon antagonistic activity. Biol Chem 392:1135–1144. doi:10.1515/BC.2011.174
Eaton MD (1940) Transmission of epidemic influenza virus in mice by contact. Journal of Bacteriology 39:229–241
Edenborough KM, Gilbertson BP, Brown LE (2012) A mouse model for the study of contact-dependent transmission of influenza A virus and the factors that govern transmissibility. J Virol 86:12544–12551. doi:10.1128/JVI.00859-12
Everitt AR, Clare S, Pertel T, John SP, Wash RS, Smith SE, Chin CR, Feeley EM, Sims JS, Adams DJ, Wise HM, Kane L, Goulding D, Digard P, Anttila V, Baillie JK, Walsh TS, Hume DA, Palotie A, Xue Y, Colonna V, Tyler-Smith C, Dunning J, Gordon SB, Gen II, Investigators M, Smyth RL, Openshaw PJ, Dougan G, Brass AL, Kellam P (2012) IFITM3 restricts the morbidity and mortality associated with influenza. Nature 484:519–523. doi:10.1038/nature10921
Fan S, Deng G, Song J, Tian G, Suo Y, Jiang Y, Guan Y, Bu Z, Kawaoka Y, Chen H (2009) Two amino acid residues in the matrix protein M1 contribute to the virulence difference of H5N1 avian influenza viruses in mice. Virology 384:28–32. doi:10.1016/j.virol.2008.11.044
Fan S, Macken CA, Li C, Ozawa M, Goto H, Iswahyudi NF, Nidom CA, Chen H, Neumann G, Kawaoka Y (2013) Synergistic effect of the PDZ and p85beta-binding domains of the NS1 protein on virulence of an avian H5N1 influenza A virus. J Virol 87:4861–4871. doi:10.1128/JVI.02608-12
Ferraris O, Escuret V, Bouscambert M, Casalegno JS, Jacquot F, Raoul H, Caro V, Valette M, Lina B, Ottmann M (2012) H1N1 influenza A virus neuraminidase modulates infectivity in mice. Antiviral Res 93:374–380. doi:10.1016/j.antiviral.2012.01.008
Ferris MT, Aylor DL, Bottomly D, Whitmore AC, Aicher LD, Bell TA, Bradel-Tretheway B, Bryan JT, Buus RJ, Gralinski LE, Haagmans BL, McMillan L, Miller DR, Rosenzweig E, Valdar W, Wang J, Churchill GA, Threadgill DW, McWeeney SK, Katze MG, Pardo-Manuel de Villena F, Baric RS, Heise MT (2013) Modeling host genetic regulation of influenza pathogenesis in the collaborative cross. PLoS Pathog 9:e1003196. doi:10.1371/journal.ppat.1003196
Fonville JM, Burke DF, Lewis NS, Katzelnick LC, Russell CA (2013) Quantifying the fitness advantage of polymerase substitutions in Influenza A/H7N9 viruses during adaptation to humans. PLoS One 8:e76047. doi:10.1371/journal.pone.0076047
Forbes NE, Ping J, Dankar SK, Jia JJ, Selman M, Keleta L, Zhou Y, Brown EG (2012) Multifunctional adaptive NS1 mutations are selected upon human influenza virus evolution in the mouse. PLoS One 7:e31839. doi:10.1371/journal.pone.0031839
Francis T Jr (1934) Transmission of influenza by a filterable virus. Science 80:457–459. doi:10.1126/science.80.2081.457-a

Francis T, Magill TP (1935) Immunological studies with the virus of influenza. J Exp Med 62:505–516
Gabriel G, Abram M, Keiner B, Wagner R, Klenk HD, Stech J (2007) Differential polymerase activity in avian and mammalian cells determines host range of influenza virus. J Virol 81:9601–9604. doi:10.1128/JVI.00666-07
Gabriel G, Czudai-Matwich V, Klenk HD (2013) Adaptive mutations in the H5N1 polymerase complex. Virus Res 178:53–62. doi:10.1016/j.virusres.2013.05.010
Gabriel G, Dauber B, Wolff T, Planz O, Klenk HD, Stech J (2005) The viral polymerase mediates adaptation of an avian influenza virus to a mammalian host. Proc Natl Acad Sci U S A 102:18590–18595. doi:10.1073/pnas.0507415102
Gabriel G, Herwig A, Klenk HD (2008) Interaction of polymerase subunit PB2 and NP with importin alpha1 is a determinant of host range of influenza A virus. PLoS Pathog 4:e11. doi:10.1371/journal.ppat.0040011
Gabriel G, Klingel K, Otte A, Thiele S, Hudjetz B, Arman-Kalcek G, Sauter M, Shmidt T, Rother F, Baumgarte S, Keiner B, Hartmann E, Bader M, Brownlee GG, Fodor E, Klenk HD (2011) Differential use of importin-alpha isoforms governs cell tropism and host adaptation of influenza virus. Nat Commun 2:156. doi:10.1038/ncomms1158
Gao P, Watanabe S, Ito T, Goto H, Wells K, McGregor M, Cooley AJ, Kawaoka Y (1999) Biological heterogeneity, including systemic replication in mice, of H5N1 influenza A virus isolates from humans in Hong Kong. J Virol 73:3184–3189
Garcia-Sastre A, Egorov A, Matassov D, Brandt S, Levy DE, Durbin JE, Palese P, Muster T (1998) Influenza A virus lacking the NS1 gene replicates in interferon-deficient systems. Virology 252:324–330
Garten W, Klenk HD (1999) Understanding influenza virus pathogenicity. Trends Microbiol 7:99–100
Ge S, Wang Z (2011) An overview of influenza A virus receptors. Crit Rev Microbiol 37:157–165. doi:10.3109/1040841X.2010.536523
Gen F, Yamada S, Kato K, Akashi H, Kawaoka Y, Horimoto T (2013) Attenuation of an influenza A virus due to alteration of its hemagglutinin-neuraminidase functional balance in mice. Arch Virol 158:1003–1011. doi:10.1007/s00705-012-1577-3
Gibson HG, Connor JI (1918) A filterable virus as the cause of the early stage of the present epidemic of influenza: (A preliminary note). Br Med J 2:645–646
Goto H, Kawaoka Y (1998) A novel mechanism for the acquisition of virulence by a human influenza A virus. Proc Natl Acad Sci U S A 95:10224–10228
Govorkova EA, Gambaryan AS, Claas EC, Smirnov YA (2000) Amino acid changes in the hemagglutinin and matrix proteins of influenza a (H2) viruses adapted to mice. Acta Virol 44:241–248
Grimm D, Staeheli P, Hufbauer M, Koerner I, Martinez-Sobrido L, Solorzano A, Garcia-Sastre A, Haller O, Kochs G (2007) Replication fitness determines high virulence of influenza A virus in mice carrying functional Mx1 resistance gene. Proc Natl Acad Sci U S A 104:6806–6811. doi:10.1073/pnas.0701849104
Gubareva LV, McCullers JA, Bethell RC, Webster RG (1998) Characterization of influenza A/HongKong/156/97 (H5N1) virus in a mouse model and protective effect of zanamivir on H5N1 infection in mice. J Infect Dis 178:1592–1596. doi:JID980456
Gustin KM, Belser JA, Katz JM, Tumpey TM, Maines TR (2012) Innovations in modeling influenza virus infections in the laboratory. Trends Microbiol 20:275–281. doi:10.1016/j.tim.2012.03.006
Hale B, Steel J, Manicassamy B, Medina R, Ye J, Hickman D, Lowen A, Perez D, Garcia-Sastre A (2010) Mutations in the NS1 C-terminal tail do not enhance replication or virulence of the 2009 pandemic H1N1 influenza A virus. J Gen Virol 91:1737–1742
Hale BG, Randall RE, Ortin J, Jackson D (2008) The multifunctional NS1 protein of influenza A viruses. J Gen Virol 89:2359–2376. doi:10.1099/vir.0.2008/004606-0
Haller O (1981) Inborn resistance of mice to orthomyxoviruses. Curr Top Microbiol Immunol 92:25–52

Haller O, Stertz S, Kochs G (2007) The Mx GTPase family of interferon-induced antiviral proteins. Microbes Infect 9:1636–1643. doi:10.1016/j.micinf.2007.09.010

Hatta M, Gao P, Halfmann P, Kawaoka Y (2001) Molecular basis for high virulence of Hong Kong H5N1 influenza A viruses. Science 293:1840–1842. doi:10.1126/science.1062882

Hiromoto Y, Saito T, Lindstrom S, Nerome K (2000) Characterization of low virulent strains of highly pathogenic A/Hong Kong/156/97 (H5N1) virus in mice after passage in embryonated hens' eggs. Virology 272:429–437. doi:10.1006/viro.2000.0371

Hirst GK (1947) Studies on the Mechanism of Adaptation of Influenza Virus to Mice. J Exp Med 86:357–366

Horby P, Nguyen NY, Dunstan SJ, Baillie JK (2012) The role of host genetics in susceptibility to influenza: a systematic review. PLoS One 7:e33180. doi:10.1371/journal.pone.0033180

Horby P, Nguyen NY, Dunstan SJ, Baillie JK (2013) An updated systematic review of the role of host genetics in susceptibility to influenza. Influenza Other Respir Viruses 7(Suppl 2):37–41. doi:10.1111/irv.12079

Hu J, Zhao K, Liu X, Wang X, Chen Z, Liu X (2012) Two Highly Pathogenic Avian Influenza H5N1 Viruses of Clade 2.3.2.1 with similar genetic background but with different pathogenicity in Mice and Ducks. Transbound Emerg Dis. doi:10.1111/j.1865-1682.2012.01325.x

Huber VC (2012) Can surveillance of the influenza virus PB1-F2 gene be used to predict the severity of secondary bacterial infections? Virulence 3:523–524. doi:10.4161/viru.21811

Hudjetz B, Gabriel G (2012) Human-like PB2 627K influenza virus polymerase activity is regulated by importin-alpha1 and -alpha7. PLoS Pathog 8:e1002488. doi:10.1371/journal.ppat.1002488

Hughes AL, Packer B, Welch R, Chanock SJ, Yeager M (2005) High level of functional polymorphism indicates a unique role of natural selection at human immune system loci. Immunogenetics 57:821–827. doi:10.1007/s00251-005-0052-7

Hutchinson EC, Fodor E (2012) Nuclear import of the influenza A virus transcriptional machinery. Vaccine 30:7353–7358. doi:10.1016/j.vaccine.2012.04.085

Ibricevic A, Pekosz A, Walter MJ, Newby C, Battaile JT, Brown EG, Holtzman MJ, Brody SL (2006) Influenza virus receptor specificity and cell tropism in mouse and human airway epithelial cells. J Virol 80:7469–7480. doi:10.1128/JVI.02677-05

Ilyushina NA, Khalenkov AM, Seiler JP, Forrest HL, Bovin NV, Marjuki H, Barman S, Webster RG, Webby RJ (2010) Adaptation of pandemic H1N1 influenza viruses in mice. J Virol. doi:10.1128/JVI.00159-10

Imai M, Kawaoka Y (2012) The role of receptor binding specificity in interspecies transmission of influenza viruses. Curr Opin Virol 2:160–167. doi:10.1016/j.coviro.2012.03.003

Ison MG, Mishin VP, Braciale TJ, Hayden FG, Gubareva LV (2006) Comparative activities of oseltamivir and A-322278 in immunocompetent and immunocompromised murine models of influenza virus infection. J Infect Dis 193:765–772. doi:10.1086/500464

Iverson AR, Boyd KL, McAuley JL, Plano LR, Hart ME, McCullers JA (2011) Influenza virus primes mice for pneumonia from Staphylococcus aureus. J Infect Dis 203:880–888. doi:10.1093/infdis/jiq113

Jackson D, Hossain MJ, Hickman D, Perez DR, Lamb RA (2008) A new influenza virus virulence determinant: the NS1 protein four C-terminal residues modulate pathogenicity. Proc Natl Acad Sci U S A 105:4381–4386. doi:10.1073/pnas.0800482105

Jiao P, Tian G, Li Y, Deng G, Jiang Y, Liu C, Liu W, Bu Z, Kawaoka Y, Chen H (2008) A single-amino-acid substitution in the NS1 protein changes the pathogenicity of H5N1 avian influenza viruses in mice. J Virol 82:1146–1154. doi:10.1128/JVI.01698-07

Job ER, Deng YM, Barfod KK, Tate MD, Caldwell N, Reddiex S, Maurer-Stroh S, Brooks AG, Reading PC (2013) Addition of glycosylation to influenza A virus hemagglutinin modulates antibody-mediated recognition of H1N1 2009 pandemic viruses. J Immunol 190:2169–2177. doi:10.4049/jimmunol.1202433

Josset L, Belser JA, Pantin-Jackwood MJ, Chang JH, Chang ST, Belisle SE, Tumpey TM, Katze MG (2012) Implication of inflammatory macrophages, nuclear receptors, and interferon regulatory factors in increased virulence of pandemic 2009 H1N1 influenza A virus after host adaptation. J Virol 86:7192–7206. doi:10.1128/JVI.00563-12

Kash JC, Tumpey TM, Proll SC, Carter V, Perwitasari O, Thomas MJ, Basler CF, Palese P, Taubenberger JK, Garcia-Sastre A, Swayne DE, Katze MG (2006) Genomic analysis of increased host immune and cell death responses induced by 1918 influenza virus. Nature 443:578–581. doi:10.1038/nature05181

Kato H, Takeuchi O, Sato S, Yoneyama M, Yamamoto M, Matsui K, Uematsu S, Jung A, Kawai T, Ishii KJ, Yamaguchi O, Otsu K, Tsujimura T, Koh CS, Reis e Sousa C, Matsuura Y, Fujita T, Akira S (2006) Differential roles of MDA5 and RIG-I helicases in the recognition of RNA viruses. Nature 441:101–105. doi:10.1038/nature04734

Katz JM, Lu X, Tumpey TM, Smith CB, Shaw MW, Subbarao K (2000) Molecular correlates of influenza A H5N1 virus pathogenesis in mice. J Virol 74:10807–10810

Kaverin NV, Rudneva IA, Ilyushina NA, Varich NL, Lipatov AS, Smirnov YA, Govorkova EA, Gitelman AK, Lvov DK, Webster RG (2002) Structure of antigenic sites on the haemagglutinin molecule of H5 avian influenza virus and phenotypic variation of escape mutants. J Gen Virol 83:2497–2505

Keleta L, Ibricevic A, Bovin NV, Brody SL, Brown EG (2008) Experimental evolution of human influenza virus H3 hemagglutinin in the mouse lung identifies adaptive regions in HA1 and HA2. J Virol 82:11599–11608. doi:10.1128/JVI.01393-08

Kilander A, Rykkvin R, Dudman SG, Hungnes O (2010) Observed association between the HA1 mutation D222G in the 2009 pandemic influenza A(H1N1) virus and severe clinical outcome, Norway 2009–2010. Euro Surveill 15(9):6–8

Kilbourne ED, Schulman JL, Schild GC, Schloer G, Swanson J, Bucher D (1971) Correlated studies of a recombinant influenza-virus vaccine. I. Derivation and characterization of virus and vaccine. J Infect Dis 124:449–462

Kim J, Hatta M, Watanabe S, Neumann G, Watanabe T, Kawaoka Y (2010) Role of host-specific amino acids in the pathogenicity of avian H5N1 influenza viruses in mice. J Gen Virol 91:1284–1289

Klenk HD, Garten W (1994) Host cell proteases controlling virus pathogenicity. Trends Microbiol 2:39–43

Kobasa D, Takada A, Shinya K, Hatta M, Halfmann P, Theriault S, Suzuki H, Nishimura H, Mitamura K, Sugaya N, Usui T, Murata T, Maeda Y, Watanabe S, Suresh M, Suzuki T, Suzuki Y, Feldmann H, Kawaoka Y (2004) Enhanced virulence of influenza A viruses with the haemagglutinin of the 1918 pandemic virus. Nature 431:703–707. doi:10.1038/nature02951

Koerner I, Matrosovich MN, Haller O, Staeheli P, Kochs G (2012) Altered receptor specificity and fusion activity of the haemagglutinin contribute to high virulence of a mouse-adapted influenza A virus. J Gen Virol 93:970–979. doi:10.1099/vir.0.035782-0

Korth MJ, Tchitchek N, Benecke AG, Katze MG (2013) Systems approaches to influenza-virus host interactions and the pathogenesis of highly virulent and pandemic viruses. Semin Immunol 25:228–239. doi:10.1016/j.smim.2012.11.001

Krenn BM, Egorov A, Romanovskaya-Romanko E, Wolschek M, Nakowitsch S, Ruthsatz T, Kiefmann B, Morokutti A, Humer J, Geiler J, Cinatl J, Michaelis M, Wressnigg N, Sturlan S, Ferko B, Batishchev OV, Indenbom AV, Zhu R, Kastner M, Hinterdorfer P, Kiselev O, Muster T, Romanova J (2011) Single ha2 mutation increases the infectivity and immunogenicity of a live attenuated H5N1 intranasal influenza vaccine candidate lacking NS1. PLoS One 6:e18577. doi:10.1371/journal.pone.0018577

Krumbholz A, Philipps A, Oehring H, Schwarzer K, Eitner A, Wutzler P, Zell R (2011) Current knowledge on PB1-F2 of influenza A viruses. Med Microbiol Immunol 200:69–75. doi:10.1007/s00430-010-0176-8

Le Goffic R, Bouguyon E, Chevalier C, Vidic J, Da Costa B, Leymarie O, Bourdieu C, Decamps L, Dhorne-Pollet S, Delmas B (2010) Influenza A virus protein PB1-F2 exacerbates IFN-beta expression of human respiratory epithelial cells. J Immunol 185:4812–4823. doi:10.4049/jimmunol.0903952

Leymarie O, Jouvion G, Herve PL, Chevalier C, Lorin V, Lecardonnel J, Da Costa B, Delmas B, Escriou N, Le Goffic R (2013) Kinetic characterization of PB1-F2-mediated immunopathology during highly pathogenic avian H5N1 influenza virus infection. PLoS One 8:e57894. doi:10.1371/journal.pone.0057894

Li J, Ishaq M, Prudence M, Xi X, Hu T, Liu Q, Guo D (2009) Single mutation at the amino acid position 627 of PB2 that leads to increased virulence of an H5N1 avian influenza virus during adaptation in mice can be compensated by multiple mutations at other sites of PB2. Virus Res 144:123–129. doi:10.1016/j.virusres.2009.04.008

Li S, Schulman J, Itamura S, Palese P (1993) Glycosylation of neuraminidase determines the neurovirulence of influenza A/WSN/33 virus. J Virol 67:6667–6673

Li Z, Chen H, Jiao P, Deng G, Tian G, Li Y, Hoffmann E, Webster RG, Matsuoka Y, Yu K (2005) Molecular basis of replication of duck H5N1 influenza viruses in a mammalian mouse model. J Virol 79:12058–12064. doi:10.1128/JVI.79.18.12058-12064.2005

Liedmann S, Hrincius ER, Anhlan D, McCullers JA, Ludwig S, Ehrhardt C (2013) new virulence determinants contribute to the enhanced immune response and reduced virulence of an influenza A virus A/PR8/34 variant. J Infect Dis. doi:10.1093/infdis/jit463

Lindenmann J (1964) Inheritance of resistance to influenza virus in mice. Proc Soc Exp Biol Med 116:506–509

Long JX, Peng DX, Liu YL, Wu YT, Liu XF (2008) Virulence of H5N1 avian influenza virus enhanced by a 15-nucleotide deletion in the viral nonstructural gene. Virus Genes 36:471–478. doi:10.1007/s11262-007-0187-8

Lowen AC, Mubareka S, Tumpey TM, Garcia-Sastre A, Palese P (2006) The guinea pig as a transmission model for human influenza viruses. Proc Natl Acad Sci U S A 103:9988–9992. doi:10.1073/pnas.0604157103

Lu X, Tumpey TM, Morken T, Zaki SR, Cox NJ, Katz JM (1999) A mouse model for the evaluation of pathogenesis and immunity to influenza A (H5N1) viruses isolated from humans. J Virol 73:5903–5911

Maines T, Jayaraman A, Belser J, Wadford D, Pappas C, Zeng H, Gustin K, Pearce M, Viswanathan K, Shriver Z, Raman R, Cox N, Sasisekharan R, Katz J, Tumpey T (2009) Transmission and pathogenesis of swine-origin 2009 A(H1N1) influenza viruses in ferrets and mice. Science 325:484–487

Maines TR, Lu XH, Erb SM, Edwards L, Guarner J, Greer PW, Nguyen DC, Szretter KJ, Chen LM, Thawatsupha P, Chittaganpitch M, Waicharoen S, Nguyen DT, Nguyen T, Nguyen HH, Kim JH, Hoang LT, Kang C, Phuong LS, Lim W, Zaki S, Donis RO, Cox NJ, Katz JM, Tumpey TM (2005) Avian influenza (H5N1) viruses isolated from humans in Asia in 2004 exhibit increased virulence in mammals. J Virol 79:11788–11800. doi:10.1128/JVI.79.18.11788-11800.2005

Mancini N, Solforosi L, Clementi N, De Marco D, Clementi M, Burioni R (2011) A potential role for monoclonal antibodies in prophylactic and therapeutic treatment of influenza. Antiviral Res 92:15–26. doi:10.1016/j.antiviral.2011.07.013

Manz B, Brunotte L, Reuther P, Schwemmle M (2012) Adaptive mutations in NEP compensate for defective H5N1 RNA replication in cultured human cells. Nat Commun 3:802. doi:10.1038/ncomms1804

Manz B, Schwemmle M, Brunotte L (2013) Adaptation of avian influenza A virus polymerase in mammals to overcome the host species barrier. J Virol 87:7200–7209. doi:10.1128/JVI.00980-13

Matsuoka Y, Lamirande EW, Subbarao K (2009a) The mouse model for influenza. Curr Protoc Microbiol Chap. 15: Unit 15G 3. doi: 10.1002/9780471729259.mc15g03s13

Matsuoka Y, Swayne DE, Thomas C, Rameix-Welti MA, Naffakh N, Warnes C, Altholtz M, Donis R, Subbarao K (2009b) Neuraminidase stalk length and additional glycosylation of the hemagglutinin influence the virulence of influenza H5N1 viruses for mice. J Virol 83:4704–4708. doi:10.1128/JVI.01987-08

McAuley JL, Hornung F, Boyd KL, Smith AM, McKeon R, Bennink J, Yewdell JW, McCullers JA (2007) Expression of the 1918 influenza A virus PB1-F2 enhances the pathogenesis of viral and secondary bacterial pneumonia. Cell Host Microbe 2:240–249. doi:10.1016/j.chom.2007.09.001

McAuley JL, Tate MD, MacKenzie-Kludas CJ, Pinar A, Zeng W, Stutz A, Latz E, Brown LE, Mansell A (2013) Activation of the NLRP3 inflammasome by IAV virulence protein PB1-F2 contributes to severe pathophysiology and disease. PLoS Pathog 9:e1003392. doi:10.1371/journal.ppat.1003392

McAuley JL, Zhang K, McCullers JA (2010) The effects of influenza A virus PB1-F2 protein on polymerase activity are strain specific and do not impact pathogenesis. J Virol 84:558–564. doi:10.1128/JVI.01785-09

Medina RA, Garcia-Sastre A (2011) Influenza A viruses: new research developments. Nat Rev Microbiol 9:590–603. doi:10.1038/nrmicro2613

Medina RA, Stertz S, Manicassamy B, Zimmermann P, Sun X, Albrecht RA, Uusi-Kerttula H, Zagordi O, Belshe RB, Frey SE, Tumpey TM, Garcia-Sastre A (2013) Glycosylations in the globular head of the hemagglutinin protein modulate the virulence and antigenic properties of the H1N1 influenza viruses. Sci Transl Med 5: 187ra70. doi: 10.1126/scitranslmed.3005996

Meunier I, von Messling V (2012) PB1-F2 modulates early host responses but does not affect the pathogenesis of H1N1 seasonal influenza virus. J Virol 86:4271–4278. doi:10.1128/JVI.07243-11

Mok CK, Lee HH, Lestra M, Nicholls JM, Chan MC, Sia SF, Zhu H, Poon LL, Guan Y, Peiris JS (2014) amino acid substitutions in polymerase basic protein 2 gene contribute to the pathogenicity of the novel A/H7N9 influenza virus in mammalian hosts. J Virol 88:3568–3576. doi:10.1128/JVI.02740-13

Mount AM, Belz GT (2010) Mouse models of viral infection: influenza infection in the lung. Methods Mol Biol 595:299–318. doi:10.1007/978-1-60761-421-0_20

Munster VJ, de Wit E, van Riel D, Beyer WE, Rimmelzwaan GF, Osterhaus AD, Kuiken T, Fouchier RA (2007) The molecular basis of the pathogenicity of the Dutch highly pathogenic human influenza A H7N7 viruses. J Infect Dis 196:258–265. doi:10.1086/518792

Narasaraju T, Sim MK, Ng HH, Phoon MC, Shanker N, Lal SK, Chow VT (2009) Adaptation of human influenza H3N2 virus in a mouse pneumonitis model: insights into viral virulence, tissue tropism and host pathogenesis. Microbes Infect 11:2–11. doi:10.1016/j.micinf.2008.09.013

Noda T, Kawaoka Y (2010) Structure of influenza virus ribonucleoprotein complexes and their packaging into virions. Rev Med Virol 20:380–391. doi:10.1002/rmv.666

O'Donnell CD, Subbarao K (2011) The contribution of animal models to the understanding of the host range and virulence of influenza A viruses. Microbes Infect 13:502–515. doi:10.1016/j.micinf.2011.01.014

Obenauer JC, Denson J, Mehta PK, Su X, Mukatira S, Finkelstein DB, Xu X, Wang J, Ma J, Fan Y, Rakestraw KM, Webster RG, Hoffmann E, Krauss S, Zheng J, Zhang Z, Naeve CW (2006) Large-scale sequence analysis of avian influenza isolates. Science 311:1576–1580. doi:10.1126/science.1121586

Ohuchi M, Ohuchi R, Feldmann A, Klenk HD (1997) Regulation of receptor binding affinity of influenza virus hemagglutinin by its carbohydrate moiety. J Virol 71:8377–8384

Ozawa M, Basnet S, Burley LM, Neumann G, Hatta M, Kawaoka Y (2011) Impact of amino acid mutations in PB2, PB1-F2, and NS1 on the replication and pathogenicity of pandemic (H1N1) 2009 influenza viruses. J Virol 85:4596–4601. doi:10.1128/JVI.00029-11

Pappas C, Aguilar PV, Basler CF, Solorzano A, Zeng H, Perrone LA, Palese P, Garcia-Sastre A, Katz JM, Tumpey TM (2008) Single gene reassortants identify a critical role for PB1, HA, and NA in the high virulence of the 1918 pandemic influenza virus. Proc Natl Acad Sci U S A 105:3064–3099. doi:10.1073/pnas.0711815105

Pasricha G, Mishra AC, Chakrabarti AK (2013) Comprehensive global amino acid sequence analysis of PB1F2 protein of influenza A H5N1 viruses and the influenza A virus subtypes responsible for the 20th-century pandemics. Influenza Other Respi Viruses 7:497–505. doi:10.1111/j.1750-2659.2012.00400.x

Paterson D, Fodor E (2012) Emerging roles for the influenza A virus nuclear export protein (NEP). PLoS Pathog 8:e1003019. doi:10.1371/journal.ppat.1003019

Pekosz A, Newby C, Bose PS, Lutz A (2009) Sialic acid recognition is a key determinant of influenza A virus tropism in murine trachea epithelial cell cultures. Virology 386:61–67. doi:10.1016/j.virol.2009.01.005

Pepin KM, Lass S, Pulliam JR, Read AF, Lloyd-Smith JO (2010) Identifying genetic markers of adaptation for surveillance of viral host jumps. Nat Rev Microbiol 8:802–813. doi:10.1038/nrmicro2440

Pica N, Iyer A, Ramos I, Bouvier NM, Fernandez-Sesma A, Garcia-Sastre A, Lowen AC, Palese P, Steel J (2011) The DBA.2 mouse is susceptible to disease following infection with a broad, but limited, range of influenza A and B viruses. J Virol 85:12825–12829. doi:10.1128/JVI.05930-11

Ping J, Dankar SK, Forbes NE, Keleta L, Zhou Y, Tyler S, Brown EG (2010) PB2 and hemagglutinin mutations are major determinants of host range and virulence in mouse-adapted influenza A virus. J Virol 84:10606–10618. doi:10.1128/JVI.01187-10

Ping J, Keleta L, Forbes NE, Dankar S, Stecho W, Tyler S, Zhou Y, Babiuk L, Weingartl H, Halpin RA, Boyne A, Bera J, Hostetler J, Fedorova NB, Proudfoot K, Katzel DA, Stockwell TB, Ghedin E, Spiro DJ, Brown EG (2011) Genomic and protein structural maps of adaptive evolution of human influenza A virus to increased virulence in the mouse. PLoS One 6:e21740. doi:10.1371/journal.pone.0021740

Price GE, Lo CY, Misplon JA, Epstein SL (2014) Mucosal immunization with a candidate universal influenza vaccine reduces virus transmission in a mouse model. J Virol. doi:10.1128/JVI.03101-13

Qi L, Davis AS, Jagger BW, Schwartzman LM, Dunham EJ, Kash JC, Taubenberger JK (2012) Analysis by single-gene reassortment demonstrates that the 1918 influenza virus is functionally compatible with a low-pathogenicity avian influenza virus in mice. J Virol 86:9211–9220. doi:10.1128/JVI.00887-12

Qi L, Kash JC, Dugan VG, Wang R, Jin G, Cunningham RE, Taubenberger JK (2009) Role of sialic acid binding specificity of the 1918 influenza virus hemagglutinin protein in virulence and pathogenesis for mice. J Virol 83:3754–3761. doi:10.1128/JVI.02596-08

Rajsbaum R, Albrecht RA, Wang MK, Maharaj NP, Versteeg GA, Nistal-Villan E, Garcia-Sastre A, Gack MU (2012) Species-specific inhibition of RIG-I ubiquitination and IFN induction by the influenza A virus NS1 protein. PLoS Pathog 8:e1003059. doi:10.1371/journal.ppat.1003059

Raut S, Hurd J, Blandford G, Heath RB, Cureton RJ (1975) The pathogenesis of infections of the mouse caused by virulent and avirulent variants of an influenza virus. J Med Microbiol 8:127–136

Reading PC, Pickett DL, Tate MD, Whitney PG, Job ER, Brooks AG (2009) Loss of a single N-linked glycan from the hemagglutinin of influenza virus is associated with resistance to collectins and increased virulence in mice. Respir Res 10:117. doi:10.1186/1465-9921-10-117

Reading PC, Tate MD, Pickett DL, Brooks AG (2007) Glycosylation as a target for recognition of influenza viruses by the innate immune system. Adv Exp Med Biol 598:279–292. doi:10.1007/978-0-387-71767-8_20

Reperant LA, Kuiken T, Osterhaus AD (2012) Adaptive pathways of zoonotic influenza viruses: from exposure to establishment in humans. Vaccine 30:4419–4434. doi:10.1016/j.vaccine.2012.04.049

Rossman JS, Lamb RA (2011) Influenza virus assembly and budding. Virology 411:229–236. doi:10.1016/j.virol.2010.12.003

Rudneva IA, Ilyushina NA, Timofeeva TA, Webster RG, Kaverin NV (2005) Restoration of virulence of escape mutants of H5 and H9 influenza viruses by their readaptation to mice. J Gen Virol 86:2831–2838. doi:10.1099/vir.0.81185-0

Rudneva IA, Kaverin NV, Varich NL, Gitelman AK, Makhov AM, Klimenko SM, Zhdanov VM (1986) Studies on the genetic determinants of influenza virus pathogenicity for mice with the use of reassortants between mouse-adapted and non-adapted variants of the same virus strain. Arch Virol 90:237–248

Schmolke M, Manicassamy B, Pena L, Sutton T, Hai R, Varga ZT, Hale BG, Steel J, Perez DR, Garcia-Sastre A (2011) Differential contribution of PB1-F2 to the virulence of highly pathogenic H5N1 influenza A virus in mammalian and avian species. PLoS Pathog 7:e1002186. doi:10.1371/journal.ppat.1002186

Schulman JL (1967) Experimental transmission of influenza virus infection in mice IV. Relationship of transmissibility of different strains of virus and recovery of airborne virus in the environment of infector mice. J Exp Med 125:479–488

Schulman JL (1968) The use of an animal model to study transmission of influenza virus infection. Am J Public Health Nations Health 58:2092–2096

Schulman JL, Kilbourne ED (1962) Airborne transmission of influenza virus infection in mice. Nature 195:1129–1130

Schulman JL, Kilbourne ED (1963a) Experimental transmission of influenza virus infection in mice i. the period of transmissibility. J Exp Med 118:257–266

Schulman JL, Kilbourne ED (1963b) Experimental Transmission of Influenza Virus Infection in Mice II. Some Factors Affecting the Incidence of Transmitted Infection. J Exp Med 118:267–275

Seyer R, Hrincius ER, Ritzel D, Abt M, Mellmann A, Marjuki H, Kuhn J, Wolff T, Ludwig S, Ehrhardt C (2012) Synergistic adaptive mutations in the hemagglutinin and polymerase acidic protein lead to increased virulence of pandemic 2009 H1N1 influenza A virus in mice. J Infect Dis 205:262–271. doi:10.1093/infdis/jir716

Shilov AA, Sinitsyn BV (1994) Changes in its hemagglutinin during the adaptation of the influenza virus to mice and their role in the acquisition of virulent properties and resistance to serum inhibitors. Vopr Virusol 39:153–157

Shinya K, Ebina M, Yamada S, Ono M, Kasai N, Kawaoka Y (2006) Avian flu: influenza virus receptors in the human airway. Nature 440:435–436. doi:10.1038/440435a

Shinya K, Hamm S, Hatta M, Ito H, Ito T, Kawaoka Y (2004) PB2 amino acid at position 627 affects replicative efficiency, but not cell tropism, of Hong Kong H5N1 influenza A viruses in mice. Virology 320:258–266. doi:10.1016/j.virol.2003.11.030

Shinya K, Watanabe S, Ito T, Kasai N, Kawaoka Y (2007) Adaptation of an H7N7 equine influenza a virus in mice. J Gen Virol 88:547–553. doi:10.1099/vir.0.82411-0

Shope RE (1935) The infection of mice with swine influenza virus. J Exp Med 62:561–572

Sidwell RW, Smee DF (2000) In vitro and in vivo assay systems for study of influenza virus inhibitors. Antiviral Res 48:1–16

Smee DF, Wandersee MK, Checketts MB, O'Keefe BR, Saucedo C, Boyd MR, Mishin VP, Gubareva LV (2007) Influenza A (H1N1) virus resistance to cyanovirin-N arises naturally during adaptation to mice and by passage in cell culture in the presence of the inhibitor. Antivir Chem Chemother 18:317–327

Smeenk CA, Brown EG (1994) The influenza virus variant A/FM/1/47-MA possesses single amino acid replacements in the hemagglutinin, controlling virulence, and in the matrix protein, controlling virulence as well as growth. J Virol 68:530–534

Smeenk CA, Wright KE, Burns BF, Thaker AJ, Brown EG (1996) Mutations in the hemagglutinin and matrix genes of a virulent influenza virus variant, A/FM/1/47-MA, control different stages in pathogenesis. Virus Res 44:79–95

Smirnov YA, Lipatov AS, Van Beek R, Gitelman AK, Osterhaus AD, Claas EC (2000) Characterization of adaptation of an avian influenza A (H5N2) virus to a mammalian host. Acta Virol 44:1–8

Smith W, Andrewes CH, Laidlaw PP (1933) A virus obtained from influenza patients. The Lancet 222:66–68

Song MS, Cho YH, Park SJ, Pascua PN, Baek YH, Kwon HI, Lee OJ, Kong BW, Kim H, Shin EC, Kim CJ, Choi YK (2013a) Early regulation of viral infection reduces inflammation and rescues mx-positive mice from lethal avian influenza infection. Am J Pathol 182:1308–1321. doi:10.1016/j.ajpath.2012.12.022

Song MS, Hee Baek Y, Kim EH, Park SJ, Kim S, Lim GJ, Kwon HI, Pascua PN, Decano AG, Lee BJ, Kim YI, Webby RJ, Choi YK (2013b) Increased virulence of neuraminidase inhibitor-resistant pandemic H1N1 virus in mice: potential emergence of drug-resistant and virulent variants. Virulence 4:489–493. doi:10.4161/viru.25952

Song MS, Pascua PN, Lee JH, Baek YH, Park KJ, Kwon HI, Park SJ, Kim CJ, Kim H, Webby RJ, Webster RG, Choi YK (2011) Virulence and genetic compatibility of polymerase reassortant viruses derived from the pandemic (H1N1) 2009 influenza virus and circulating influenza A viruses. J Virol 85:6275–6286. doi:10.1128/JVI.02125-10

Spesock A, Malur M, Hossain MJ, Chen LM, Njaa BL, Davis CT, Lipatov AS, York IA, Krug RM, Donis RO (2011) The virulence of 1997 H5N1 influenza viruses in the mouse model is increased by correcting a defect in their NS1 proteins. J Virol 85:7048–7058. doi:10.1128/JVI.00417-11

Srivastava B, Blazejewska P, Hessmann M, Bruder D, Geffers R, Mauel S, Gruber AD, Schughart K (2009) Host genetic background strongly influences the response to influenza a virus infections. PLoS One 4:e4857. doi:10.1371/journal.pone.0004857

Staeheli P, Grob R, Meier E, Sutcliffe JG, Haller O (1988) Influenza virus-susceptible mice carry Mx genes with a large deletion or a nonsense mutation. Mol Cell Biol 8:4518–4523

Steel J, Lowen AC, Mubareka S, Palese P (2009) Transmission of influenza virus in a mammalian host is increased by PB2 amino acids 627K or 627E/701N. PLoS Pathog 5:e1000252. doi:10.1371/journal.ppat.1000252

Steidle S, Martinez-Sobrido L, Mordstein M, Lienenklaus S, Garcia-Sastre A, Staheli P, Kochs G (2010) Glycine 184 in nonstructural protein NS1 determines the virulence of influenza A virus strain PR8 without affecting the host interferon response. J Virol 84:12761–12770. doi:10.1128/JVI.00701-10

Stieneke-Grober A, Vey M, Angliker H, Shaw E, Thomas G, Roberts C, Klenk HD, Garten W (1992) Influenza virus hemagglutinin with multibasic cleavage site is activated by furin, a subtilisin-like endoprotease. EMBO J 11:2407–2414

Stuart-Harris CH (1939) A neutrotropic strain of human influenza virus. The Lancet 233:497–499

Subbarao EK, London W, Murphy BR (1993) A single amino acid in the PB2 gene of influenza A virus is a determinant of host range. J Virol 67:1761–1764

Suguitan AL Jr, Matsuoka Y, Lau YF, Santos CP, Vogel L, Cheng LI, Orandle M, Subbarao K (2012) The multibasic cleavage site of the hemagglutinin of highly pathogenic A/Vietnam/1203/2004 (H5N1) avian influenza virus acts as a virulence factor in a host-specific manner in mammals. J Virol 86:2706–2714. doi:10.1128/JVI.05546-11

Sun X, Jayaraman A, Maniprasad P, Raman R, Houser KV, Pappas C, Zeng H, Sasisekharan R, Katz JM, Tumpey TM (2013a) N-linked glycosylation of the hemagglutinin protein influences virulence and antigenicity of the 1918 pandemic and seasonal H1N1 influenza A viruses. J Virol 87:8756–8766. doi:10.1128/JVI.00593-13

Sun X, Tse LV, Ferguson AD, Whittaker GR (2010) Modifications to the hemagglutinin cleavage site control the virulence of a neurotropic H1N1 influenza virus. J Virol 84:8683–8690. doi:10.1128/JVI.00797-10

Sun Y, Tan Y, Wei K, Sun H, Shi Y, Pu J, Yang H, Gao GF, Yin Y, Feng W, Perez DR, Liu J (2013b) Amino acid 316 of hemagglutinin and the neuraminidase stalk length influence virulence of H9N2 influenza virus in chickens and mice. J Virol 87:2963–2968. doi:10.1128/JVI.02688-12

Tanaka H, Park CH, Ninomiya A, Ozaki H, Takada A, Umemura T, Kida H (2003) Neurotropism of the 1997 Hong Kong H5N1 influenza virus in mice. Vet Microbiol 95:1–13. doi:S0378113503001329

Tate MD, Brooks AG, Reading PC (2011) Specific sites of N-linked glycosylation on the hemagglutinin of H1N1 subtype influenza A virus determine sensitivity to inhibitors of the innate immune system and virulence in mice. J Immunol 187:1884–1894. doi:10.4049/jimmunol.1100295

Tian J, Qi W, Li X, He J, Jiao P, Zhang C, Liu GQ, Liao M (2012) A single E627K mutation in the PB2 protein of H9N2 avian influenza virus increases virulence by inducing higher glucocorticoids (GCs) level. PLoS One 7:e38233. doi:10.1371/journal.pone.0038233

Trammell RA, Liberati TA, Toth LA (2012) Host genetic background and the innate inflammatory response of lung to influenza virus. Microbes Infect 14:50–58. doi:10.1016/j.micinf.2011.08.008

Trammell RA, Toth LA (2008) Genetic susceptibility and resistance to influenza infection and disease in humans and mice. Expert Rev Mol Diagn 8:515–529. doi:10.1586/14737159.8.4.515

Tripp RA, Tompkins SM (2009) Animal models for evaluation of influenza vaccines. Curr Top Microbiol Immunol 333:397–412. doi:10.1007/978-3-540-92165-3_19

Tumpey TM, Basler CF, Aguilar PV, Zeng H, Solorzano A, Swayne DE, Cox NJ, Katz JM, Taubenberger JK, Palese P, Garcia-Sastre A (2005) Characterization of the reconstructed 1918 Spanish influenza pandemic virus. Science 310:77–780. doi:10.1126/science.1119392

Tumpey TM, Garcia-Sastre A, Taubenberger JK, Palese P, Swayne DE, Basler CF (2004) Pathogenicity and immunogenicity of influenza viruses with genes from the 1918 pandemic virus. Proc Natl Acad Sci U S A 101:3166–3171. doi:10.1073/pnas.0308391100

Uraki R, Kiso M, Shinya K, Goto H, Takano R, Iwatsuki-Horimoto K, Takahashi K, Daniels RS, Hungnes O, Watanabe T, Kawaoka Y (2013) Virulence determinants of pandemic a (H1N1)2009 influenza virus in a mouse model. J Virol 87:2226–2233. doi:10.1128/JVI.01565-12

van der Laan JW, Herberts C, Lambkin-Williams R, Boyers A, Mann AJ, Oxford J (2008) Animal models in influenza vaccine testing. Expert Rev Vaccines 7:783–793. doi:10.1586/14760584.7.6.783

Varga ZT, Grant A, Manicassamy B, Palese P (2012) Influenza virus protein PB1-F2 inhibits the induction of type i interferon by binding to MAVS and decreasing mitochondrial membrane potential. J Virol 86:8359–8366. doi:10.1128/JVI.01122-12

Varga ZT, Ramos I, Hai R, Schmolke M, Garcia-Sastre A, Fernandez-Sesma A, Palese P (2011) The influenza virus protein PB1-F2 inhibits the induction of type i interferon at the level of the MAVS adaptor protein. PLoS Pathog 7:e1002067. doi:10.1371/journal.ppat.1002067

Vigerust DJ, Ulett KB, Boyd KL, Madsen J, Hawgood S, McCullers JA (2007) N-linked glycosylation attenuates H3N2 influenza viruses. J Virol 81:8593–8600. doi:10.1128/JVI.00769-07

Wang CC, Chen JR, Tseng YC, Hsu CH, Hung YF, Chen SW, Chen CM, Khoo KH, Cheng TJ, Cheng YS, Jan JT, Wu CY, Ma C, Wong CH (2009) Glycans on influenza hemagglutinin affect receptor binding and immune response. Proc Natl Acad Sci U S A 106:18137–18142. doi:10.1073/pnas.0909696106

Wang J, Sun Y, Xu Q, Tan Y, Pu J, Yang H, Brown EG, Liu J (2012) Mouse-adapted H9N2 influenza A virus PB2 protein M147L and E627K mutations are critical for high virulence. PLoS One 7:e40752. doi:10.1371/journal.pone.0040752

Wang TT, Tan GS, Hai R, Pica N, Petersen E, Moran TM, Palese P (2010) Broadly protective monoclonal antibodies against H3 influenza viruses following sequential immunization with different hemagglutinins. PLoS Pathog 6:e1000796. doi:10.1371/journal.ppat.1000796

Ward AC (1995) Specific changes in the M1 protein during adaptation of influenza virus to mouse. Arch Virol 140:383–389

Ward AC (1996) Neurovirulence of influenza A virus. J Neurovirol 2:139–151

Ward AC (1997) Virulence of influenza A virus for mouse lung. Virus Genes 14:187–194

Watanabe T, Tisoncik-Go J, Tchitchek N, Watanabe S, Benecke AG, Katze MG, Kawaoka Y (2013) 1918 Influenza virus hemagglutinin (HA) and the viral RNA polymerase complex enhance viral pathogenicity, but only HA induces aberrant host responses in mice. J Virol 87:5239–5254. doi:10.1128/JVI.02753-12

Webster RG, Rott R (1987) Influenza virus A pathogenicity: the pivotal role of hemagglutinin. Cell 50:665–666

Whittaker GR, Kann M, Helenius A (2000) Viral entry into the nucleus. Annu Rev Cell Dev Biol 16:627–651. doi:10.1146/annurev.cellbio.16.1.627

Wu R, Zhang H, Yang K, Liang W, Xiong Z, Liu Z, Yang X, Shao H, Zheng X, Chen M, Xu D (2009) Multiple amino acid substitutions are involved in the adaptation of H9N2 avian influenza virus to mice. Vet Microbiol 138:85–91. doi:10.1016/j.vetmic.2009.03.010

Wyde PR, Cate TR (1978) Cellular changes in lungs of mice infected with influenza virus: characterization of the cytotoxic responses. Infect Immun 22:423–429

Wyde PR, Peavy DL, Cate TR (1978) Morphological and cytochemical characterization of cells infiltrating mouse lungs after influenza infection. Infect Immun 21:140–146

Xu C, Hu WB, Xu K, He YX, Wang TY, Chen Z, Li TX, Liu JH, Buchy P, Sun B (2012) Amino acids 473V and 598P of PB1 from an avian-origin influenza A virus contribute to polymerase activity, especially in mammalian cells. J Gen Virol 93:531–540. doi:10.1099/vir.0.036434-0

Xu L, Bao L, Deng W, Zhu H, Chen T, Lv Q, Li F, Yuan J, Xiang Z, Gao K, Xu Y, Huang L, Li Y, Liu J, Yao Y, Yu P, Yong W, Wei Q, Zhang L, Qin C (2013) The mouse and ferret models for studying the novel avian-origin human influenza A (H7N9) virus. Virol J 10:253. doi:10.1186/1743-422X-10-253

Xu L, Bao L, Li F, Lv Q, Ma Y, Zhou J, Xu Y, Deng W, Zhan L, Zhu H, Ma C, Shu Y, Qin C (2011a) Adaption of seasonal H1N1 influenza virus in mice. PLoS One 6:e28901. doi:10.1371/journal.pone.0028901

Xu L, Bao L, Zhou J, Wang D, Deng W, Lv Q, Ma Y, Li F, Sun H, Zhan L, Zhu H, Ma C, Shu Y, Qin C (2011b) Genomic polymorphism of the pandemic A (H1N1) influenza viruses correlates with viral replication, virulence, and pathogenicity in vitro and in vivo. PLoS One 6:e20698. doi:10.1371/journal.pone.0020698

Yamada S, Hatta M, Staker BL, Watanabe S, Imai M, Shinya K, Sakai-Tagawa Y, Ito M, Ozawa M, Watanabe T, Sakabe S, Li C, Kim JH, Myler PJ, Phan I, Raymond A, Smith E, Stacy R, Nidom CA, Lank SM, Wiseman RW, Bimber BN, O'Connor DH, Neumann G, Stewart LJ, Kawaoka Y (2010) Biological and structural characterization of a host-adapting amino acid in influenza virus. PLoS Pathog 6:e1001034. doi:10.1371/journal.ppat.1001034

Yen HL, Aldridge JR, Boon AC, Ilyushina NA, Salomon R, Hulse-Post DJ, Marjuki H, Franks J, Boltz DA, Bush D, Lipatov AS, Webby RJ, Rehg JE, Webster RG (2009) Changes in H5N1 influenza virus hemagglutinin receptor binding domain affect systemic spread. Proc Natl Acad Sci U S A 106:286–291. doi:10.1073/pnas.0811052106

Zamarin D, Ortigoza MB, Palese P (2006) Influenza A virus PB1-F2 protein contributes to viral pathogenesis in mice. J Virol 80:7976–7983. doi:10.1128/JVI.00415-06

Zaraket H, Bridges OA, Russell CJ (2013) The pH of Activation of the Hemagglutinin Protein Regulates H5N1 Influenza Virus Replication and Pathogenesis in Mice. J Virol. doi:10.1128/JVI.03110-12

Zell R, Krumbholz A, Eitner A, Krieg R, Halbhuber KJ, Wutzler P (2007) Prevalence of PB1-F2 of influenza A viruses. J Gen Virol 88:536–546. doi:10.1099/vir.0.82378-0

Zhang H, Li X, Guo J, Li L, Chang C, Li Y, Bian C, Xu K, Chen H, Sun B (2014) The PB2 E627K mutation contributes to the high polymerase activity and enhanced replication of H7N9 influenza virus. J Gen Virol. doi:10.1099/vir.0.061721-0

Zhang Y, Sun Y, Sun H, Pu J, Bi Y, Shi Y, Lu X, Li J, Zhu Q, Gao GF, Yang H, Liu J (2012) a single amino acid at the hemagglutinin cleavage site contributes to the pathogenicity and neurovirulence of H5N1 influenza virus in mice. J Virol. doi:10.1128/JVI.07142-11

Zhang Y, Zhu J, Li Y, Bradley KC, Cao J, Chen H, Jin M, Zhou H (2013) Glycosylation on hemagglutinin affects the virulence and pathogenicity of pandemic H1N1/2009 influenza a virus in mice. PLoS One 8:e61397. doi:10.1371/journal.pone.0061397

Zhang Z, Hu S, Li Z, Wang X, Liu M, Guo Z, Li S, Xiao Y, Bi D, Jin H (2011) Multiple amino acid substitutions involved in enhanced pathogenicity of LPAI H9N2 in mice. Infect Genet Evol 11:1790–1797. doi:10.1016/j.meegid.2011.07.025

Zheng B, Chan KH, Zhang AJ, Zhou J, Chan CC, Poon VK, Zhang K, Leung VH, Jin DY, Woo PC, Chan JF, To KK, Chen H, Yuen KY (2010) D225G mutation in hemagglutinin of pandemic influenza H1N1 (2009) virus enhances virulence in mice. Exp Biol Med (Maywood) 235:981–988. doi:10.1258/ebm.2010.010071

Zhou B, Pearce MB, Li Y, Wang J, Mason RJ, Tumpey TM, Wentworth DE (2013) Asparagine substitution at PB2 residue 701 enhances the replication, pathogenicity, and transmission of the 2009 pandemic H1N1 influenza A virus. PLoS One 8:e67616. doi:10.1371/journal.pone.0067616

Zhou H, Yu Z, Hu Y, Tu J, Zou W, Peng Y, Zhu J, Li Y, Zhang A, Yu Z, Ye Z, Chen H, Jin M (2009) The special neuraminidase stalk-motif responsible for increased virulence and pathogenesis of H5N1 influenza A virus. PLoS One 4:e6277. doi:10.1371/journal.pone.0006277

Zielecki F, Semmler I, Kalthoff D, Voss D, Mauel S, Gruber AD, Beer M, Wolff T (2010) Virulence determinants of avian H5N1 influenza A virus in mammalian and avian hosts: role of the C-terminal ESEV motif in the viral NS1 protein. J Virol 84:10708–10718. doi:10.1128/JVI.00610-10

Mammalian Models for the Study of H7 Virus Pathogenesis and Transmission

Jessica A. Belser and Terrence M. Tumpey

Abstract Mammalian models, most notably the mouse and ferret, have been instrumental in the assessment of avian influenza virus pathogenicity and transmissibility, and have been used widely to characterize the molecular determinants that confer H5N1 virulence in mammals. However, while H7 influenza viruses have typically been associated with conjunctivitis and/or mild respiratory disease in humans, severe disease and death is also possible, as underscored by the recent emergence of H7N9 viruses in China. Despite the public health need to understand the pandemic potential of this virus subtype, H7 virus pathogenesis and transmission has not been as extensively studied. In this review, we discuss the heterogeneity of H7 subtype viruses isolated from humans, and the characterization of mammalian models to study the virulence of H7 subtype viruses associated with human infection, including viruses of both high and low pathogenicity and following multiple inoculation routes. The use of the ferret transmission model to assess the influence of receptor binding preference among contemporary H7 influenza viruses is described. These models have enabled the study of preventative and therapeutic agents, including vaccines and antivirals, to reduce disease burden, and have permitted a greater appreciation that not all highly pathogenic influenza viruses are created equal.

Contents

J. A. Belser · T. M. Tumpey (✉)
Influenza Division, MS G-16, National Center for Immunization and Respiratory Diseases, Centers for Disease Control and Prevention, 1600 Clifton Rd. NE, Atlanta, GA 30333, USA
e-mail: tft9@cdc.gov

Current Topics in Microbiology and Immunology (2014) 385: 275–305
DOI: 10.1007/82_2014_383

Published Online: 5 July 2014

1 Introduction

The emergence in Spring 2013 of a low pathogenic avian influenza (LPAI) H7N9 virus in Southeast Asia, capable of causing severe human disease and death, resulted in rapid global recognition of the pandemic potential of H7 viruses. H7N9 viruses are second only to H5N1 viruses in total confirmed human cases and fatalities from avian influenza viruses, though different epidemiologic profiles make it difficult to draw extensive parallels between these viruses (Cowling et al. 2013), necessitating increased investigation of H7 subtype virus virulence. However, this was not the first instance of human infection caused by an H7 subtype virus; prior to this outbreak, there were >100 confirmed or presumed human infections with H7 viruses in both Europe and the Americas (Belser et al. 2009a). Similar to H5N1, human-to-human transmission of H7 viruses has been rare but documented (Abdel-Ghafar et al. 2008; Koopmans et al. 2004; Qi et al. 2013).

There are numerous features of viruses within the H7 subtype associated with human infection that set this subtype apart from other avian influenza viruses with pandemic potential. While H5 viruses have caused the greatest number of confirmed human infections among all avian influenza virus subtypes, with a great deal of clade-specific diversity among viruses associated with human infection, all cases have resulted from viruses possessing only the N1 neuraminidase (NA) (Abdel-Ghafar et al. 2008). Similarly, human infections with H6 and H9 viruses have been restricted to the N1 and N2 NA subtypes, respectively (Arzey et al. 2012; Peiris et al. 1999; Yuan et al. 2013); H10 viruses with N7 and N8 subtypes have both been detected in humans (Arzey et al. 2012; WHO 2013). In contrast, human disease has resulted from H7 viruses possessing four NA subtypes, indicating a greater compatibility of the H7 hemagglutinin (HA) with several different NA configurations and suggesting that there is not a clear association of NA subtypes with virulent H7 viruses (Fig. 1) (Belser et al. 2009a). Furthermore, while human infection with H5, H6, and H9 viruses has been limited to Eurasian lineage strains, among all avian-origin viruses H7 viruses represent the subtype with the

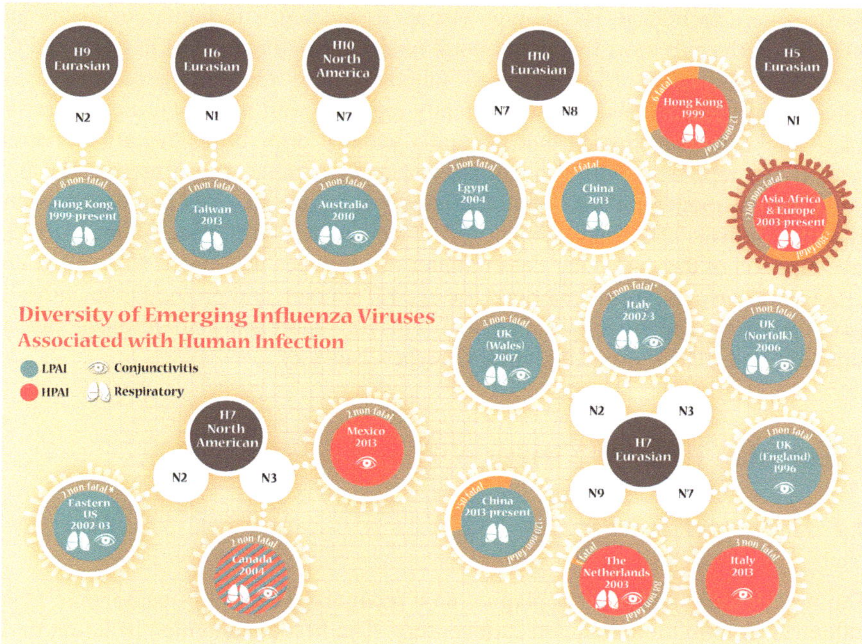

Fig. 1 Diversity of emerging influenza viruses associated with human infection. Confirmed or presumed (serologic evidence only) cases of avian influenza virus infection in humans since 1996. *, includes cases with serologic evidence only. *Red border* surrounding H5N1 human cases from 2003 to present indicates that there is substantial clade-specific and geographic diversity among this virus subtype not represented in this image (Abdel-Ghafar et al. 2008)

greatest number of documented human infections from both Eurasian and North American lineages. The diversity of these H7 avian influenza viruses resulting in human disease, encompassing both highly pathogenic (HPAI) and LPAI viruses and with select strains possessing atypical molecular determinants of virulence, also sets this virus subtype apart from other avian subtypes with pandemic potential. These features underscore the difficulty of studying the subtype as a whole, frequently requiring the inclusion of several virus isolates when establishing experimental designs to eliminate strain-specific differences inherent among this virus subtype.

Due to the pronounced heterogeneity among H7 subtype viruses associated with human infection, there is an urgent public health need to develop mammalian models for the study of pathogenicity and transmissibility, and to assess lineage-specific and strain-specific diversity of viruses within this subtype. Furthermore, the frequent detection of conjunctivitis with occasional and generally mild respiratory illness caused by H7 viruses in humans necessitates the development of models which reflect the multiple exposure routes exploited by this virus subtype. In this review, we discuss the pathogenicity and transmissibility of Eurasian and

North American lineage H7 subtype viruses associated with human exposure and infection in the two most frequently employed mammalian models in this research, the mouse and ferret. Mice are well suited to study the pathogenesis of avian influenza viruses, as these viruses typically replicate well in the murine respiratory tract, unlike human viruses that generally require prior host adaptation. Ferrets support the replication of both human and avian strains, and represent the best small animal model for the coincident study of influenza virus pathogenesis and transmissibility. The development of these models has greatly enabled the investigation of the molecular determinants of virulence associated with HPAI and LPAI Eurasian and North American lineage H7 viruses, and provided the ability to assess existing and novel vaccine and antiviral approaches to mitigate human infection and illness from this virus subtype (Table 1).

2 Eurasian Lineage Viruses

2.1 Historical H7 Viruses

The first descriptions of fowl plague, now recognized as highly pathogenic avian influenza, date to the late nineteenth century. The first documented isolation of an influenza virus was of the H7N7 subtype in 1902 from an outbreak of fowl plague virus (FPV) in Italy, predating the first isolation of a human influenza virus by thirty years (Horimoto and Kawaoka 2001). Documented human infection with a fowl plague virus dates to 1959, with additional human cases infrequently detected, typically following laboratory or occupational exposure (Belser et al. 2009a). While belonging to a separate phylogenetic clade, these historical H7 influenza viruses exhibit closest sequence similarity to the current Eurasian lineage (Bulach et al. 2010). Eurasian lineage H7 viruses isolated throughout the 20th century have exhibited great strain-specific diversity; most viruses examined have been found to infect mice without prior adaptation, with select H7N1 and H7N3 viruses causing lethal disease in this species (Joseph et al. 2007). Interestingly, many of these viruses elicit broadly cross-reactive antibodies to contemporary Eurasian and North American viruses (Joseph et al. 2007).

Equine influenza viruses of the H7N7 subtype were first isolated in 1956 (Timoney 1996). Despite only causing mild respiratory symptoms in horses, equine H7N7 influenza viruses were found to be lethal in mice without prior adaptation, with the capacity for enhanced neurovirulence in this species following murine adaptation (Christensen et al. 2000; Kawaoka 1991; Shinya et al. 2005, 2007). This is likely due to the presence of a multibasic amino acid HA0 cleavage site among these viruses, a molecular determinant of virulence frequently detected in viruses that exhibit a high pathogenicity phenotype in mice. While they have not been isolated from horses since 1978, serologic evidence supporting the continued circulation of equine H7N7 viruses in horses has been reported (Abd El-Rahim and Hussein 2004; Mancini et al. 2012; Timoney 1996; Webster 1993).

Table 1 Virulence of H7 viruses in mammalian models following intranasal inoculation

Location	Year(s)	IVPI[a]	Subtype	Mouse virulence	Ferret virulence	References
Pakistan	1995–2002	HPAI	H7N3	Lethality detected among selected strains	Moderate disease, no lethality reported	(Aamir et al. 2009; Abbas et al. 2010)
Italy	1999–2000	HPAI	H7N1	Lethality detected among selected strains	Moderate disease, no lethality reported	(Cox et al. 2009; Rigoni et al. 2007; Whiteley et al. 2007)
Northeast US	1994–2006	LPAI	H7N3	Mild to moderate disease, no lethality reported	Mild to moderate disease, no lethality reported	(Belser et al. 2007; Joseph et al. 2007)
Chile	2002	HPAI[b]	H7N3	Mild to moderate disease, no lethality reported	Not examined	(Belser et al. 2013a; Joseph et al. 2007)
The Netherlands	2003	HPAI	H7N7	Lethality detected among selected strains	Lethality detected among selected strains	(Belser et al. 2007; de Wit et al. 2005; Ekiert et al. 2011; Joseph et al. 2007)
Canada	2004	HPAI[b]	H7N3	Lethality detected among selected strains	Moderate disease, no lethality reported	(Belser et al. 2007, 2008; Joseph et al. 2007, 2008; Song et al. 2009)
Mexico	2012	HPAI	H7N3	Lethality detected among selected strains	Moderate disease, no lethality reported	(Belser et al. 2013a)
China	2013–present	LPAI	H7N9	Lethality detected among selected strains	Lethality detected among selected strains	(Belser et al. 2013b; Mok et al. 2013; Watanabe et al. 2013; Zhang et al. 2013; Zhu et al. 2013)

[a] IVPI, intravenous pathogenicity index

[b] LPAI viruses were also isolated from these outbreaks and tested in these models

2.2 H7N7 Viruses

The first well-documented case of direct transmission of an H7 subtype avian influenza virus from an avian to a human host occurred in 1996 in England (Kurtz et al. 1996). The causative strain, a LPAI H7N7 virus, possessed a close phylogenetic relationship with H7 viruses isolated from poultry in the area (Banks et al. 1998). Transmission likely occurred following ocular exposure to infected ducks, and the exposed woman developed conjunctivitis; no subsequent human cases were detected (Belser et al. 2009a). The virus isolated from the human case replicated in the respiratory tract of inoculated mice without prior adaptation but did not cause severe disease in this model (Belser et al. 2009b). An outbreak of HPAI H7N7 virus in Italy resulting in three confirmed cases of human conjunctivitis represents the most recent detection of this virus (ProMED-mail 2013); research investigating the virulence of these isolates in mammalian models is ongoing.

In 2003, an outbreak of HPAI H7N7 virus in the Netherlands necessitated the culling of >30 million birds, and resulted in 89 virologically confirmed human cases, representing the largest instance of HPAI H7 subtype infection in humans to date (Fouchier et al. 2004; Koopmans et al. 2004). The majority of human infections presented as conjunctivitis, though respiratory symptoms were occasionally detected; there was one fatality, due to acute respiratory distress syndrome (Fouchier et al. 2004). Two molecular features of viruses isolated from this outbreak were similar to those identified in HPAI H5N1 viruses: the presence of a multibasic amino acid HA cleavage site and, in the virus isolated from the fatal case (A/Netherlands/219/03, NL/219), the presence of a lysine at position 627 in PB2 (Fouchier et al. 2004). With the exception of E627K, many of the known molecular markers of virulence and human adaptation present among human isolates were detected in poultry isolates before potential avian-to-human transmission events (Jonges et al. 2011), underscoring the need for early intervention during outbreaks of influenza virus in poultry.

Viruses isolated from the 2003 Netherlands outbreak were found to be highly infectious in mice following intranasal inoculation, replicating efficiently throughout the murine respiratory tract and demonstrating the capacity for systemic spread to extrapulmonary tissues, including the brain (Belser et al. 2007; de Wit et al. 2005; Joseph et al. 2007). Most viruses from this outbreak caused substantial morbidity in mice before eventual recovery, with select viruses (including the fatal case isolate NL/219) possessing a lethal dose comparable to H5N1 viruses which exhibit a high pathogenicity phenotype in this species (Belser et al. 2007; de Wit et al. 2005; Joseph et al. 2007). HPAI H7N7 viruses from 2003 also replicated to high titer following intranasal inoculation in ferrets, exhibiting systemic spread to extrapulmonary tissues and causing fever, weight loss, nasal discharge, and transient lymphopenia (Belser et al. 2007). Similar to mice, NL/219 virus exhibited a high pathogenicity phenotype in ferrets, with ferrets losing >20% of their initial body weight and exhibiting neurological symptoms during the course of infection, necessitating euthanasia.

Due to the high incidence of conjunctivitis among infected individuals, several studies were conducted to assess the ocular tropism of HPAI H7N7 viruses in mammalian models. Underscoring the ability of H7 viruses to both use the eye as a site of virus replication and as a portal of entry to establish a productive respiratory infection, virus was detected in ocular and respiratory tissues of both mice and ferrets when inoculated by the ocular route with isolates from the Netherlands outbreak (Belser et al. 2009c, 2012a). Furthermore, NL/219 virus maintained a lethal phenotype in mice (but not ferrets) following ocular inoculation. Notably, HPAI H7N7 viruses were detected in ferret ocular tissue following traditional intranasal inoculation, indicating that influenza viruses can spread from nasal to ocular tissues more readily than previously considered (Belser et al. 2012a, 2013c).

Limited human-to-human transmission was reported during the Netherlands outbreak (Koopmans et al. 2004), raising the need to model transmissibility of HPAI H7N7 viruses. NL/219 virus was not capable of transmission to naïve contacts in the ferret model, similar to HPAI H5N1 viruses. However, a virus isolated from a conjunctivitis case (A/Netherlands/230/03, NL/230) transmitted in a ferret direct contact model following either intranasal or ocular inoculation of ferrets, though airborne transmission was not detected (Belser et al. 2008, 2012a).

2.3 *H7N1, H7N2, and H7N3 Viruses*

H7N1, H7N2, and H7N3 influenza viruses have been responsible for frequent poultry outbreaks in Europe and Asia in recent years (Alexander 2007; Brown 2010). While reduced in scope compared with the Netherlands outbreak in 2003, these epornitics have resulted in large culling operations leading to human occupational exposure and occasional infection (Fig. 1) (2007; Nguyen-Van-Tam et al. 2006; Puzelli et al. 2005). Suspected and confirmed human cases have been reported following several of these outbreaks, with disease ranging from mild to severe (Dudley 2008; Eames et al. 2010; Nguyen-Van-Tam et al. 2006; Puzelli et al. 2005).

Numerous poultry outbreaks of LPAI (H7N1, H7N3) and HPAI (H7N1) viruses have been reported in Italy since 1999. Viruses isolated from the 1999–2000 HPAI H7N1 outbreak in Italy generally demonstrated similar pathogenicity in mice as HPAI H7N7 viruses from the Netherlands, with frequent detection of extrapulmonary spread and lethality reported among select isolates (Rigoni et al. 2007; Whiteley et al. 2007). However, while systemic spread of virus was detected in ferrets (albeit later in infection, with virus detected day 6 p.i. and not day 3 p.i. as compared with HPAI H7N7 viruses), the lethal phenotype observed with selected HPAI H7N1 viruses does not appear to be maintained in this model (Cox et al. 2009; Whiteley et al. 2007). Virus was not isolated from any exposed individuals from these outbreaks, though a percentage (<4 %) of tested poultry workers exhibited seroconversion to H7 virus, albeit at very low titers (Di Trani et al. 2012; Puzelli et al. 2005).

The first introduction of LPAI H7N3 virus to Pakistan occurred in 1995, with viruses acquiring a HPAI phenotype within a year of circulation in chickens (Abbas et al. 2010). Over the following decade, H7N3 viruses caused sporadic outbreaks in geographically distinct regions of the country, suggesting a virus reservoir in poultry. HPAI H7N3 viruses isolated from Pakistan between 1995 and 2002 replicated efficiently in the lungs of inoculated mice, but did not spread to extrapulmonary tissues; lethality in this model was observed among the most recently isolated strains (Aamir et al. 2009). Similar to HPAI H7N1 viruses, ferrets inoculated by the intranasal or ocular route with an HPAI H7N3 virus associated with lethality in mice exhibited moderate signs of illness but did not succumb to infection. It is important to note that while the majority of mammalian pathotyping of H7 viruses occurs with strains associated with human infection or large poultry outbreaks, viruses that are detected during routine surveillance have been shown to replicate efficiently in mammalian models. As an example, a LPAI H7N2 virus from northern China isolated in 2002 was found to replicate efficiently in both pulmonary and extrapulmonary tissues in mice (Li et al. 2006), underscoring the ability of H7 avian influenza viruses to replicate in mammals without prior adaptation and emphasizing the continuous need for surveillance of galliform poultry for strains which could possess pandemic potential.

Conjunctivitis is a frequent symptom among suspected and virologically confirmed cases of H7N1, H7N2, and H7N3 virus infection (Dudley 2008; Puzelli et al. 2005). Similar to HPAI H7N7 viruses, virus was recovered from eye swabs of intranasally (HPAI H7N1) or ocularly (HPAI H7N3) inoculated ferrets, though the detection of other avian and human influenza viruses recovered from ferret ocular samples indicates that this ability is not specific to H7 subtype viruses (Aamir et al. 2009; Belser et al. 2012a; Cox et al. 2009). While mice inoculated by the ocular route with influenza viruses typically do not display macroscopic ocular signs of disease, ocular symptoms were nonetheless detected among HPAI H7N1 and H7N3 viruses following intranasal inoculation in mice (Aamir et al. 2009; Cox et al. 2009; Belser et al. 2009c). These studies further support the ability of Eurasian lineage H7 viruses to employ the ocular system during in vivo mammalian infection.

Unfortunately, mammalian pathotyping of avian H7 subtype viruses associated with all human outbreaks has not been performed, and the transmissibility of these viruses has not been well examined in mammalian models. HPAI H7N1 viruses were reported to transmit in a direct contact mouse model (Rigoni et al. 2010), though there is a paucity of data establishing mice as a reliable transmission model for influenza viruses. While human-to-human transmission has not been confirmed among the outbreaks discussed here, contact tracing of suspected cases from an outbreak of LPAI H7N2 virus in Wales indicates that limited independent chains of person-to-person transmission cannot be ruled out (Eames et al. 2010). Future studies are needed to better identify the capacity of these viruses to transmit between mammals, as studied previously using viruses from the Netherlands outbreak.

2.4 LPAI H7N9 Viruses

Eurasian lineage LPAI H7N9 viruses had been previously detected during surveillance activities, but not until 2013 were these viruses associated with human infection (Gerloff et al. 2013; Gonzalez-Reiche et al. 2012; Perez-Ramirez et al. 2010). Human infections, totaling over 440 confirmed cases with over 160 fatalities, have been characterized by severe pulmonary disease and acute respiratory distress syndrome (Gao et al. 2013b). The majority of human infections have resulted from exposure to infected poultry, though evidence for limited human-to-human transmission has been reported (Li et al. 2013; Qi et al. 2013). Interestingly, the older median age among H7N9 cases compared with H5N1 cases, and the relatively few reports of mild cases among H7N9-infected individuals, indicates potential differences in susceptibility to severe illness for these avian viruses (Cowling et al. 2013).

The unexpected severe disease following human infection with a LPAI virus warranted swift assessment of the virulence of H7N9 viruses in mammalian models. Mice inoculated by the intranasal route with the first human isolates A/Anhui/1/13 (Anhui/1) and A/Shanghai/1/13 generally showed severe morbidity and comparable lethality as HPAI H5N1 and H7N7 viruses, though strain-specific differences in virulence have been detected and not all human isolates possess a lethal phenotype in this model (Baranovich et al. 2013; Belser et al. 2013b; Mok et al. 2013; Watanabe et al. 2013; Zhang et al. 2013). While extrapulmonary spread of virus is frequently associated with a lethal outcome in infected mice, viral titers in the brain were only sporadically reported following H7N9 infection (Belser et al. 2013b; Zhang et al. 2013). In contrast, avian isolates tested from this outbreak (which lack a lysine at position 627 in PB2) were not highly pathogenic in mice (Zhang et al. 2013). This is in accord with other genetically related LPAI H7N9 viruses isolated in prior years from avian species which possess comparable infectivity and replicative ability in murine respiratory tract tissues but did not maintain the high virulence observed with human isolates in this species (Belser et al. 2013b; Mok et al. 2013; Watanabe et al. 2013).

Ferrets were used extensively to characterize the pathogenicity of LPAI H7N9 viruses isolated from humans. Intranasally inoculated ferrets exhibited modest signs of disease, including weight loss, fever, sneezing and lethargy, but generally did not maintain the lethal phenotype observed in the mouse model (Belser et al. 2013b; Watanabe et al. 2013; Zhu et al. 2013). H7N9 viruses replicated efficiently throughout the ferret respiratory tract; virus was frequently but not uniformly detected in the lungs, with moderate inflammation and inflammatory infiltrates detected in bronchioles and alveoli (Belser et al. 2013b; Watanabe et al. 2013; Zhu et al. 2013). Lymphopenia and leukopenia, often detected among severe human cases, were also observed in infected ferrets, though levels of circulating white blood cells generally returned to pre-infection levels following the acute phase of infection (Belser et al. 2013b; Gao et al. 2013a; Xu et al. 2013b). While nonhuman primates are not typically utilized for the study of H7 influenza viruses, cynomolgus macaques inoculated with Anhui/1 virus in one study exhibited generally comparable levels of pathogenicity as observed in ferrets (Watanabe et al. 2013). Taken together, these studies found that LPAI H7N9 viruses exhibit similar disease in the ferret as HPAI H7N7 viruses, but cause a less severe infection compared to highly virulent HPAI H5N1 viruses (Maines et al. 2005).

The association of LPAI H7N9 viruses with severe respiratory disease in humans is in disaccord with the mild ocular and/or respiratory disease most typically observed following infection with H7 viruses (Belser et al. 2009a). However, varying the inoculation route was found to modulate H7N9 virulence, as intratracheal inoculation of ferrets with Anhui/1 virus lead to heightened morbidity and mortality compared to traditional intranasal inoculation (Kreijtz et al. 2013). While LPAI H7N9 viruses have not exhibited an ocular tropism in humans, swollen eyes were reported at the onset of mild respiratory illness in one H7N9 virus-positive poultry worker wearing respiratory but not ocular protection during the time of occupational exposure (Lv et al. 2013), suggesting that the virus may have employed the ocular route to cause human disease. In support of this, virus was detected in the respiratory tract of mice inoculated by the ocular route, as well as in the eye washes of ferrets inoculated by the intranasal route (Belser et al. 2013b). Collectively these results indicate that, similar to other influenza viruses, H7N9 virus can utilize multiple exposure routes to mount a productive infection in mammals, with severity of disease influenced by the route of infection.

The presence among several H7N9 isolates of amino acid changes known to increase binding to α2-6 linked sialic acids, notably Q226L in the HA protein, suggested that these viruses were capable of enhanced transmissibility compared to other avian-origin viruses (Gao et al. 2013b). H7N9 isolates were found to transmit efficiently between either ferrets or guinea pigs when placed in direct contact, a capacity not generally present among avian influenza viruses (Belser et al. 2013b; Gabbard et al. 2013; Maines et al. 2006; Zhu et al. 2013). However, while the incidence of H7N9 virus transmission between ferrets in the absence of direct or indirect contact (i.e. respiratory droplet or airborne transmission) was greater than that observed among H7N9 viruses isolated from avian species, transmission was overall less efficient than seasonal or pandemic influenza viruses in this model, with delayed kinetics of virus spread to contact ferrets when transmission did occur (Belser et al. 2013b; Maines et al. 2006, 2009; Richard et al. 2013; Watanabe et al. 2013; Xu et al. 2013a; Zhang et al. 2013; Zhu et al. 2013). While selected studies have observed efficient transmission of H7N9 viruses by respiratory droplets, serial passage of H7N9 virus in ferrets did not enhance virus transmissibility, further indicating that, despite the efficient transmissibility detected in a direct contact model, these viruses are not yet capable of sustained airborne mammalian transmission and would require additional mammalian adaptation to achieve this property (Richard et al. 2013; Zhang et al. 2013).

3 North American Lineage Viruses

3.1 Historical H7 Viruses

North American lineage H7 viruses, similar to those within the Eurasian lineage, have been detected numerous times throughout the twentieth century, most frequently from avian species. The majority of these viruses examined have been found to infect mice without prior adaptation, though rarely have they been associated with lethal disease in this species following intranasal inoculation (Goff et al. 2013; Joseph et al. 2007). In 1979–1980, an H7N7 virus caused the death of ~500 seals on the New England coast, subsequently resulting in five cases of conjunctivitis among laboratory workers with occupational exposure to the virus (one virologically confirmed, four suspect cases) (Lang et al. 1981; Webster et al. 1981). This virus only caused mild respiratory disease in mice, and did not infect experimentally inoculated ferrets or rats (Scheiblauer et al. 1995). However, mammalian passage of a laboratory variant of this virus was capable of causing lethal disease in mice and ferrets, indicating that the H7N7 virus could acquire neurotropic and pathogenic properties following host adaptation.

3.2 LPAI H7 Viruses in the United States

LPAI H7N2 virus was first isolated from northeastern United States live bird markets (LBM) in 1994, and persisted for over a decade before the virus was eradicated from those markets in 2006 (Senne 2010; Senne et al. 2003). During this time, the largest outbreak of LPAI in the United States occurred in 2002 in Virginia and the surrounding area, due to an H7N2 virus with high genetic sequence identity to the LBM circulating strains (Senne 2007). These viruses were associated with one confirmed case of human infection (A/New York/107/03, NY/107) and one suspected case (among 80 tested poultry workers) with occupational exposure who seroconverted to H7N2 virus during the Virginia outbreak (CDC 2004; Ostrowsky et al. 2012). Despite the elimination of H7N2 viruses from the northeastern United States LBM system, H7 subtype viruses continue to circulate across the United States, as H7N3, H7N7, and H7N9 subtype viruses have been isolated from wild birds and galliform poultry over the past decade (Pasick et al. 2012; Senne 2007, 2010). While these viruses have not caused any confirmed human cases, seroprevalence studies have reported statistically significant elevated titers against H7 influenza virus in veterinarians compared to healthy control subjects, suggesting that occupational exposure may be occurring (Myers et al. 2007).

LPAI H7 viruses from North America have typically not caused severe disease in experimentally inoculated mammalian models. In mice, virus replicates efficiently in both the nose and lung following intranasal inoculation, with extrapulmonary spread generally not observed (Belser et al. 2007, 2013a; Joseph et al.

2007). However, LPAI H7 viruses are highly infectious in this model, and morbidity can range from mild to moderate depending on the strain used to infect. Similar results are detected in the ferret model, though viral titers are generally lower in the lung compared with the nasal turbinates and trachea (Belser et al. 2007, 2008, 2013a). Ferret morbidity is generally mild and transient, with few lymphohematopoietic perturbations and an absence of extrapulmonary virus spread.

Similar to Eurasian lineage H7 viruses, many LPAI H7 viruses from the North American lineage can employ the mammalian eye as a site of primary virus replication as well as a portal of entry to spread to the respiratory tract (Belser et al. 2009c, 2012, 2013a). Ocular inoculation of mice with LPAI H7N3 and H7N9 viruses resulted in detectable virus titers in both eye and respiratory tract tissues in the absence of significant morbidity, as did ferrets inoculated by the ocular route with the LPAI H7N2 virus NY/107 (Belser et al. 2012a, 2013a). These findings indicate that avian isolates not associated with human disease are still capable of using the eye as a portal of entry to cause a productive mammalian infection.

Unlike the majority of Eurasian lineage viruses (with the exception of LPAI H7N9 viruses), selected contemporary North American viruses possess the ability to bind both α2-3 and α2-6 linked sialic acids, as will be discussed in more detail in the subsequent section. As a binding preference for α2-6 linked sialic acids is a feature of human and pandemic influenza viruses, continued evaluation of the transmissibility of North American H7 viruses has become essential to monitor the true pandemic potential of these viruses. Notably, several LPAI H7N2 viruses have been found capable of transmission to naïve contacts in the presence of direct contact. NY/107 virus was found to transmit efficiently in a direct contact model following either intranasal or ocular inoculation of ferrets (Belser et al. 2008, 2012a). A chicken LPAI H7N2 virus from 2003, but not a turkey isolate from 2002, was also capable of limited transmission in a direct contact model, despite both viruses exhibiting a similar receptor-binding profile as NY/107 virus, indicating that receptor binding preference alone does not confer a transmissible phenotype (Belser et al. 2008). However, no wild-type LPAI H7 North American strains have been found capable of transmission between ferrets by the airborne route to date.

3.3 HPAI H7N3 Viruses

HPAI H7N3 viruses have caused numerous poultry outbreaks in both North and South America (Lopez-Martinez et al. 2013; Senne 2007, 2010). Unlike Eurasian HPAI H5N1 and H7N7 viruses which typically possess traditional multibasic amino acid HA cleavage sites, these HPAI H7N3 viruses all appear to have arisen from LPAI precursor viruses by nonhomologous recombination between the HA and other viral genes or host cell genetic sequences (Berhane et al. 2009; Maurer-Stroh et al. 2013; Pasick et al. 2005; Suarez et al. 2004). To date, H7 viruses are the only known subtype to use this mechanism as a way to acquire a virulent phenotype. Several of these epornitics have resulted in human infection, underscoring the need to study the virulence of these viruses which pose a threat to both avian and human health and to elucidate why H7 subtype viruses are most susceptible to this form of recombination. Unfortunately, viruses from these unique outbreaks have been infrequently studied in mammalian models, limiting our understanding of how the H7 HA cleavage site influences viral pathogenicity.

An outbreak of LPAI H7N3 virus occurred in Chile in May 2002, representing the first isolation of H7 influenza virus in poultry or wild birds in the continent of South America (Suarez et al. 2004). Recombination between the HA and nucleoprotein genes of the LPAI virus resulted in the generation of a HPAI virus within a month; no human cases were reported during the 7 months of eradication efforts (Max et al. 2007). Despite exhibiting differential infectivity and virulence in poultry, both HPAI and LPAI viruses from this outbreak replicated efficiently in the lungs of mice in the absence of substantial morbidity or mortality (Belser et al. 2013a; Jones and Swayne 2004; Joseph et al. 2007).

In 2004, an H7N3 virus detected in British Columbia, Canada acquired a high pathogenicity phenotype in chickens following a nonhomologous recombination event between the HA and matrix genes (Hirst et al. 2004). This outbreak necessitated the culling of over 19 million birds, and resulted in 55 suspected and two virologically confirmed human infections (Tweed et al. 2004). While both human cases presented with conjunctivitis and influenza-like illness following direct conjunctival contact with poultry, virus isolated from one human case was found to be LPAI whereas the other human isolate (A/Canada/504/04, Can/504) was found to be HPAI (Hirst et al. 2004; Tweed et al. 2004); one additional basic amino acid was present in the cleavage region of the HPAI virus compared with the LPAI strain (Hirst et al. 2004). Similar to the H7N3 viruses from Chile, both HPAI and LPAI H7N3 viruses associated with human infection, as well as a closely related LPAI chicken isolate from this outbreak and a mallard LPAI H7N3 virus isolated in 2001, replicated efficiently in the lungs of mice but did not spread to extrapulmonary tissues following intranasal inoculation (Belser et al. 2007; Joseph et al. 2008; Song et al. 2009). However, a HPAI chicken isolate from this outbreak was capable of lethality and extrapulmonary spread to the brain following intranasal murine inoculation, indicating strain-specific differences in virulence among these HPAI H7N3 viruses (Belser et al. 2008; Joseph et al. 2007). Both HPAI and LPAI chicken isolates from the 2004 H7N3 outbreak replicated efficiently in ferret upper and lower respiratory tract tissues (Joseph et al. 2008). Unlike mice, ferrets inoculated by the intranasal route with Can/504 virus exhibited substantial weight loss before recovery, indicating species-specific differences in mammalian virulence among HPAI H7N3 viruses from this outbreak. A subsequent outbreak of HPAI H7N3 in Saskatchewan, Canada in 2007 also emerged from a nonhomologous recombination event with a genetic sequence of eukaryotic origin, but did not result in confirmed human infection (Pasick et al. 2010); the causative virus has not been extensively studied in mammalian models.

The most recent outbreak of HPAI H7N3 virus occurred in Jalisco, Mexico in 2012, resulting in the culling of over 22 million birds and the detection of two confirmed human cases among poultry workers, both presenting with conjunctivitis (Kreijtz et al. 2013; Lopez-Martinez et al. 2013). Similar to other HPAI H7N3 viruses from North America, isolates from this outbreak also possess an extended HA cleavage site, with the insertion likely acquired from host 28S ribosomal RNA (Lopez-Martinez et al. 2013; Maurer-Stroh et al. 2013). Mice inoculated with virus isolated from a human conjunctivitis case (A/Mexico/InDRE7218/12, (Mex/7218)) exhibited greater morbidity and mortality compared with previously examined North American HPAI H7N3 viruses, causing severe morbidity and lethal disease in mice intranasally inoculated with high (10^6 EID_{50}) doses of virus (Belser et al. 2013a). Similar to mice, Mex/7218 virus replicated to high titer throughout the ferret respiratory tract following intranasal inoculation, including the lung, but did not spread systemically to multiple organs. However, extrapulmonary spread of virus was detected in the brain and gastrointestinal tract of infected ferrets, although all ferrets survived viral challenge (Belser et al. 2013a).

Conjunctivitis was documented among all human infections with North American H7N3 viruses, underscoring a need to examine the pathogenicity of these viruses following ocular exposure. Mice inoculated by the ocular route with HPAI H7N3 human isolates from Canada and Mexico did not exhibit significant morbidity or mortality, but did possess detectable virus in both ocular and respiratory tract tissues post-inoculation (Belser et al. 2009c, 2013a). Similarly, ferrets inoculated by the ocular route with Can/504 virus became productively infected in the absence of substantial morbidity, with virus detected in samples collected from both the respiratory tract and eye (Belser et al. 2012a).

Just as North American lineage LPAI H7N2 viruses have exhibited limited transmissibility in a direct contact ferret model, LPAI and HPAI H7N3 viruses from several of these outbreaks have also demonstrated this property. Low levels of the HPAI H7N3 virus Can/504 isolated from a human conjunctivitis case, were transmitted to ferrets in a direct contact model (Belser et al. 2008). The HPAI H7N3 virus Mex/7218, also isolated from a conjunctivitis case, efficiently transmitted to naïve ferrets in a direct contact model, though a genetically related North American LPAI H7N9 virus did not (Belser et al. 2013a). Strikingly, transmission in the presence of direct contact was also reported with a mallard LPAI H7N3 virus in the absence of prior host adaptation, underscoring the need for continued surveillance and evaluation of H7 viruses from wild birds (Song et al. 2009). However, similar to other North American H7 viruses, airborne transmission of H7N3 viruses has not been reported to date.

4 Molecular and Host Determinants of Pathogenicity and Transmissibility

4.1 Molecular Correlates of Pathogenicity

The majority of research studies investigating molecular determinants associated with mammalian virulence of avian influenza viruses have been conducted with H5 and not H7 subtype viruses. However, it appears that many features associated with H5N1 virulence are maintained in H7 subtype viruses, especially with regard to the virus HA. Similar to results obtained using H5N1 strains, reassortant viruses bearing surface glycoproteins derived from a virulent H7 virus on the backbone of a nonlethal H7 virus exhibit moderately enhanced virulence without increases in lethality in mammalian species (Munster et al. 2007; Salomon et al. 2006). The introduction of glycosylation sites proximal to the receptor binding pocket of the HA has been shown to facilitate adaptation of H7 and H5 viruses from wild birds to poultry and modulate virulence in mammalian hosts (Banks and Plowright 2003; Belser and Tumpey 2013; de Wit et al. 2010). Building upon prior studies conducted with H5N1 viruses has thus enabled greater precision in the identification of molecular markers of H7 subtype virulence.

Previous studies on H5N1 mammalian virulence have indicated that while a multibasic amino acid HA cleavage site is necessary (Hatta et al. 2001; Horimoto and Kawaoka 1994; Suguitan et al. 2012), this molecular feature alone is not sufficient for a virulent phenotype in mammals (Maines et al. 2005). In this vein, studies with H7 viruses have demonstrated that the presence of a multibasic amino acid cleavage site does not necessarily confer a virulent phenotype in either avian or mammalian species (Belser et al. 2007; Senne et al. 1996). While the composition of HA cleavage sites among Eurasian lineage LPAI and HPAI H7 viruses generally resemble that of H5N1 viruses, there is a greater diversity among North American H7 viruses. Insertion of additional basic amino acids at the cleavage site of a LPAI H7N2 virus, but not mutation of residues already present at the cleavage site, resulted in viruses which were highly pathogenic in chickens, demonstrating the capacity of an H7 virus to acquire a pathogenic phenotype, in contrast to H5 viruses which can acquire a virulent phenotype by either mutation or insertion (Lee et al. 2006). Molecular modeling of the insertional event resulting in a HPAI H7N3 virus isolated from Canada in 2004 revealed an enlarged loop formation resulting in greater accessibility of the HA cleavage site, likely leading to increased cleavage by furin-like proteases resulting in enhanced pathogenicity (Hirst et al. 2004). Future study is needed to better understand the contribution and mechanism of HA cleavage site insertions as they pertain to mammalian pathogenesis, especially among North American lineage viruses.

The presence of a lysine at position 627 in PB2 has long been identified as a critical factor of mammalian adaptation and virulence among influenza viruses, including HPAI H5N1 viruses (Hatta et al. 2001; Subbarao et al. 1993). As mentioned in preceding sections, several studies have demonstrated a similar role for this position among H7 subtype viruses. An E627K substitution in a HPAI H7N7 virus isolated from the Netherlands was identified as a major determinant of mammalian pathogenicity, with viruses bearing this mutation exhibiting increased viral load, extrapulmonary spread to the brain, and heightened mortality in mice; this enhanced virulence was absent in mice inoculated with virus bearing the reciprocal K627E mutation (de Jong et al. 2013; Munster et al. 2007). Similarly, the presence of a lysine at this position in LPAI H7N9 human isolates and a mouse-adapted H7N7 equine virus resulted in increased PB2 polymerase activity, viral loads, and disease severity in mice (Mok et al. 2014; Shinya et al. 2007; Zhang et al. 2013, 2014). Enhanced virulence of HPAI H7N7 viruses bearing E627K was also observed in ferrets compared with virus bearing glutamic acid at this position (de Jong et al. 2013). However, H7N9 viruses bearing either a glutamic acid or lysine at position 627 exhibited comparable transmissibility in the ferret model, indicating E627K does not solely confer an enhanced transmissibility phenotype (Zhang et al. 2013). These studies are supported by in vitro assays in human cell types revealing that the E627K mutation confers enhanced replicative ability and polymerase activity compared with K627E (de Wit et al. 2010). While the presence of E627K yields heightened virulence in mice, it should be noted that select viruses lacking this mutation are still capable of extrapulmonary spread and murine lethality (Aamir et al. 2009; Rigoni et al. 2007), demonstrating the polygenic nature of mammalian virulence and underscoring that PB2 position 627 is only one of several molecular determinants that contribute to mammalian pathogenicity. In support of this, additional mutations in the H7 viral polymerase independent of position 627 have been demonstrated to modulate virulence in a mouse model (Gabriel et al. 2005, 2007; Yamayoshi et al. 2013). Disruption of protein–protein binding between PA and PB1 was also shown to attenuate murine virulence of an H7N7 virus (Manz et al. 2011).

The influenza non-structural (NS1) protein possesses several varied functions which contribute to block type-1 interferon-mediated host antiviral responses (Garcia-Sastre 2001). A mouse-adapted H7N7 virus with a complete deletion of the NS1 gene was nonpathogenic in wild-type mice, suggesting that enhanced synthesis of IFN in infected mouse lungs led to virus attenuation in this species (Kochs et al. 2007). Comparison of NS1 proteins encoded by different subtypes in vitro revealed that an H7 NS1 induced lower levels of proinflammatory cytokines, chemokines, and levels of apoptosis compared to an H5 NS1 (Lam et al. 2011); this is in agreement with in vivo studies which detect proinflammatory cytokines and chemokines in respiratory tract tissues of mice and ferrets inoculated by the intranasal route with H7 subtype viruses, albeit at lower levels than virulent HPAI H5N1 viruses (Belser et al. 2007, 2013c). However, a recombinant H7N7 virus bearing the NS gene from a fatal human case did not enhance the virulence of an H7N7 virus isolated from a conjunctivitis case, indicating that this gene does not represent a major determinant of virulence among H7 strains (Munster et al. 2007).

Dysregulation of innate host immune responses and other host factors following H5N1 virus infection in mammals has been shown to contribute to the high pathogenicity phenotype of this virus subtype (Belser and Tumpey 2013), but there are only limited studies conducted in the context of mammalian H7 virus infection. H7N7 virus infection of different inbred laboratory mouse strains resulted in divergent survival outcomes, demonstrating a role for the genetic background of the host in H7 mammalian virulence (Srivastava et al. 2009). The interferon-induced resistance factor Mx1 has been shown to contribute to interferon-mediated protection against both avian and human influenza viruses, including the H7 subtype, as viruses which cause lethal disease in standard laboratory mouse strains (which lack a functional *Mx1* gene) do not maintain a lethal phenotype in mice carrying a wild-type functional *Mx1* gene (Koerner et al. 2007; Tumpey et al. 2007). However, Mx1-positive mice lacking functional toll-like receptor (TLR)-7 signaling or which were depleted of plasmacytoid dendritic cells exhibit a decrease in resistance to a mouse-adapted HPAI H7N7 virus isolate of seal origin (SC35M), indicating TLR7-mediated detection of H7 virus via plasmacytoid dendritic cells represents a source of interferon which confers antiviral protection in the host (Kaminski et al. 2012). Lymphopenia has been reported following H7 virus infection in mice and ferrets (Belser et al. 2007); studies using SC35M virus found that the high polymerase activity of this strain was responsible for the severe lymphocyte depletion and impaired immune response observed in SC35M virus-infected mice but not in the non-mouse-adapted counterpart (Gabriel et al. 2009). Collectively, these studies indicate that while several molecular determinants of avian influenza virulence are maintained between H5 and H7 viruses, there are numerous ways in which H7 subtype viruses differentially interact with the mammalian host.

4.2 Receptor Binding and Correlates of Transmissibility

The distribution of host cell glycoconjugates containing terminal sialic acid moities, which serve as the receptors for all influenza viruses, greatly contributes to the tissue tropism and pandemic potential of influenza viruses in humans. Avian influenza viruses, such as HPAI H5N1 viruses, typically exhibit preferential binding to sialic acids linked to galactose by an α2-3 linkage; in humans, these receptors are generally located in the lower respiratory tract. In contrast, human influenza viruses preferentially bind to α2-6 linked sialic acids, which are present in greater abundance in the epithelia of the human upper respiratory tract (Matrosovich et al. 2000). Thus, the receptor binding site of the HA governs the ability of influenza viruses to bind sialic acids in an α2-3 linked conformation, α2-6 linked conformation, or both. It is generally believed that a switch in receptor-binding specificity from an avian-like to a human-like preference is necessary for a virus to achieve sustained human-to-human transmission and cause a pandemic, though an increase in α2-6 binding preference alone is not sufficient for efficient airborne mammalian transmission of avian influenza viruses (Maines et al. 2011). Analysis of contemporary Eurasian and North American H7 influenza viruses revealed that despite maintaining a strong preference for α2-3 linked sialic acids, many possess

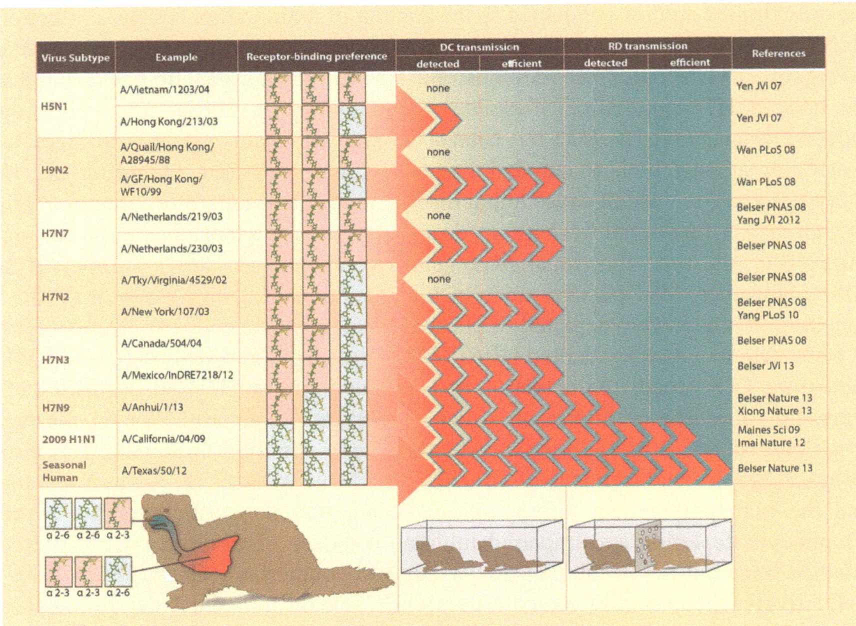

Fig. 2 Contribution of receptor binding preference on virus transmissibility in the ferret model. Virus receptor binding is depicted as maintaining a strong avian binding preference (three α2-3 icons), maintaining an avian binding preference with detectable binding to human receptors (two α2-3 icons, one α2-6 icon), enhanced binding to human receptors while maintaining binding to avian receptors (two α2-6 icons, one α2-3 icon) or strong human binding preference (three α2-6 icons) and is not meant to be quantitative. "DC transmission" indicates one inoculated ferret and one naïve ferret co-housed in the same cage, sharing food, water, and bedding; "RD transmission" indicates one inoculated ferret and one naïve ferret housed in adjacent cages separated by a perforated side-wall (3–5 mm in diameter) allowing air exchange only in the absence of direct or indirect contact

weak but detectable binding to α2-6 linked sialic acids (Gambaryan et al. 2012). The ferret model is most commonly used to model the transmissibility of human and avian influenza viruses in a laboratory setting due to similar lung physiology, receptor distribution, and virus attachment patterns between ferrets and humans (Fig. 2) (Belser et al. 2009b).

With the notable exception of LPAI H7N9 viruses, Eurasian lineage H7 viruses associated with human infection have generally maintained an avian-like α2-3 sialic acid receptor binding preference, similar to the binding preference observed with HPAI H5N1 viruses isolated from humans (Belser et al. 2008; Gambaryan et al. 2008, 2012; Yang et al. 2012). As discussed in previous sections, these Eurasian lineage H7 viruses have generally not transmitted well between ferrets in direct contact or respiratory droplet models. However, divergent patterns of transmissibility in the ferret model and attachment patterns in respiratory tract mammalian cells between viruses isolated from the H7N7 2003 Netherlands outbreak (Fig. 2) point to subtle differences in receptor binding not identified in glycan arrays (Belser et al. 2008; de Wit et al. 2010), highlighting the complexity between closely related strains and warranting further study of additional determinants of influenza virus receptor binding specificity.

Numerous contemporary North American H7 subtype viruses possess increased binding to α2-6 linked sialosides while maintaining strong binding to α2-3 linked sialosides, with select strains further showing reduced binding to α2-3 linked sialosides similar to human influenza viruses (Belser et al. 2008; Gambaryan et al. 2008; Yang et al. 2010). Viruses which possess increased binding to α2-6 linked sialic acids have transmitted with moderate to high efficiency in a ferret direct contact model, though strain-specific differences remain (Belser et al. 2008). Furthermore, limited transmission in a direct contact model has been reported among H7 viruses which maintain an avian α2-3 sialic acid binding preference (Song et al. 2009), indicating that additional factors contribute to the transmissibility of H7 viruses.

HPAI H7N9 viruses isolated from China possess the highest affinity for α2-6 linked sialic acids among all avian viruses associated with human infection, although unlike human influenza viruses, maintain strong binding to α2-3 linked sialic acids (Tharakaraman et al. 2013; Xiong et al. 2013). As mentioned earlier, this enhanced binding to human receptors is likely aided by the presence of a glutamine to leucine switch at position 226 of the HA among H7N9 viruses, a substitution previously associated with human adaptation (Gao et al. 2013b). However, strain-specific differences in both receptor-binding preference and transmission efficiency have been reported among H7N9 human isolates (Belser et al. 2013b; Watanabe et al. 2013; Zhang et al. 2013), further underscoring the polygenic nature of both of these properties. Decreased NA activity of the H7N9 virus compared to a human pandemic virus, and reduced virus replication efficiency at 33 °C compared to pandemic viruses, were also suggested to contribute to the reduced airborne transmissibility of this virus (Belser et al. 2013b; Xiong et al. 2013).

5 Use of Mammalian Models to Develop Vaccines and Antivirals Against H7 Viruses

As discussed throughout this review, H7 influenza viruses associated with human infection exhibit a wide range of pathogenic and transmissible phenotypes, posing unique challenges in designing appropriate strategies to both reduce disease severity and prevent infection. Mammalian models have proven invaluable in assessing the immunogenicity and efficacy of novel candidate vaccines against this virus subtype. While pre-existing influenza virus-specific $CD8^+$ T-cell memory may provide some degree of protection against virulent viruses such as H7 (Christensen et al. 2000; Zhong et al. 2010), there is a clear need to develop effective vaccines and antivirals to protect a serologically naïve human population in the event of an H7 outbreak or pandemic. The poor immunogenicity observed with vaccine candidates against avian influenza viruses in general, and H7 viruses in particular, further highlights the importance of generation and testing of vaccine approaches to this virus subtype as well as improved means to identify appropriate correlates of protection to measure H7 vaccine effectiveness in mammalian species (Krammer and Cox 2013).

Numerous inactivated vaccine candidates have been generated against both Eurasian and North American H7 strains in pandemic preparedness efforts, generally showing promise in mammalian models despite variable immunogenicity. A β-propiolactone-inactivated whole virus vaccine against a Eurasian H7N1 virus was found to elicit humoral antibody responses in mice following vaccination (Hovden et al. 2009). Formalin-inactivated whole virus vaccines against both North American and Eurasian H7 viruses were found to be effective in mice following heterologous virus challenge, restricting pulmonary virus replication and conferring protection from lethal challenge, even in the absence of an adjuvant (Jadhao et al. 2008; Pappas et al. 2007). However, the addition of adjuvant was found to greatly enhance the protection of an H7N7 classical subunit preparation in mice (de Wit et al. 2005), an important feature given the generally poor induction of neutralizing antibodies by this subtype in mammals and potential obstacles involved in eliciting broadly cross-reactive antibodies that can neutralize viruses from both Eurasian and North American lineages (Joseph et al. 2007).

Several live attenuated influenza vaccine (LAIV) candidates against H7 viruses have also been developed and evaluated in mammalian models for safety, immunogenicity, and efficacy. Both North American H7N3 and Eurasian H7N7 LAIV candidates were found to protect mice and ferrets from homologous and heterologous H7 viruses (Joseph et al. 2008; Min et al. 2010). Furthermore, sera from H7N3 and H7N7 LAIV immunized ferrets was found to cross-react against Anhui/1 virus, suggesting that these vaccine candidates may offer some degree of protection against H7N9 strains (Talaat et al. 2009; Xu et al. 2013c); based on results obtained from these studies in mammalian models, both of these candidate vaccines have been used in clinical trials in humans (Xu et al. 2013c).

While the inactivated and LAIV candidates discussed above have relied on production techniques previously established for the generation of seasonal influenza virus vaccines, alternate approaches are also under investigation. In recent years, there has been renewed interest in producing egg-free inactivated influenza vaccines, especially for viruses with pandemic potential. H7N1 vaccines have been generated in cell culture (Cox et al. 2009), and those that have undergone further evaluation have been found to elicit protective antibody responses against homologous and heterologous viruses in both mice and ferrets (Legastelois et al. 2007; Whiteley et al. 2007). Non-infectious recombinant virus-like particles (VLPs) have been generated with both Eurasian and North American lineage H7 HAs (either alone or co-expressed with other subtypes) and have been evaluated for their ability to induce serum antibody responses and confer protection against viral challenge in both mice and ferrets (Smith et al. 2013; Szecsi et al. 2006; Tretyakova et al. 2013). The use of well-characterized mammalian models for H7 virus infection has allowed for the evaluation of vaccine approaches against conserved HA epitopes, as mice vaccinated with chimeric HA constructs expressing an H3 stalk vaccine developed cross-reactive antibodies which protected against murine heterosubtypic challenge with H7 viruses (Krammer et al. 2013; Margine et al. 2013). Further demonstrating the ability of antibodies raised against H7 viruses to exhibit cross-protection, mice immunized with a recombinant Newcastle disease virus expressing a North American H7 HA were protected against a lethal heterologous H7 challenge. This was also true of mice immunized with a recombinant vaccine expressing a Eurasian H7 HA on the surface of baculovirus subsequently challenged with a murine-passaged homologous virus exhibiting enhanced virulence (Goff et al. 2013; Rajesh Kumar et al. 2013). Future study will allow for a better evaluation of these novel approaches to effectively elicit protective antibody and cell-mediated immune responses toward H7 influenza viruses in humans.

In the absence of a well-matched vaccine, such as the early stages of an outbreak or pandemic, antivirals represent the first line of defense against influenza viruses. Unfortunately, the sensitivity of H7 viruses to currently available antiviral drugs has not been extensively examined, with a paucity of studies performed in in vivo models. Similar to many avian and human influenza viruses, the HPAI H7N7 NL/219 virus was found to be resistant to the M2 inhibitor amantadine in a mouse model (Ilyushina et al. 2007). The neuraminidase inhibitor oseltamivir was found to be effective against H7N9 virus in mice in a dose-dependent manner, with mice protected from death and the emergence of drug-resistant variants (Baranovich et al. 2013). However, the presence of a lysine at position 292 of the NA (detected among select H7N9 clinical isolates) resulted in oseltamivir resistance without a loss of virulence or transmissibility in vivo (Hai et al. 2013). Oseltamivir has been prescribed during H7 influenza outbreaks resulting in conjunctivitis (Koopmans et al. 2004; Tweed et al. 2004); in support of this, mice inoculated by the ocular route with HPAI H7N7 Eurasian lineage or H7N3 North American lineage viruses receiving prophylactic oseltamivir treatment exhibited reductions in viral titer in both respiratory and ocular tissues (Belser et al. 2012b). Interference with host cell signaling pathways may also represent a possible avenue to attenuate virulence in vivo, as mice treated with morpholino oligomers or inhibitors of the JNK or NF-κB pathways exhibited reduced viral loads in the lungs following H7N7 virus infection (Gabriel et al. 2008; Haasbach et al. 2013; Nacken et al. 2012). Serine protease inhibitors have furthermore shown antiviral activity in mice following H7N7 challenge (Bahgat et al. 2011). The successful use of convalescent plasma for the treatment of human H5N1 infection has indicated that passive immunotherapy warrants further investigation as a therapeutic agent (Zhou et al. 2007); monoclonal antibodies with neutralizing activity against H7 viruses have increased survival and limited virus replication in mice following lethal H7 virus challenge, indicating that antibody-based therapy, administered either prophylactically or as a treatment, could provide an additional approach to contain the spread of a pandemic H7 virus in the absence of an effective vaccine (Ekiert et al. 2011; He et al. 2013).

6 Conclusions

Influenza viruses within the H7 subtype have caused a broad spectrum of human infection, from conjunctivitis and mild influenza-like illness to severe disease, acute respiratory distress syndrome, and death. The viruses responsible for these human infections have similarly demonstrated a wide range of virulence and transmissible phenotypes in mammalian models (Table 1). This diversity highlights that there is no one "representative" H7 strain and that laboratory study of this subtype necessitates the inclusion of multiple isolates to identify and control for strain-specific differences. Just as mammalian models for the study of H7 virus pathogenesis and transmission have become more refined to better address specific questions pertinent to this virus subtype, such as investigating the apparent ocular tropism frequently associated with this virus subtype in humans, the development and evaluation of both existing and nascent vaccines and antiviral treatments must be attuned to the different manifestations of disease possible following human infection with this subtype. There is a need for continued analysis of the clinical relevance of mutations present among H7 viruses which confer a resistant phenotype in vitro, especially as they pertain to viral fitness and transmissibility (Sleeman et al. 2013). The frequent interspecies transmission of Eurasian and North American H7 viruses from aquatic birds to poultry, and detection of a poultry-like receptor binding site among contemporary viruses within this subtype, indicates that this subtype readily infects gallinaceous poultry and underscores both the need for continued surveillance of this virus subtype and preparation for future cases of human infection (Gambaryan et al. 2012). Collectively, these studies show the continued requirement for the study of H7 viruses and the concurrent reminder to use personal protective equipment that includes both respiratory and eye protection in all instances of potential laboratory or occupational exposure to LPAI and HPAI H7 viruses (Morgan et al. 2009).

Acknowledgments The authors thank Alissa Eckert for graphical assistance. The findings and conclusions in this report are those of the authors and do not necessarily reflect the views of the funding agency.

References

(2007) Avian influenza A/(H7N2) outbreak in the United Kingdom. Euro Surveill 12(5): E070532.1

Aamir UB, Naeem K, Ahmed Z, Obert CA, Franks J, Krauss S, Seiler P, Webster RG (2009) Zoonotic potential of highly pathogenic avian H7N3 influenza viruses from Pakistan. Virology 390:212–220

Abbas MA, Spackman E, Swayne DE, Ahmed Z, Sarmento L, Siddique N, Naeem K, Hameed A, Rehmani S (2010) Sequence and phylogenetic analysis of H7N3 avian influenza viruses isolated from poultry in Pakistan 1995-2004. Virol J 7:137

Abd El-Rahim IH, Hussein M (2004) An epizootic of equine influenza in Upper Egypt in 2000. Rev Sci Tech 23:921–930

Abdel-Ghafar AN, Chotpitayasunondh T, Gao Z, Hayden FG, Nguyen DH, de Jong MD, Naghdaliyev A, Peiris JS, Shindo N, Soeroso S et al (2008) Update on avian influenza A (H5N1) virus infection in humans. N Engl J Med 358:261–273

Alexander DJ (2007) Summary of avian influenza activity in Europe, Asia, Africa, and Australasia, 2002–2006. Avian Dis 51:161–166

Arzey GG, Kirkland PD, Arzey KE, Frost M, Maywood P, Conaty S, Hurt AC, Deng YM, Iannello P, Barr I et al (2012) Influenza virus A (H10N7) in chickens and poultry abattoir workers, Australia. Emerg Infect Dis 18:814–816

Bahgat MM, Blazejewska P, Schughart K (2011) Inhibition of lung serine proteases in mice: a potentially new approach to control influenza infection. Virol J 8:27

Banks J, Plowright L (2003) Additional glycosylation at the receptor binding site of the hemagglutinin (HA) for H5 and H7 viruses may be an adaptation to poultry hosts, but does it influence pathogenicity? Avian Dis 47:942–950

Banks J, Speidel E, Alexander DJ (1998) Characterisation of an avian influenza A virus isolated from a human—is an intermediate host necessary for the emergence of pandemic influenza viruses? Arch Virol 143:781–787

Baranovich T, Burnham AJ, Marathe BM, Armstrong J, Guan Y, Shu Y, Peiris JM, Webby RJ, Webster RG, Govorkova EA (2013) The neuraminidase inhibitor oseltamivir is effective against A/Anhui/1/2013 (H7N9) influenza virus in a mouse model of acute respiratory distress syndrome. J Infect Dis

Belser JA, Blixt O, Chen LM, Pappas C, Maines TR, Van Hoeven N, Donis R, Busch J, McBride R, Paulson JC et al (2008) Contemporary North American influenza H7 viruses possess human receptor specificity: Implications for virus transmissibility. Proc Natl Acad Sci USA 105:7558–7563

Belser JA, Bridges CB, Katz JM, Tumpey TM (2009a) Past, present, and possible future human infection with influenza virus A subtype H7. Emerg Infect Dis 15:859–865

Belser JA, Davis CT, Balish A, Edwards LE, Zeng H, Maines TR, Gustin KM, Martinez IL, Fasce R, Cox NJ et al (2013a) Pathogenesis, transmissibility, and ocular tropism of a highly pathogenic avian influenza A (H7N3) virus associated with human conjunctivitis. J Virol 87:5746–5754

Belser JA, Gustin KM, Maines TR, Pantin-Jackwood MJ, Katz JM, Tumpey TM (2012a) Influenza virus respiratory infection and transmission following ocular inoculation in ferrets. PLoS Pathog 8:e1002569

Belser JA, Gustin KM, Pearce MB, Maines TR, Zeng H, Pappas C, Sun X, Carney PJ, Villanueva JM, Stevens J et al (2013b) Pathogenesis and transmission of avian influenza A (H7N9) virus in ferrets and mice. Nature 501:556–559

Belser JA, Lu X, Maines TR, Smith C, Li Y, Donis RO, Katz JM, Tumpey TM (2007) Pathogenesis of avian influenza (H7) virus infection in mice and ferrets: enhanced virulence of Eurasian H7N7 viruses isolated from humans. J Virol 81:11139–11147

Belser JA, Maines TR, Gustin KM, Katz JM, Tumpey TM (2013c) Kinetics of viral replication and induction of host responses in ferrets differs between ocular and intranasal routes of inoculation. Virology 438:56–60

Belser JA, Sleeman K, Pearce MB, Katz JM, Gubareva LV, Tumpey TM (2012b) Oseltamivir inhibits H7 influenza virus replication in mice inoculated by the ocular route. Antimicrob Agents Chemother 56:1616–1618

Belser JA, Szretter KJ, Katz JM, Tumpey TM (2009b) Use of animal models to understand the pandemic potential of highly pathogenic avian influenza viruses. Adv Virus Res 73:55–97

Belser JA, Tumpey TM (2013) H5N1 pathogenesis studies in mammalian models. Virus Res

Belser JA, Wadford DA, Xu J, Katz JM, Tumpey TM (2009c) Ocular infection of mice with influenza A (H7) viruses: a site of primary replication and spread to the respiratory tract. J Virol 83:7075–7084

Berhane Y, Hisanaga T, Kehler H, Neufeld J, Manning L, Argue C, Handel K, Hooper-McGrevy K, Jonas M, Robinson J et al (2009) Highly pathogenic avian influenza virus A (H7N3) in domestic poultry, Saskatchewan, Canada, 2007. Emerg Infect Dis 15:1492–1495

Brown IH (2010) Summary of avian influenza activity in Europe, Asia, and Africa, 2006–2009. Avian Dis 54:187–193

Bulach D, Halpin R, Spiro D, Pomeroy L, Janies D, Boyle DB (2010) Molecular analysis of H7 avian influenza viruses from Australia and New Zealand: genetic diversity and relationships from 1976 to 2007. J Virol 84:9957–9966

CDC (2004) Update: influenza activity—United States, 2003–2004 season. MMWR Morb Mortal Wkly Rep 53:284–287

Christensen JP, Doherty PC, Branum KC, Riberdy JM (2000) Profound protection against respiratory challenge with a lethal H7N7 influenza A virus by increasing the magnitude of CD8(+) T-cell memory. J Virol 74:11690–11696

Cowling BJ, Jin L, Lau EH, Liao Q, Wu P, Jiang H, Tsang TK, Zheng J, Fang VJ, Chang Z et al (2013) Comparative epidemiology of human infections with avian influenza A H7N9 and H5N1 viruses in China: a population-based study of laboratory-confirmed cases. Lancet 382:129–137

Cox RJ, Major D, Hauge S, Madhun AS, Brokstad KA, Kuhne M, Smith J, Vogel FR, Zambon M, Haaheim LR et al (2009) A cell-based H7N1 split influenza virion vaccine confers protection in mouse and ferret challenge models. Influenza Other Respi Viruses 3:107–117

de Jong RM, Stockhofe-Zurwieden N, Verheij ES, de Boer-Luijtze EA, Ruiter SJ, de Leeuw OS, Cornelissen LA (2013) Rapid emergence of a virulent PB2 E627K variant during adaptation of highly pathogenic avian influenza H7N7 virus to mice. Virol J 10:276

de Wit E, Munster VJ, Spronken MI, Bestebroer TM, Baas C, Beyer WE, Rimmelzwaan GF, Osterhaus AD, Fouchier RA (2005) Protection of mice against lethal infection with highly pathogenic H7N7 influenza A virus by using a recombinant low-pathogenicity vaccine strain. J Virol 79:12401–12407

de Wit E, Munster VJ, van Riel D, Beyer WE, Rimmelzwaan GF, Kuiken T, Osterhaus AD, Fouchier RA (2010) Molecular determinants of adaptation of highly pathogenic avian influenza H7N7 viruses to efficient replication in the human host. J Virol 84:1597–1606

Di Trani L, Porru S, Bonfanti L, Cordioli P, Cesana BM, Boni A, Di Carlo AS, Arici C, Donatelli I, Tomao P et al (2012) Serosurvey against H5 and H7 avian influenza viruses in Italian poultry workers. Avian Dis 56:1068–1071

Dudley JP (2008) Public health and epidemiological considerations for avian influenza risk mapping and risk assessment. Ecol Soc 13:21

Eames KT, Webb C, Thomas K, Smith J, Salmon R, Temple JM (2010) Assessing the role of contact tracing in a suspected H7N2 influenza a outbreak in humans in Wales. BMC Infect Dis 10:141

Ekiert DC, Friesen RH, Bhabha G, Kwaks T, Jongeneelen M, Yu W, Ophorst C, Cox F, Korse HJ, Brandenburg B et al (2011) A highly conserved neutralizing epitope on group 2 influenza A viruses. Science 333:843–850

Fouchier RA, Schneeberger PM, Rozendaal FW, Broekman JM, Kemink SA, Munster V, Kuiken T, Rimmelzwaan GF, Schutten M, Van Doornum GJ et al (2004) Avian influenza A virus (H7N7) associated with human conjunctivitis and a fatal case of acute respiratory distress syndrome. Proc Natl Acad Sci USA 101:1356–1361

Gabbard JD, Dlugolenski D, Van Riel D, Marshall N, Galloway SE, Howerth EW, Campbell PJ, Jones C, Johnson S, Byrd-Leotis L et al (2013) Novel H7N9 influenza virus shows low infectious dose, high growth and efficient contact transmission in the guinea pig model. J Virol

Gabriel G, Abram M, Keiner B, Wagner R, Klenk HD, Stech J (2007) Differential polymerase activity in avian and mammalian cells determines host range of influenza virus. J Virol 81:9601–9604

Gabriel G, Dauber B, Wolff T, Planz O, Klenk HD, Stech J (2005) The viral polymerase mediates adaptation of an avian influenza virus to a mammalian host. Proc Natl Acad Sci USA 102:18590–18595

Gabriel G, Klingel K, Planz O, Bier K, Herwig A, Sauter M, Klenk HD (2009) Spread of infection and lymphocyte depletion in mice depends on polymerase of influenza virus. Am J Pathol 175:1178–1186

Gabriel G, Nordmann A, Stein DA, Iversen PL, Klenk HD (2008) Morpholino oligomers targeting the PB1 and NP genes enhance the survival of mice infected with highly pathogenic influenza A H7N7 virus. J Gen Virol 89:939–948

Gambaryan AS, Matrosovich TY, Philipp J, Munster VJ, Fouchier RA, Cattoli G, Capua I, Krauss SL, Webster RG, Banks J et al (2012) Receptor-binding profiles of H7 subtype influenza viruses in different host species. J Virol 86:4370–4379

Gambaryan AS, Tuzikov AB, Pazynina GV, Desheva JA, Bovin NV, Matrosovich MN, Klimov AI (2008) 6-sulfo sialyl Lewis X is the common receptor determinant recognized by H5, H6, H7 and H9 influenza viruses of terrestrial poultry. Virol J 5:85

Gao HN, Lu HZ, Cao B, Du B, Shang H, Gan JH, Lu SH, Yang YD, Fang Q, Shen YZ et al (2013a) Clinical findings in 111 cases of influenza A (H7N9) virus infection. N Engl J Med 368:2277–2285

Gao R, Cao B, Hu Y, Feng Z, Wang D, Hu W, Chen J, Jie Z, Qiu H, Xu K et al (2013b) Human infection with a novel avian-origin influenza A (H7N9) virus. N Engl J Med 368:1888–1897

Garcia-Sastre A (2001) Inhibition of interferon-mediated antiviral responses by influenza A viruses and other negative-strand RNA viruses. Virology 279:375–384

Gerloff NA, Jones J, Simpson N, Balish A, Elbadry MA, Baghat V, Rusev I, de Mattos CC, de Mattos CA, Zonkle LE et al (2013) A high diversity of eurasian lineage low pathogenicity avian influenza A viruses circulate among wild birds sampled in Egypt. PLoS One 8:e68522

Goff PH, Krammer F, Hai R, Seibert CW, Margine I, Garcia-Sastre A, Palese P (2013) Induction of cross-reactive antibodies to novel H7N9 influenza virus by recombinant Newcastle disease virus expressing a North American lineage H7 subtype hemagglutinin. J Virol 87:8235–8240

Gonzalez-Reiche AS, Morales-Betoulle ME, Alvarez D, Betoulle JL, Muller ML, Sosa SM, Perez DR (2012) Influenza a viruses from wild birds in Guatemala belong to the North American lineage. PLoS One 7:e32873

Haasbach E, Reiling SJ, Ehrhardt C, Droebner K, Ruckle A, Hrincius ER, Leban J, Strobl S, Vitt D, Ludwig S et al (2013) The NF-kappaB inhibitor SC75741 protects mice against highly pathogenic avian influenza A virus. Antiviral Res 99:336–344

Hai R, Schmolke M, Leyva-Grado VH, Thangavel RR, Margine I, Jaffe EL, Krammer F, Solorzano A, Garcia-Sastre A, Palese P et al (2013) Influenza A(H7N9) virus gains neuraminidase inhibitor resistance without loss of in vivo virulence or transmissibility. Nat Commun 4:2854

Hatta M, Gao P, Halfmann P, Kawaoka Y (2001) Molecular basis for high virulence of Hong Kong H5N1 influenza A viruses. Science 293:1840–1842

He F, Kumar SR, Syed Khader SM, Tan Y, Prabakaran M, Kwang J (2013) Effective intranasal therapeutics and prophylactics with monoclonal antibody against lethal infection of H7N7 influenza virus. Antiviral Res 100:207–214

Hirst M, Astell CR, Griffith M, Coughlin SM, Moksa M, Zeng T, Smailus DE, Holt RA, Jones S, Marra MA et al (2004) Novel avian influenza H7N3 strain outbreak, British Columbia. Emerg Infect Dis 10:2192–2195

Horimoto T, Kawaoka Y (1994) Reverse genetics provides direct evidence for a correlation of hemagglutinin cleavability and virulence of an avian influenza A virus. J Virol 68:3120–3128

Horimoto T, Kawaoka Y (2001) Pandemic threat posed by avian influenza A viruses. Clin Microbiol Rev 14:129–149

Hovden AO, Brokstad KA, Major D, Wood J, Haaheim LR, Cox RJ (2009) A pilot study of the immune response to whole inactivated avian influenza H7N1 virus vaccine in mice. Influenza Other Respi Viruses 3:21–28

Ilyushina NA, Govorkova EA, Russell CJ, Hoffmann E, Webster RG (2007) Contribution of H7 haemagglutinin to amantadine resistance and infectivity of influenza virus. J Gen Virol 88:1266–1274

Jadhao SJ, Achenbach J, Swayne DE, Donis R, Cox N, Matsuoka Y (2008) Development of Eurasian H7N7/PR8 high growth reassortant virus for clinical evaluation as an inactivated pandemic influenza vaccine. Vaccine 26:1742–1750

Jones YL, Swayne DE (2004) Comparative pathobiology of low and high pathogenicity H7N3 Chilean avian influenza viruses in chickens. Avian Dis 48:119–128

Jonges M, Bataille A, Enserink R, Meijer A, Fouchier RA, Stegeman A, Koch G, Koopmans M (2011) Comparative analysis of avian influenza virus diversity in poultry and humans during a highly pathogenic avian influenza A (H7N7) virus outbreak. J Virol 85:10598–10604

Joseph T, McAuliffe J, Lu B, Jin H, Kemble G, Subbarao K (2007) Evaluation of replication and pathogenicity of avian influenza a H7 subtype viruses in a mouse model. J Virol 81:10558–10566

Joseph T, McAuliffe J, Lu B, Vogel L, Swayne D, Jin H, Kemble G, Subbarao K (2008) A live attenuated cold-adapted influenza A H7N3 virus vaccine provides protection against homologous and heterologous H7 viruses in mice and ferrets. Virology 378:123–132

Kaminski MM, Ohnemus A, Cornitescu M, Staeheli P (2012) Plasmacytoid dendritic cells and Toll-like receptor 7-dependent signalling promote efficient protection of mice against highly virulent influenza A virus. J Gen Virol 93:555–559

Kapczynski DR, Pantin-Jackwood M, Guzman SG, Ricardez Y, Spackman E, Bertran K, Suarez DL, Swayne DE (2013) Characterization of the 2012 highly pathogenic avian influenza H7N3 virus isolated from poultry in an outbreak in Mexico: pathobiology and vaccine protection. J Virol 87:9086–9096

Kawaoka Y (1991) Equine H7N7 influenza A viruses are highly pathogenic in mice without adaptation: potential use as an animal model. J Virol 65:3891–3894

Kochs G, Koerner I, Thiel L, Kothlow S, Kaspers B, Ruggli N, Summerfield A, Pavlovic J, Stech J, Staeheli P (2007) Properties of H7N7 influenza A virus strain SC35M lacking interferon antagonist NS1 in mice and chickens. J Gen Virol 88:1403–1409

Koerner I, Kochs G, Kalinke U, Weiss S, Staeheli P (2007) Protective role of beta interferon in host defense against influenza A virus. J Virol 81:2025–2030

Koopmans M, Wilbrink B, Conyn M, Natrop G, van der Nat H, Vennema H, Meijer A, van Steenbergen J, Fouchier R, Osterhaus A et al (2004) Transmission of H7N7 avian influenza A virus to human beings during a large outbreak in commercial poultry farms in the Netherlands. Lancet 363:587–593

Krammer F, Cox RJ (2013) The emergence of H7N9 viruses: a chance to redefine correlates of protection for influenza virus vaccines. Expert Rev Vaccines 12:1369–1372

Krammer F, Margine I, Hai R, Flood A, Hirsh A, Tsvetnitsky V, Chen D, Palese P (2013) H3 stalk-based chimeric hemagglutinin influenza virus constructs protect mice from H7N9 challenge. J Virol

Kreijtz JH, Kroeze EJ, Stittelaar KJ, de Waal L, van Amerongen G, van Trierum S, van Run P, Bestebroer T, Kuiken T, Fouchier RA, et al (2013) Low pathogenic avian influenza A(H7N9) virus causes high mortality in ferrets upon intratracheal challenge: A model to study intervention strategies. Vaccine

Kurtz J, Manvell RJ, Banks J (1996) Avian influenza virus isolated from a woman with conjunctivitis. Lancet 348:901–902

Lam WY, Yeung AC, Chan PK (2011) Apoptosis, cytokine and chemokine induction by non-structural 1 (NS1) proteins encoded by different influenza subtypes. Virol J 8:554

Lang G, Gagnon A, Geraci JR (1981) Isolation of an influenza A virus from seals. Arch Virol 68:189–195

Lee CW, Lee YJ, Senne DA, Suarez DL (2006) Pathogenic potential of North American H7N2 avian influenza virus: a mutagenesis study using reverse genetics. Virology 353:388–395

Legastelois I, Garcia-Sastre A, Palese P, Tumpey TM, Maines TR, Katz JM, Vogel FR, Moste C (2007) Preparation of genetically engineered A/H5N1 and A/H7N1 pandemic vaccine viruses by reverse genetics in a mixture of Vero and chicken embryo cells. Influenza Other Respi Viruses 1:95–104

Li Q, Zhou L, Zhou M, Chen Z, Li F, Wu H, Xiang N, Chen E, Tang F, Wang D, et al (2013) Preliminary report: epidemiology of the avian influenza a (H7N9) outbreak in China. N Engl J Med

Li Y, Li C, Liu L, Wang H, Wang C, Tian G, Webster RG, Yu K, Chen H (2006) Characterization of an avian influenza virus of subtype H7N2 isolated from chickens in northern China. Virus Genes 33:117–122

Lopez-Martinez I, Balish A, Barrera-Badillo G, Jones J, Nunez-Garcia TE, Jang Y, Aparicio-Antonio R, Azziz-Baumgartner E, Belser JA, Ramirez-Gonzalez JE, et al (2013) Highly pathogenic avian influenza A(H7N3) virus in poultry workers, Mexico, 2012. Emerg Infect Dis 19

Lv H, Han J, Zhang P, Lu Y, Wen D, Cai J, Liu S, Sun J, Yu Z, Zhang H, et al (2013) Mild illness in avian influenza A(H7N9) virus-infected poultry worker, Huzhou, China, April 2013. Emerg Infect Dis 19

Maines TR, Chen LM, Matsuoka Y, Chen H, Rowe T, Ortin J, Falcon A, Nguyen TH, le Mai Q, Sedyaningsih ER et al (2006) Lack of transmission of H5N1 avian-human reassortant influenza viruses in a ferret model. Proc Natl Acad Sci U S A 103:12121–12126

Maines TR, Chen LM, Van Hoeven N, Tumpey TM, Blixt O, Belser JA, Gustin KM, Pearce MB, Pappas C, Stevens J et al (2011) Effect of receptor binding domain mutations on receptor binding and transmissibility of avian influenza H5N1 viruses. Virology 413:139–147

Maines TR, Jayaraman A, Belser JA, Wadford DA, Pappas C, Zeng H, Gustin KM, Pearce MB, Viswanathan K, Shriver ZH et al (2009) Transmission and pathogenesis of swine-origin 2009 A(H1N1) influenza viruses in ferrets and mice. Science 325:484–487

Maines TR, Lu XH, Erb SM, Edwards L, Guarner J, Greer PW, Nguyen DC, Szretter KJ, Chen LM, Thawatsupha P et al (2005) Avian influenza (H5N1) viruses isolated from humans in Asia in 2004 exhibit increased virulence in mammals. J Virol 79:11788–11800

Mancini DA, Mendonca RM, Pereira AS, Kawamoto AH, Vannucchi CI, Pinto JR, Mori E, Mancini Filho J (2012) Influenza viruses in adult dogs raised in rural and urban areas in the state of Sao Paulo, Brazil. Rev Inst Med Trop Sao Paulo 54:311–314

Manz B, Gotz V, Wunderlich K, Eisel J, Kirchmair J, Stech J, Stech O, Chase G, Frank R, Schwemmle M (2011) Disruption of the viral polymerase complex assembly as a novel approach to attenuate influenza A virus. J Biol Chem 286:8414–8424

Margine I, Krammer F, Hai R, Heaton NS, Tan GS, Andrews SA, Runstadler JA, Wilson PC, Albrecht RA, Garcia-Sastre A et al (2013) Hemagglutinin Stalk-Based Universal Vaccine Constructs Protect against Group 2 Influenza A Viruses. J Virol 87:10435–10446

Matrosovich M, Tuzikov A, Bovin N, Gambaryan A, Klimov A, Castrucci MR, Donatelli I, Kawaoka Y (2000) Early alterations of the receptor-binding properties of H1, H2, and H3 avian influenza virus hemagglutinins after their introduction into mammals. J Virol 74:8502–8512

Maurer-Stroh S, Lee RT, Gunalan V, Eisenhaber F (2013) The highly pathogenic H7N3 avian influenza strain from July 2012 in Mexico acquired an extended cleavage site through recombination with host 28S rRNA. Virol J 10:139

Max V, Herrera J, Moreira R, Rojas H (2007) Avian influenza in Chile: a successful experience. Avian Dis 51:363–365

Min JY, Vogel L, Matsuoka Y, Lu B, Swayne D, Jin H, Kemble G, Subbarao K (2010) A live attenuated H7N7 candidate vaccine virus induces neutralizing antibody that confers protection from challenge in mice, ferrets, and monkeys. J Virol 84:11950–11960

Mok CK, Lee HH, Chan MC, Sia SF, Lestra M, Nicholls JM, Zhu H, Guan Y, Peiris JM (2013) Pathogenicity of the novel A/H7N9 influenza virus in mice. MBio 4

Mok CK, Lee HH, Lestra M, Nicholls JM, Chan CW, Sia SF, Zhu H, Poon LL, Guan Y, Peiris JM (2014) Amino-acid substitutions in polymerase basic protein 2 gene contributes to the pathogenicity of the novel A/H7N9 influenza virus in mammalian hosts. J Virol
Morgan O, Kuhne M, Nair P, Verlander NQ, Preece R, McDougal M, Zambon M, Reacher M (2009) Personal protective equipment and risk for avian influenza (H7N3). Emerg Infect Dis 15:59–62
Munster VJ, de Wit E, van Riel D, Beyer WE, Rimmelzwaan GF, Osterhaus AD, Kuiken T, Fouchier RA (2007) The molecular basis of the pathogenicity of the Dutch highly pathogenic human influenza A H7N7 viruses. J Infect Dis 196:258–265
Myers KP, Setterquist SF, Capuano AW, Gray GC (2007) Infection due to 3 avian influenza subtypes in United States veterinarians. Clin Infect Dis 45:4–9
Nacken W, Ehrhardt C, Ludwig S (2012) Small molecule inhibitors of the c-Jun N-terminal kinase (JNK) possess antiviral activity against highly pathogenic avian and human pandemic influenza A viruses. Biol Chem 393:525–534
Nguyen-Van-Tam JS, Nair P, Acheson P, Baker A, Barker M, Bracebridge S, Croft J, Ellis J, Gelletlie R, Gent N et al (2006) Outbreak of low pathogenicity H7N3 avian influenza in UK, including associated case of human conjunctivitis. Euro Surveill 11(E060504):2
Ostrowsky B, Huang A, Terry W, Anton D, Brunagel B, Traynor L, Abid S, Johnson G, Kacica M, Katz J et al (2012) Low pathogenic avian influenza A (H7N2) virus infection in immunocompromised adult, New York, USA, 2003. Emerg Infect Dis 18:1128–1131
Pappas C, Matsuoka Y, Swayne DE, Donis RO (2007) Development and evaluation of an Influenza virus subtype H7N2 vaccine candidate for pandemic preparedness. Clin Vaccine Immunol 14:1425–1432
Pasick J, Berhane Y, Hisanaga T, Kehler H, Hooper-McGrevy K, Handel K, Neufeld J, Argue C, Leighton F (2010) Diagnostic test results and pathology associated with the 2007 Canadian H7N3 highly pathogenic avian influenza outbreak. Avian Dis 54:213–219
Pasick J, Handel K, Robinson J, Copps J, Ridd D, Hills K, Kehler H, Cottam-Birt C, Neufeld J, Berhane Y et al (2005) Intersegmental recombination between the haemagglutinin and matrix genes was responsible for the emergence of a highly pathogenic H7N3 avian influenza virus in British Columbia. J Gen Virol 86:727–731
Pasick J, Pedersen J, Hernandez MS (2012) Avian influenza in North America, 2009-2011. Avian Dis 56:845–848
Peiris M, Yuen KY, Leung CW, Chan KH, Ip PL, Lai RW, Orr WK, Shortridge KF (1999) Human infection with influenza H9N2. Lancet 354:916–917
Perez-Ramirez E, Gerrikagoitia X, Barral M, Hofle U (2010) Detection of low pathogenic avian influenza viruses in wild birds in Castilla-La Mancha (south central Spain). Vet Microbiol 146:200–208
ProMED-mail (2013) Avian Influenza (90): ITALY (EMILIA-ROMAGNA) Poultry, HPAI H7N7, Human, Public health. ProMED mail
Puzelli S, Di Trani L, Fabiani C, Campitelli L, De Marco MA, Capua I, Aguilera JF, Zambon M, Donatelli I (2005) Serological analysis of serum samples from humans exposed to avian H7 influenza viruses in Italy between 1999 and 2003. J Infect Dis 192:1318–1322
Qi X, Qian YH, Bao CJ, Guo XL, Cui LB, Tang FY, Ji H, Huang Y, Cai PQ, Lu B et al (2013) Probable person to person transmission of novel avian influenza A (H7N9) virus in Eastern China, 2013: epidemiological investigation. BMJ 347:f4752
Rajesh Kumar S, Syed Khader SM, Kiener TK, Szyporta M, Kwang J (2013) Intranasal immunization of baculovirus displayed hemagglutinin confers complete protection against mouse adapted highly pathogenic H7N7 reassortant influenza virus. PLoS One 8:e63856
Richard M, Schrauwen EJ, de Graaf M, Bestebroer TM, Spronken MI, van Boheemen S, de Meulder D, Lexmond P, Linster M, Herfst S et al (2013) Limited airborne transmission of H7N9 influenza A virus between ferrets. Nature 501:560–563
Rigoni M, Shinya K, Toffan A, Milani A, Bettini F, Kawaoka Y, Cattoli G, Capua I (2007) Pneumo- and neurotropism of avian origin Italian highly pathogenic avian influenza H7N1 isolates in experimentally infected mice. Virology 364:28–35

Rigoni M, Toffan A, Viale E, Mancin M, Cilloni F, Bertoli E, Salomoni A, Marciano S, Milani A, Zecchin B et al (2010) The mouse model is suitable for the study of viral factors governing transmission and pathogenesis of highly pathogenic avian influenza (HPAI) viruses in mammals. Vet Res 41:66

Salomon R, Franks J, Govorkova EA, Ilyushina NA, Yen HL, Hulse-Post DJ, Humberd J, Trichet M, Rehg JE, Webby RJ et al (2006) The polymerase complex genes contribute to the high virulence of the human H5N1 influenza virus isolate A/Vietnam/1203/04. J Exp Med 203:689–697

Scheiblauer H, Kendal AP, Rott R (1995) Pathogenicity of influenza A/Seal/Mass/1/80 virus mutants for mammalian species. Arch Virol 140:341–348

Senne DA (2007) Avian influenza in North and South America, 2002–2005. Avian Dis 51:167–173

Senne DA (2010) Avian influenza in North and South America, the Caribbean, and Australia, 2006–2008. Avian Dis 54:179–186

Senne DA, Panigrahy B, Kawaoka Y, Pearson JE, Suss J, Lipkind M, Kida H, Webster RG (1996) Survey of the hemagglutinin (HA) cleavage site sequence of H5 and H7 avian influenza viruses: amino acid sequence at the HA cleavage site as a marker of pathogenicity potential. Avian Dis 40:425–437

Senne DA, Suarez DL, Pedersen JC, Panigrahy B (2003) Molecular and biological characteristics of H5 and H7 avian influenza viruses in live-bird markets of the northeastern United States, 1994–2001. Avian Dis 47:898–904

Shinya K, Suto A, Kawakami M, Sakamoto H, Umemura T, Kawaoka Y, Kasai N, Ito T (2005) Neurovirulence of H7N7 influenza A virus: brain stem encephalitis accompanied with aspiration pneumonia in mice. Arch Virol 150:1653–1660

Shinya K, Watanabe S, Ito T, Kasai N, Kawaoka Y (2007) Adaptation of an H7N7 equine influenza A virus in mice. J Gen Virol 88:547–553

Sleeman K, Guo Z, Barnes J, Shaw M, Stevens J, Gubareva LV (2013) R292K substitution and drug susceptibility of influenza A(H7N9) viruses. Emerg Infect Dis 19:1521–1524

Smith GE, Flyer DC, Raghunandan R, Liu Y, Wei Z, Wu Y, Kpamegan E, Courbron D, Fries LF 3rd, Glenn GM (2013) Development of influenza H7N9 virus like particle (VLP) vaccine: Homologous A/Anhui/1/2013 (H7N9) protection and heterologous A/chicken/Jalisco/CPA1/2012 (H7N3) cross-protection in vaccinated mice challenged with H7N9 virus. Vaccine 31:4305–4313

Song H, Wan H, Araya Y, Perez DR (2009) Partial direct contact transmission in ferrets of a mallard H7N3 influenza virus with typical avian-like receptor specificity. Virol J 6:126

Srivastava B, Blazejewska P, Hessmann M, Bruder D, Geffers R, Mauel S, Gruber AD, Schughart K (2009) Host genetic background strongly influences the response to influenza a virus infections. PLoS One 4:e4857

Suarez DL, Senne DA, Banks J, Brown IH, Essen SC, Lee CW, Manvell RJ, Mathieu-Benson C, Moreno V, Pedersen JC et al (2004) Recombination resulting in virulence shift in avian influenza outbreak, Chile. Emerg Infect Dis 10:693–699

Subbarao EK, London W, Murphy BR (1993) A single amino acid in the PB2 gene of influenza A virus is a determinant of host range. J Virol 67:1761–1764

Suguitan AL Jr, Matsuoka Y, Lau YF, Santos CP, Vogel L, Cheng LI, Orandle M, Subbarao K (2012) The multibasic cleavage site of the hemagglutinin of highly pathogenic A/Vietnam/1203/2004 (H5N1) avian influenza virus acts as a virulence factor in a host-specific manner in mammals. J Virol 86:2706–2714

Szecsi J, Boson B, Johnsson P, Dupeyrot-Lacas P, Matrosovich M, Klenk HD, Klatzmann D, Volchkov V, Cosset FL (2006) Induction of neutralising antibodies by virus-like particles harbouring surface proteins from highly pathogenic H5N1 and H7N1 influenza viruses. Virol J 3:70

Talaat KR, Karron RA, Callahan KA, Luke CJ, DiLorenzo SC, Chen GL, Lamirande EW, Jin H, Coelingh KL, Murphy BR et al (2009) A live attenuated H7N3 influenza virus vaccine is well tolerated and immunogenic in a Phase I trial in healthy adults. Vaccine 27:3744–3753

Tharakaraman K, Jayaraman A, Raman R, Viswanathan K, Stebbins NW, Johnson D, Shriver Z, Sasisekharan V, Sasisekharan R (2013) Glycan receptor binding of the influenza A virus H7N9 hemagglutinin. Cell 153:1486–1493
Timoney PJ (1996) Equine influenza. Comp Immunol Microbiol Infect Dis 19:205–211
Tretyakova I, Pearce MB, Florese R, Tumpey TM, Pushko P (2013) Intranasal vaccination with H5, H7 and H9 hemagglutinins co-localized in a virus-like particle protects ferrets from multiple avian influenza viruses. Virology 442:67–73
Tumpey TM, Szretter KJ, Van Hoeven N, Katz JM, Kochs G, Haller O, Garcia-Sastre A, Staeheli P (2007) The Mx1 gene protects mice against the pandemic 1918 and highly lethal human H5N1 influenza viruses. J Virol 81:10818–10821
Tweed SA, Skowronski DM, David ST, Larder A, Petric M, Lees W, Li Y, Katz J, Krajden M, Tellier R et al (2004) Human illness from avian influenza H7N3, British Columbia. Emerg Infect Dis 10:2196–2199
Watanabe T, Kiso M, Fukuyama S, Nakajima N, Imai M, Yamada S, Murakami S, Yamayoshi S, Iwatsuki-Horimoto K, Sakoda Y et al (2013) Characterization of H7N9 influenza A viruses isolated from humans. Nature 501:551–555
Webster RG (1993) Are equine 1 influenza viruses still present in horses? Equine Vet J 25:537–538
Webster RG, Geraci J, Petursson G, Skirnisson K (1981) Conjunctivitis in human beings caused by influenza A virus of seals. N Engl J Med 304:911
Whiteley A, Major D, Legastelois I, Campitelli L, Donatelli I, Thompson CI, Zambon MC, Wood JM, Barclay WS (2007) Generation of candidate human influenza vaccine strains in cell culture—rehearsing the European response to an H7N1 pandemic threat. Influenza Other Respi Viruses 1:157–166
WHO (2013) Avian influenza A (H10N8)
Xiong X, Martin SR, Haire LF, Wharton SA, Daniels RS, Bennett MS, McCauley JW, Collins PJ, Walker PA, Skehel JJ et al (2013) Receptor binding by an H7N9 influenza virus from humans. Nature 499:496–499
Xu L, Bao L, Deng W, Dong L, Zhu H, Chen T, Lv Q, Li F, Yuan J, Xiang Z, et al (2013a) Novel avian-origin human influenza A(H7N9) can be transmitted between ferrets via respiratory droplets. J Infect Dis
Xu L, Bao L, Deng W, Zhu H, Chen T, Lv Q, Li F, Yuan J, Xiang Z, Gao K et al (2013b) The mouse and ferret models for studying the novel avian-origin human influenza A (H7N9) virus. Virol J 10:253
Xu Q, Chen Z, Cheng X, Xu L, Jin H (2013c) Evaluation of Live Attenuated H7N3 and H7N7 Vaccine Viruses for Their Receptor Binding Preferences, Immunogenicity in Ferrets and Cross Reactivity to the Novel H7N9 Virus. PLoS One 8:e76884
Yamayoshi S, Yamada S, Fukuyama S, Murakami S, Zhao D, Uraki R, Watanabe T, Tomita Y, Macken C, Neumann G, et al (2013) Virulence-Affecting Amino Acid Changes in the PA Protein of H7N9 Influenza A Viruses. J Virol
Yang H, Carney PJ, Donis RO, Stevens J (2012) Structure and receptor complexes of the hemagglutinin from a highly pathogenic H7N7 influenza virus. J Virol 86:8645–8652
Yang H, Chen LM, Carney PJ, Donis RO, Stevens J (2010) Structures of receptor complexes of a North American H7N2 influenza hemagglutinin with a loop deletion in the receptor binding site. PLoS Pathog 6:e1001081
Yuan J, Zhang L, Kan X, Jiang L, Yang J, Guo Z, Ren Q (2013) Origin and molecular characteristics of a novel 2013 avian influenza A(H6N1) virus causing human infection in Taiwan. Clin Infect Dis 57:1367–1368
Zhang H, Li X, Guo J, Li L, Chang C, Li Y, Bian C, Xu K, Chen H, Sun B (2014) The PB2 E627K mutation contributes to the high polymerase activity and enhanced replication of H7N9 influenza virus. J Gen Virol
Zhang Q, Shi J, Deng G, Guo J, Zeng X, He X, Kong H, Gu C, Li X, Liu J et al (2013) H7N9 influenza viruses are transmissible in ferrets by respiratory droplet. Science 341:410–414

Zhong W, Liu F, Dong L, Lu X, Hancock K, Reinherz EL, Katz JM, Sambhara S (2010) Significant impact of sequence variations in the nucleoprotein on CD8 T cell-mediated cross-protection against influenza A virus infections. PLoS One 5:e10583

Zhou B, Zhong N, Guan Y (2007) Treatment with convalescent plasma for influenza A (H5N1) infection. N Engl J Med 357:1450–1451

Zhu H, Wang D, Kelvin DJ, Li L, Zheng Z, Yoon SW, Wong SS, Farooqui A, Wang J, Banner D et al (2013) Infectivity, transmission, and pathology of human-isolated H7N9 influenza virus in ferrets and pigs. Science 341:183–186

Pathogenesis and Vaccination of Influenza A Virus in Swine

Daniela S. Rajao, Tavis K. Anderson, Phillip C. Gauger and Amy L. Vincent

Abstract Swine influenza is an acute respiratory disease of pigs caused by influenza A virus (IAV) and characterized by fever followed by lethargy, anorexia, and serous nasal discharge. The disease progresses rapidly and may be complicated when associated with other respiratory pathogens. IAV is one of the most prevalent respiratory pathogens of swine, resulting in substantial economic burden to pork producers. In the past 10–15 years, a dramatic evolution of the IAV in U.S. swine has occurred, resulting in the co-circulation of many antigenically distinct IAV strains, derived from 13 phylogenetically distinct hemagglutinin clusters of H1 and H3 viruses. Vaccination is the most common strategy to prevent influenza in pigs, however, the current diverse IAV epidemiology poses a challenge for the production of efficacious and protective vaccines. A concern regarding the use of traditional inactivated vaccines is the possibility of inducing vaccine-associated enhanced respiratory disease (VAERD) when vaccine virus strains are mismatched with the infecting strain. In this review, we discuss the current epidemiology and pathogenesis of swine influenza in the United States, different vaccines platforms with potential to control influenza in pigs, and the factors associated with vaccine-associated disease enhancement.

D.S. Rajao · A.L. Vincent (✉)
Virus and Prion Diseases of Livestock Research Unit, USDA-ARS,
1920 Dayton Avenue, PO Box 70, Ames, IA 50010, USA
e-mail: amy.vincent@ars.usda.gov

T.K. Anderson
Department of Biology, Georgia Southern University, PO Box 8042
Statesboro, GA 30460, USA

P.C. Gauger
Department of Veterinary Microbiology and Preventive Medicine,
2180 Veterinary Medicine, Iowa State University, Ames, IA 50010, USA

Current Topics in Microbiology and Immunology (2014) 385: 307–326 307
DOI: 10.1007/82_2014_391

Published Online: 18 July 2014

Contents

1 Introduction

Swine influenza is an acute respiratory viral disease caused by influenza A virus (IAV) of the Orthomyxovirus family that decreases health and welfare of pigs and results in a significant economic loss for the swine industry worldwide. Clinical signs of influenza illness in pigs can display a range of severity, but often occurs as mild respiratory disease with high morbidity and rapid recovery, with rare fatal cases. However, the disease has substantial economic burden as a result of weight loss, reduced weight gain and, in some cases, reproductive failure in infected sows due to high fevers. Further, when associated with other respiratory pathogens, as part of the porcine respiratory disease complex, it can lead to complicated pneumonia and severe clinical signs. IAV is one of the most important respiratory pathogens of swine, and its high prevalence in swine herds worldwide directly correlates to the economic impact of the disease.

In addition, swine influenza poses a threat to public health and to the control of influenza in humans. Influenza viruses can transmit between pigs and humans, as observed during the 2009 pandemic (H1N1pdm09), when a virus generated by reassortment between two established lineages of swine viruses became globally widespread and reached a pandemic level in humans. The virus then quickly transmitted from humans to swine. Because pigs are susceptible to both avian and human influenza viruses (CDC 2012; Myers et al. 2007), they have been referred to as the "mixing vessel" of IAV with the potential to generate novel viruses (Scholtissek et al. 1985). This can occur when infection with two or more strains leads to development of swine, avian, and/or human reassortant viruses that can then be transmitted between pigs and to other species (Brown 2000). However, humans and other mammals can also be directly infected with avian and swine viruses, thus can potentially serve as mixing vessel hosts as well (Myers et al. 2007; Smith et al. 2009b). Recent surveillance efforts have increased the amount of available sequence data for IAV in swine, revealing a vast diversity of IAV circulating in pig populations worldwide, underscoring that pigs are an important host in the overall epidemiology of influenza and to the generation of novel reassorted viruses. Thus, the swine reservoir of reassorted viruses, particularly those with gene segments from prior human seasonal IAV combined with novel or substantially drifted HA and NA, poses a risk for future incursions of swine viruses into the human population.

Since 1998, a dramatic evolutionary expansion in IAV diversity in U.S. swine has occurred, with frequent reassortment events between endemic swine-adapted viruses, resulting in the co-circulation of many antigenically and genetically distinct IAV strains. This diversity has complicated the control of swine influenza. During the 1990s, vaccination became a common practice for U.S. swine producers to reduce the disease burden of swine influenza, but vaccine efficacy is compromised by the rapid evolution of influenza viruses, resulting in suboptimal protection against distantly related strains. Thus, improved strain selection or improved vaccine platforms are needed to use in conjunction with management practices to aid in comprehensive intervention strategies to effectively control IAV in swine.

2 Influenza A Virus in Pigs

2.1 Current Epidemiology of Influenza A Virus in Swine

Respiratory disease caused by IAV was first documented in pigs coincident with the 1918 "Spanish flu" pandemic in humans (Koen 1919). The causative agent was subsequently isolated in 1930 and identified as an H1N1 IAV (Shope 1931); and though the origin of this pandemic is unresolved (Smith et al. 2009a), it is clear that the H1N1 lineages evolving from the 1918 Spanish flu and circulating in pigs and humans before 2009 were evolutionarily related (Van Reeth et al. 2012). These H1N1 viruses, classified as classical-swine H1N1 (cH1N1), persisted in the United States swine population for nearly 80 years and remained relatively genetically and antigenically stable (Van Reeth et al. 2012; Vincent et al. 2008b). However, in the late 1990s, two novel reassortant genotypes were detected in swine (Olsen 2002; Zhou et al. 1999). One of the genotypes was a double human-swine reassortant H3N2 virus that was unsuccessful in establishing itself as stable lineage. The second was a triple-reassortant H3N2 virus that swept through the swine population. It contained hemagglutinin (HA), neuraminidase (NA), and polymerase basic 1 (PB1) gene segments derived from seasonal human H3N2, polymerase basic 2 (PB2) and polymerase acidic (PA) gene segments derived from avian IAV, and nucleoprotein (NP), matrix (M), and nonstructural (NS) gene segments from the classical H1N1 swine IAV (Zhou et al. 1999). Subsequently, the triple-reassortant H3N2 viruses reassorted with classical H1N1 and resulted in new genotypes of H1N1 and H1N2 viruses (Karasin et al. 2000, 2002). Of these reassortment events, the majority included the H1 and/or N1 segments while

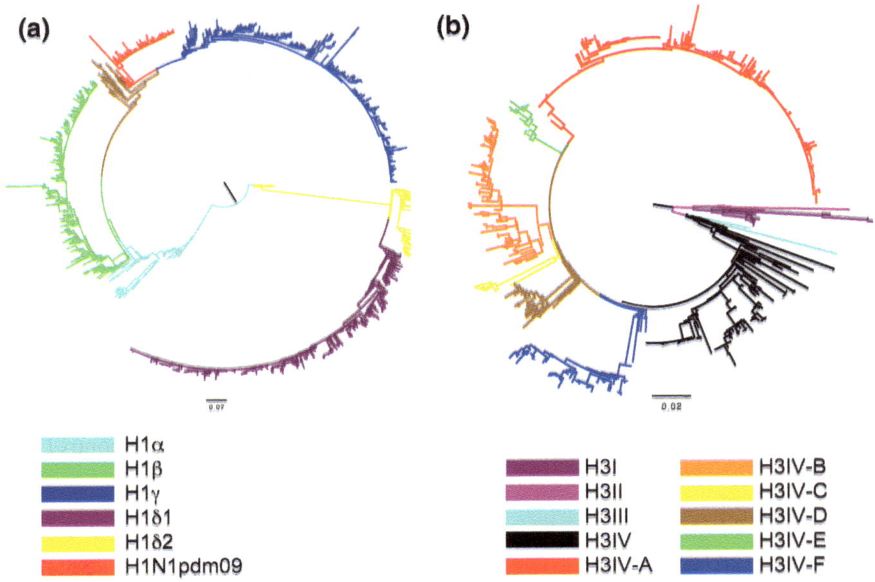

Fig. 1 The major H1 and H3 genetic lineages co-circulating in United States swine populatons. **a** Phylogenetic relationships of 2543 H1 sequences from 1930–2013; and **b** phylogenetic relationships of 607 H3 sequences from 1977–2013 (data available upon request). The phylogenies were constructed using maximum likelihood methods with a GTR + Γ model of molecular evolution implemented in RAxML v.7.7.8. North American swine influenza A virus subtypes are derived from either classical H1 swine-lineage or human seasonal lineage H1 viruses (**a**) or a triple reassortant H3N2 (**b**). This has resulted in six distinct H1 genetic clusters: H1α; H1β; H1γ; H1δ1; H1δ2; and H1N1pdm09. The H3N2 viruses are represented by three lineages, with cluster I, cluster II, cluster III; and a recent and rapidly diversifying cluster IV with emerging sub-clusters A–F

maintaining an intact triple reassortant internal gene (TRIG) constellation containing swine (M, NP, and NS), avian (PB2 and PA), and human (PB1) influenza virus genes (Vincent et al. 2008b). In addition to these dramatic reassortment events, the evolution of swine IAV has been driven by the accumulation of mutations leading to changes in viral surface proteins (antigenic drift) (de Jong et al. 1999) and resulting in diversity of circulating viruses in the United States (Anderson et al. 2013; Kitikoon et al. 2013b) (Fig. 1) and globally (Vincent et al. 2013).

In the early twenty-first century in North America, newly introduced lineages of swine H1N1 and H1N2 viruses were detected. These viruses contained either the HA, NA or both genes derived from human seasonal IAV (Vincent et al. 2009b), and now represent nearly 50 % of the circulating swine IAV (Anderson et al. 2013). The HA of these viruses were genetically and antigenically distinct from those of classical swine-lineage H1 viruses, but their TRIG genes were similar to those found in contemporary triple-reassortant viruses. The growing diversity of

IAV in swine in the U.S. at that time led to phylogenetic "cluster" terminology supported by antigenic data: viruses with HA genes most similar to human seasonal H1 viruses were termed δ-cluster, and those more similar to the classical H1N1 viruses were termed α-, β-, or γ-cluster (Lorusso et al. 2011, 2013; Vincent et al. 2009b). The δ-cluster was further divided into δ-1 and δ-2 subclusters with the recognition that the HA genes that formed this cluster likely emerged from at least two separate introductions of human seasonal-lineage viruses and subsequently diversified into two distinct phylogenetic clades (Lorusso et al. 2011). The δ-1 and δ-2 HA lineages paired with neuraminadase N2 genes derived from the human H3N2 seasonal-lineage (Nelson et al. 2011), N1 genes from the classical swine-lineage, or human H1N1 seasonal-lineage N1 genes (Anderson et al. 2013). In total, there are at least ten phylogenetically distinct HA genetic clusters among the co-circulating IAV in North American swine populations: H1α, H1β, and H1γ; H1δ1 and H1δ2; the H1pdm09; and H3 cluster IV viruses that are now subdivided into emerging clusters A–F (Anderson et al. 2013; Kitikoon et al. 2013b; Lorusso et al. 2013) (Fig. 1).

The observed diversity over the past century of IAV in swine populations around the world can be attributed primarily to three distinct processes: relatively frequent transmission from humans to swine, followed by antigenic shift and drift in pigs, then relatively infrequent transmission from swine to humans (Nelson et al. 2012). This bidirectional transmission dynamic has contributed to the more than 10 genetically and antigenically distinct lineages in North America, and a similarly diverse population of human-, avian-, and classical swine-lineage subtypes circulating in Europe and Asia. A direct consequence of the exchange between human and swine lineages of IAV is the swine-origin pandemic in 2009 (Garten et al. 2009), and viruses of potential zoonotic or pandemic risk exemplified by the detection of variant H3N2 (H3N2v) viruses in humans in the United States in 2011–2013 (Epperson et al. 2013).

2.2 Pathogenesis and Host Response in Swine

Influenza in pigs is typically characterized as an acute, high morbidity-low mortality disease, very similar to what is observed for seasonal influenza in humans, therefore pigs serve as models for the study of human influenza pathogenesis (Khatri et al. 2010) in addition to the study of swine IAV in its natural host. The acute phase of the disease begins in 1–2 days post infection with the onset of fever followed by lethargy, anorexia, and serous nasal discharge. The disease can progress to conjunctivitis, tachypnea and development of cough and dyspnea. In uncomplicated cases, clinical signs may abate in the order of appearance with resolution of disease in 7–10 days. Swine of all ages are susceptible to infection, with clinical signs less pronounced in nursing pigs with passive maternal antibody and sows (Janke 2013). IAV targets epithelial cells lining upper and lower respiratory tracts of pigs, including nasal mucosa, tonsils, trachea, bronchi, bronchioles, and alveoli (Nelli et al. 2010), although swine IAV typically shows a preference for the lungs (Khatri et al. 2010). Virus excretion and transmission occurs exclusively via the respiratory route through nasal or oral secretions (Van Reeth et al. 2012; Vincent et al. 2009a). Virus can be detected in nasal secretions by 1–3 days after infection and virus excretion typically lasts for 5–7 days (Janke 2013; Jo et al. 2007; Vincent et al. 2010b). Traditionally, swine influenza was a seasonal disease with variable kinetics of a herd outbreak following a pattern that could be drawn out for several weeks as it rolled through the herd, depending on the size and immune status of the population at risk. In some large swine production facilities with a continuous flow of pigs of different ages moving in and out of the barn, endemic influenza infection may result in year-round presence of IAV in the herd. Although current disease outbreaks tend to peak in seasonal patterns, IAV has recently been shown to be routinely detected year-round in herds with and without obvious signs of illness (Janke 2013).

Protective immunity against infection with IAV involves both innate and adaptive immune responses (Doherty et al. 2006). After infection, the innate immune response is important for the direct lysis of infected cells via natural killer cells or by the production of pro-inflammatory cytokines (Cox et al. 2004; La Gruta et al. 2007), however, these same mechanisms can also contribute to lung epithelial damage. Production of pro-inflammatory and chemotactic cytokines, such as TNF-α, IL-1β, IL-6, and type I interferons (IFN), contribute to the recruitment of neutrophils, monocytes, macrophages, and lymphocytes to the site of infection and are directly correlated with viral titers in the lungs and the clinicopathologic effects of influenza in pigs (Jo et al. 2007; Van Reeth 2000; Van Reeth et al. 1998, 2002). The innate response signals the migration of cells and subsequent stimulation of the adaptive arm of the immune system.

The adaptive immune response to IAV in swine includes mucosal and systemic humoral and cell-mediated immunity (CMI); however, most studies investigating the immune response to influenza virus have been conducted in mice. The humoral immunity produced after natural exposure or vaccination plays a significant role in reducing or preventing subsequent infection. Antibodies are mainly directed to the HA, NA, M, and NP viral proteins, however only antibodies to the HA can neutralize viral infectivity, and these are measured by hemagglutination inhibition (HI) or virus neutralization (VN) assays (Cox et al. 2004). These neutralizing antibodies are correlated with clinical protection against challenge virus (Bikour et al. 1996); however, this is likely only true when the priming HA inducing the HI titer is antigenically closely related to the HA of the challenge virus. The antibody response at the mucosa is also important for preventing or clearing the respiratory tract of IAV, and levels of mucosal IgA and IgG have been detected in the nasal cavity of pigs that did not have a detectable anamnestic serum antibody response (Larsen et al. 2000). Additionally, numerous studies have demonstrated significant protection against infection and clinical disease in the absence of detectable HI titers, such as when using live attenuated influenza vaccines in pigs (Loving et al. 2012; Vincent et al. 2012), suggesting that absence of HI titers does not always correlate with lack of protection. The CMI response, which includes CD4+ T helper cells and CD8+ cytotoxic T cells, plays an important role in IAV clearance and recovery after natural infection (Larsen et al. 2000; Platt et al. 2011). T cells mediating CMI against influenza can target the surface glycoproteins as well as the more conserved internal proteins, such as the NP (reviewed in Thomas et al. 2006), and may provide broader heterologous protection.

3 Vaccination Against Influenza in Swine

3.1 Currently Available Vaccines for Use in Swine

Vaccination against influenza is routinely used by U.S. swine producers and veterinarians to reduce clinical disease, however infection and transmission may still occur. Commercial swine influenza vaccines licensed in the U.S. are oil-in-water adjuvanted, bivalent or multivalent vaccines composed of inactivated whole viruses. Vaccination usually consists of two intramuscular injections, 2–4 weeks apart (Vincent et al. 2008b). Approximately 70 % of breeding herd producers vaccinated their stocks in 2007 according to the USDA survey (USDA 2007). Sows are usually vaccinated 3–6 weeks before farrowing to protect their litters through the transfer of maternal derived antibodies (MDA). The presence of passive MDA at the time of infection results in reduction of clinical disease when piglets show sufficient serum antibody titers to the challenge virus (Renshaw 1975). However, the presence of MDA at the time of vaccination can negatively interfere with vaccine efficacy (Kitikoon et al. 2006, 2013a; Loeffen et al. 2003), by suppressing active IgM, IgG, and HI serum antibody responses upon primary and secondary infections and suppressing nasal IgA response as well (Loeffen et al. 2003). MDA can last up to 14 weeks of age, and thus are a major obstacle for the use of killed vaccines in growing pigs.

Intramuscular vaccination with killed vaccines can induce high levels of serum antibody to the IAV HA, but limited mucosal antibodies (Heinen et al. 2001). Whole inactivated virus (WIV) vaccine protection depends on the priming antigen and challenge virus being closely related or antigenically matched. Vaccination with commercial WIV vaccines has been shown to partially protect against clinical signs and reduce nasal viral excretion when pigs are challenged with closely related viruses (Kitikoon et al. 2006, 2013a; Lee et al. 2007; Loving et al. 2013; Macklin et al. 1998). However, when infected with antigenically distinct viruses, these vaccines are less protective (Van Reeth et al. 2003; Vincent et al. 2008a, 2010a). Although genetic proximity of the HA gene to the challenge strain may be inferred to predict vaccine protection, other factors are involved. When comparing the efficacy of four commercial swine IAV vaccines against a European H1N1 virus, Kyriakis et al. (2010) found that the vaccine containing the highest sequence homology with the challenge virus did not offer protection, emphasizing that genetic homology is not the only predictor of vaccine efficacy and cross-protection. Antigenic cross-reactivity as well as immunogenicity needs to be taken into account when evaluating vaccine efficacy to a challenge strain; additionally, studies have shown that single amino acid mutations were sufficient to induce changes near the receptor-binding site and resulted in escape from protective antibodies (Lewis et al. 2011; Li et al. 2011). The continual co-circulation of antigenically diverse swine IAV greatly complicates control by traditional vaccines. As a result, the use of herd-specific autogenous killed vaccines increased in recent years and represented half of the vaccine doses produced for swine in the U.S. (Ma and Richt 2010). Another concern with the use of killed adjuvanted IAV vaccines in the current diverse swine IAV epidemiology is the phenomenon of vaccine-associated enhanced respiratory disease (VAERD), characterized by severe respiratory disease in heterologous challenged pigs following vaccination with mismatched WIV (Gauger et al. 2011), discussed in detail below.

3.2 Experimental Vaccines

New generations of experimental vaccine platforms have been studied as alternatives to improve the suboptimal protection observed for currently available inactivated swine influenza vaccines. Experimental studies using live-attenuated IAV (LAIV) vaccines in swine have repeatedly demonstrated that these vaccines are safe and result in superior protection against heterologous infections (Kappes et al. 2012; Loving et al. 2012; Vincent et al. 2007). Swine IAV with attenuated replication properties have been obtained through molecular approaches, including truncation of the NS1 gene (Solorzano et al. 2005), temperature-sensitive mutations in polymerase genes (PB1 and PB2) (Pena et al. 2011), and modification of the HA cleavage site to an elastase-sensitive motif (Masic et al. 2009). Intratracheal (IT) inoculation of pigs with the H3N2 TRIG NS1-truncated LAIV elicited neutralizing serum antibodies and mucosal IgA and resulted in complete protection against homologous challenge and partial protection against a heterosubtypic H1N1 virus (Richt et al. 2006). Later, Vincent et al. (2007) showed that the intranasal (IN) route primed a mucosal antibody response, and the use of IN vaccination with H3N2 NS-1-truncated LAIV in pigs induced a virus-specific T-cell response and provided partial cross-protection against a heterosubtypic H1N1 challenge (Kappes et al. 2012). The temperature sensitive (ts), cold-adapted LAIV induced CMI through the expansion of virus-specific IFN-γ- secreting cells and increases in total T cells and CD4/CD8 double-positive memory T cells (Loving et al. 2012). Vaccination of pigs with the tsLAIV resulted in sterilizing immunity upon challenge with homologous virus (Pena et al. 2011). The elastase-dependent LAIV was shown to completely protect pigs against homologous infection and resulted in partial protection from heterosubtypic challenge with two IT vaccine administrations (Masic et al. 2009).

Expression of IAV proteins in viral vectors is another appealing alternative to WIV because engineered vectors can be produced in a short period of time and can be administered IN to induce mucosal, humoral and/or CMI (Tutykhina et al. 2011).

A replication-defective human adenovirus serotype 5 vector (Ad5) expressing IAV HA and NP proteins showed complete protection against a closely related challenge strain in pigs, and significant protection when expressing HA alone (Wesley et al. 2004). In another study in pigs, a single intranasal dose of an Ad5-vectored HA induced mucosal IgA and primed a cross-reactive IAV-specific IFN-γ response, providing protection against the homologous virus and partial protection against the heterologous virus (Braucher et al. 2012). Additionally, vectored vaccines have been shown to prime the immune system even in the presence of MDA. Piglets with MDA primed with the Ad5 vector expressing HA and NP and then vaccinated with a commercial killed vaccine were protected against the homologous challenge, while there was no protection in pigs with MDA that were not primed with the vectored vaccine prior to the WIV (Wesley and Lager 2006). Recently, an alphavirus derived replicon particle (RP) vaccine expressing H3N2 IAV HA RNA was approved for use in swine, and has been shown to induce protective immunity against homologous challenge, although it did not protect in the presence of MDA (Bosworth et al. 2010). The same platform expressing H1N1pdm09 HA protein was also protective against homologous infection, and the vaccine expressing an H3N2-derived NP gene reduced nasal shedding and viral replication following H1N1 challenge in pigs (Vander Veen et al. 2012, 2013).

DNA vaccines may also offer advantages over conventional WIV vaccines, and have been shown to elicit both humoral and cellular immune responses and result in broader protection (Kim and Jacob 2009; Ma and Richt 2010; Macklin et al. 1998). In pig studies, DNA vaccines have been shown to induce a strong humoral response against swine IAV, resulting in reduction of viral load in the lungs (Gorres et al. 2011; Larsen and Olsen 2002; Macklin et al. 1998). Recombinant vaccines were also tested as potential alternative platforms for swine IAV vaccination. A previous study showed that recombinant M2 protein vaccine constructs induced an antibody response, but resulted in poor protection against H1N1 challenge (Heinen et al. 2002). Other studies showed reduction in respiratory signs when pigs were vaccinated with a recombinant M2 vaccine (Kitikoon et al. 2009) or reduction of viral replication and shedding after recombinant HA vaccination (Loeffen et al. 2011).

3.3 Vaccine-Associated Enhanced Respiratory Disease in Swine

Infection or vaccination with a number of viruses has been associated with cases of enhanced systemic or respiratory disease in humans. Severe cases of dengue virus infection, termed dengue hemorrhagic fever (DHF), occur mostly after secondary infection with a different serotype (Rothman 2010). This is proposed to be due to the mechanism of antibody-dependent enhancement (ADE), by which pre-existing antibodies elicited in the primary infection bind to the new virus serotype but do not neutralize infection. Viral replication is enhanced after entry into Fc receptor-positive macrophages, resulting in DHF in susceptible individuals (Halstead and O'Rourke 1977). The immune response to inactivated vaccines has also been implicated in cases of severe respiratory disease in humans and other mammals. The use of an inactivated whole-virus vaccine against respiratory syncytial virus (RSV) in naïve infants was associated with severe lower respiratory tract disease upon subsequent natural RSV infection, leading to the death of two vaccinated children (Kim et al. 1969). The RSV vaccine enhancement was suggested to be a result of RSV-specific antibody response with poor neutralizing activity (Chin et al. 1969; Murphy et al. 1986). During the 2009 H1N1 influenza pandemic, severe and fatal cases of infection were predominantly observed in middle-aged adults (Chowell et al. 2009), an age group not typically affected during seasonal influenza. This shift in the age group was associated, among other factors, with the presence of low avidity, nonprotective antibodies, and complement activation resulting from previous vaccination and/or exposure to seasonal influenza viruses (Monsalvo et al. 2011). Inactivated seasonal influenza vaccines were correlated with an increased risk of influenza-like-illness in H1N1pdm09 infected humans, suggesting vaccine-induced immune responses may have contributed to a more severe disease outcome (Skowronski et al. 2010). In a separate study, high titers of non-neutralizing antibodies against the H1N1pdm09 vaccine antigen were identified in the serum of patients with laboratory-confirmed, severe H1N1pdm09 influenza illness (To et al. 2012).

Vaccine-associated enhancement of influenza respiratory disease in swine was suggested in a study with a DNA construct expressing a fusion protein consisting of the extracellular domain of the M2 protein (M2e) and nucleoprotein (NP) (Heinen et al. 2002). The M2eNP DNA vaccine induced an antibody and lymphoproliferative immune response to the virus although subsequent challenge with an H1N1 resulted in severe clinical respiratory disease in all M2eNP immunized pigs and death of three pigs compared to the nonvaccinated and challenged group. The results of this study suggested both non-neutralizing antibodies to M2e and T helper (Th) cells directed against the NP were involved in the disease enhancement. However, lung lesions were not assessed in this report, and the mechanism of enhancement was not further characterized.

Enhancement of influenza disease was subsequently reported when pigs were vaccinated with a bivalent commercial H1N1/H3N2 vaccine in the presence of MDA and challenged with a cH1N1 virus (A/Swine/Iowa/40776/92) with limited serological HI cross-reactivity against the vaccine strain (Kitikoon et al. 2006). MDA-positive, vaccinated and challenged pigs demonstrated prolonged clinical signs and significantly elevated macroscopic pneumonia compared to the non-vaccinated challenged group (Kitikoon et al. 2006). Furthermore, enhancement of lung lesions and broader viral distribution was also shown in the lungs of pigs vaccinated with an α-cluster cH1N1 (A/Swine/Iowa/15/1930) adjuvanted WIV and challenged with a homosubtypic, but antigenically distinct γ-cluster H1N2 virus with no evidence of serological cross-reactivity in the HI assay (Vincent et al. 2008a). Subsequent studies reproduced enhancement of IAV disease as a result of

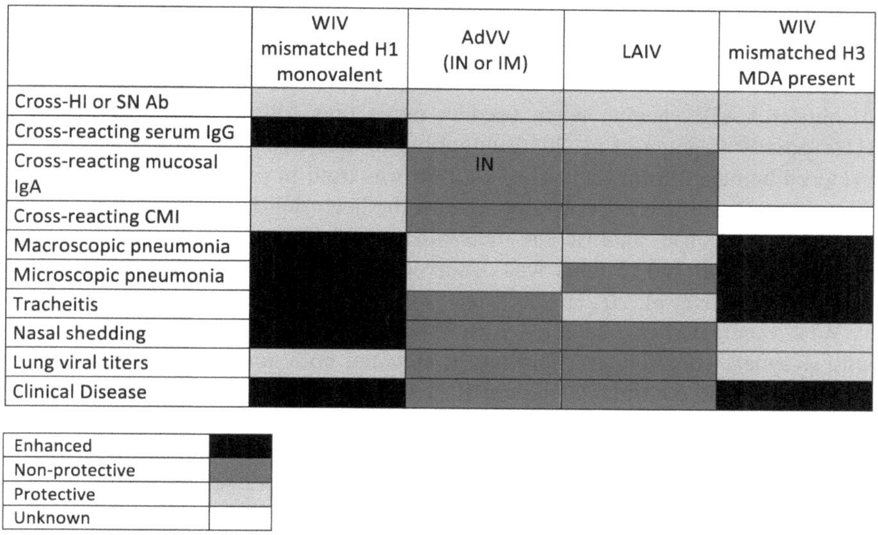

Fig. 2 Vaccine models using different combinations of vaccine platforms and challenge viruses and relative involvement in vaccine-associated enhanced respiratory disease (VAERD). Relative impact was *color coded* for parameters tested in pigs challenged with the homosubtypic heterologous virus after vaccination with either δ1-H1N2 or H1N1pdm09 whole inactivated virus (WIV) vaccine; human adenovirus serotype 5 vector expressing IAV HA (Ad5-HA); live attenuated influenza virus (LAIV) vaccine; or WIV in the presence of homologous maternal derived antibodies (MDA). *Black* indicates parameters that were associated with enhancement; *dark grey* indicates parameters associate with vaccine failure without enhancement; *light grey* indicates parameters that were associated with vaccine protection; *white* denotes parameters that were not evaluated

vaccine/challenge mismatch and the resulting clinical outcome in swine was termed vaccine-associated enhanced respiratory disease (VAERD) (Gauger et al. 2011).

VAERD has been reproduced under numerous combinations of factors (Fig. 2), but the common denominator suggests the phenomenon is associated with the use of oil-in-water adjuvanted, inactivated WIV followed by challenge with a homosubtypic but antigenically distinct IAV (Braucher et al. 2012; Gauger et al. 2011; Vincent et al. 2012). Of relevance to current influenza epidemiology, the vaccination of pigs with a δ1 cluster H1N2 (A/Sw/MN/02011/2008) WIV vaccine and challenge with the H1N1pdm09 resulted in VAERD (Gauger et al. 2011, 2012, 2013). Reports also described the reverse vaccine/virus combination, using H1N1pdm09 WIV vaccine and δ1-H1N2 challenge virus, causing VAERD after intranasal challenge (Braucher et al. 2012). The use of H3N2 viruses for vaccine and challenge with limited serologic cross-reactivity also resulted in VAERD when pigs were vaccinated in the presence of MDA (Vincent et al. 2012).

In contrast, alternative IAV vaccine platforms, such as modified live virus using mucosal delivery, were shown to avoid the risk of VAERD. Moreover, vaccination with a cluster I H3N2 LAIV attenuated by truncation of the NS1 protein in pigs with matching MDA resulted in decreased nasal shedding and viral replication after heterologous cluster II H3N2 infection, while avoiding VAERD development (Vincent et al. 2012). Similar results were observed for a temperature sensitive δ1-cluster LAIV vaccine when used in naïve pigs followed by challenge with H1N1pdm09 (Gauger et al. 2014, unpublished data). Additionally, when an Ad5 vectored hemagglutinin (Ad5-HA) vaccine was used in pigs in a single intranasal dose, it conferred partial protection against the heterologous challenge and primed an IFN-γ response against the heterologous challenge virus while avoiding VAERD, as opposed to what was observed in WIV-vaccinated pigs in the same study (Braucher et al. 2012).

WIV vaccinated pigs affected with VAERD exhibit severe respiratory disease with dyspnea, coughing, and prolonged elevated body temperatures compared to naïve challenged controls (Gauger et al. 2011). Pigs with VAERD display a greater magnitude of purple-red colored multifocal to coalescing lung consolidation (Fig. 3a), with some cases of localized hemorrhage and bullous emphysema (Gauger et al. 2011). Histopathological lesions were more severe and widespread than typically observed with uncomplicated IAV infection in swine, and occurred as rapidly as 1 day post infection (dpi) (Gauger et al. 2012). Initial lesions demonstrate various amounts of hemorrhage, edema, and a suppurative inflammatory response that progress to a predominantly lymphocytic interstitial infiltrate by 5 dpi. The microscopic changes include prominent peribronchiolar lymphocytic cuffing, severe necrotizing to proliferative bronchiolitis (Fig. 3c), and moderate lymphohistiocytic interstitial pneumonia (Gauger et al. 2012). The trachea is also more affected in pigs with VAERD, showing lymphoplasmacytic and necrosuppurative tracheitis with regional loss of cilia and severe lymphoplasmacytic infiltration of the tracheal submucosa (Gauger et al. 2012).

Elevated virus titers in bronchoalveolar lavage fluid (BALF) of VAERD-affected pigs were observed during the early post-infection period (1 and 2 dpi) compared to nonvaccinated and challenged pigs (Gauger et al. 2012). These *in vivo* findings support the *in vitro* evidence of enhanced viral-fusion caused by non-neutralizing antibodies specific to the conserved HA2 domain of the H1N1pdm09 HA protein described in a recent study (Khurana et al. 2013). Although virus titers in the lungs and nasal secretions of VAERD-affected pigs have consistently been shown to be lower than nonvaccinated challenged animals at 5 dpi, the percentage of shedding pigs and group mean titers remain higher than pigs protected by matched WIV or mismatched LAIV, consistent with the lack of protection typical of VAERD (Braucher et al. 2012; Gauger et al. 2011; Vincent et al. 2012). Nonetheless, VAERD-affected pigs recovered from disease by 21 days after infection in the absence of complication by co-infection with other pathogens (Gauger et al. 2011; Vincent et al. 2012). In addition, pigs affected with VAERD mounted a robust neutralizing antibody response by 21 days to the heterologous challenge virus, with elevated adaptive cytokine levels in the lungs, suggesting VAERD does not interfere negatively with the immune response to the challenge virus in spite of the exacerbated disease and inflammatory response and is not related to the concept of original antigenic sin (Gauger et al. 2013).

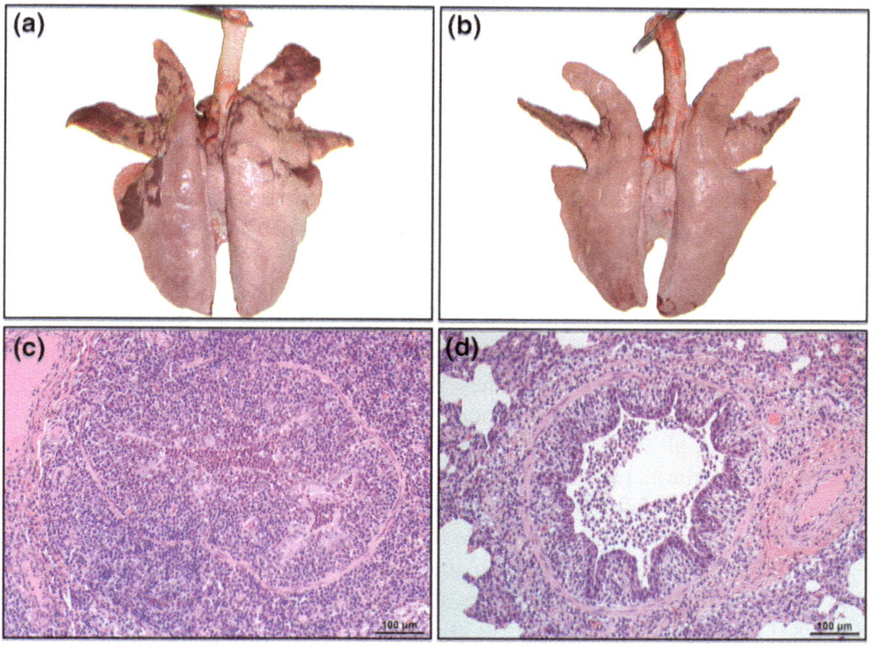

Fig. 3 Representative macroscopic and microscopic lung lesions of pigs with and without prior vaccination challenged with H1N1pdm09 virus. VAERD-affected pigs vaccinated with δ1-H1N2 WIV demonstrate a greater percentage of multifocal, *purple-colored* cranioventral and diaphragmatic pulmonary consolidation (**a**) compared to nonvaccinated H1N1pdm09-infected pigs (**b**). Severe necrotizing to proliferative bronchiolitis and prominent peribronchiolar lymphocytic cuffing in pigs with enhanced pneumonia (**c**) in comparison to nonvaccinated, challenged pigs (**d**). Hematoxylin and eosin (200×)

The mechanism responsible for VAERD remains unknown; however, studies conducted to date suggest disease enhancement may be dependent on nonprotective but cross-reactive antibodies induced by the adjuvanted WIV vaccine in the absence of a cross-protective mucosal and/or cell-mediated immune response against antigenically distinct IAV in swine (Khurana et al. 2013). Antibody response to WIV was strongly suggested to be sufficient to initiate the onset of VAERD in a recent experiment in pigs with passive MDA derived from WIV vaccinated dams (Rajao et al. 2013). Passive antibodies from WIV vaccinated sows were sufficient to induce enhancement in challenged, unvaccinated three-week-old piglets, albeit at an intermediate level compared to WIV vaccinated pigs demonstrating VAERD. This suggested that although passive antibodies may have been sufficient to incite VAERD, other aspects of the active immune response are involved in the magnitude of the pathologic changes observed in WIV-induced VAERD.

A consistent predisposing factor of VAERD is the presence of serum and mucosal IgG antibodies that cross-react with the heterologous challenge virus in whole-virus ELISA in the absence of neutralizing or HI antibodies (Braucher et al. 2012;

Gauger et al. 2011; Vincent et al. 2012). These non-neutralizing but cross-reactive antibodies may contribute in several proposed mechanisms responsible for the VAERD phenomenon (Crowe 2013). These include activation of the classical complement cascade mediated through immune-complex formation, antibody dependent cell-mediated cytotoxicity (ADCC), or antibody-mediated increased virus uptake into cells that support virus replication or increase the production of pro-inflammatory cytokines (Gauger et al. 2011; Khurana et al. 2013). Elevated concentrations of IL-1β, IL-8, and TNF-α proteins were observed in pigs with VAERD, manifested as early as 1 dpi, and additional pro-inflammatory cytokines were elevated by 5 dpi (Gauger et al. 2012). In contrast to the elevated pro-inflammatory response, the antiviral IFN-α protein levels were paradoxically lacking in VAERD affected pigs at 1 dpi, despite the presence of high virus titers in the nose and lungs. Collectively, dysregulation of pro-inflammatory and anti-viral cytokines seems to be a common characteristic of VAERD-affected pigs compared to nonvaccinated, challenged pigs (Braucher et al. 2012; Gauger et al. 2011, 2012). The cytokine imbalances are likely to have an important role in the increased pathology associated with VAERD through generalized inflammation, neutrophil recruitment and degranulation, vasodilation, and an absent or abrogated innate protection against IAV infection.

Limited cross-protection between inactivated vaccines and antigenically distinct IAV circulating in swine poses the question of risk of VAERD in field settings (Gauger et al. 2011; Vincent et al. 2008a, 2010b). The extent of antigenic diversity between vaccine antigen and challenge virus necessary to induce VAERD is unknown. The δ1-cluster H1N2 and H1N1pdm09 viruses used in this VAERD model displayed complete loss in HI cross-reactivity and shared only 71 % HA1 amino acid homology (Khurana et al. 2013). Although WIV vaccines may protect against homologous IAV, they typically lack the ability to induce robust mucosal IgA (Gauger et al. 2011; Vincent et al. 2008a) or cell-mediated immune responses (Platt et al. 2011; Vincent et al. 2008a). Previous reports describing VAERD in WIV-vaccinated swine have also emphasized the lack of IgA in the upper and lower respiratory tract or deficient cell-mediated immune responses correlated with protection in LAIV-vaccinated swine (Braucher et al. 2012; Gauger et al. 2011, 2014). Further research is needed to elucidate the mechanisms involved in VAERD and the vaccine platforms or methods of delivery that may enhance cross-protection and avoid VAERD. Live-attenuated influenza and vectored vaccine products have demonstrated superior efficacy compared to inactivated preparations and may be appropriate for averting undesirable vaccine-associated adverse immune responses and enhanced disease in swine.

4 Conclusions

Homologous vaccination remains the most feasible and effective means of preventing swine IAV infection; however, the rapid evolution and the immense diversity of IAV currently circulating in the swine population greatly challenge the development of effectively matched vaccines. Viral surveillance and genetic analysis in conjunction with antigenic evaluation are important tools for vaccine strain selection and can be used to improve the antigenic coverage among swine IAV vaccines. Updating vaccine formulations to include contemporary circulating strains is essential to provide broader protection and overcome the lack of cross-protection observed for traditional vaccines. Although current WIV vaccines have disadvantages when it comes to heterologous protection and possibility of disease enhancement, no vaccine is flawless. The choice of vaccines and immunization program depends on the herd's IAV epidemiology, immunological status, and age of animals, among other factors. When placed in the right situation and in combination with other control measures, WIV can be highly effective. However, other vaccine platforms offer improved heterologous protection under many conditions, and should continue to be evaluated for their ability to provide broader and more balanced cross-protection.

References

Anderson TK, Nelson MI, Kitikoon P, Swenson SL, Korslund JA, Vincent AL (2013) Population dynamics of cocirculating swine influenza A viruses in the United States from 2009 to 2012. Influenza Other Respir Viruses 7(Suppl 4):42–51

Bikour MH, Cornaglia E, Elazhary Y (1996) Evaluation of a protective immunity induced by an inactivated influenza H3N2 vaccine after an intratracheal challenge of pigs. Can J Vet Res 60:312–314

Bosworth B, Erdman MM, Stine DL, Harris I, Irwin C, Jens M et al (2010) Replicon particle vaccine protects swine against influenza. Comp Immunol Microbiol Infect Dis 33:e99–e103

Braucher DR, Henningson JN, Loving CL, Vincent AL, Kim E, Steitz J et al (2012) Intranasal vaccination with replication-defective adenovirus type 5 encoding influenza virus hemagglutinin elicits protective immunity to homologous challenge and partial protection to heterologous challenge in pigs. Clin Vaccine Immunol 19:1722–1729

Brown IH (2000) The epidemiology and evolution of influenza viruses in pigs. Vet Microbiol 74:29–46

CDC [Centers for Disease Control and Prevention] (2012) Influenza A (H3N2) variant virus outbreaks. http://www.cdc.gov/flu/swineflu/h3n2v-outbreak.htm. Accessed 24 June 2013

Chin J, Magoffin RL, Shearer LA, Schieble JH, Lennette EH (1969) Field evaluation of a respiratory syncytial virus vaccine and a trivalent parainfluenza virus vaccine in a pediatric population. Am J Epidemiol 89:449–463

Chowell G, Bertozzi SM, Colchero MA, Lopez-Gatell H, Alpuche-Aranda C, Hernandez M et al (2009) Severe respiratory disease concurrent with the circulation of H1N1 influenza. N Engl J Med 361:674–679

Cox RJ, Brokstad KA, Ogra P (2004) Influenza virus: immunity and vaccination strategies. Comparison of the immune response to inactivated and live, attenuated influenza vaccines. Scand J Immunol 59:1–15

Crowe JE Jr (2013) Universal flu vaccines: primum non nocere. Sci Trans Med 5:200fs234

de Jong JC, van Nieuwstadt AP, Kimman TG, Loeffen WL, Bestebroer TM, Bijlsma K et al (1999) Antigenic drift in swine influenza H3 haemagglutinins with implications for vaccination policy. Vaccine 17:1321–1328

Doherty PC, Turner SJ, Webby RG, Thomas PG (2006) Influenza and the challenge for immunology. Nat Immunol 7:449–455

Epperson S, Jhung M, Richards S, Quinlisk P, Ball L, Moll M et al (2013) Human infections with influenza A(H3N2) variant virus in the United States, 2011-2012. Clin Infect Dis 57(Suppl 1):S4–S11
Garten RJ, Davis CT, Russell CA, Shu B, Lindstrom S, Balish A et al (2009) Antigenic and genetic characteristics of swine-origin 2009 A(H1N1) influenza viruses circulating in humans. Science 325:197–201
Gauger PC, Vincent AL, Loving CL, Lager KM, Janke BH, Kehrli ME Jr et al (2011) Enhanced pneumonia and disease in pigs vaccinated with an inactivated human-like (delta-cluster) H1N2 vaccine and challenged with pandemic 2009 H1N1 influenza virus. Vaccine 29:2712–2719
Gauger PC, Vincent AL, Loving CL, Henningson JN, Lager KM, Janke BH et al (2012) Kinetics of lung lesion development and pro-inflammatory cytokine response in pigs with vaccine-associated enhanced respiratory disease induced by challenge with pandemic (2009) A/H1N1 influenza virus. Vet Pathol 49:900–912
Gauger PC, Loving CL, Lager KM, Janke BH, Kehrli ME Jr, Roth JA et al (2013) Vaccine-associated enhanced respiratory disease does not interfere with the adaptive immune response following challenge with pandemic A/H1N1 2009. Viral Immunol 26:314–321
Gauger PC, Loving CL, Lorusso A, Perez D, Kehrli ME Jr, Roth JA et al (2014) Live attenuated influenza A virus vaccine protects against A(H1N1)pdm09 heterologous challenge without vaccine associated enhanced respiratory disease (manuscript submitted for publication)
Gorres JP, Lager KM, Kong WP, Royals M, Todd JP, Vincent AL et al (2011) DNA vaccination elicits protective immune responses against pandemic and classic swine influenza viruses in pigs. Clin Vaccine Immunol 18:1987–1995
Halstead SB, O'Rourke EJ (1977) Dengue viruses and mononuclear phagocytes. I. Infection enhancement by non-neutralizing antibody. J Exp Med 146:201–217
Heinen PP, van Nieuwstadt AP, de Boer-Luijtze EA, Bianchi AT (2001) Analysis of the quality of protection induced by a porcine influenza A vaccine to challenge with an H3N2 virus. Vet Immunol Immunopathol 82:39–56
Heinen PP, Rijsewijk FA, de Boer-Luijtze EA, Bianchi AT (2002) Vaccination of pigs with a DNA construct expressing an influenza virus M2-nucleoprotein fusion protein exacerbates disease after challenge with influenza A virus. J Gen Virol 83:1851–1859
Janke BH (2013) Clinicopathological features of swine influenza. Curr Top Microbiol Immunol 370:69–83
Jo SK, Kim HS, Cho SW, Seo SH (2007) Pathogenesis and inflammatory responses of swine H1N2 influenza viruses in pigs. Virus Res 129:64–70
Kappes MA, Sandbulte MR, Platt R, Wang C, Lager KM, Henningson JN et al (2012) Vaccination with NS1-truncated H3N2 swine influenza virus primes T cells and confers cross-protection against an H1N1 heterosubtypic challenge in pigs. Vaccine 30:280–288
Karasin AI, Schutten MM, Cooper LA, Smith CB, Subbarao K, Anderson GA et al (2000) Genetic characterization of H3N2 influenza viruses isolated from pigs in North America, 1977–1999: evidence for wholly human and reassortant virus genotypes. Virus Res 68:71–85
Karasin AI, Landgraf J, Swenson S, Erickson G, Goyal S, Woodruff M et al (2002) Genetic characterization of H1N2 influenza A viruses isolated from pigs throughout the United States. J Clin Microbiol 40:1073–1079
Khatri M, Dwivedi V, Krakowka S, Manickam C, Ali A, Wang L et al (2010) Swine influenza H1N1 virus induces acute inflammatory immune responses in pig lungs: a potential animal model for human H1N1 influenza virus. J Virol 84:11210–11218
Khurana S, Loving CL, Manischewitz J, King LR, Gauger PC, Henningson J et al (2013) Vaccine-induced anti-HA2 antibodies promote virus fusion and enhance influenza virus respiratory disease. Sci Trans Med 5:200ra114
Kim HW, Canchola JG, Brandt CD, Pyles G, Chanock RM, Jensen K et al (1969) Respiratory syncytial virus disease in infants despite prior administration of antigenic inactivated vaccine. Am J Epidemiol 89:422–434

Kim JH, Jacob J (2009) DNA vaccines against influenza viruses. Curr Top Microbiol Immunol 333:197–210
Kitikoon P, Nilubol D, Erickson BJ, Janke BH, Hoover TC, Sornsen SA et al (2006) The immune response and maternal antibody interference to a heterologous H1N1 swine influenza virus infection following vaccination. Vet Immunol Immunopathol 112:117–128
Kitikoon P, Vincent AL, Janke BH, Erickson B, Strait EL, Yu S et al (2009) Swine influenza matrix 2 (M2) protein contributes to protection against infection with different H1 swine influenza virus (SIV) isolates. Vaccine 28:523–531
Kitikoon P, Gauger PC, Anderson TK, Culhane MR, Swenson S, Loving CL et al (2013a) Swine influenza virus vaccine serologic cross-reactivity to contemporary US swine H3N2 and efficacy in pigs infected with an H3N2 similar to 2011–2012 H3N2v. Influenza Other Respir Viruses 7(Suppl 4):32–41
Kitikoon P, Nelson MI, Killian ML, Anderson TK, Koster L, Culhane MR et al (2013b) Genotype patterns of contemporary reassorted H3N2 virus in US swine. J Gen Virol 94:1236–1241
Koen JS (1919) A practical method for field diagnosis of swine diseases. Am J Vet Med 14:468–470
Kyriakis CS, Gramer MR, Barbe F, Van Doorsselaere J, Van Reeth K (2010) Efficacy of commercial swine influenza vaccines against challenge with a recent European H1N1 field isolate. Vet Microbiol 144:67–74
La Gruta NL, Kedzierska K, Stambas J, Doherty PC (2007) A question of self-preservation: immunopathology in influenza virus infection. Immunol Cell Biol 85:85–92
Larsen DL, Karasin A, Zuckermann F, Olsen CW (2000) Systemic and mucosal immune responses to H1N1 influenza virus infection in pigs. Vet Microbiol 74:117–131
Larsen DL, Olsen CW (2002) Effects of DNA dose, route of vaccination, and coadministration of porcine interleukin-6 DNA on results of DNA vaccination against influenza virus infection in pigs. Am J Vet Res 63:653–659
Lee JH, Gramer MR, Joo HS (2007) Efficacy of swine influenza A virus vaccines against an H3N2 virus variant. Can J Vet Res 71:207–212
Lewis NS, Daly JM, Russell CA, Horton DL, Skepner E, Bryant NA et al (2011) Antigenic and genetic evolution of equine influenza A (H3N8) virus from 1968 to 2007. J Virol 85:12742–12749
Li Z, Liu Z, Ma C, Zhang L, Su Y, Gao GF et al (2011) Identification of amino acids in highly pathogenic avian influenza H5N1 virus hemagglutinin that determine avian influenza species specificity. Arch Virol 156:1803–1812
Loeffen WL, Heinen PP, Bianchi AT, Hunneman WA, Verheijden JH (2003) Effect of maternally derived antibodies on the clinical signs and immune response in pigs after primary and secondary infection with an influenza H1N1 virus. Vet Immunol Immunopathol 92:23–35
Loeffen WL, de Vries RP, Stockhofe N, van Zoelen-Bos D, Maas R, Koch G et al (2011) Vaccination with a soluble recombinant hemagglutinin trimer protects pigs against a challenge with pandemic (H1N1) 2009 influenza virus. Vaccine 29:1545–1550
Lorusso A, Vincent AL, Harland ML, Alt D, Bayles DO, Swenson SL et al (2011) Genetic and antigenic characterization of H1 influenza viruses from United States swine from 2008. J Gen Virol 92:919–930
Lorusso A, Vincent AL, Gramer ME, Lager KM, Ciacci-Zanella JR (2013) Contemporary epidemiology of North American lineage triple reassortant influenza A viruses in pigs. Curr Top Microbiol Immunol 370:113–132
Loving CL, Vincent AL, Pena L, Perez DR (2012) Heightened adaptive immune responses following vaccination with a temperature-sensitive, live-attenuated influenza virus compared to adjuvanted, whole-inactivated virus in pigs. Vaccine 30:5830–5838
Loving CL, Lager KM, Vincent AL, Brockmeier SL, Gauger PC, Anderson TK et al (2013) Efficacy of inactivated and live-attenuated influenza virus vaccines in pigs against infection and transmission of emerging H3N2 similar to the 2011–2012 H3N2v. J Virol 87(17):9895–8903
Ma W, Richt JA (2010) Swine influenza vaccines: current status and future perspectives. Anim Health Res Rev 11:81–96

Macklin MD, McCabe D, McGregor MW, Neumann V, Meyer T, Callan R et al (1998) Immunization of pigs with a particle-mediated DNA vaccine to influenza A virus protects against challenge with homologous virus. J Virol 72:1491–1496
Masic A, Booth JS, Mutwiri GK, Babiuk LA, Zhou Y (2009) Elastase-dependent live attenuated swine influenza A viruses are immunogenic and confer protection against swine influenza A virus infection in pigs. J Virol 83:10198–10210
Monsalvo AC, Batalle JP, Lopez MF, Krause JC, Klemenc J, Hernandez JZ et al (2011) Severe pandemic 2009 H1N1 influenza disease due to pathogenic immune complexes. Nat Med 17:195–199
Murphy BR, Prince GA, Walsh EE, Kim HW, Parrott RH, Hemming VG et al (1986) Dissociation between serum neutralizing and glycoprotein antibody responses of infants and children who received inactivated respiratory syncytial virus vaccine. J Clin Microbiol 24:197–202
Myers KP, Olsen CW, Gray GC (2007) Cases of swine influenza in humans: a review of the literature. Clin Infect Dis 44:1084–1088
Nelli RK, Kuchipudi SV, White GA, Perez BB, Dunham SP, Chang KC (2010) Comparative distribution of human and avian type sialic acid influenza receptors in the pig. BMC Vet Res 6:4
Nelson MI, Lemey P, Tan Y, Vincent A, Lam TT, Detmer S et al (2011) Spatial dynamics of human-origin H1 influenza A virus in North American swine. PLoS Pathog 7:e1002077
Nelson MI, Gramer MR, Vincent AL, Holmes EC (2012) Global transmission of influenza viruses from humans to swine. J Gen Virol 93:2195–2203
Olsen CW (2002) The emergence of novel swine influenza viruses in North America. Virus Res 85:199–210
Pena L, Vincent AL, Ye J, Ciacci-Zanella JR, Angel M, Lorusso A et al (2011) Modifications in the polymerase genes of a swine-like triple-reassortant influenza virus to generate live attenuated vaccines against 2009 pandemic H1N1 viruses. J Virol 85:456–469
Platt R, Vincent AL, Gauger PC, Loving CL, Zanella EL, Lager KM et al (2011) Comparison of humoral and cellular immune responses to inactivated swine influenza virus vaccine in weaned pigs. Vet Immunol Immunopathol 142:252–257
Rajao DS, Vincent AL, Loving CL, Sandbulte MR, Gauger PC, Kitikoon P (2013) Heterologous challenge of weaned piglets in the presence of maternal derived antibodies results in vaccine-associated enhanced respiratory disease. In: Conference of research workers in animal diseases. CRWAD, Chicago, p 090
Renshaw HW (1975) Influence of antibody-mediated immune suppression on clinical, viral, and immune responses to swine influenza infection. Am J Vet Res 36:5–13
Richt JA, Lekcharoensuk P, Lager KM, Vincent AL, Loiacono CM, Janke BH et al (2006) Vaccination of pigs against swine influenza viruses by using an NS1-truncated modified live-virus vaccine. J Virol 80:11009–11018
Rothman AL (2010) Cellular immunology of sequential dengue virus infection and its role in disease pathogenesis. Curr Top Microbiol Immunol 338:83–98
Scholtissek C, Burger H, Kistner O, Shortridge KF (1985) The nucleoprotein as a possible major factor in determining host specificity of influenza H3N2 viruses. Virology 147:287–294
Shope RE (1931) Swine influenza: III. Filtration experiments and etiology. J Exp Med 54:373–385
Skowronski DM, De Serres G, Crowcroft NS, Janjua NZ, Boulianne N, Hottes TS et al (2010) Association between the 2008–2009 seasonal influenza vaccine and pandemic H1N1 illness during Spring-Summer 2009: four observational studies from Canada. PLoS Med 7:e1000258
Smith GJ, Bahl J, Vijaykrishna D, Zhang J, Poon LL, Chen H et al (2009a) Dating the emergence of pandemic influenza viruses. Proc Natl Acad Sci USA 106:11709–11712
Smith GJ, Vijaykrishna D, Bahl J, Lycett SJ, Worobey M, Pybus OG et al (2009b) Origins and evolutionary genomics of the 2009 swine-origin H1N1 influenza A epidemic. Nature 459:1122–1125

Solorzano A, Webby RJ, Lager KM, Janke BH, Garcia-Sastre A, Richt JA (2005) Mutations in the NS1 protein of swine influenza virus impair anti-interferon activity and confer attenuation in pigs. J Virol 79:7535–7543

Thomas PG, Keating R, Hulse-Post DJ, Doherty PC (2006) Cell-mediated protection in influenza infection. Emerg Infect Dis 12:48–54

To KK, Zhang AJ, Hung IF, Xu T, Ip WC, Wong RT et al (2012) High titer and avidity of nonneutralizing antibodies against influenza vaccine antigen are associated with severe influenza. Clin Vaccine Immunol 19:1012–1018

Tutykhina IL, Logunov DY, Shcherbinin DN, Shmarov MM, Tukhvatulin AI, Naroditsky BS et al (2011) Development of adenoviral vector-based mucosal vaccine against influenza. J Mol Med (Berl) 89:331–341

USDA [United States Department of Agriculture] (2007) Swine 2006, Part II: reference of swine health and health management practices in the United States, 2006

Van Reeth K, Nauwynck H, Pensaert M (1998) Bronchoalveolar interferon-alpha, tumor necrosis factor-alpha, interleukin-1, and inflammation during acute influenza in pigs: a possible model for humans? J Infect Dis 177:1076–1079

Van Reeth K (2000) Cytokines in the pathogenesis of influenza. Vet Microbiol 74:109–116

Van Reeth K, Van Gucht S, Pensaert M (2002) Correlations between lung proinflammatory cytokine levels, virus replication, and disease after swine influenza virus challenge of vaccination-immune pigs. Viral Immunol 15:583–594

Van Reeth K, Van Gucht S, Pensaert M (2003) Investigations of the efficacy of European H1N1- and H3N2-based swine influenza vaccines against the novel H1N2 subtype. Vet Rec 153:9–13

Van Reeth K, Brown IH, Olsen CW (2012) Influenza virus. In: Zimmerman JJ, Karriker LA, Ramirez A et al (eds) Diseases of swine, 10th edn. Wiley-Blackwell, Ames, pp 557–571

Vander Veen RL, Loynachan AT, Mogler MA, Russell BJ, Harris DL, Kamrud KI (2012) Safety, immunogenicity, and efficacy of an alphavirus replicon-based swine influenza virus hemagglutinin vaccine. Vaccine 30:1944–1950

Vander Veen RL, Mogler MA, Russell BJ, Loynachan AT, Harris DL, Kamrud KI (2013) Haemagglutinin and nucleoprotein replicon particle vaccination of swine protects against the pandemic H1N1 2009 virus. Vet Rec 173:344

Vincent A, Awada L, Brown I, Chen H, Claes F, Dauphin G et al (2013) Review of influenza A virus in swine worldwide: a call for increased surveillance and research. Zoonoses Public Health 61:4–17

Vincent AL, Ma W, Lager KM, Janke BH, Webby RJ, Garcia-Sastre A et al (2007) Efficacy of intranasal administration of a truncated NS1 modified live influenza virus vaccine in swine. Vaccine 25:7999–8009

Vincent AL, Lager KM, Janke BH, Gramer MR, Richt JA (2008a) Failure of protection and enhanced pneumonia with a US H1N2 swine influenza virus in pigs vaccinated with an inactivated classical swine H1N1 vaccine. Vet Microbiol 126:310–323

Vincent AL, Ma W, Lager KM, Janke BH, Richt JA (2008b) Swine influenza viruses a North American perspective. Adv Virus Res 72:127–154

Vincent AL, Lager KM, Harland M, Lorusso A, Zanella E, Ciacci-Zanella JR et al (2009a) Absence of 2009 pandemic H1N1 influenza A virus in fresh pork. PLoS One 4:e8367

Vincent AL, Ma W, Lager KM, Gramer MR, Richt JA, Janke BH (2009b) Characterization of a newly emerged genetic cluster of H1N1 and H1N2 swine influenza virus in the United States. Virus Genes 39:176–185

Vincent AL, Ciacci-Zanella JR, Lorusso A, Gauger PC, Zanella EL, Kehrli ME Jr et al (2010a) Efficacy of inactivated swine influenza virus vaccines against the 2009 A/H1N1 influenza virus in pigs. Vaccine 28:2782–2787

Vincent AL, Lager KM, Faaberg KS, Harland M, Zanella EL, Ciacci-Zanella JR et al (2010b) Experimental inoculation of pigs with pandemic H1N1 2009 virus and HI cross-reactivity with contemporary swine influenza virus antisera. Influenza Other Respi Viruses 4:53–60

Vincent AL, Ma W, Lager KM, Richt JA, Janke BH, Sandbulte MR et al (2012) Live attenuated influenza vaccine provides superior protection from heterologous infection in pigs with maternal antibodies without inducing vaccine-associated enhanced respiratory disease. J Virol 86:10597–10605

Wesley RD, Tang M, Lager KM (2004) Protection of weaned pigs by vaccination with human adenovirus 5 recombinant viruses expressing the hemagglutinin and the nucleoprotein of H3N2 swine influenza virus. Vaccine 22:3427–3434

Wesley RD, Lager KM (2006) Overcoming maternal antibody interference by vaccination with human adenovirus 5 recombinant viruses expressing the hemagglutinin and the nucleoprotein of swine influenza virus. Vet Microbiol 118:67–75

Zhou NN, Senne DA, Landgraf JS, Swenson SL, Erickson G, Rossow K et al (1999) Genetic reassortment of avian, swine, and human influenza A viruses in American pigs. J Virol 73:8851–8856

Secondary Bacterial Infections in Influenza Virus Infection Pathogenesis

Amber M. Smith and Jonathan A. McCullers

Abstract Influenza is often complicated by bacterial pathogens that colonize the nasopharynx and invade the middle ear and/or lung epithelium. Incidence and pathogenicity of influenza-bacterial coinfections are multifactorial processes that involve various pathogenic virulence factors and host responses with distinct site- and strain-specific differences. Animal models and kinetic models have improved our understanding of how influenza viruses interact with their bacterial co-pathogens and the accompanying immune responses. Data from these models indicate that considerable alterations in epithelial surfaces and aberrant immune responses lead to severe inflammation, a key driver of bacterial acquisition and infection severity following influenza. However, further experimental and analytical studies are essential to determining the full mechanistic spectrum of different viral and bacterial strains and species and to finding new ways to prevent and treat influenza-associated bacterial coinfections. Here, we review recent advances regarding transmission and disease potential of influenza-associated bacterial infections and discuss the current gaps in knowledge.

Contents

A.M. Smith (✉) · J.A. McCullers
Department of Infectious Diseases, St. Jude Children's Research Hospital, Memphis, TN 38105, USA
e-mail: amber.smith@stjude.org

J.A. McCullers
Department of Pediatrics, University of Tennessee Health Science Center, Memphis, TN 38103, USA

Current Topics in Microbiology and Immunology (2014) 385: 327–356
DOI: 10.1007/82_2014_394

Published Online: 16 July 2014

1 Introduction

Pneumonia is a leading cause of death in the United States and worldwide [(Centers for Disease Control Deaths and Mortality); World Health Organization]. Multiple respiratory viruses and bacteria can cause an infection that leads to severe pneumonia, and it is now recognized that a high proportion of community-acquired pneumonia is caused by coinfections. Pathogens including influenza viruses, parainfluenza viruses, respiratory syncytial virus (RSV), human metapneumovirus (HMPV), *Streptococcus pneumoniae* (pneumococcus), *Staphylococcus aureus*, and group A streptococcus (*S. pyogenes*, or GAS), and others, alone or in various combinations, cause millions of ambulatory care visits for pneumonia and thousands of deaths each year in the United States. The resulting economic burden is greater than 17 billion dollars (File and Marrie 2010). In addition, otitis media is the leading reason for visits to a pediatrician (2.4 million visits per year) (Centers for Disease Control Deaths and Mortality, Centers for Disease Control Estimated Burden of Acute Otits Externa)], further increasing the health care cost of these pathogens. Although otitis media has classically been considered a bacterial disease, an increasing amount of evidence suggests that viral infections are a common cause and a great deal of acute otitis media (AOM) results from coinfections with two or more pathogens (Heikkinen 2000).

Of the multiple viruses and bacteria that participate in coinfections of the lung and middle ear, one of the most important is the influenza virus. Although influenza is a major public health threat on its own, bacterial coinfections complicating influenza contribute greatly by exacerbating disease severity. Detailed descriptions of fatal cases date as far back as the eighteenth century (Laennec 1923) indicating that viral-bacterial coinfections have been recognized as being prevalent for hundreds of years. Since then, further study has taken place. The most infamous event was during the "Spanish Flu" pandemic in 1918–1919 where more than

95 % of the 50+ million deaths were complicated by a bacterial coinfection (Morens et al. 2008). Although significant improvements regarding health care have been made, new pathogenic strains emerge and complications from bacterial coinfections continue. Approximately 50–70 % of severe or fatal cases in the 1957 H2N2 and 1968 H3N2 pandemics and nearly one-third of those in the 2009 H1N1 pandemic had bacterial complications (Louria et al. 1959; Weinberger et al. 2011). Furthermore, when a bacterial coinfection was identified, mortality was high despite appropriate antibiotic use in the majority of cases (Domínguez-Cherit 2009; Kumar 2009; Jain et al. 2009; Palacios et al. 2009). Today, it is well recognized that bacterial pneumonia complicates disease initiated by respiratory pathogens like influenza viruses.

Pneumococcus remains the most frequently identified bacterial pathogen associated with influenza infections and the most common cause of community-acquired pneumonia (CAP) despite use of the pneumococcal conjugate vaccine (PCV) in children and adults (Nelson et al. 2008). However, over the last decade, *S. aureus* dominated influenza-associated childhood fatalities in the US and accounted for ~75 % of deaths from bacterial coinfections. *S. aureus* has likely become a more common cause of fulminant coinfections due to the emergence in some countries of the methicillin-resistant clonotypes USA300 and USA400 (MRSA). It is unclear why these strains are more likely lead to secondary pneumonia with influenza than other circulating strains. There is currently no vaccine for *S. aureus*. Group A streptococcus only occasionally complicates viral infections (Chaussee et al. 2011) and, when present, falls behind pneumococcus and *S. aureus* in prevalence.

There is little systematic surveillance of bacterial coinfections during seasonal influenza, but this continued threat to public health has led to increased research on the co-pathogenesis of pneumonia due to influenza viruses and bacterial pathogens [reviewed in (Short et al. 2012a; Bosch et al. 2013; Metzger and Sun 2013; McCullers 2014)]. This research has significantly improved the current state of knowledge of influenza coinfections through the use of animal models and, more recently, through the use of theoretical models. Key questions regarding transmission, invasion, and pathogenicity remain unanswered. Identifying how a bacterial coinfection renders mild influenza infections fatal is key to effectively combating pneumonia and preparing for future influenza pandemics.

2 Animal Models to Study Influenza-Bacterial Coinfections

During the 1918 influenza pandemic, the armed forces of several countries made detailed accounts of infectious disease-related illnesses since their efforts during World War I were severely impacted (Brundage and Shanks 2007; Shanks et al. 2010). This led to the first animal studies confirming that bacteria contribute to disease during influenza virus infections by using filtered and unfiltered human sputum (Wherry and Butterfield 1921). These experiments were followed in 1931

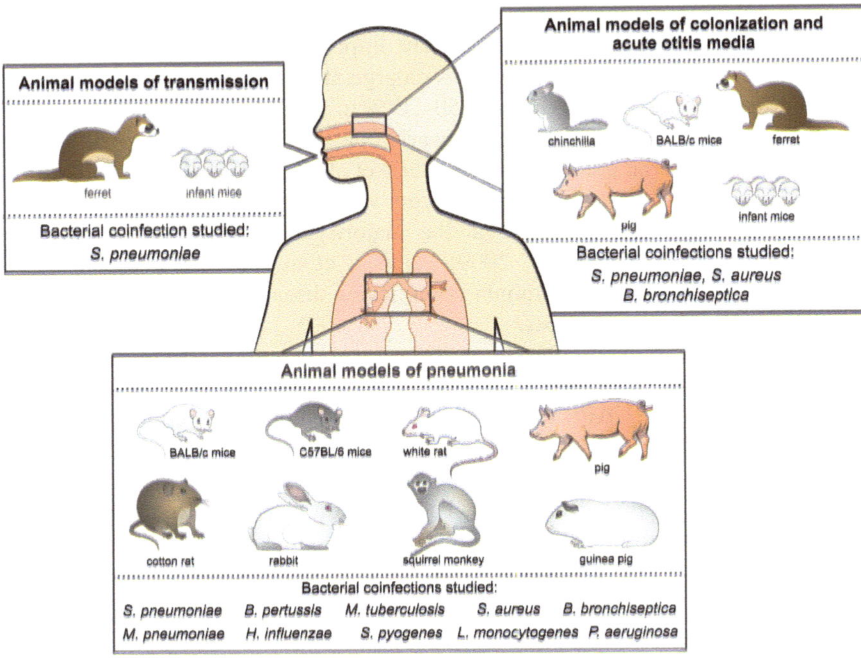

Fig. 1 Animal models of influenza-associated bacterial coinfections. Several different animal models have been used to study the effect that influenza viruses have on bacterial transmission and colonization and on invasive diseases, such as acute otitis media and pneumonia (Wherry and Butterfield 1921; Shope 1931; Francis and de Torregrosa 1945; Berendt et al. 1975; Rarey et al. 1987; Hajek et al. 1999; Hirano et al. 1999; McCullers and Rehg 2002; Okamoto et al. 2003; Seki et al. 2004; Peltola et al. 2006; Montgomery et al. 2008; Small et al. 2010; Diavatopoulos et al. 2010b; Lee et al. 2010; Jamieson et al. 2010; McCullers et al. 2010; Loving et al. 2010; Iverson et al. 2011; Kudva et al. 2011; Ayala et al. 2011; Chaussee et al. 2011; Short et al. 2011; Mina et al. 2013; McHugh et al. 2013; Redford et al. 2014)

by Shope, who conducted controlled experiments in pigs with a swine influenza virus and *Haemophilus influenzae suis* (Shope 1931), and in the 1940s by Francis and Torregrosa, who used a mouse model with the mouse-adapted influenza A/Puerto Rico/8/1934 virus together with pneumococcus, *S. aureus*, or *H. influenzae* (Francis and de Torregrosa 1945). Since then, a variety of animal models have been used to study coinfections (Fig. 1).

The sequential viral-bacterial mouse model of pneumonia, which we characterized in detail in 2002, remains the most useful and well-defined system for investigating coinfections, particularly given the lack of comprehensive data from natural infections in humans. In the initial model, sublethal doses of PR8 and of a type 2 laboratory strain of pneumococcus (D39) reproducibly caused severe secondary bacterial pneumonia when given intranasally in BALB/c mice (McCullers and Rehg 2002). The influenza virus infection had to precede the bacterial challenge to observe synergistic disease. An interval of 3–14 days between inoculations with the organisms resulted in the most severe disease, and peak severity occurred when pneumococcus was given 7 days postviral infection. Simultaneous administration of the two pathogens had only additive effects on morbidity, rather than the synergistic effects observed during the sequential infection. This model was later improved by engineering pneumococcal strains to express luciferase, which allows for quantitative bioluminescent imaging to track progression of the infection in live mice (McCullers and Bartmess 2003).

Multiple strains of influenza, including the 2009 H1N1 pandemic virus, can prime mice for secondary pneumonia (Wanzeck et al. 2011), but the doses necessary to have comparable results differ in a strain-dependent fashion. In addition, other viruses (e.g., rhinovirus, adenovirus, coronavirus, parainfluenza virus, HMPV, and RSV) have been used within the same model framework (reviewed in (Bosch et al. 2013). A variety of clinical outcomes can be modeled with different pneumococcal strains, including pneumonia with and without bacteremia, sepsis with secondary seeding of organs leading to pneumonia, otitis media, and sinusitis (Peltola et al. 2005; McCullers et al. 2007; Smith et al. 2007). Furthermore, multiple bacterial species can synergize with influenza viruses to cause disease (Fig. 1).

The mouse model for influenza-bacterial coinfections has several limitations. For example, viruses that replicate well in mice are required to produce robust and reproducible effects, a limitation that affects the certainty with which conclusions can be extrapolated to humans. This is mitigated somewhat by using different species of mice, including the C57BL/6 strain, which behaves similarly to the BALB/c strain (Karlstrom et al. 2011), and the DBA/2 strain, which is highly permissive to a variety of human influenza strains (Alymova et al. 2011). In addition, a ferret model can be used to confirm results found using the mouse model or to answer questions about strain-related differences in pathogenesis since ferrets are susceptible to most human viruses and exhibit a disease course similar to humans (Peltola et al. 2006; McCullers et al. 2010).

Chinchillas and weanling ferrets can also be infected with a variety of pneumococcal strains (Hajek et al. 1999; McCullers et al. 2010), although the disease manifestations do not map precisely to the mouse model. There are limited data with other viruses and bacteria in the ferret model, but unpublished experience from our laboratory has shown that *S. aureus* will not cause respiratory infections in ferrets even when the animals are preinfected with influenza. Another limitation of the mouse model is the poor transmission potential for respiratory viruses or bacteria between mice, thus requiring the use of alternate models such as neonatal mice (Diavatopoulos et al. 2010a) or ferrets (McCullers et al. 2010) for transmission studies.

Early animal models of AOM utilized the chinchilla due to their large and accessible middle ear spaces (Hajek et al. 1999). These studies demonstrated that the greatest incidence of AOM occurred in animals receiving bacteria 4–8 days following influenza (Hajek et al. 1999), similar to the data concerning timing of pneumonia. More recently, juvenile and infant mouse models have been developed so that diseases of young adults and children, respectively, can be mimicked (McCullers et al. 2007; Diavatopoulos et al. 2010a). Similar models are used to investigate the effects that influenza viruses have on bacterial colonization (Tong et al. 2001; Nakamura et al. 2011).

Studying viral-bacterial interactions in animal models has significantly increased our knowledge about the transmission and pathogenicity of coinfections. However, age, gender, weight, and exposure to anesthesia all contribute to susceptibility to infection in animals in these models, so extreme care must be taken in pathogenesis studies to control all these variables. In addition, studies must carefully select pathogen strains, inoculum sizes, and the sequence and timing of infections since all influence the progression of bacterial pneumonia following influenza.

3 Effect of Influenza Virus Infection on Pneumococcal Transmission

Influenza viruses readily transmit from person to person via small or large respiratory droplets from a sneeze or cough. Successful transmission and infection typically begins 1 day prior to developing symptoms, which can last up to 7 days in adults and 21 days or more in children (World Health Organization Writing Group et al. 2006). While influenza viruses can spread by large droplets up to six feet away, pneumococcal transmission is thought to require close contact of individuals. Recent evidence, however, suggests that this distance can be lengthened if the individual is virus infected. In fact, epidemiological studies found connections between upper respiratory tract (URT) infections, likely of viral origin, and an increase in bacterial transmission and carriage prevalence (Gwaltney et al. 1975; García-Rodríguez and Fresnadillo Martínez 2002; Pettigrew et al. 2008; Murphy et al. 2009; Ansaldi et al. 2012).

Influenza virus' impact on pneumococcal transmission was recently illustrated in the ferret model where transmission events and recipient acquisition increased while the distance necessary for successful bacterial acquisition decreased (McCullers et al. 2010). Both bacterial titers and disease severity intensified in the contact ferrets. This relationship was further examined in the infant mouse model, where influenza virus replication and nasopharyngeal bacterial growth were deemed essential for pneumococcal transmission between littermates (Diavatopoulos et al. 2010b; Short et al. 2012b).

These outcomes may not be seen with all influenza-pneumococcal pairings since all observed effects were both viral and bacterial strain dependent. For instance, H3N2 influenza viruses enhance pneumococcal sinusitis and AOM and induce bacterial colonization and disease more frequently than H1N1 or influenza B viruses (Peltola et al. 2006; Short et al. 2013b). Similarly, colonization and AOM development were greater with pneumococcal serotype 19F compared to serotype 7F (McCullers et al. 2010).

Although the precise mechanisms responsible for enhancing the transmission profile that influenza viruses provide pneumococci are currently unknown, it is likely due to an increase in pathogen density and frequency of secretion events (e.g., sneezing and coughing) in the infected individual combined with a decrease in immunity and resistance from natural barriers breaking down in the person who is newly exposed.

4 Mechanisms of Interaction Between Influenza Viruses and Bacterial Pathogens

During an influenza virus infection, the respiratory tract environment is primed for efficient bacterial invasion. Natural physiological barriers are compromised and a heightened state of inflammation is reached. Numerous factors dictate whether an individual develops a mild or serious infection. The time between exposure to the virus and the bacteria and the pathogen strain and inoculum size all influence influenza coinfection pathogenesis. In addition, many of the virulence factors expressed by each viral and bacterial pathogen act in strain-specific and site-specific manners and can favor different outcomes. The extensive, and growing, list of possible mechanisms (Fig. 2) emphasizes the need to understand how each interacts and how to effectively combat the disease.

4.1 Influenza Virus Effects on Physiologic Barriers to Bacterial Invasion

As an influenza virus infection progresses, respiratory tract damage accumulates and primes the damaged and undamaged areas for bacterial colonization due to disrupted mechanical clearance mechanisms and exposed receptors. Airway damage from overexuberant inflammatory responses and disruption of specific immune responses to viral pathogens leave the airways suitable for invasion by bacterial pathogens (reviewed in Short et al. 2012a; Bosch et al. 2013; Metzger and Sun 2013; McCullers 2014).

The host depends on the mucociliary apparatus in the lung and nasal passages to clear invading pathogens, but viral insults can damage the respiratory epithelium and inhibit this mode of removal (Pittet et al. 2010). Receptors [e.g., plate-activating factor receptor (PAFr) (Cundell et al. 1995; Miller et al. 2007)] permissive to attachment of bacterial invaders become exposed in these inflamed areas, as shown by autopsy studies in humans and in vivo infections in mice (Giles and Shuttleworth 1957; Oseasohn et al. 1959; Herzog et al. 1959; Plotkowski et al. 1986, 1993; Louie et al. 2009). Additional adhesion sites in the lung appear as the viral lesions begin to heal. Pneumococcus, *H. influenzae*, and *S. aureus*, in particular, all use bacterial adhesions to bind exposed laminin, type I and IV collagen, and fibrin/fibrinogen deposition in areas of incomplete healing (Fainstein et al. 1980).

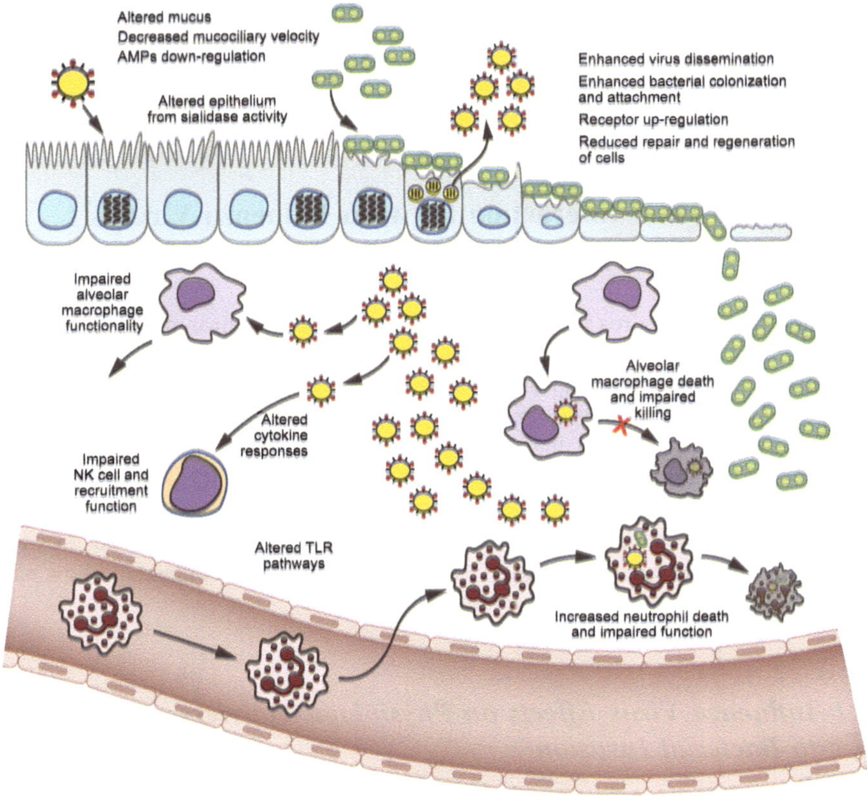

Fig. 2 Influenza-bacterial interaction during coinfections. Numerous alterations of the respiratory epithelium and host immune responses occur during influenza virus infection that predisposes a host to coinfection with bacterial pathogens. As influenza virus infects and kills host cells, epithelial surfaces become exposed and permissive to bacterial attachment. Physical barriers (e.g., mucociliary transport) are damaged, pathogen detection is decreased, antimicrobial peptides (AMPs) are downregulated, receptors are upregulated, virus production is enhanced, bacterial transepithelial migration is permitted, and repair mechanisms are lost. Several host responses are also dampened, altered, or removed. Alveolar macrophages, neutrophils, dendritic cells, and NK cells have altered cytokine profiles and become impaired and/or depleted. These changes result in a heightened inflammatory environment with decreased bacterial surveillance and eradication

Injured or differentiating cells also provide new sites on apical receptors [e.g., asialylated glycans or integrins) for both *S. aureus* and *Pseudomonas aeruginosa* (reviewed in Puchelle et al. 2006).

This increased attachment within the lung, trachea, and nasopharyngeal surfaces may be mediated, at least in part, by viral neuraminidase (NA) activity (Hirano et al. 1999; Peltola et al. 2005), which facilitates bacterial adherence by exposing host cell receptors and providing decoy receptors when sialylated mucins are disrupted. Some bacteria, like pneumococcus, use their own NAs to access receptors and cleave sialic acids to avoid host defenses and prevent mucociliary clearance, replacing, or complementing antecedent viral infections (Camara et al. 1991).

Although decreased mucociliary transport impacts bacterial access to the middle ear, receptor-mediated mechanisms may not be as relevant in AOM. In neonatal mice infected with influenza virus, bacteria can be seen localizing to inflammatory infiltrates, rather than to the epithelium (Short et al. 2011), suggesting that different mechanisms are driving the enhanced bacterial replication. It has been hypothesized that viral-mediated destruction still has a role, but in the context of nutrient availability rather than receptor upregulation (Short et al. 2011).

The influenza glycoprotein hemagglutinin (HA), which binds sialic acid residues on host cell surfaces and aids viral internalization, has an indirect effect on influenza-pneumococcal synergy within the middle ear (Short et al. 2013b). The HA specificity is sufficient to produce differential viral replication and bacterial localization. Here, H3 viruses have higher replicative ability than H1 viruses, even across various NAs, but the effect is site specific and does not depend on cell tropisms (Short et al. 2013a). It is likely that similar mechanisms dictate each type of infection with both the HA and NA having specific roles, but the interactions remain complex. It does, however, help to explain the differential outcomes that different influenza strains have on pneumococcal coinfection.

4.1.1 Bacterial Invasion of the Nasopharynx and Middle Ear

Various bacterial species frequently colonize the nasopharynx and reach a balance with the mucosal immune response such that they are not harmful to the host. In this state, bacteria exist in either biofilms or move between intracellular and extracellular states. Most often, colonizing strains remain in the upper airways since movement to the lower respiratory tract is inhibited by physical barriers (e.g., mucociliary mechanisms) and by immune responses (e.g., resident immune cells, complement, and mucosal antibodies). In this manner, the breakdown of physical barriers and disruption of host responses can result in bacteria emerging from biofilms (Marks et al. 2013).

Most invasive infections and pneumonia occur within a short period of time after a new strain is acquired from the environment rather than from a long-term carriage isolate disseminating to other sites in the body. This is likely due to colonizing strains being limited by systemic immunity, such as pathogen-specific serum IgG, which prevents successful invasion of the lower airways but may tolerate carriage [reviewed in (McCullers et al. 2010)]. In the mouse model, bacteria are delivered directly to the lung, which mimics a direct-inhalation scenario. It is unclear whether this scenario is physiologically relevant in humans, or whether some period of colonization of the nasopharynx must occur prior to invasion and dissemination [reviewed in (McCullers 2014)]. The inoculum size and volume, and the length of anesthesia all influence how much bacteria reach the lower airways and thus the disease model being studied. Both the ferret and mouse AOM models, on the other hand, use colonization as a prerequisite for pneumonia or AOM, so a more natural infection for this site can be instigated.

Influenza viruses and live-attenuated influenza vaccines (LAIV) can result in prolonged bacterial colonization and enhanced bacterial replication within the nasopharynx in both mice and ferrets (Peltola et al. 2006; Nakamura et al. 2011; Short et al. 2012b; Mina et al. 2013). This may be mediated by type I interferon responses and bacterial toxins, such as pneumococcal pneumolysin (Nakamura et al. 2011). Nasopharyngeal colonization can result in bacterial migration to the middle ear via the Eustachian tube. In mice AOM models, animals become colonized with pneumococcus within 72 h after inoculation and can experience recurrent episodes when virus infected (McCullers et al. 2007). Of those that developed AOM ($\sim$70 %), resolution occurred within 48 h but colonization persisted for nearly 30 days (McCullers et al. 2007). In the chinchilla model, animals infected with influenza virus experience negative middle ear pressure and eardrum inflammation associated with epithelial damage and cellular and mucosal debris accumulation in the Eustachian tube (Giebink et al. 1987). Similarly, both virus-mediated inflammation and hearing loss are observed in neonatal mice and ferrets (Rarey et al. 1987; Short et al. 2011). However, in contrast to the findings in chinchillas, minimal Eustachian tube damage is observed and bacterial localization specificity suggests that invasion techniques observed in other models are unlikely to be relevant to bacterial AOM (Short et al. 2011). Nevertheless, the increased pathology may support bacterial replication in the middle ear.

4.1.2 Bacterial Invasion of the Lung

If bacteria are successful in migrating to the lungs, a combination of increasing bacterial burden and an accompanying, intense inflammatory response may result in the host developing pneumonia. The early stages of pneumonia are marked by capillary congestion and fluid in the alveolar regions, which provides a medium where pneumococci can readily grow [reviewed in (McCullers 2001)]. As blood vessels become permeable, inflammatory cells are allowed to enter the lung, receptors become upregulated and bacteria easily adhere to, invade, and kill epithelial cells. The combined effects result in significant inflammation, a hallmark of pneumonia.

During coinfections, the host is in a relative state of immune dysregulation with heightened inflammatory and anti-inflammatory responses [reviewed in (Short et al. 2012a; Bosch et al. 2013; Metzger and Sun 2013; McCullers 2014)] likely due to expression of various pathogenic factors. Bacterial cytotoxins, like the pneumococcal pneumolysin (Tuomanen et al. 1995; Kadioglu et al. 2008), *S. aureus* panton-valentine leukocidin (PVL) (Niemann et al. 2012) and *B. pertussis* toxin (PT) (Ayala et al. 2011), are known to influence host inflammation and may work in concert with viral cytotoxins. These bacterial factors may intensify the cell death and inflammatory signaling resulting from pores formed by the influenza cytotoxic protein, PB1-F2 (Chen et al. 2001).

The PB1-F2 protein of some influenza viruses increases pathologic effects by causing cell death, increasing viral replication, and altering inflammatory responses to primary viral infections and to bacterial coinfections (Conenello et al. 2007; McAuley et al. 2007, 2010a, b; Smith et al. 2011a, 2013). PB1-F2 can act in a proapoptotic fashion due to its mitochondrial targeting sequence and ability to form pores when interacting with membrane-based proteins (Chen et al. 2001; Gibbs et al. 2003; Chanturiya et al. 2004; Zamarin et al. 2005; Danishuddin et al. 2010; McAuley et al. 2010a). This likely results in the death of epithelial cells and immune cells, which may balance the high replicative ability and support rapid spread through cell monolayers thereby contributing to virulence in vivo (Zamarin et al. 2006; McAuley et al. 2010a, b; Smith et al. 2011a; Varga et al. 2011, 2012). Kinetic analyses suggest that this mechanism impacts viral loads during the later stages of the influenza infection, but is overshadowed by more prominent mechanisms during secondary bacterial infections (Smith et al. 2011a, 2013).

These cellular effects have been mapped to a specific set of amino acids in the C-terminal end of the protein, which are found in most of the early twentieth century H1N1 strains (McAuley et al. 2010a). Although rare (Hai et al. 2010), a serine at position 66 (i.e., '66S polymorphism') impacts virulence with highly pathogenic strains with full-length PB1-F2s (e.g., 1918 H1N1, H5N1) but not less pathogenic strains with truncated PB1-F2s (e.g., 2009 H1N1) (Conenello et al. 2007, 2011; Hai et al. 2010; Varga et al. 2011, 2012). The 66S polymorphism facilitates binding of PB1-F2 to the mitochondrial antiviral-signaling (MAVS) protein adaptor protein and subsequent inhibition of interferon production (Varga et al. 2011, 2012). As a result, viral virulence in primary infection and secondary bacterial infection models is severely exacerbated.

The most relevant PB1-F2 mechanism may be its ability to modulate the immune response during influenza infections and coinfections. The high proinflammatory activity of PB1-F2 intensifies disease in animal models, particularly with respect to induction and severity of bacterial coinfections (McAuley et al. 2007, 2010a; Alymova et al. 2011; Weeks-Gorospe et al. 2012), and is marked by a large influx of immune cells and cytokine storm (Conenello et al. 2007; McAuley et al. 2007, 2010a). Pathogenic PB1-F2s, such as that from the 1918 pandemic strain, elevate neutrophils and macrophages and contribute to the pathologic tissue destruction observed during bacterial coinfections (McAuley et al. 2007). This is likely due to regulation of the type I interferon response (le Goffic et al. 2010; Conenello et al. 2011; Varga et al. 2011, 2012) and apoptotic monocytes infected with influenza (Chen et al. 2001; Gibbs et al. 2003; Zamarin et al. 2005). Specific molecular signatures that facilitate this inflammatory environment have been identified (McAuley et al. 2007, 2010b). Amino acids 62L, 75R, 79R, and 82L in the C-terminal portion of PB1-F2 of select strains are positively associated with inflammation and hypercytokinemia in infected animals and negatively correlated with survival. The precise mechanism and contribution of these signatures singly or in combination is unclear.

Pathogenicity of the 1918 H1N1 pandemic strain was likely impacted by possession all four inflammatory signatures together with the 66S polymorphism. While the 1957 H2N2 and 1968 H3N2 pandemic strains also had all four amino acids, both lacked the 66S polymorphism. Subsequent circulating strains became truncated in the H1N1 lineage around 1948 (McAuley et al. 2010a) and mutations in the H3N2 lineage resulted in loss of the inflammatory signatures by the 1980s. In fact, an antibacterial phenotype emerged in the H3N2 lineage such that viruses and invading bacteria compete via PB1-F2 expression rather than synergize (Alymova et al. 2011; Weeks-Gorospe et al. 2012). The expression of these signatures and the length of PB1-F2 proteins vary widely within influenza viruses. In general, viruses with only the 66S signature or a full set of the polymorphisms strongly support bacterial pathogens, while truncation results in an intermediate phenotype (Zell et al. 2007) and a single substitution at position 82 confers anti-inflammatory properties (Weeks-Gorospe et al. 2012). The range of phenotypes highlights the intricate and complicated nature of influenza infections and coinfections and links specific molecular signatures to pathogenicity. It is important to note that each of these effects was demonstrated in animal models using specific viruses [e.g., H3N2 (Alymova et al. 2011), swine H1N1, H1N2, H3N2 (Weeks-Gorospe et al. 2012)]. The H5N1 viruses are of particular interest for further study since they have little diversity in their PB1-F2, are typically full length, and are more likely to possess the full inflammatory panel with the exception of the 66S polymorphism (Smith and McCullers 2013).

4.2 Influenza Virus Effects on Host Immune Responses

A key prerequisite for bacterial invasion into respiratory epithelium is the induction of inflammation. Several immune responses are activated and act to control bacterial pathogens that invade the lung (Joyce et al. 2009; Koppe et al. 2012). A robust initial response is sufficient to immobilize bacterial invaders before full establishment and uncontrolled growth puts the host in a harmful inflammatory state. The degree of attack and the initial replicative ability within mice are dose-dependent and occur in the lung only when alveolar macrophages become overwhelmed with bacteria (Smith et al. 2011b).

For small inocula of bacteria, resident macrophages provide the first line of defense and result in rapid elimination of bacterial pathogens while maintaining homeostasis, which is represented by a low inflammatory state (Knapp et al. 2003; Dockrell et al. 2003 Smith et al. 2003, 2011b). However, given a large invasion or a compromised host state (e.g., influenza virus infected), bacterial outgrowth occurs and an inflammatory response is launched. Neutrophils appear first and are followed shortly by inflammatory macrophages (Jonsson et al. 1985; Fillion et al. 2001; Knapp et al. 2003). The inflammatory influx in the lungs results from bacterial recognition by antigen presenting cells (APCs), and subsequent cytokine and chemokine production. Bacterial phagocytosis by these cells is only efficient if ample complement proteins are available to opsonize the pathogen or if type-specific antibody is made available by B-cells.

Respiratory viruses compromise many aspects of the early detection and response to bacterial pathogens like pneumococcus or *S. aureus* [reviewed in (Short et al. 2012a; Bosch et al. 2013; Metzger and Sun 2013; McCullers 2014)]. Viruses and bacteria also activate many of the same cytokines, inflammatory cells, and pattern recognition receptors (e.g., TLR4) that can synergize during coinfections and generate inflammation (Navarini et al. 2006; Joyce et al. 2009; Karlstrom et al. 2011; Kuri et al. 2013). Interference of immune responses occurs through various manners, such as by viral expression of multifunctional proteins like the influenza virus NS-1 (Hale et al. 2008) and PB1-F2 (McAuley et al. 2007, 2010a). Depending on the stage of influenza, the innate, cellular, and anergic responses may differentially synergize.

4.2.1 Inflammation in Pneumonia

Since phagocytic cells are critical in creating a bactericidal environment, it is not surprising that these cells are impacted by viral and bacterial mechanisms when secondary infections occur. The activity of neutrophils and macrophages is dampened along with their cytokine production as natural killer (NK) cells become impaired during influenza virus infection and undergo additional suppression during coinfections as a result of heightened TNF-α expression (Small et al. 2010). Further functional suppression of these cells occurs when type I interferons (IFN-α,β) place epithelial cells in antiviral states and alter their chemotactic functions (Joyce et al. 2009; Shahangian et al. 2009; le Goffic et al. 2010; Conenello et al. 2011; Nakamura et al. 2011; Tian et al. 2012; Li et al. 2012). For instance, neutrophil chemoattractants KC and MIP-2 (Shahangian et al. 2009) and macrophage chemoattractant CCL2 (Nakamura et al. 2011) all become downregulated, which inhibits recruitment of immune cells leading to inefficient bacterial clearance. IFN-α,β may also decrease Th-17 cytokines IL-17, IL-22, and IL-23 during *S. aureus* coinfection, which increases inflammation and decreases viral and bacterial clearance (Kudva et al. 2011).

Production of interferon-γ increases during influenza resolution and can downregulate bacterial scavenger receptors (e.g., MARCO) on macrophages leaving phagocytic cells suppressed and cytokine profiles altered (Didierlaurent et al. 2008; Sun and Metzger 2008). Additional proinflammatory [e.g., IL-1β, TNF-α, IL-6, and IL-12 (Seki et al. 2004; Smith et al. 2007; Shahangian et al. 2009; Nakamura et al. 2011; McHugh et al. 2013)] and anti-inflammatory cytokines [e.g., IL-10 (van der Sluijs et al. 2004)] become inflated and further compound downstream events like macrophage and neutrophil recruitment and dendritic cell function during influenza-pneumococcal coinfection (Sun and Metzger 2008; Shahangian et al. 2009; Wu et al. 2011; Nakamura et al. 2011; Kuri et al. 2013). Even viral and bacterial pathogens themselves can induce apoptosis of phagocytic cells (Colamussi et al. 1999; Engelich et al. 2001; Kobayashi et al. 2003, 2010; McNamee and Harmsen 2006), thus leaving the infected areas severely damaged.

Many of these responses require time to activate, thus the susceptibility of a host to bacterial pathogens following influenza virus infection indicates an effect that may be fully realized during viral preinfection. Indeed, novel analyses of coinfection kinetics identified and detailed the dominant mechanism driving influenza-pneumococcal synergy as a direct viral-dependent reduction in bacterial phagocytosis by alveolar macrophages (Smith et al. 2013). We predicted that this phagocytosis is reduced by 85–90 % at day 7 of the influenza virus infection. We later confirmed that this was a major driver of influenza-pneumococcal synergy with a mouse model by labeling and tracking these cells before and during influenza infections (Ghoneim et al. 2013). Our experiments showed that the resident macrophage population declines as influenza progresses, suggesting that influenza virus directly depletes these cells rather than simply reducing their function. Furthermore, bacterial outgrowth correlated to the level of depletion, which offers new insight into why the timing of bacterial infection has a profound impact on disease outcome (Ghoneim et al. 2013).

As the vigorous antiviral inflammatory response begins to subside, a new state of innate immune activation that may alter responsiveness to new pathogenic insults is reached. The lung becomes repopulated with resident alveolar macrophages as recruited macrophages proliferate and differentiate. In an attempt to return the lung to homeostasis, wound-healing processes coordinate an anti-inflammatory response characterized by IL-10 (van der Sluijs et al. 2004; Hussell and Cavanagh 2009) and suppress pathogen recognition systems [reviewed in (Metzger and Sun 2013)].

During the recovery phase, the host becomes immunologically desensitized both locally and systemically (van der Sluijs et al. 2004; Didierlaurent et al. 2008), which can last for several weeks and prolongs the opportunity for bacterial invasion. The degree and length of this suppression is viral strain-dependent (Ludewick et al. 2011), and occurs through diverse mechanisms. For instance, alveolar macrophages with high expression of homeostatic moieties such as CD200R, a regulatory anti-inflammatory ligand (Barclay et al. 2002; Minas and Liversidge 2006; Snelgrove et al. 2008; Jiang-Shieh et al. 2010), become desensitized when expression of CD200 on apoptotic immune cells increases and open the airways to bacterial invasion (Goulding et al. 2011). In conjunction, absence of CD200R in mice inhibits bacterial outgrowth and prevents migration of bacteria to exogenous sites, such as the blood, in influenza-infected mice (Goulding et al. 2011). Elevated glucocorticoid levels also cause sustained immunosuppression, as was demonstrated in a model of *Listeria monocytogenes* coinfection (Jamieson et al. 2010).

4.2.2 Inflammation in Acute Otitis Media

The inflammatory nature of influenza-pneumococcal coinfections extends to middle ear invasions. Middle ear inflammation, regardless of pathogenic origin, is sufficient to induce AOM. Influenza viruses can initiate this inflammation (Abramson et al. 1981, 1982; Short et al. 2013b), cause hearing loss and instigate bacterial growth within the ear cavity in a viral strain-dependent, but cell tropism independent, manner (Short et al. 2013a).

Inflammation in AOM is characterized by an influx of neutrophils and expression of key proinflammatory genes (i.e., pro-IL-1β, IL-1α, and CXCL2) (Abramson et al. 1981, 1982; Short et al. 2011, 2013b). Chinchilla studies indicate that immunosuppression, rather than inflammation, is the key mechanism contributing to enhanced pneumococcal replication since influenza viruses can inhibit neutrophils and render clearance ineffective (Abramson et al. 1981, 1982). However, it has also been hypothesized that the enhanced bacterial growth results from nutrients becoming available in areas damaged by this response (Short et al. 2013b). More experiments are clearly necessary to elucidate the underlying relationship between viral replication and inflammation in AOM.

4.3 Bacterial Effects on Influenza Virus Clearance

The mechanisms discussed thus far have detailed how the virus affects host responses to invading bacteria. The relationship is somewhat complementary, however, since rebounds in viral load and reduced viral clearance are consistently observed in animal models (McCullers and Rehg 2002; Iverson et al. 2011; Weeks-Gorospe et al. 2012; Smith et al. 2013). The rapidity of the viral reply suggests a fast-acting mechanism, which may occur through direct interactions of the two pathogens, bacterial interference with antiviral immunity, or virulence factors synergizing (Smith et al. 2013). Kinetic studies suggest that pneumococci directly interact with influenza-infected epithelial cells to cause a sudden release of virus (Smith et al. 2013), but experimental studies have not been crafted to confirm this prediction.

The precise effects of bacterial virulence factors are likely type specific as *S. aureus* may be capable of cleaving influenza hemagglutinin to enhance invasion of host cells and thus impact viral load (Tashiro et al. 1987). It is feasible that additional, although unknown, bacterial virulence factors have immune-modulatory effects and could interfere with viral clearance, such as T-cell-mediated infected cell clearance or other innate immune components. Determining if bacterial gene products from various species have similar effects during infection and, if so, how they complement viral factors is an important, but largely unexplored, area. Another interesting area for investigation in animal models is the idea that the respiratory and gastrointestinal microbiome affects the development of antiviral responses to pathogens (Ichinohe et al. 2011; Licciardi et al. 2012; Abt et al. 2012; Wang et al. 2013). Understanding how commensal species influence host immune status may help explain the heterogeneity in responses to pathogenic invasions.

5 Treatment and Prevention

The outcome of influenza virus coinfection is often severe despite appropriate vaccination and treatment. In addition, antiviral and antimicrobial resistance is increasing (Musher et al. 2002; Levy and Marshall 2004; Hayden 2006) and many treatment options have the potential to cause adverse effects on the host (McCullers and English 2008; Karlstrom et al. 2011). With the high prevalence of viral-bacterial coinfections in some situations, discovering treatment options that can prevent or treat both the influenza virus infection and the secondary bacterial infection are of utmost importance.

5.1 Antiviral Treatment

Several antiviral drugs targeted against various influenza virus components have been or are currently being developed [reviewed in (Hayden 2013)]. Neuraminidase inhibitors (NAIs) are one class of drugs that have become the pillar of influenza treatment in recent years. NAIs act to block virus from budding out of infected cells, thereby preventing the spread of virus to neighboring cells (Moscona 2005).

NAI therapy can prevent secondary bacterial pneumonia in animals infected with influenza. Mice given NAI treatment within 72 h postinfluenza infection or prophylactically experienced improved survival due to delayed development of and progression of pneumonia with NAI treatment, decreased viral loads, and reduced secondary bacterial infections. Treatment in the later stages of influenza virus infection (i.e., 5 days) reduces bacterial invasion without any impact on viral loads, signifying that mechanisms independent of replication inhibition are in play (McCullers 2005). Their action may directly, or indirectly, lessen the effects of viral virulence factors, prevent receptor exposure, reduce use of sialic acids as catabolic substrates, and/or activate immunological components in a viral strain-dependent manner [reviewed in (McCullers 2011)].

Although NAI resistance remains problematic and may be reduced with combination therapy, treatment with multiple NAIs can inhibit antiviral efficacy (Duval et al. 2010). NAIs in combination with antibiotics, on the other hand, can facilitate recovery from influenza and alter the coinfection pathogenesis in infected animals (McCullers 2004). Other antivirals [e.g., peramivir and laninamivir (NAIs), favipiravir (RNA polymerase inhibitor)] that are not yet licensed may provide benefit to coinfected hosts, but these have yet to be tested in animal coinfection models.

It is important to note that NAIs specifically target influenza virus NA and do not inhibit bacterial NAs with clinically relevant doses (Nishikawa et al. 2012). While NAIs may reduce incidence of bacterial pneumonia, and thus antibiotic requirements, NAs derived from invasive or commensal bacteria may antagonize their effectiveness (Nishikawa et al. 2012). Thus, NAI therapy could have differential bacterial-dependent effects as well.

5.2 Antibacterial Treatment

Unlike antivirals, which interrupt disease progression by preventing viral spread, antibiotics work to eliminate pathogens directly. Some antibiotics, however, kill bacteria through mechanisms that can have harmful repercussions. For instance, therapy with cell wall active agents (e.g., ampicillin), the mainstay of treatment of community-acquired pneumonia in children (Bradley et al. 2011), causes significant inflammation and lung injury in animal models (Karlström et al. 2009). The characteristic inflammation in secondary bacterial infections is due to immune cells responding to the release of bacterial components, such as cell wall components, during lysis (Karlstrom et al. 2011). Thus, alternative treatments that eliminate pathogens while preserving host integrity are desirable.

Antibiotics that reduce neutrophil influx or cytokines and thus circumvent the inflammatory tissue damage are beneficial in coinfected animals (Karlstrom et al. 2011; Liu et al. 2013). In particular, protein synthesis inhibitors (e.g., clindamycin) and macrolides (e.g., azithromycin) have anti-inflammatory properties in addition to bactericidal activity, thereby clinically curing mice with influenza-associated pneumococcal pneumonia (Karlström et al. 2009). Nevertheless, antibiotic treatment alone is suboptimal.

Anti-inflammatory agents, such as corticosteroids (e.g., dexamethasone), in conjunction with antibiotic therapy improve beta-lactam-induced immunopathology and mortality in animals with severe pneumonia. However, giving dexamethasone prophylactically during influenza infections negatively impacts adaptive immunity and results in reduced viral clearance (Ghoneim and McCullers 2013). Thus, use of immune-modulatory approaches may be best reserved for severe infections where inflammation is driving poor outcomes but avoided in primary viral infections where detrimental effects on the host response may result. Given the benefit of anti-inflammatory treatment and the importance of inflammation in viral-bacterial coinfections, better success may occur if specific inflammatory pathways or pathogenic factors are targeted [reviewed in (McCullers 2011)].

5.3 Vaccination

Vaccination remains fundamental to prevention of influenza and bacterial infections. Data from animal models indicate that vaccinating against influenza viruses effectively circumvents bacterial associated pneumonia (Huber et al. 2010; Chaussee et al. 2011; Mina et al. 2013) but may support colonization and replication in the URT (Mina et al. 2013). An important caveat of current vaccines against influenza viruses is that partial protection of related strains may not be sufficient to alleviate bacterial complications. On the other hand, antibacterial vaccines are important to block the specific bacteria being targeted and to prevent clinically severe influenza infections by reducing the coinfection component.

Vaccinating animals against some bacteria (e.g., pneumococcus, *H. influenzae*) prevents invasive diseases caused by these pathogens, but protection is limited to vaccine-specific types and efficacy may be lost in influenza virus infected animals (Mina et al. 2013). In addition, vaccines against other coinfecting bacteria, such as *S. aureus,* are not currently available. Interestingly, vaccination with a live-attenuated *B. pertussis* vaccine can protect against lethal challenges with influenza virus by controlling cytokine responses that lead to virus-mediated inflammation (Li et al. 2010). Although the vaccine has not yet been approved for use in humans, it is promising and may benefit as a prophylactic agent against infection with influenza viruses.

6 Kinetic Modeling

Unraveling the relationships between pathogen replication and interactions and the resulting airway alterations and inflammation that are driving coinfection host pathology and disease is complicated. Kinetic models are a robust means of analyzing experimental results and explaining biological phenomena without testing every scenario experimentally. They have proven valuable in the identification and characterization of mechanisms driving influenza virus infections, pneumococcal infections, and bacterial coinfection establishment and severity.

6.1 Modeling Influenza Virus Infections

A growing body of work modeling in vivo influenza virus infections has improved our knowledge about the viral life cycle, viral control by the host, pathogenic differences in strains, and efficacy of antiviral treatment [reviewed in (Smith and Ribeiro 2010; Beauchemin and Handel 2011; Smith and Perelson 2011)]. These models have characterized the spread of virus during early infection and yielded estimates of strain-specific viral infection and production rates, infected cell life spans, and infectious virus half-life, all of which are not amenable to experimental investigation.

Most of these studies model data from humans or large animals where only nasal wash titers are available and, thus, are restricted to studying nasopharyngeal infections (Baccam et al. 2006; Saenz et al. 2010; Canini and Carrat 2010; Pawelek et al. 2012). A few models, however, take advantage of data collected in the mouse model, including pathogen and immunological measurements, to study invasive lung infections (Handel et al. 2010; Miao et al. 2010; Smith et al. 2011a). Viral load dynamics can be accurately modeled using target-cell limitation (Fig. 3), through undefined mechanisms, as the primary means of viral control while excluding specific immune responses (Baccam et al. 2006; Smith et al. 2011a). It is important to note that these models do not discount the fact that

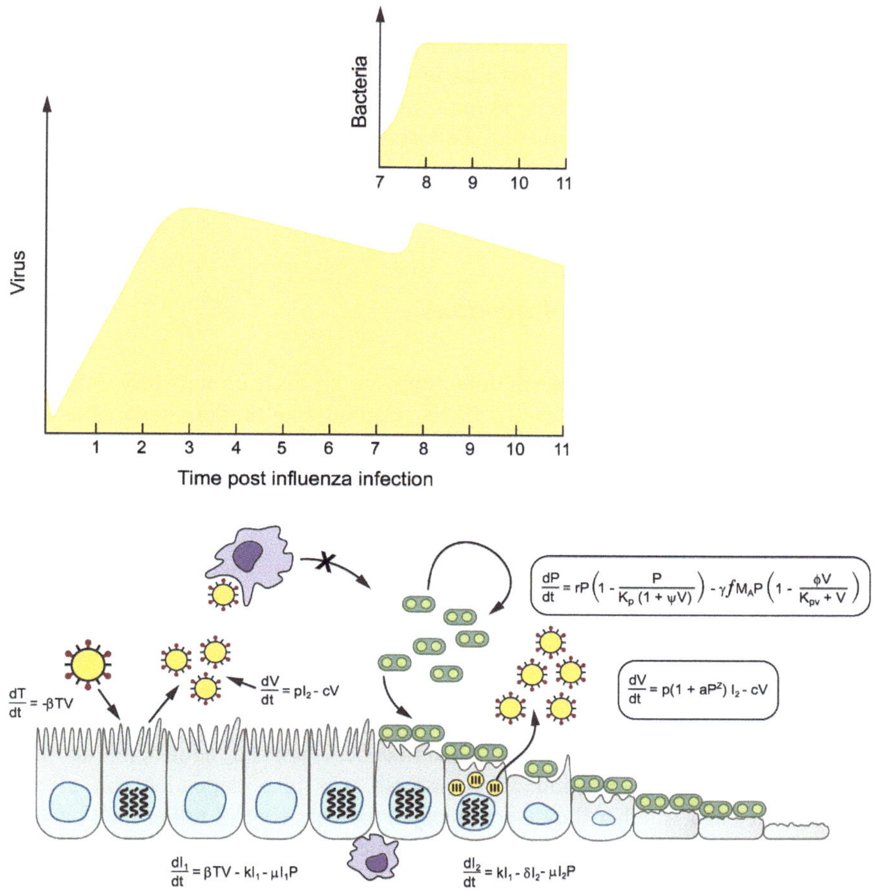

Fig. 3 Kinetics of influenza-pneumococcal coinfection. Model schematic and equations that result in the observed kinetics of influenza virus infection followed by pneumococcus given 7 days postinfluenza infection (Smith et al. 2013). During primary influenza, susceptible epithelial (target) cells (*T*) become infected at a rate βV per cell. Infected cells (I_1) first undergo an eclipse phase at rate *k* per cell prior to entering a state (I_2) in which virus is produced. Productively infected cells are lost, through apoptosis, viral cytopathic effects, or removal by immune cells, at a rate δ per cell. Virus (*V*) is produced at rate *p* per cell, which is significantly increased by bacterial presence (aP^z) (boxed), and cleared at rate *c*. Invading pneumococci (*P*) proliferate at maximum rate *r* with a tissue capacity K_p CFU/ml. Bacteria are cleared via phagocytosis by alveolar macrophages (M_A) at rate γf per cell, which is significantly reduced by virus presence ($\phi V/(K_{PV} + V)$) (boxed). With this kinetic description, viral titers increase exponentially, peak and begin to decline prior to bacterial invasion. Once bacteria are present, a viral rebound occurs and bacteria grow exponentially before reaching a maximum capacity. The potential increase in bacterial adherence to virus-infected cells and any accompanying cell death has little effect are excluded here

immunological factors may drive influenza virus pathogenesis, but that this information can simply not be extracted from viral loads (Smith et al. 2010; Miao et al. 2011). Quantifying the effect that host factors have on viral replication has been restricted by the limited amount of data detailing the innate immune responses [reviewed in (Smith and Perelson 2011)]. As more data arise, new quantitative descriptions of influenza virus kinetics will be developed and will undoubtedly aid experimental interpretation.

6.2 Modeling Bacterial Infections

Kinetic models depicting bacterial infections are an exciting new tool being used to study pathogenesis (Smith et al. 2011b). Capitalizing on data obtainable in the mouse model system, we characterized the dose-dependent innate immune control of a pneumococcal invasion and quantified the contributions of alveolar macrophages, neutrophils, inflammatory macrophages, cytokines, and damage to bacterial pathogenesis (Smith et al. 2011b).

Model analysis revealed the exact thresholds for bacterial establishment, growth, and eradication with alveolar macrophages playing a central role. The dose-dependent invasive ability of pneumococci was solely dependent on the number and phagocytic ability of resident macrophages initially present. Thus, any alterations to resident cells, such as death from an antecedent viral infection, would result in immediate pathogenic invasion. While the rapid neutrophil influx could facilitate bacterial removal, pneumococcal-induced neutrophil apoptosis hindered complete eradication. This process was also dependent on alveolar macrophages and whether they were engaged damage control rather than bacterial clearance. Inflammatory macrophages had little effect on clearance but still contributed to respiratory tract damage. Through this model, we successfully captured the biochemical, cellular, immunological interactions of pneumococci with the host and identify the critical processes driving pathogenesis.

6.3 Modeling Coinfections

Modeling the interactions of two pathogens requires the combination of previously developed single infection models. Thus far, the only model depicting a coinfection is one that we formulated for influenza-pneumococcal coinfection using the models discussed above (Fig. 3) (Smith et al. 2013). We quantitated the enhanced bacterial growth and viral rebound and evaluated prior hypotheses about the interaction between influenza, pneumococcus, and the host.

Careful model development and analyses showed that any enhanced bacterial adherence to epithelial cells, with respect to both invasion and cell death, was negligible compared to the viral-induced impairment of alveolar macrophages.

While the mechanism for this prediction is not available through initial modeling efforts, it did pinpoint the process driving influenza-pneumococcal synergy that should be subject to further examination in the laboratory (Smith et al. 2013). In fact, the decrease in phagocytosis by alveolar macrophage was later determined to be a result of influenza virus directly killing these cells (Ghoneim et al. 2013). Remarkably, both our models (Smith et al. 2013) and experiments (Ghoneim et al. 2013) agreed that these cells diminished to 85–90 % of their baseline level within 7 days.

Receptor-mediated mechanisms may still drive the synergy, although not in the context of enhanced invasion. Our model predicts that bacterial interaction with virus-infected epithelial cells releases virus and thus increases viral loads post-bacterial invasion. Bacterial proteases or NAs may liberate virus in the same manner as viral NA. *S. aureus* proteases can activate influenza virus HA cleavage and enhance viral invasion into host cells (Tashiro et al. 1987). Furthermore, some commensal bacteria [e.g., *S. mitis* (Nonaka et al. 1983; Beighton and Whiley 1990), certain *S. pneumoniae* (Scanlon et al. 1989), *Actinomyces naeslundii* and *A. viscosus* (Moncla and Braham 1989), *Porphyromonas gingivalis* (Moncla et al. 1990), and *S. oralis* (Homer et al. 1996)] that secrete NAs or exogenous NA can rescue influenza virus replication if viral NA is missing or inhibited (Liu and Air 1993; Hughes et al. 2000; Nishikawa et al. 2012). Thus, it is feasible that other NA possessing bacterial species can act in the manner predicted by our model. To uncover the underlying mechanism, a combination of in vitro and in vivo experiments with viruses and bacteria that exhibit differential expression of NA is necessary (Smith et al. 2013).

7 Concluding Remarks and Key Research Questions

It is becoming better appreciated that pneumonia is frequently caused by coinfecting pathogens. Viral-mediated mechanisms are also important in other invasive infections, such as otitis media. The underlying relationship between viral and bacterial density, inflammation, and the host microbiome during influenza coinfections is exceptionally complex. Even with the growing body of work detailing various aspects of viral-bacterial coinfections, determining the precise contributions of each interrelated factor is challenging. Furthermore, studying coinfections has become problematic due to the numerous site-, pathogen-, time-, and host-specific variations to consider. Thus, it is necessary to employ the next generation of analyses using a mixture of animal models and kinetic models with the goal of obtaining results translatable to infections in humans.

Some important areas for consideration include determining how different bacterial virulence factors leverage the environment set forth by influenza viruses to cause disease, how timing of sequential infections impacts each of the aforementioned mechanisms, how the synergistic relationship is facilitated by host genetics, and how each of these factors differ between the nasopharyngeal, middle ear, and lung niches and between different viral and bacterial species. Our understanding of coinfection biology should increase as new and different data emerge. With such data, treatment options suitable for clinical practice are permissive to investigation as we focus on preparation for the next influenza pandemic.

Acknowledgments This work was supported by NIH grant AI100946 (AMS) and ALSAC (JAM). We thank Betsy Williford and Klo Spelshouse (SJCRH Biomedical Communications) for assistance with figure illustration.

References

Abramson JS, Giebink GS, Mills EL, Quie PG (1981) Polymorphonuclear leukocyte dysfunction during influenza virus infection in chinchillas. J Infect Dis 143:836–845. doi:10.1093/infdis/143.6.836

Abramson JS, Giebink GS, Quie PG (1982) Influenza A virus-induced polymorphonuclear leukocyte dysfunction in the pathogenesis of experimental pneumococcal otitis media. Infect Immun 36:289–296

Abt MC, Osborne LC, Monticelli LA et al (2012) Commensal bacteria calibrate the activation threshold of innate antiviral immunity. Immunity 37:158–170. doi:10.1016/j.immuni.2012.04.011

Alymova IV, Green AM, van de Velde N et al (2011) Immunopathogenic and antibacterial effects of H3N2 influenza A virus PB1-F2 map to amino acid residues 62, 75, 79, and 82. J Virol 85:12324–12333. doi:10.1128/JVI.05872-11

Ansaldi F, de Florentiis D, Parodi V et al (2012) Bacterial carriage and respiratory tract infections in subjects > or = 60 years during an influenza season: implications for the epidemiology of community acquired pneumonia and influenza vaccine effectiveness. J Prev Med Hyg 53:94–97

Ayala VI, Teijaro JR, Farber DL et al (2011) Bordetella pertussis infection exacerbates influenza virus infection through pertussis toxin-mediated suppression of innate immunity. PLoS ONE 6:e19016. doi:10.1371/journal.pone.0019016

Baccam P, Beauchemin C, Macken CA et al (2006) Kinetics of influenza a virus infection in humans. J Virol 80:7590–7599. doi:10.1128/JVI.01623-05

Barclay AN, Wright GJ, Brooke G, Brown MH (2002) CD200 and membrane protein interactions in the control of myeloid cells. Trends Immunol 23:285–290. doi:10.1016/S1471-4906(02)02223-8

Beauchemin CA, Handel A (2011) A review of mathematical models of influenza A infections within a host or cell culture: lessons learned and challenges ahead. BMC Public Health 11:S7. doi:10.1186/1471-2458-11-S1-S7

Beighton D, Whiley RA (1990) Sialidase activity of the "Streptococcus milleri group" and other viridans group Streptococci. J Clin Microbiol 28:1431–1433

Berendt RF, Long GG, Walker JS (1975) Influenza alone and in sequence with pneumonia due to Streptococcus pneumoniae in the squirrel monkey. J Infect Dis 132:689–693. doi:10.1093/infdis/132.6.689

Bosch AATM, Biesbroek G, Trzcinski K et al (2013) Viral and bacterial interactions in the upper respiratory tract. PLoS Pathog 9:e1003057. doi:10.1371/journal.ppat.1003057

Bradley JS, Byington CL, Shah SS et al (2011) The management of community-acquired pneumonia in infants and children older than 3 months of age: clinical practice guidelines by the pediatric infectious diseases society and the infectious diseases society of America. Clin Infect Dis 53:e25–e76. doi:10.1093/cid/cir531

Brundage JF, Shanks GD (2007) What really happened during the 1918 influenza pandemic? The importance of bacterial secondary infections. J Infect Dis 196:1717–1718. doi:10.1086/522355

Camara M, Mitchell TJ, Andrew PW, Boulnois GJ (1991) Streptococcus pneumoniae produces at least two distinct enzymes with neuraminidase activity: cloning and expression of a second neuraminidase gene in Escherichia coli. Infect Immun 59:2856–2858

Canini L, Carrat F (2010) Population modeling of influenza A/H1N1 virus kinetics and symptom dynamics. J Virol 85:2764–2770. doi:10.1128/JVI.01318-10

Centers for Disease Control Deaths and Mortality

Centers for Disease Control Estimated Burden of Acute Otits Externa

Chanturiya AN, Basanez G, Schubert U et al (2004) PB1-F2, an influenza A virus-encoded proapoptotic mitochondrial protein, creates variably sized pores in planar lipid membranes. J Virol 78:6304–6312. doi:10.1128/JVI.78.12.6304-6312.2004

Chaussee MS, Sandbulte HR, Schuneman MJ et al (2011) Inactivated and live, attenuated influenza vaccines protect mice against influenza: Streptococcus pyogenes super-infections. Vaccine 29:3773–3781. doi:10.1016/j.vaccine.2011.03.031

Chen W, Calvo PA, Malide D et al (2001) A novel influenza A virus mitochondrial protein that induces cell death. Nat Med 7:1306–1312. doi:10.1038/nm1201-1306

Colamussi ML, White MR, Crouch E, Hartshorn KL (1999) Influenza A virus accelerates neutrophil apoptosis and markedly potentiates apoptotic effects of bacteria. Blood 93:2395–2403

Conenello GM, Tisoncik JR, Rosenzweig E et al (2011) A single N66S mutation in the PB1-F2 protein of influenza A virus increases virulence by inhibiting the early interferon response in vivo. J Virol 85:652–662. doi:10.1128/JVI.01987-10

Conenello GM, Zamarin D, Perrone LA et al (2007) A single mutation in the PB1-F2 of H5N1 (HK/97) and 1918 influenza A viruses contributes to increased virulence. PLoS Pathog 3:e141. doi:10.1371/journal.ppat.0030141

Cundell DR, Gerard NP, Gerard C et al (1995) Streptococcus pneumoniae anchor to activated human cells by the receptor for platelet-activating factor. Nature 377:435–438. doi:10.1038/377435a0

Danishuddin M, Khan SN, Khan AU (2010) Molecular interactions between mitochondrial membrane proteins and the C-terminal domain of PB1-F2: an in silico approach. J Mol Model 16:535–541. doi:10.1007/s00894-009-0555-5

Diavatopoulos DA, Short KR, Price JT et al (2010a) Influenza A virus facilitates Streptococcus pneumoniae transmission and disease. FASEB J 24:1789–1798. doi:10.1096/fj.09-146779

Diavatopoulos DA, Short KR, Price JT et al (2010b) Influenza A virus facilitates Streptococcus pneumoniae transmission and disease. FASEB J 24:1789–1798. doi:10.1096/fj.09-146779

Didierlaurent A, Goulding J, Patel S et al (2008) Sustained desensitization to bacterial toll-like receptor ligands after resolution of respiratory influenza infection. J Exp Med 205:323–329. doi:10.1084/jem.20070891

Dockrell DH, Marriott HM, Prince LR et al (2003) Alveolar macrophage apoptosis contributes to pneumococcal clearance in a resolving model of pulmonary infection. J Immunol (Baltimore Md 1950) 171:5380–5388

Domínguez-Cherit G (2009) Critically ill patients with 2009 influenza A(H1N1) in Mexico. JAMA 302:1880. doi:10.1001/jama.2009.1536

Duval X, van der Werf S, Blanchon T et al (2010) Efficacy of Oseltamivir-Zanamivir combination compared to each monotherapy for seasonal influenza: a randomized placebo-controlled trial. PLoS Med 7:e1000362. doi:10.1371/journal.pmed.1000362

Engelich G, White M, Hartshorn KL (2001) Neutrophil survival is markedly reduced by incubation with influenza virus and Streptococcus pneumoniae: role of respiratory burst. J Leukoc Biol 69:50–56

Fainstein V, Musher DM, Cate TR (1980) Bacterial adherence to pharyngeal cells during viral infection. J Infect Dis 141:172–176. doi:10.1093/infdis/141.2.172

File TM Jr, Marrie TJ (2010) Burden of community-acquired pneumonia in North American adults. Postgrad Med 122:130–141. doi:10.3810/pgm.2010.03.2130
Fillion I, Ouellet N, Simard M et al (2001) Role of chemokines and formyl peptides in pneumococcal pneumonia-induced monocyte/macrophage recruitment. J Immunol (Baltimore Md 1950) 166:7353–7361
Francis T, de Torregrosa MV (1945) Combined infection of mice with H. influenzae and influenza virus by the intranasal route. J Infect Dis 76:70–77. doi:10.2307/30085688
García-Rodríguez JA, Fresnadillo Martínez MJ (2002) Dynamics of nasopharyngeal colonization by potential respiratory pathogens. J Antimicrob Chemother 50 Suppl S2:59–73
Ghoneim HE, McCullers JA (2013) Adjunctive corticosteroid therapy improves lung immunopathology and survival during severe secondary pneumococcal pneumonia in mice. J Infect Dis. doi:10.1093/infdis/jit653
Ghoneim HE, Thomas PG, McCullers JA (2013) Depletion of alveolar macrophages during influenza infection facilitates bacterial superinfections. J Immunol 191:1250–1259. doi:10.4049/jimmunol.1300014
Gibbs JS, Malide D, Hornung F et al (2003) The influenza A virus PB1-F2 protein targets the inner mitochondrial membrane via a predicted basic amphipathic helix that disrupts mitochondrial function. J Virol 77:7214–7224. doi:10.1128/JVI.77.13.7214-7224.2003
Giebink G, Ripley M, Wright P (1987) Eustachian tube histopathology during experimental influenza A virus infection in the chinchilla. Ann Otol Rhinol Laryngol 96:199–206
Giles C, Shuttleworth EM (1957) Postmortem findings in 46 influenza deaths. Lancet 273:1224–1225
le Goffic R, Bouguyon E, Chevalier C et al (2010) Influenza A virus protein PB1-F2 exacerbates IFN-expression of human respiratory epithelial cells. J Immunol 185:4812–4823. doi:10.4049/jimmunol.0903952
Goulding J, Godlee A, Vekaria S et al (2011) Lowering the threshold of lung innate immune cell activation alters susceptibility to secondary bacterial superinfection. J Infect Dis 204:1086–1094. doi:10.1093/infdis/jir467
Gwaltney JM, Sande MA, Austrian R, Hendley JO (1975) Spread of Streptococcus pneumoniae in families. II. Relation of transfer of S. pneumoniae to incidence of colds and serum antibody. J Infect Dis 132:62–68. doi:10.1093/infdis/132.1.62
Hai R, Schmolke M, Varga ZT et al (2010) PB1-F2 expression by the 2009 pandemic H1N1 influenza virus has minimal impact on virulence in animal models. J Virol 84:4442–4450. doi:10.1128/JVI.02717-09
Hajek DM, Yuan Z, Quartey MK, Giebink GS (1999) Otitis media: the chinchilla model. In: Zak O, Sande MA (eds) Handbook of animal models of infection: experimental models in antimicrobial chemotherapy Academic, San Diego, CA, pp 389–401
Hale BG, Randall RE, Ortin J, Jackson D (2008) The multifunctional NS1 protein of influenza A viruses. J Gen Virol 89:2359–2376. doi:10.1099/vir.0.2008/004606-0
Handel A, Longini IM, Antia R (2010) Towards a quantitative understanding of the within-host dynamics of influenza A infections. J R Soc Interface 7:35–47. doi:10.1098/rsif.2009.0067
Hayden FG (2006) Antiviral resistance in influenza viruses—implications for management and pandemic response. N Engl J Med 354:785–788. doi:10.1056/NEJMp068030
Hayden FG (2013) Newer influenza antivirals, biotherapeutics and combinations: novel influenza antivirals and therapeutics. Influenza Other Respir Viruses 7:63–75. doi:10.1111/irv.12045
Heikkinen T (2000) The role of respiratory viruses in otitis media. Vaccine 19:S51–S55. doi:10.1016/S0264-410X(00)00278-4
Herzog H, Staub H, Richterich R (1959) Gas-analytical studies in severe pneumonia; observations during the 1957 influenza epidemic. Lancet 1:593–597
Hirano T, Kurono Y, Ichimiya I et al (1999) Effects of influenza A virus on lectin-binding patterns in murine nasopharyngeal mucosa and on bacterial colonization. Otolaryngol Head Neck Surg 121:616–621. doi:10.1016/S0194-5998(99)70068-9
Homer KA, Kelley S, Hawkes J et al (1996) Metabolism of glycoprotein-derived sialic acid and N-acetylglucosamine by Streptococcus oralis. Microbiol Read Engl 142(Pt 5):1221–1230

Huber VC, Peltola V, Iverson AR, McCullers JA (2010) Contribution of Vaccine-induced immunity toward either the HA or the NA component of influenza viruses limits secondary bacterial complications. J Virol 84:4105–4108. doi:10.1128/JVI.02621-09

Hughes MT, Matrosovich M, Rodgers ME et al (2000) Influenza A viruses lacking sialidase activity can undergo multiple cycles of replication in cell culture, eggs, or mice. J Virol 74:5206–5212. doi:10.1128/JVI.74.11.5206-5212.2000

Hussell T, Cavanagh MM (2009) The innate immune rheostat: influence on lung inflammatory disease and secondary bacterial pneumonia. Biochem Soc Trans 37:811. doi:10.1042/BST0370811

Ichinohe T, Pang IK, Kumamoto Y et al (2011) Microbiota regulates immune defense against respiratory tract influenza A virus infection. Proc Nat Acad Sci U.S.A 108:5354–5359. doi:10.1073/pnas.1019378108

Iverson AR, Boyd KL, McAuley JL et al (2011) Influenza virus primes mice for pneumonia from Staphylococcus aureus. J Infect Dis 203:880–888. doi:10.1093/infdis/jiq113

Jain S, Kamimoto L, Bramley AM et al (2009) Hospitalized patients with 2009 H1N1 influenza in the United States, April–June 2009. N Engl J Med 361:1935–1944. doi:10.1056/NEJMoa0906695

Jamieson AM, Yu S, Annicelli CH, Medzhitov R (2010) Influenza virus-induced glucocorticoids compromise innate host defense against a secondary bacterial infection. Cell Host Microbe 7:103–114. doi:10.1016/j.chom.2010.01.010

Jiang-Shieh Y-F, Chien H-F, Chang C-Y et al (2010) Distribution and expression of CD200 in the rat respiratory system under normal and endotoxin-induced pathological conditions. J Anat 216:407–416. doi:10.1111/j.1469-7580.2009.01190.x

Jonsson S, Musher DM, Chapman A et al (1985) Phagocytosis and killing of common bacterial pathogens of the lung by human alveolar macrophages. J Infect Dis 152:4–13. doi:10.1093/infdis/152.1.4

Joyce EA, Popper SJ, Falkow S (2009) Streptococcus pneumoniae nasopharyngeal colonization induces type I interferons and interferon-induced gene expression. BMC Genomics 10:404. doi:10.1186/1471-2164-10-404

Kadioglu A, Weiser JN, Paton JC, Andrew PW (2008) The role of Streptococcus pneumoniae virulence factors in host respiratory colonization and disease. Nat Rev Microbiol 6:288–301. doi:10.1038/nrmicro1871

Karlström Å, Boyd KL, English BK, McCullers JA (2009) Treatment with protein synthesis inhibitors improves outcomes of secondary bacterial pneumonia after influenza. J Infect Dis 199:311–319. doi:10.1086/596051

Karlstrom A, Heston SM, Boyd KL et al (2011) Toll-like receptor 2 mediates fatal immunopathology in mice during treatment of secondary pneumococcal pneumonia following influenza. J Infect Dis 204:1358–1366. doi:10.1093/infdis/jir522

Knapp S, Leemans JC, Florquin S et al (2003) Alveolar macrophages have a protective antiinflammatory role during murine pneumococcal pneumonia. Am J Respir Crit Care Med 167:171–179. doi:10.1164/rccm.200207-698OC

Kobayashi SD, Braughton KR, Palazzolo-Ballance AM et al (2010) Rapid neutrophil destruction following phagocytosis of Staphylococcus aureus. J. Innate Immun 2:560–575. doi:10.1159/000317134

Kobayashi SD, Braughton KR, Whitney AR et al (2003) Bacterial pathogens modulate an apoptosis differentiation program in human neutrophils. Proc Nat Acad Sci U.S.A 100:10948–10953. doi:10.1073/pnas.1833375100

Koppe U, Suttorp N, Opitz B (2012) Recognition of Streptococcus pneumoniae by the innate immune system: Innate immune recognition of Streptococcus pneumoniae. Cell Microbiol 14:460–466. doi:10.1111/j.1462-5822.2011.01746.x

Kudva A, Scheller EV, Robinson KM et al (2011) Influenza A inhibits Th17-mediated host defense against bacterial pneumonia in mice. J Immunol (Baltimore Md 1950) 186:1666–1674. doi:10.4049/jimmunol.1002194

Kumar A (2009) Critically ill patients with 2009 influenza A(H1N1) infection in Canada. JAMA 302:1872. doi:10.1001/jama.2009.1496

Kuri T, Smed Sörensen A, Thomas S et al (2013) Influenza A virus-mediated priming enhances cytokine secretion by human dendritic cells infected with *Streptococcus pneumoniae*: Influenza virus and pneumococcal co-infection of human DCs. Cell Microbiol 15:1385–1400. doi:10.1111/cmi.12122

Laennec RTH (1923) Translation of selected passages from De l'auscultation mediate. Bale & Danielsson, London

Lee LN, Dias P, Han D et al (2010) A mouse model of lethal synergism between influenza virus and haemophilus influenzae. Am J Pathol 176:800–811. doi:10.2353/ajpath.2010.090596

Levy SB, Marshall B (2004) Antibacterial resistance worldwide: causes, challenges and responses. Nat Med 10:S122–S129. doi:10.1038/nm1145

Li R, Lim A, Phoon MC et al (2010) Attenuated bordetella pertussis protects against highly pathogenic influenza a viruses by dampening the cytokine storm. J Virol 84:7105–7113. doi:10.1128/JVI.02542-09

Li W, Moltedo B, Moran TM (2012) Type I interferon induction during influenza virus infection increases susceptibility to secondary Streptococcus pneumoniae infection by negative regulation of $\gamma\delta$ T cells. J Virol 86:12304–12312. doi:10.1128/JVI.01269-12

Licciardi PV, Toh ZQ, Dunne E et al (2012) Protecting against ppneumococcal disease: critical interactions between probiotics and the airway microbiome. PLoS Pathog 8:e1002652. doi:10.1371/journal.ppat.1002652

Liu C, Air GM (1993) Selection and characterization of a neuraminidase-minus mutant of influenza virus and its rescue by cloned neuraminidase genes. Virology 194:403–407. doi:10.1006/viro.1993.1276

Liu X, He Y, Xiao K et al (2013) Effect of linezolid on clinical severity and pulmonary cytokines in a murine model of influenza A and Staphylococcus aureus coinfection. PLoS ONE 8:e57483. doi:10.1371/journal.pone.0057483

Louie J, Jean C, Chen TH et al (2009) Bacterial coinfections in lung tissue specimens from fatal cases of 2009 pandemic influenza A (H1N1)—United States, May-August 2009. Morb Mortal Wkly Rep 58:1071–1074

Louria DB, Blumenfeld HL, Ellis JT et al (1959) Studies on influenza in the pandemic of 1957-1958. II. Pulmonary Complications Of Influenza. J Clin Invest 38:213–265. doi:10.1172/JCI103791

Loving CL, Brockmeier SL, Vincent AL et al (2010) Influenza virus coinfection with Bordetella bronchiseptica enhances bacterial colonization and host responses exacerbating pulmonary lesions. Microb Pathog 49:237–245. doi:10.1016/j.micpath.2010.06.004

Ludewick HP, Aerts L, Hamelin M-E, Boivin G (2011) Long-term impairment of Streptococcus pneumoniae lung clearance is observed after initial infection with influenza A virus but not human metapneumovirus in mice. J Gen Virol 92:1662–1665. doi:10.1099/vir.0.030825-0

Marks LR, Davidson BA, Knight PR, Hakansson AP (2013) Interkingdom signaling induces streptococcus pneumoniae biofilm dispersion and transition from asymptomatic colonization to disease. mBio 4:e00438–13. doi:10.1128/mBio.00438-13

McAuley JL, Chipuk JE, Boyd KL et al (2010a) PB1-F2 proteins from H5N1 and 20th century pandemic influenza viruses cause immunopathology. PLoS Pathog 6:e1001014. doi:10.1371/journal.ppat.1001014

McAuley JL, Hornung F, Boyd KL et al (2007) Expression of the 1918 influenza A virus PB1-F2 enhances the pathogenesis of viral and secondary bacterial pneumonia. Cell Host Microbe 2:240–249. doi:10.1016/j.chom.2007.09.001

McAuley JL, Zhang K, McCullers JA (2010b) The effects of influenza A virus PB1-F2 protein on polymerase activity are strain specific and do not impact pathogenesis. J Virol 84:558–564. doi:10.1128/JVI.01785-09

McCullers JA (2001) Molecular pathogenesis of pneumococcal pneumonia. Front Biosci 6:d877. doi:10.2741/McCullers

McCullers JA (2005) Antiviral therapy of influenza. Expert Opin Investig Drugs 14:305–312. doi:10.1517/13543784.14.3.305
McCullers JA (2011) Preventing and treating secondary bacterial infections with antiviral agents. Antivir Ther 16:123–135. doi:10.3851/IMP1730
McCullers JA (2014) The co-pathogenesis of influenza viruses with bacteria in the lung. Nat Rev Microbiol 12:252–262. doi:10.1038/nrmicro3231
McCullers JA (2004) Effect of antiviral treatment on the outcome of secondary bacterial pneumonia after influenza. J Infect Dis 190:519–526. doi:10.1086/421525
McCullers JA, Bartmess KC (2003) Role of neuraminidase in lethal synergism between influenza virus and Streptococcus pneumoniae. J Infect Dis 187:1000–1009. doi:10.1086/368163
McCullers JA, English BK (2008) Improving therapeutic strategies for secondary bacterial pneumonia following influenza. Future Microbiol 3:397–404. doi:10.2217/17460913.3.4.397
McCullers JA, Karlström Å, Iverson AR et al (2007) Novel strategy to prevent otitis media caused by colonizing Streptococcus pneumoniae. PLoS Pathog 3:e28. doi:10.1371/journal.ppat.0030028
McCullers JA, McAuley JL, Browall S et al (2010) Influenza enhances susceptibility to natural acquisition of and disease due to Streptococcus pneumoniae in ferrets. J Infect Dis 202:1287–1295. doi:10.1086/656333
McCullers JA, Rehg JE (2002) Lethal synergism between influenza virus and streptococcus pneumoniae: characterization of a mouse model and the role of platelet-activating factor receptor. J Infect Dis 186:341–350. doi:10.1086/341462
McHugh KJ, Mandalapu S, Kolls JK et al (2013) A novel outbred mouse model of 2009 pandemic influenza and bacterial co-infection severity. PLoS ONE 8:e82865. doi:10.1371/journal.pone.0082865
McNamee LA, Harmsen AG (2006) Both influenza-induced neutrophil dysfunction and neutrophil-independent mechanisms contribute to increased susceptibility to a secondary Streptococcus pneumoniae infection. Infect Immun 74:6707–6721. doi:10.1128/IAI.00789-06
Metzger DW, Sun K (2013) Immune dysfunction and bacterial coinfections following influenza. J Immunol 191:2047–2052. doi:10.4049/jimmunol.1301152
Miao H, Hollenbaugh JA, Zand MS et al (2010) Quantifying the early immune response and adaptive immune response kinetics in mice infected with influenza A virus. J Virol 84:6687–6698. doi:10.1128/JVI.00266-10
Miao H, Xia X, Perelson AS, Wu H (2011) On identifiability of nonlinear ODE models and applications in viral dynamics. SIAM Rev 53:3–39. doi:10.1137/090757009
Miller ML, Gao G, Pestina T et al (2007) Hypersusceptibility to invasive pneumococcal infection in experimental sickle cell disease involves platelet-activating factor receptor. J Infect Dis 195:581–584. doi:10.1086/510626
Mina MJ, Klugman KP, McCullers JA (2013) Liveattenuated influenza vaccine, but not pneumococcal conjugate vaccine, protects against increased density and duration of pneumococcal carriage after influenza infection in pneumococcal colonized mice. J Infect Dis 208:1281–1285. doi:10.1093/infdis/jit317
Minas K, Liversidge J (2006) Is the CD200/CD200 receptor interaction more than just a myeloid cell inhibitory signal? Crit Rev Immunol 26:213–230
Moncla BJ, Braham P (1989) Detection of sialidase (neuraminidase) activity in Actinomyces species by using 2'-(4-methylumbelliferyl) alpha-D-N-acetylneuraminic acid in a filter paper spot test. J Clin Microbiol 27:182–184
Moncla BJ, Braham P, Hillier SL (1990) Sialidase (neuraminidase) activity among gram-negative anaerobic and capnophilic bacteria. J Clin Microbiol 28:422–425
Montgomery CP, Boyle-Vavra S, Adem PV et al (2008) Comparison of virulence in community-associated methicillin-resistant Staphylococcus aureus pulsotypes USA300 and USA400 in a rat model of pneumonia. J Infect Dis 198:561–570. doi:10.1086/590157
Morens DM, Taubenberger JK, Fauci AS (2008) Predominant role of bacterial pneumonia as a cause of death in pandemic influenza: implications for pandemic influenza preparedness. J Infect Dis 198:962–970. doi:10.1086/591708

Moscona A (2005) Neuraminidase inhibitors for influenza. N Engl J Med 353:1363–1373. doi:10.1056/NEJMra050740
Murphy TF, Bakaletz LO, Smeesters PR (2009) Microbial interactions in the respiratory tract. Pediatr Infect Dis J 28:S121–S126. doi:10.1097/INF.0b013e3181b6d7ec
Musher DM, Dowell ME, Shortridge VD et al (2002) Emergence of macrolide resistance during treatment of pneumococcal pneumonia. N Engl J Med 346:630–631. doi:10.1056/NEJM200202213460820
Nakamura S, Davis KM, Weiser JN (2011) Synergistic stimulation of type I interferons during influenza virus coinfection promotes Streptococcus pneumoniae colonization in mice. J Clin Invest 121:3657–3665. doi:10.1172/JCI57762
Navarini AA, Recher M, Lang KS et al (2006) Increased susceptibility to bacterial superinfection as a consequence of innate antiviral responses. Proc Nat Acad Sci U.S.A 103:15535–15539. doi:10.1073/pnas.0607325103
Nelson JC, Jackson M, Yu O et al (2008) Impact of the introduction of pneumococcal conjugate vaccine on rates of communityacquired pneumonia in children and adults. Vaccine 26:4947–4954. doi:10.1016/j.vaccine.2008.07.016
Niemann S, Ehrhardt C, Medina E et al (2012) Combined action of influenza virus and Staphylococcus aureus panton-valentine leukocidin provokes severe lung epithelium damage. J Infect Dis 206:1138–1148. doi:10.1093/infdis/jis468
Nishikawa T, Shimizu K, Tanaka T et al (2012) Bacterial neuraminidase rescues influenza virus replication from inhibition by a neuraminidase inhibitor. PLoS ONE 7:e45371. doi:10.1371/journal.pone.0045371
Nonaka H, Ishikawa Y, Otsuka M et al (1983) Purification and some properties of neuraminidase isolated from the culture medium of oral bacterium Streptococcus mitis ATCC 9811. J Dent Res 62:792–797
Okamoto S, Kawabata S, Nakagawa I et al (2003) Influenza A virus-infected hosts boost an invasive type of Streptococcus pyogenes infection in mice. J Virol 77:4104–4112
Oseasohn R, Adelson L, Kaji M (1959) Clinicopathologic study of thirty-three fatal cases of Asian influenza. N Engl J Med 260:509–518. doi:10.1056/NEJM195903122601101
Palacios G, Hornig M, Cisterna D et al (2009) Streptococcus pneumoniae coinfection is correlated with the severity of H1N1 pandemic influenza. PLoS ONE 4:e8540. doi:10.1371/journal.pone.0008540
Pawelek KA, Huynh GT, Quinlivan M et al (2012) Modeling within-host dynamics of influenza virus infection including immune responses. PLoS Comput Biol 8:e1002588. doi:10.1371/journal.pcbi.1002588
Peltola VT, Boyd KL, McAuley JL et al (2006) Bacterial sinusitis and otitis media following influenza virus infection in ferrets. Infect Immun 74:2562–2567. doi:10.1128/IAI.74.5.2562-2567.2006
Peltola VT, Murti KG, McCullers JA (2005) Influenza virus neuraminidase contributes to secondary bacterial pneumonia. J Infect Dis 192:249–257. doi:10.1086/430954
Pettigrew MM, Gent JF, Revai K et al (2008) Microbial interactions during upper respiratory tract infections. Emerg Infect Dis 14:1584–1591. doi:10.3201/eid1410.080119
Pittet LA, Hall-Stoodley L, Rutkowski MR, Harmsen AG (2010) Influenza virus infection decreases tracheal mucociliary velocity and clearance of Streptococcus pneumoniae. Am J Respir Cell Mol Biol 42:450–460. doi:10.1165/rcmb.2007-0417OC
Plotkowski MC, Bajolet-Laudinat O, Puchelle E (1993) Cellular and molecular mechanisms of bacterial adhesion to respiratory mucosa. Eur Respir J 6:903–916
Plotkowski MC, Puchelle E, Beck G et al (1986) Adherence of type I Streptococcus pneumoniae to tracheal epithelium of mice infected with influenza A/PR8 virus. Am Rev Respir Dis 134:1040–1044
Puchelle E, Zahm J-M, Tournier J-M, Coraux C (2006) Airway epithelial repair, regeneration, and remodeling after injury in chronic obstructive pulmonary disease. Proc Am Thorac Soc 3:726–733. doi:10.1513/pats.200605-126SF

Rarey KE, DeLacure MA, Sandridge SA, Small PA (1987) Effect of upper respiratory infection on hearing in the ferret model. Am J Otolaryngol 8:161–170. doi:10.1016/S0196-0709(87)80040-6

Redford PS, Mayer-Barber KD, McNab FW et al (2014) Influenza A virus impairs control of mycobacterium tuberculosis coinfection through a type I interferon receptor-dependent pathway. J Infect Dis 209:270–274. doi:10.1093/infdis/jit424

Saenz RA, Quinlivan M, Elton D et al (2010) Dynamics of influenza virus infection and pathology. J Virol 84:3974–3983. doi:10.1128/JVI.02078-09

Scanlon KL, Diven WF, Glew RH (1989) Purification and properties of Streptococcus pneumoniae neuraminidase. Enzyme 41:143–150

Seki M, Yanagihara K, Higashiyama Y et al (2004) Immunokinetics in severe pneumonia due to influenza virus and bacteria coinfection in mice. Eur Respir J 24:143–149. doi:10.1183/09031936.04.00126103

Shahangian A, Chow EK, Tian X et al (2009) Type I IFNs mediate development of postinfluenza bacterial pneumonia in mice. J Clin Invest 119:1910–1920. doi:10.1172/JCI35412

Shanks GD, Mackenzie A, McLaughlin R et al (2010) Mortality risk factors during the 1918-1919 influenza pandemic in the Australian army. J Infect Dis 201:1880–1889. doi:10.1086/652868

Shope RE (1931) Swine influenza: I. Experimental transmission and pathology. J Exp Med 54:349–359. doi:10.1084/jem.54.3.349

Short KR, Diavatopoulos DA, Thornton R et al (2011) Influenza virus induces bacterial and nonbacterial otitis media. J Infect Dis 204:1857–1865. doi:10.1093/infdis/jir618

Short KR, Habets MN, Hermans PW, Diavatopoulos DA (2012a) Interactions between Streptococcus pneumoniae and influenza virus: a mutually beneficial relationship? Future Microbiol 7:609–624. doi:10.2217/fmb.12.29

Short KR, Habets MN, Payne J et al (2013a) Influenza A virusinduced bacterial otitis media is independent of virus tropism for α2,6-linked sialic acid. Virol J 10:128. doi:10.1186/1743-422X-10-128

Short KR, Reading PC, Brown LE et al (2013b) Influenza-induced inflammation drives pneumococcal otitis media. Infect Immun 81:645–652. doi:10.1128/IAI.01278-12

Short KR, Reading PC, Wang N et al (2012b) Increased nasopharyngeal bacterial titers and local inflammation facilitate transmission of Streptococcus pneumoniae. mBio 3:e00255–12. doi:10.1128/mBio.00255-12

van der Sluijs KF, van Elden LJR, Nijhuis M et al (2004) IL-10 is an important mediator of the enhanced susceptibility to pneumococcal pneumonia after influenza infection. J Immunol (Baltimore Md 1950) 172:7603–7609

Small C-L, Shaler CR, McCormick S et al (2010) Influenza infection leads to increased susceptibility to subsequent bacterial superinfection by impairing NK cell responses in the lung. J Immunol 184:2048–2056. doi:10.4049/jimmunol.0902772

Smith AM, Adler FR, McAuley JL et al (2011a) Effect of 1918 PB1-F2 expression on influenza A virus infection kinetics. PLoS Comput Biol 7:e1001081. doi:10.1371/journal.pcbi.1001081

Smith AM, Adler FR, Perelson AS (2010) An accurate two-phase approximate solution to an acute viral infection model. J Math Biol 60:711–726. doi:10.1007/s00285-009-0281-8

Smith AM, Adler FR, Ribeiro RM et al (2013) Kinetics of coinfection with influenza A virus and Streptococcus pneumoniae. PLoS Pathog 9:e1003238. doi:10.1371/journal.ppat.1003238

Smith AM, McCullers JA (2013) Molecular signatures of virulence in the PB1-F2 proteins of H5N1 influenza viruses. Virus Res 178:146–150. doi:10.1016/j.virusres.2013.02.012

Smith AM, McCullers JA, Adler FR (2011b) Mathematical model of a three-stage innate immune response to a pneumococcal lung infection. J Theor Biol 276:106–116. doi:10.1016/j.jtbi.2011.01.052

Smith AM, Perelson AS (2011) Influenza A virus infection kinetics: quantitative data and models. Wiley Interdiscip Rev Syst Biol Med 3:429–445. doi:10.1002/wsbm.129

Smith AM, Ribeiro RM (2010) Modeling the viral dynamics of influenza A virus infection. Crit Rev Immunol 30:291–298. doi:10.1615/CritRevImmunol.v30.i3.60

Smith MW, Schmidt JE, Rehg JE et al (2007) Induction of pro- and anti-inflammatory molecules in a mouse model of pneumococcal pneumonia after influenza. Comp Med 57:82–89
Snelgrove RJ, Goulding J, Didierlaurent AM et al (2008) A critical function for CD200 in lung immune homeostasis and the severity of influenza infection. Nat Immunol 9:1074–1083. doi:10.1038/ni.1637
Sun K, Metzger DW (2008) Inhibition of pulmonary antibacterial defense by interferon-γ during recovery from influenza infection. Nat Med 14:558–564. doi:10.1038/nm1765
Tashiro M, Ciborowski P, Klenk H-D et al (1987) Role of Staphylococcus protease in the development of influenza pneumonia. Nature 325:536–537. doi:10.1038/325536a0
Tian X, Xu F, Lung WY et al (2012) Poly I: C enhances susceptibility to secondary pulmonary infections by gram-positive bacteria. PLoS ONE 7:e41879. doi:10.1371/journal.pone.0041879
Tong HH, Weiser JN, James MA, DeMaria TF (2001) Effect of influenza A virus infection on nasopharyngeal colonization and otitis media induced by transparent or opaque phenotype variants of Streptococcus pneumoniae in the chinchilla model. Infect Immun 69:602–606. doi:10.1128/IAI.69.1.602-606.2001
Tuomanen EI, Austrian R, Masure HR (1995) Pathogenesis of pneumococcal infection. N Engl J Med 332:1280–1284. doi:10.1056/NEJM199505113321907
Varga ZT, Grant A, Manicassamy B, Palese P (2012) Influenza virus protein PB1-F2 inhibits the induction of type I interferon by binding to MAVS and decreasing mitochondrial membrane potential. J Virol 86:8359–8366. doi:10.1128/JVI.01122-12
Varga ZT, Ramos I, Hai R et al (2011) The influenza virus protein PB1-F2 inhibits the induction of type I interferon at the level of the MAVS adaptor protein. PLoS Pathog 7:e1002067. doi:10.1371/journal.ppat.1002067
Wang J, Li F, Sun R et al (2013) Bacterial colonization dampens influenza-mediated acute lung injury via induction of M2 alveolar macrophages. Nat Commun. doi:10.1038/ncomms3106
Wanzeck K, Boyd KL, McCullers JA (2011) Glycan shielding of the influenza virus hemagglutinin contributes to immunopathology in mice. Am J Respir Crit Care Med 183:767–773. doi:10.1164/rccm.201007-1184OC
Weeks-Gorospe JN, Hurtig HR, Iverson AR et al (2012) Naturally occurring swine influenza A virus PB1-F2 phenotypes that contribute to superinfection with gram-positive respiratory pathogens. J Virol 86:9035–9043. doi:10.1128/JVI.00369-12
Weinberger DM, Simonsen L, Jordan R et al (2011) Impact of the 2009 influenza pandemic on pneumococcal pneumonia hospitalizations in the United States. J Infect Dis 205:458–465. doi:10.1093/infdis/jir749
Wherry WB, Butterfield CT (1921) Inhalation experiments on influenza and pneumonia, and on the importance of spray-borne bacteria in respiratory infections. Public Health Rep (1896–1970) 36:1443. doi:10.2307/4576030
World Health Organization Global Burden of Disease
World Health Organization Writing Group, Bell D, Nicoll A et al (2006) Non-pharmaceutical interventions for pandemic influenza, international measures. Emerg Infect Dis 12:81–87. doi:10.3201/eid1201.051370
Wu Y, Mao H, Ling M-T et al (2011) Successive influenza virus infection and Streptococcus pneumoniae stimulation alter human dendritic cell function. BMC Infect Dis 11:201. doi:10.1186/1471-2334-11-201
Zamarin D, García-Sastre A, Xiao X et al (2005) Influenza virus PB1-F2 protein induces cell death through mitochondrial ANT3 and VDAC1. PLoS Pathog 1:e4. doi:10.1371/journal.ppat.0010004
Zamarin D, Ortigoza MB, Palese P (2006) Influenza A virus PB1-F2 protein contributes to viral pathogenesis in mice. J Virol 80:7976–7983. doi:10.1128/JVI.00415-06
Zell R, Krumbholz A, Eitner A et al (2007) Prevalence of PB1-F2 of influenza A viruses. J Gen Virol 88:536–546. doi:10.1099/vir.0.82378-0

Part V
Evolution, Antigenic Variation

Evolution and Ecology of Influenza A Viruses

Sun-Woo Yoon, Richard J. Webby and Robert G. Webster

Abstract Wild aquatic bird populations have long been considered the natural reservoir for influenza A viruses with virus transmission from these birds seeding other avian and mammalian hosts. While most evidence still supports this dogma, recent studies in bats have suggested other reservoir species may also exist. Extensive surveillance studies coupled with an enhanced awareness in response to H5N1 and pandemic 2009 H1N1 outbreaks is also revealing a growing list of animals susceptible to infection with influenza A viruses. Although in a relatively stable host–pathogen interaction in aquatic birds, antigenic, and genetic evolution of influenza A viruses often accompanies interspecies transmission as the virus adapts to a new host. The evolutionary changes in the new hosts result from a number of processes including mutation, reassortment, and recombination. Depending on host and virus these changes can be accompanied by disease outbreaks impacting wildlife, veterinary, and public health.

Contents

S.-W. Yoon · R. J. Webby · R. G. Webster (✉)
Division of Virology, Department of Infectious Diseases,
St Jude Children's Research Hospital, Memphis, TN, USA
e-mail: robert.webster@stjude.org

S.-W. Yoon
Viral Infectious Disease Research Center,
Korea Research Institute of Bioscience and Biotechnology,
Daejeon 305-806, South Korea

Current Topics in Microbiology and Immunology (2014) 385: 359–375

DOI: 10.1007/82_2014_396

Published Online: 3 July 2014

1 Introduction

The known natural reservoirs of influenza viruses are the aquatic birds of the world with mounting evidence for a role of bats. Sixteen hemagglutinin (HA) and nine neuraminidase (NA) subtypes are found in aquatic birds (Fouchier et al. 2005; Webster et al. 1992) and two HA and two NA's have been described in bats (Tong et al. 2013). The influenza viruses of domestic poultry and mammals including humans have all evolved directly or indirectly from influenza viruses in the aquatic bird reservoirs. The processes involved in influenza virus evolution are acquisition of mutations, reassortment, and less frequently nonhomologous recombination. The negative sense RNA genome of the influenza virus has no proof reading mechanism during replication and is extremely error prone giving rise to a high rate of mutation. The eight segments of RNA in influenza viruses permit genomic mixing when two or more influenza viruses infect a cell. Nonhomologous recombination has to date only been detected in the genesis of highly pathogenic (HP) H7 influenza viruses (Suarez et al. 2004) and is presumably confined to H7 viruses.

In aquatic birds, influenza viruses replicate predominantly in the intestinal tract (Webster et al. 1978) at a body temperature of 42 °C and have an HA binding preference for α2,3 sialic acid (SA) receptors on the host cell. In contrast, influenza viruses that have become endemic in humans replicate predominantly in the upper respiratory tract at 35–37 °C with an HA binding preference for α2,6 SA receptors. Thus, multiple evolutionary changes are required for the establishment of stable lineages in other hosts. In this chapter, we will consider the ecology of influenza viruses in their reservoir hosts and the different hosts where influenza viruses have established permanent lineages and what we know about the evolutionary processes involved.

2 Hosts of Influenza A Viruses

2.1 *Wild Birds*

Ecological studies on influenza viruses in wild birds in multiple sites in the world have established the following general principles:

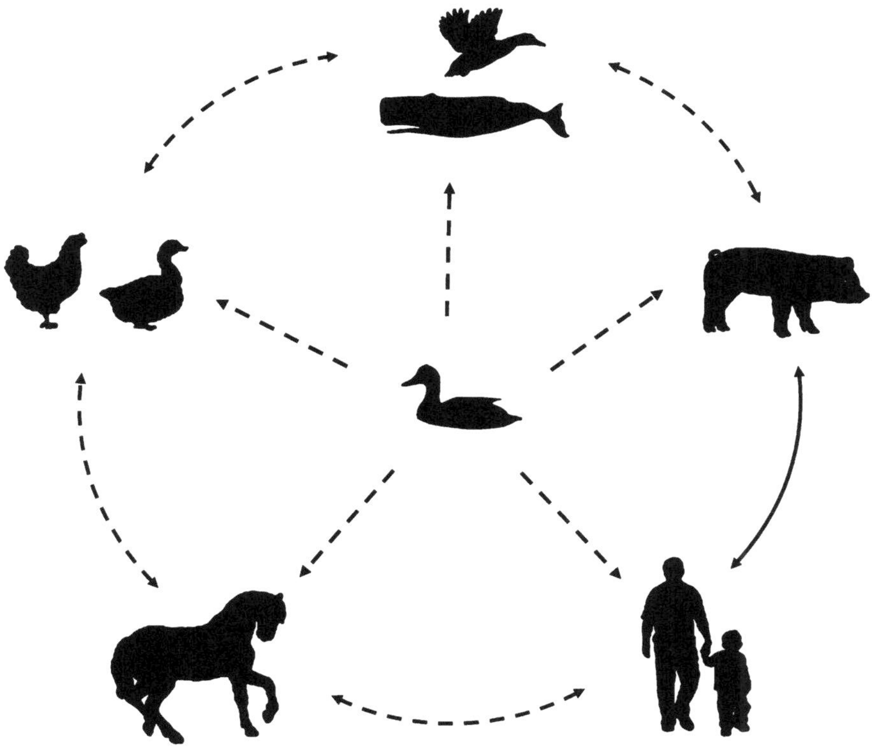

Fig. 1 Original "Wheel of Influenza." Aquatic birds are the natural reservoir for all subtypes of influenza A viruses. Transmission of influenza A viruses is known between pigs and humans (*solid line*)

(i) *Natural reservoirs in aquatic birds*: Anseriformes (ducks, geese, swans) and Charadriiformes (shorebirds, gulls, alcids) are the aquatic bird species that are the ultimate source of influenza virus infecting domestic avian species and mammals (Fouchier et al. 2005; Webster et al. 1992). The simplified original diagram (Fig. 1) gives the overall concept and has been expanded with additional information (Fig. 2). Although influenza viruses have been isolated from over 100 aquatic wild bird species (Olsen et al. 2006), dabbling ducks, especially mallards (*Anas platyrhynchos*), are the most frequent source of influenza virus in wild birds (Krauss et al. 2004; Munster et al. 2007). Influenza viruses have been isolated from many other birds species but it is unclear whether they are endemic in these species or are transitorily present.

(ii) *Pathogenicity*: In their reservoir species, all influenza virus subtypes are low pathogenic (LP) and cause inapparent disease. Two of the 16 subtypes (H5 and H7) can evolve into HP virus but in aquatic birds all 16 HA

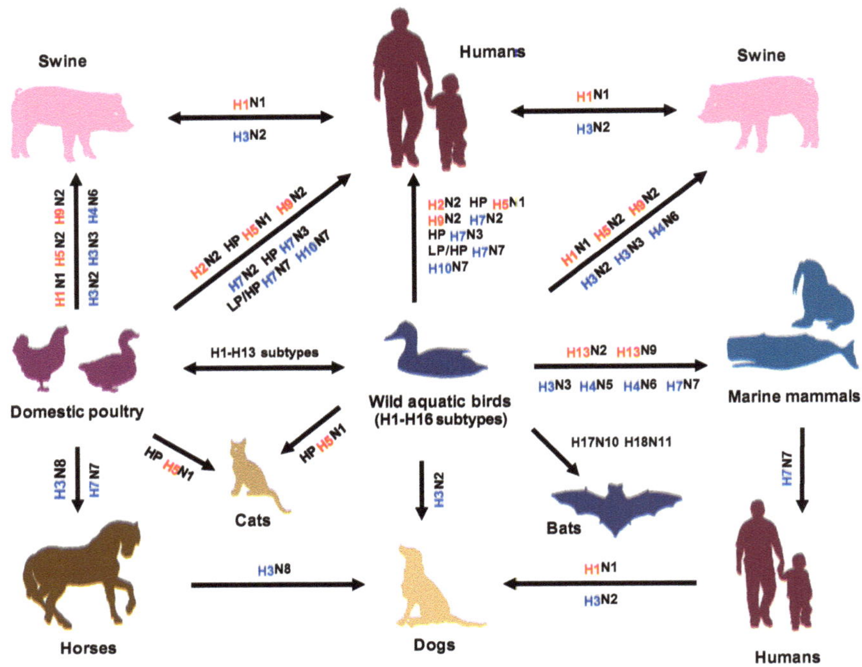

Fig. 2 Cross-species transmission of influenza A viruses. Aquatic birds are naturally the principal reservoirs for influenza A viruses including horses, cats, dogs, marine animals, domestic poultry, pigs, bats, and humans. Group 1 HA subtypes are represents in *red* and group 2 HA subtypes are represents in *blue*

subtypes including H5 and H7 influenza viruses occur as LP subtypes. There is some disagreement with this statement when applied to the HP H5N3 isolated from common terns in South Africa in 1961 (Becker 1966) and to the highly pathogenic Asian H5N1 viruses. While there is no doubt that HP H5 and H7 influenza viruses have been isolated from wild aquatic birds, the probability is that wild aquatic birds spread HP H5N1 influenza viruses between infected domestic poultry but these HP H5N1 influenza viruses are not perpetuated from one generation of aquatic birds to the next during breeding. The failure of HP H5N1 to spread to the Americas argues against perpetuation of HP H5N1 in wild aquatic birds for millions of migratory waterfowl migrate every year from Eurasia to Alaska. Although each of the HA subtypes cause inapparent disease in aquatic birds, there is some "cost" to the birds. Detailed studies in mallards and Bewicks swans (*Cygnus columbianus bewickii*) show reduced body mass and physical condition but no apparent disease (Latorre-Margalef et al. 2009; van Gils et al. 2007).

(iii) *Intestinal replication*: In aquatic birds, influenza viruses replicate predominantly in the intestinal tract and are spread fecal-orally through water. However, some subtypes (H3 and H4) also replicate equally well in the respiratory tract. Thus during wild bird surveillance, influenza viruses can be isolated from both ends of the bird. During spread in domestic avian species, the H5, H6, H7, and H9 influenza virus subtypes have acquired the property of preferential replication in the respiratory tract—a trait that facilitates spread in high density poultry houses.

(iv) *Geographical separation*: There is geographical separation of influenza viruses into distinct clades. The 16 HA subtypes of influenza A viruses can be separated phylogenetically into major geographically separable clades. The principal clades recognized to date are the Eurasian and American clades with information supporting additional separate clades in Australia and South America (Hansbro et al. 2010; Pereda et al. 2008). There is a continuous interaction between influenza viruses in these clades but there is sufficient separation for the evolution of phylogenetically distinct evolutionary groups.

(v) *Evolutionary stasis?* It has been noted that influenza viruses in their natural host reservoirs show limited evolution at the amino acid level (Webster et al. 1992). This has been interpreted to mean that the viruses are in equilibrium with their hosts and that the continued random mutations that occur provide no additional advantages. However, when influenza viruses spread to other hosts they can show rapid evolution with emergence of antigenic variants and evolution of separate clades. This is already demonstrated for H5, H7 and H9 influenza viruses when multiple antigenic variants and emergence of separate clades has occurred (WHO/FAO/OIE 2014).

(vi) *Interspecies transmission*: Interspecies transmission of influenza viruses occurs most frequently between wild and domestic waterfowl. The conditions for exchange between wild and domestic ducks in ponds and rice paddies throughout Asia are optimal (Shortridge 1992). Exchange between waterfowl-both wild and domestic and between waterfowl and terrestrial poultry (chicken, quail, etc) occurs in backyard poultry farms where multiple species are raised together and in live poultry markets that provide optimal conditions for Interspecies transmission and genetic mixing of influenza viruses. Live poultry markets also provide optimal conditions for the transfer of influenza viruses from poultry to mammals, including humans. This was dramatically demonstrated when the closure of poultry markets in China in 2013 reduced the number of human H7N9 cases to near zero (Webster et al. 2013) with an equally dramatic increase in human cases when the live poultry markets were reopened. The available information supports the notion that interspecies transmission of influenza viruses occur relatively frequently but the establishment of stable transmissible lineages are rare (Fouchier and Guan 2013). Thus, while multiple subtypes of influenza A viruses transmit to terrestrial poultry (see H6N2 and H9N2 below) and establish stable lineages, only H1, H2, and H3 subtypes have to date established stable lineages in mammals (with the exception of bats).

2.2 Domestic Birds

Since the majority of influenza A subtypes have been isolated from wild ducks, it is not surprising that the majority of influenza subtypes will replicate in domestic ducks. Only the H14 and H15 subtypes that are isolated mainly from gulls tend not to replicate in ducks. In domestic ducks, all of the subtypes including H5 and H7 are LP—unless coinfected with other disease agents. In general, domestic ducks show inapparent infection with no disease signs. When terrestrial poultry is considered, quail are unique in that they will support the replication of the vast majority of influenza viruses, again with inapparent infection with no disease signs and have been considered an intermediate host possessing both α2,3 and α2,6 SA receptors (Perez et al. 2003). Quail, along with aquatic birds, was banned from live poultry markets in Hong Kong for they were considered "mixing vessels" between aquatic and terrestrial poultry and were the source of the G1 lineage of H9N2 influenza viruses (Guan et al. 1999). The other avian species highly susceptible to influenza infection are turkeys, which will support the replication of the majority of avian and mammalian influenza viruses including the 2009 pandemic H1N1 virus. However, despite the ability to replicate influenza viruses remarkably few influenza virus subtypes have established permanent lineages in terrestrial poultry-exceptions include H6N1 and H9N2 influenza viruses which have established stable lineages in chickens. Two of the 16 HA subtypes, H5 and H7, are unique and can evolve into highly pathogenic variants that cause up to 100 % mortality in gallinaceous poultry and sporadically spread to humans with up to 60 % mortality. Since entire chapters in this book will deal with HP H5N1, LP H7N9, and H9N2 influenza viruses, these viruses will not be considered in this section.

2.3 Bats

While the aquatic birds of the world have been identified as the ultimate source of influenza viruses that directly or indirectly infect terrestrial poultry and mammals, the identification of genetic material from novel influenza viruses in bats (Tong et al. 2013) raises the possibility that additional reservoirs of influenza virus remain to be discovered. To date, the two subtypes of bat influenza virus (H17N10, H18N11) have been detected by PCR but they have not been cultured and do not attach to sialic acid residues (Tong et al. 2013). These viruses have been detected in apparently healthy bats, and their role in influenza ecology remains to be resolved.

2.4 Pigs

Swine, along with humans and horses, are the best understood mammalian influenza virus reservoirs. More akin to chickens than to wild birds, swine are hosts to only a limited number of endemic virus subtypes, these being H1N1, H3N2, and H1N2. Although limited in subtype, swine influenza viruses are for the most part geographically restricted, and there is substantial diversity between the viruses that do circulate in different regions of the world (Vincent et al. 2014). With few exceptions swine influenza viruses are endemic wherever large numbers of swine are raised. In these populations, it is a year-round disease with defined peaks of activity in winter months, at least in temperate regions. The severity of disease can range from asymptomatic to severe depending on a number of factors from virus strain to comorbidities to the presence of other pathogens. While the impact of the disease may vary from region to region, in the United States swine influenza has been ranked in the top three respiratory health challenges to the industry with direct cost estimates between $3.23 and $10.31 per head (United States Department of Agriculture ARS 2014). In the United Kingdom, the impact of swine influenza has been estimated at £7 per pig (Kay et al. 1994).

While swine have been postulated to be the mixing vessel for pandemic human influenza strains due to the presence of both α2,3 and α2,6 SA's in their respiratory tracts (Ito et al. 1998), evidence suggests that the net flow of viruses is from humans to swine rather than the other way around. Using a phylogenetic analysis, Nelson and colleagues highlighted this by detecting 49 separate human-to-swine transmissions of the 2009 pandemic H1N1 virus during 2009–2011 and an additional 23 transmissions of other seasonal human strains from 1990 to 2011 (Nelson et al. 2012). While many of these transmission events were likely dead end, at least eight human-origin viruses (or virus genes) have circulated for extended periods in swine clearly demonstrating the impact of human influenza viruses on the ecology and genetic diversity of influenza viruses in swine. In contrast, while there have been many reports of sporadic infection of avian viruses in swine, only one of the dominant circulating lineages, that being the Eurasian avian H1N1 virus, has originated from this source.

A recent hallmark of the evolution of swine influenza viruses has been reassortment. While it is clear that swine influenza viruses evolve through genetic and antigenic drift (Lorusso et al. 2011; Vijaykrishna et al. 2011), this drift is less pronounced than is typically associated with human viruses (de Jong et al. 2007). Reassortment, however, is prevalent within swine populations. The reverse zoonotic transmission of the 2009 pandemic H1N1 viruses from humans to pigs has demonstrated this. Subsequent to the human-to-pig transmission of the 2009 pandemic viruses, multiple reassortants between it and endemic swine viruses were rapidly detected in Asia (Vijaykrishna et al. 2010), the Americas (Ducatez et al. 2011; Nfon et al. 2011), and Europe (Howard et al. 2011; Moreno et al. 2011). While it is unclear how many of these reassortants have been maintained, there appear to be limited barriers to reassortment in swine. Analysis of viruses isolated from long-term surveillance activities in swine slaughterhouses in Hong Kong have also suggested that reassortment can be accompanied by antigenic changes although the mechanisms involved are yet to be resolved (Vijaykrishna et al. 2011).

As discussed, many swine virus lineages are descendants of human viruses. Extended circulation of similar viruses in human and swine over time results in substantial antigenic divergence and a lack of human population immunity to the swine viruses; most famously demonstrated by the 2009 pandemic (Garten et al. 2009). A review of the literature in 2006 identified 50 cases of likely zoonotic swine influenza virus infection (Myers et al. 2007) although the true rate was likely substantially higher. Since 2005 there have been over 350 detections of human infection with swine influenza viruses in the United States alone (www.cdc.gov), many subsequent to 2011 and in conjunction with agricultural fairs and H3N2 viruses (Lindstrom et al. 2012). Swine influenza viruses have been repeatedly isolated from turkey populations and sporadically from wild birds (Olsen et al. 2003) although the impact on the latter populations seems minimal.

2.5 Humans

Three subtypes of influenza A viruses have been known to form long-term stable lineages in humans; H1N1, H2N2, and H3N2. Over the past 100 years, these viruses have turned over in the human population due primarily to their displacement by newly emerging antigenic shift variants. H1N1 subtype viruses were introduced into humans around 1918 but disappeared in 1957 with the emergence of the H2N2 pandemic viruses only to reappear around 1977. This lineage of H1N1 virus circulated until 2009 when the distinct 2009 pandemic H1N1 virus emerged. Similarly, the 1957 pandemic H2N2 strain circulated until the emergence of the 1968 H3N2 strain. Although the replacement of strains over time can be seen in other hosts of influenza, it is most pronounced in humans, perhaps due to some combination of herd immunity and the global as opposed to regional nature of the circulating viruses. There is a distinct seasonality to influenza in humans with it being a winter disease in temperate regions and a more year round disease with peaks of activity in rainy seasons in tropical climates.

The primary driver of evolution in human influenza viruses is antigenic drift. Herd immunity built from prior infection with the virus leads to rapid selection of variants that are antigenically distinct. While there are other hypotheses (Bahl et al. 2011), it is thought that antigenic variants often emerge from equatorial regions of Asia from where they spread to other regions (Lemey et al. 2014). The tight bottlenecks in the temperate regions of the world caused by the marked seasonality provide an avenue for rapid selection of variants with a selective advantage (Gog et al. 2003). In this way, antigenic variants emerge and rapidly displace previously circulating strains causing the typical ladder-like appearance of HA-based phylogenetic trees. Reassortment can also be detected in endemic human influenza viruses but, with some notable exceptions such as the H1N2 viruses that circulated for a period of time in 2001–2002, is generally limited to intrasubtypic reassortment (Nelson et al. 2008). This comparative low rates of reassortment in humans as opposed to avian and swine hosts can be most parsimoniously explained by the fact that within a given influenza season a single influenza A subtype tends to dominate.

The susceptibility of humans to viruses from other reservoirs is complex and relatively poorly understood. The receptor preference of avian and mammalian viruses is likely a limiting factor, but as human infection with avian viruses occurs in the absence of detectable receptor changes, it is obviously not an absolute barrier. Population immunity also likely plays a role in limiting infection and onward transmission of zoonotic viruses, and has been postulated to be a factor limiting the spread of the swine H3N2 viruses that have been causing human infection in the United States. Epidemiologic evidence also suggests that host genetics are important factors in susceptibility to influenza viruses (Horby et al. 2013), but identification of causative genetic polymorphisms has been limited to a role for IFITM3 in mediating pandemic 2009 H1N1 disease severity (Everitt et al. 2012).

2.6 Horses

Equine influenza A viruses are a common respiratory pathogen of horses caused by two different subtypes, H7N7 and H3N8. The H7N7 subtype was first isolated from a horse during an outbreak in Czechoslovakia in 1956, A/eq/Prague/56 (Sovinova et al. 1958). After the first isolation, these viruses were associated with respiratory disease in the horse population of Europe lasting for approximately two decades. H3N8 viruses were first identified from an outbreak of respiratory disease in horses in Florida in 1963, which was correlated with the international movement of horses (Waddell et al. 1963). Since the late 1970s, the H3N8 subtype has been the dominant influenza virus causing respiratory diseases in horses while H7N7 viruses have rarely been reported in the last three decades and are now thought to be extinct (Webster et al. 1992).

In the late 1980s, two distinct lineages of the H3N8 subtype were identified, American- and European-like lineages (Daly et al. 1996). These lineages were distinguished by antigenic drift in their HA and/or NA and have continued to co-circulate. Currently, the H3N8 subtype virus is prevalent worldwide (Daly et al. 1996). In the 1990s, two severe outbreaks of H3N8 viruses affected large populations of horses in Northeast China raising the question of whether the outbreaks were caused by novel reassortants. The causative agent, A/Eq/Gansu/2/94, was an H3N8 influenza A virus whose gene segments were derived entirely from classical equine influenza viruses (Guo et al. 1992) indicating that the severity was due to host susceptibility rather than virus novelty. Interestingly, current outbreaks of equine influenza viruses often follow horse movement and introduction into new environments. For example, In Australia in 2007 an outbreak of equine influenza was initiated by a subclinically infected imported horse from Japan that had responded poorly to vaccination (Bryant et al. 2010). During 2007 and 2011, cases of equine influenza were reported by China, France, Germany, Ireland, Mongolia, Sweden, the United Kingdom, and the United States.

Although the origin of equine influenza viruses still remains unknown, the available data suggest that they are possibly of avian origin (Guo et al. 1995). Sequence analysis of an H3N8 virus (A/Equine/Jilin/1/89) isolated from horses in China in 1989 (Webster et al. 1992), showed that the virus was more closely related to avian influenza viruses than to equine influenza viruses. This virus was different from circulating H3N8 viruses. Interestingly, genetic features of the avian-like H3N8 viruses were responsible for the outbreak, indicating that this virus had spread directly to horses from the avian reservoir.

In the early 2000s, equine H3N8 influenza viruses crossed the species barrier and have been associated with outbreaks of respiratory disease in dogs in North America (Crawford et al. 2005) and in the United Kingdom (Daly et al. 1996). Both outbreaks were in quarry hounds, which have close contact with horses. Evidence of sporadic cases of H3N8 influenza in dogs was reported from outbreaks of equine influenza virus in Australia in 2007; direct horse-to-dog transmission was confirmed in an experimental setting (Virmani et al. 2010). Additionally, two equine H3N8 influenza viruses were isolated from pigs in China (Tu et al. 2009). However, it is not yet known whether equine influenza viruses can be transmitted from infected horses to other species.

2.7 Dogs and Cats

Until 2004, canines were not considered a reservoir species for influenza A viruses. However, after H3N8 influenza virus was first isolated from dogs, two lineages of influenza A viruses—one of avian-origin (H3N2) and the other of equine-origin (H3N8), were described (Crawford et al. 2005; Lee et al. 2009; Li et al. 2010). In 2004, H3N8 influenza A virus of equine origin was isolated from racing greyhounds with respiratory disease (Crawford et al. 2005). Because of the lack of previous exposure to this virus, respiratory disease spread rapidly and transmitted between individual dogs (Buonavoglia and Martella 2007). Studies in Asian countries including South Korea and China have reported transmission of avian-origin canine influenza A viruses (H3N2) (Lee et al. 2009; Li et al. 2010) and determined that the virus can be experimentally transmitted directly between dogs (Song et al. 2009). In 2004, serological and virological data established infection of dogs with HP H5N1 viruses from Thailand (Songserm et al. 2006). Experimental studies by Maas et al. (2007) and Giese et al. (2008) showed that dogs were susceptible to infection with HP H5N1 viruses. These results noted that dogs may play a role in HP H5N1 virus adaptation to mammals and contribute to spread to other species. During the pandemic of 2009, serological and virological data indicated infection of dogs with the pandemic H1N1 virus and transmission between dogs can occur, albeit with low efficiency (Dundon et al. 2010).

We understand little about the transmission of canine influenza viruses to avians and mammals. Canine influenza virus was experimentally able to infect mice and guinea pigs but not ducks or chickens (Castleman et al. 2010; Tu et al. 2009). Although the canine H3N2 virus is genetically and antigenically different from strains currently circulating in humans, we should remain aware of the potential of these viruses to transmit to the human population.

In the 1970s, serologic studies indicated that clinically healthy cats could be infected with H3N2 subtype influenza A viruses (Paniker and Nair 1970). Cats infected with different subtypes such as mammalian-origin (human H2N2 and H3N2, seal H7N7) and avian-origin (H7N3) were subsequently reported (Hinshaw et al. 1981; Romvary et al. 1975). Each of these viruses replicated and induced antibody titers after intranasal inoculation of cats (Kuiken et al. 2004; Romvary et al. 1975). Further studies on feline influenza viruses were not described before HP H5N1 was isolated from domestic cats, with lethal results in East Asia (Kuiken et al. 2004; Songserm et al. 2006). The HP H5N1 virus has caused fatal infection in other felines such as leopards and tigers, species previously considered to be resistant to disease from influenza A virus infection (Thanawongnuwech et al. 2005). More recently, natural infections of domestic cats with the 2009 pandemic H1N1 virus (Sponseller et al. 2010) as well as human H3N2 and H1N1 influenza A viruses (McCullers et al. 2011) during the 2009 to 2010 influenza season were reported in the United States.

Serosurvey investigations have shown that multiple subtypes of influenza viruses have infected cats but there are no known influenza viruses that have established lineages (Hinshaw et al. 1981; Paniker and Nair 1972). Zhang et al. (2013) have suggested that domestic cats and dogs were infected with H9N2 influenza viruses which are endemic in domestic poultry in Eurasia (Peiris et al. 2001). Sporadic transmissions of H9N2 to humans have also been reported (Uyeki et al. 2002). These findings suggest that both domestic dogs and cats may represent novel bridging species for cross-species transmission of avian and human influenza viruses.

2.8 Marine Mammals

Sporadic reports of influenza A virus isolations from marine mammals have been reported. From December 1979 to October 1980, influenza virus infection in harbor seals (*Phoca vitulina*) with associated severe pneumonia was reported on the New England coast of Massachusetts (Webster et al. 1981). A mass mortality of seals (approximately 600 animals) occurred during this outbreak and H7N7 influenza A virus, A/Seal/Massachusetts/1/80, was isolated from the lungs and brains of the dead seals (Geraci et al. 1982; Webster et al. 1981). Later outbreaks (1981–1982) of influenza in seals were associated with H4N5 influenza A virus (A/seal/MA/133/82) which was isolated from the lungs of dead seals (Hinshaw et al. 1986). Similarly, H4N6 and H3N3 influenza A viruses were isolated from tissues from stranded seals but no increase in deaths was observed (Callan et al. 1995).

In 1984, H13N2 and H13N9 influenza viruses were isolated from a pilot whale (*Globicephala meleana*) along the New England coast (Hinshaw et al. 1986). In 2010, the 2009 pandemic H1N1 virus was isolated from an elephant seal (*Mirounga angustirostris*) off the central California coast (Goldstein et al. 2013); it was unclear how and when exposure occurred and there was no evidence of widespread transmission between human and marine animals. In 2011 an avian-origin H3N8 influenza A virus, A/harbor seal/Massachusetts/1/2011, was isolated from an outbreak of pneumonia in Atlantic harbor seals (*Phoca vitulina*) along the United States east coast (Anthony et al. 2012). Human influenza B virus has been also detected in harbor seals (Osterhaus et al. 2000) but this seems to have been an isolated incident. Thus, it appears that marine mammals are transitorily infected with influenza viruses of avian and human origin with no evidence of establishment of stable virus lineages.

3 Concluding Remarks

The increased demand for food protein from a growing global population has created environments that have altered the interfaces between influenza virus hosts. In combination with our increased ability to detect and diagnose virus infection, this has led to an increasing number of recognized hosts of influenza viruses and documented interspecies transmission events complicating our view of influenza virus ecology. The use of mounting viral genomic information and laboratory and field studies is, however, providing us with a better understanding of the molecular events underpinning influenza virus evolution; there is, however, much more to be understood. The role of live animal markets as a site for influenza virus evolution and zoonotic infection has been repeatedly demonstrated, and in the longer term society would benefit from their phasing out. It will be hard to change tradition, but the risks to global health far outweigh the benefits of fresh chicken.

Acknowledgments This work was supported by Contract No. HHSN266200700005C from the National Institute of Allergy and Infectious Diseases, National Institutes of Health, Department of Health and Human Services, and by ALSAC. We thank James Knowles for manuscript preparation.

References

Anthony SJ, St Leger JA, Pugliares K, Ip HS, Chan JM, Carpenter ZW, Navarrete-Macias I, Sanchez-Leon M, Saliki JT, Pedersen J, Karesh W, Daszak P, Rabadan R, Rowles T, Lipkin WI (2012) Emergence of fatal avian influenza in New England harbor seals. MBio 3:e00166-12

Bahl J, Nelson MI, Chan KH, Chen R, Vijaykrishna D, Halpin RA, Stockwell TB, Lin X, Wentworth DE, Ghedin E, Guan Y, Peiris JS, Riley S, Rambaut A, Holmes EC, Smith GJ (2011) Temporally structured metapopulation dynamics and persistence of influenza A H3N2 virus in humans. Proc Natl Acad Sci USA 108:19359–19364

Becker WB (1966) The isolation and classification of Tern virus: influenza A-Tern South Africa—1961. J Hyg (Lond) 64:309–320
Bryant NA, Paillot R, Rash AS, Medcalf E, Montesso F, Ross J, Watson J, Jeggo M, Lewis NS, Newton JR, Elton DM (2010) Comparison of two modern vaccines and previous influenza infection against challenge with an equine influenza virus from the Australian 2007 outbreak. Vet Res 41:19
Buonavoglia C, Martella V (2007) Canine respiratory viruses. Vet Res 38:355–373
Callan RJ, Early G, Kida H, Hinshaw VS (1995) The appearance of H3 influenza viruses in seals. J Gen Virol 76(Pt 1):199–203
Castleman WL, Powe JR, Crawford PC, Gibbs EP, Dubovi EJ, Donis RO, Hanshaw D (2010) Canine H3N8 influenza virus infection in dogs and mice. Vet Pathol 47:507–517
Crawford PC, Dubovi EJ, Castleman WL, Stephenson I, Gibbs EP, Chen L, Smith C, Hill RC, Ferro P, Pompey J, Bright RA, Medina MJ, Johnson CM, Olsen CW, Cox NJ, Klimov AI, Katz JM, Donis RO (2005) Transmission of equine influenza virus to dogs. Science 310:482–485
Daly JM, Lai AC, Binns MM, Chambers TM, Barrandeguy M, Mumford JA (1996) Antigenic and genetic evolution of equine H3N8 influenza A viruses. J Gen Virol 77(Pt 4):661–671
de Jong JC, Smith DJ, Lapedes AS, Donatelli I, Campitelli L, Barigazzi G, Van Reeth K, Jones TC, Rimmelzwaan GF, Osterhaus AD, Fouchier RA (2007) Antigenic and genetic evolution of swine influenza A (H3N2) viruses in Europe. J Virol 81:4315–4322
Ducatez MF, Hause B, Stigger-Rosser E, Darnell D, Corzo C, Juleen K, Simonson R, Brockwell-Staats C, Rubrum A, Wang D, Webb A, Crumpton JC, Lowe J, Gramer M, Webby RJ (2011) Multiple reassortment between pandemic (H1N1) 2009 and endemic influenza viruses in pigs, United States. Emerg Infect Dis 17:1624–1629
Dundon WG, De Benedictis P, Viale E, Capua I (2010) Serologic evidence of pandemic (H1N1) 2009 infection in dogs, Italy. Emerg Infect Dis 16:2019–2021
Everitt AR, Clare S, Pertel T, John SP, Wash RS, Smith SE, Chin CR, Feeley EM, Sims JS, Adams DJ, Wise HM, Kane L, Goulding D, Digard P, Anttila V, Baillie JK, Walsh TS, Hume DA, Palotie A, Xue Y, Colonna V, Tyler-Smith C, Dunning J, Gordon SB, Gen II, Investigators M, Smyth RL, Openshaw PJ, Dougan G, Brass AL, Kellam P (2012) IFITM3 restricts the morbidity and mortality associated with influenza. Nature 484:519–523
Fouchier RA, Guan Y (2013) Ecology and evolution of influenza viruses in wild and domestic birds, 2nd edn. Wiley, London
Fouchier RA, Munster V, Wallensten A, Bestebroer TM, Herfst S, Smith D, Rimmelzwaan GF, Olsen B, Osterhaus AD (2005) Characterization of a novel influenza A virus hemagglutinin subtype (H16) obtained from black-headed gulls. J Virol 79:2814–2822
Garten RJ, Davis CT, Russell CA, Shu B, Lindstrom S, Balish A, Sessions WM, Xu X, Skepner E, Deyde V, Okomo-Adhiambo M, Gubareva L, Barnes J, Smith CB, Emery SL, Hillman MJ, Rivailler P, Smagala J, de Graaf M, Burke DF, Fouchier RA, Pappas C, Alpuche-Aranda CM, Lopez-Gatell H, Olivera H, Lopez I, Myers CA, Faix D, Blair PJ, Yu C, Keene KM, Dotson PD Jr, Boxrud D, Sambol AR, Abid SH, St George K, Bannerman T, Moore AL, Stringer DJ, Blevins P, Demmler-Harrison GJ, Ginsberg M, Kriner P, Waterman S, Smole S, Guevara HF, Belongia EA, Clark PA, Beatrice ST, Donis R, Katz J, Finelli L, Bridges CB, Shaw M, Jernigan DB, Uyeki TM, Smith DJ, Klimov AI, Cox NJ (2009) Antigenic and genetic characteristics of swine-origin 2009 A(H1N1) influenza viruses circulating in humans. Science 325:197–201
Geraci JR, St Aubin DJ, Barker IK, Webster RG, Hinshaw VS, Bean WJ, Ruhnke HL, Prescott JH, Early G, Baker AS, Madoff S, Schooley RT (1982) Mass mortality of harbor seals: pneumonia associated with influenza A virus. Science 215:1129–1131
Giese M, Harder TC, Teifke JP, Klopfleisch R, Breithaupt A, Mettenleiter TC, Vahlenkamp TW (2008) Experimental infection and natural contact exposure of dogs with avian influenza virus (H5N1). Emerg Infect Dis 14:308–310

Gog JR, Rimmelzwaan GF, Osterhaus AD, Grenfell BT (2003) Population dynamics of rapid fixation in cytotoxic T lymphocyte escape mutants of influenza A. Proc Natl Acad Sci USA 100:11143–11147

Goldstein T, Mena I, Anthony SJ, Medina R, Robinson PW, Greig DJ, Costa DP, Lipkin WI, Garcia-Sastre A, Boyce WM (2013) Pandemic H1N1 influenza isolated from free-ranging Northern Elephant Seals in 2010 off the central California coast. PLoS One 8:e62259

Guan Y, Shortridge KF, Krauss S, Webster RG (1999) Molecular characterization of H9N2 influenza viruses: were they the donors of the "internal" genes of H5N1 viruses in Hong Kong? Proc Natl Acad Sci USA 96:9363–9367

Guo Y, Wang M, Kawaoka Y, Gorman O, Ito T, Saito T, Webster RG (1992) Characterization of a new avian-like influenza A virus from horses in China. Virology 188:245–255

Guo Y, Wang M, Zheng GS, Li WK, Kawaoka Y, Webster RG (1995) Seroepidemiological and molecular evidence for the presence of two H3N8 equine influenza viruses in China in 1993–94. J Gen Virol 76(Pt 8):2009–2014

Hansbro PM, Warner S, Tracey JP, Arzey KE, Selleck P, O'Riley K, Beckett EL, Bunn C, Kirkland PD, Vijaykrishna D, Olsen B, Hurt AC (2010) Surveillance and analysis of avian influenza viruses, Australia. Emerg Infect Dis 16:1896–1904

Hinshaw VS, Bean WJ, Geraci J, Fiorelli P, Early G, Webster RG (1986) Characterization of two influenza A viruses from a pilot whale. J Virol 58:655–656

Hinshaw VS, Webster RG, Easterday BC, Bean WJ Jr (1981) Replication of avian influenza A viruses in mammals. Infect Immun 34:354–361

Horby P, Nguyen NY, Dunstan SJ, Baillie JK (2013) An updated systematic review of the role of host genetics in susceptibility to influenza. Influenza Other Respir Viruses 7(Suppl 2):37–41

Howard WA, Essen SC, Strugnell BW, Russell C, Barass L, Reid SM, Brown IH (2011) Reassortant Pandemic (H1N1) 2009 virus in pigs, United Kingdom. Emerg Infect Dis 17:1049–1052

Ito T, Couceiro JN, Kelm S, Baum LG, Krauss S, Castrucci MR, Donatelli I, Kida H, Paulson JC, Webster RG, Kawaoka Y (1998) Molecular basis for the generation in pigs of influenza A viruses with pandemic potential. J Virol 72:7367–7373

Kay RM, Done SH, Paton DJ (1994) Effect of sequential porcine reproductive and respiratory syndrome and swine influenza on the growth and performance of finishing pigs. Vet Rec 135:199–204

Krauss S, Walker D, Pryor SP, Niles L, Chenghong L, Hinshaw VS, Webster RG (2004) Influenza A viruses of migrating wild aquatic birds in North America. Vector Borne Zoonotic Dis 4:177–189

Kuiken T, Rimmelzwaan G, van Riel D, van Amerongen G, Baars M, Fouchier R, Osterhaus A (2004) Avian H5N1 influenza in cats. Science 306:241

Latorre-Margalef N, Gunnarsson G, Munster VJ, Fouchier RA, Osterhaus AD, Elmberg J, Olsen B, Wallensten A, Haemig PD, Fransson T, Brudin L, Waldenstrom J (2009) Effects of influenza A virus infection on migrating mallard ducks. Proc Biol Sci 276:1029–1036

Lee C, Song D, Kang B, Kang D, Yoo J, Jung K, Na G, Lee K, Park B, Oh J (2009) A serological survey of avian origin canine H3N2 influenza virus in dogs in Korea. Vet Microbiol 137:359–362

Lemey P, Rambaut A, Bedford T, Faria N, Bielejec F, Baele G, Russell CA, Smith DJ, Pybus OG, Brockmann D, Suchard MA (2014) Unifying viral genetics and human transportation data to predict the global transmission dynamics of human influenza H3N2. PLoS Pathog 10:e1003932

Li S, Shi Z, Jiao P, Zhang G, Zhong Z, Tian W, Long LP, Cai Z, Zhu X, Liao M, Wan XF (2010) Avian-origin H3N2 canine influenza A viruses in Southern China. Infect Genet Evol 10:1286–1288

Lindstrom S, Garten R, Balish A, Shu B, Emery S, Berman L, Barnes N, Sleeman K, Gubareva L, Villanueva J, Klimov A (2012) Human infections with novel reassortant influenza A(H3N2)v viruses, United States, 2011. Emerg Infect Dis 18:834–837

Lorusso A, Vincent AL, Harland ML, Alt D, Bayles DO, Swenson SL, Gramer MR, Russell CA, Smith DJ, Lager KM, Lewis NS (2011) Genetic and antigenic characterization of H1 influenza viruses from United States swine from 2008. J Gen Virol 92:919–930

Maas R, Tacken M, Ruuls L, Koch G, van Rooij E, Stockhofe-Zurwieden N (2007) Avian influenza (H5N1) susceptibility and receptors in dogs. Emerg Infect Dis 13:1219–1221

McCullers JA, Van De Velde LA, Schultz RD, Mitchell CG, Halford CR, Boyd KL, Schultz-Cherry S (2011) Seroprevalence of seasonal and pandemic influenza A viruses in domestic cats. Arch Virol 156:117–120

Moreno A, Di Trani L, Faccini S, Vaccari G, Nigrelli D, Boniotti MB, Falcone E, Boni A, Chiapponi C, Sozzi E, Cordioli P (2011) Novel H1N2 swine influenza reassortant strain in pigs derived from the pandemic H1N1/2009 virus. Vet Microbiol 149:472–477

Munster VJ, Baas C, Lexmond P, Waldenstrom J, Wallensten A, Fransson T, Rimmelzwaan GF, Beyer WE, Schutten M, Olsen B, Osterhaus AD, Fouchier RA (2007) Spatial, temporal, and species variation in prevalence of influenza A viruses in wild migratory birds. PLoS Pathog 3:e61

Myers KP, Olsen CW, Gray GC (2007) Cases of swine influenza in humans: a review of the literature. Clin Infect Dis 44:1084–1088

Nelson MI, Edelman L, Spiro DJ, Boyne AR, Bera J, Halpin R, Sengamalay N, Ghedin E, Miller MA, Simonsen L, Viboud C, Holmes EC (2008) Molecular epidemiology of A/H3N2 and A/H1N1 influenza virus during a single epidemic season in the United States. PLoS Pathog 4:e1000133

Nelson MI, Gramer MR, Vincent AL, Holmes EC (2012) Global transmission of influenza viruses from humans to swine. J Gen Virol 93:2195–2203

Nfon CK, Berhane Y, Hisanaga T, Zhang S, Handel K, Kehler H, Labrecque O, Lewis NS, Vincent AL, Copps J, Alexandersen S, Pasick J (2011) Characterization of H1N1 swine influenza viruses circulating in Canadian pigs in 2009. J Virol 85:8667–8679

Olsen B, Munster VJ, Wallensten A, Waldenstrom J, Osterhaus AD, Fouchier RA (2006) Global patterns of influenza A virus in wild birds. Science 312:384–388

Olsen CW, Karasin A, Erickson G (2003) Characterization of a swine-like reassortant H1N2 influenza virus isolated from a wild duck in the United States. Virus Res 93:115–121

Osterhaus AD, Rimmelzwaan GF, Martina BE, Bestebroer TM, Fouchier RA (2000) Influenza B virus in seals. Science 288:1051–1053

Paniker CK, Nair CM (1970) Infection with A2 Hong Kong influenza virus in domestic cats. Bull World Health Organ 43:859–862

Paniker CK, Nair CM (1972) Experimental infection of animals with influenzavirus types A and B. Bull World Health Organ 47:461–463

Peiris JS, Guan Y, Markwell D, Ghose P, Webster RG, Shortridge KF (2001) Cocirculation of avian H9N2 and contemporary "human" H3N2 influenza A viruses in pigs in southeastern China: potential for genetic reassortment? J Virol 75:9679–9686

Pereda AJ, Uhart M, Perez AA, Zaccagnini ME, La Sala L, Decarre J, Goijman A, Solari L, Suarez R, Craig MI, Vagnozzi A, Rimondi A, Konig G, Terrera MV, Kaloghlian A, Song H, Sorrell EM, Perez DR (2008) Avian influenza virus isolated in wild waterfowl in Argentina: evidence of a potentially unique phylogenetic lineage in South America. Virology 378:363–370

Perez DR, Lim W, Seiler JP, Yi G, Peiris M, Shortridge KF, Webster RG (2003) Role of quail in the interspecies transmission of H9 influenza A viruses: molecular changes on HA that correspond to adaptation from ducks to chickens. J Virol 77:3148–3156

Romvary J, Rozsa J, Farkas E (1975) Infection of dogs and cats with the Hong Kong influenza A (H3N2) virus during an epidemic period in Hungary. Acta Vet Acad Sci Hung 25:255–259

Shortridge KF (1992) Pandemic influenza: A zoonosis? Semin Respir Infect 7:11–25

Song D, Lee C, Kang B, Jung K, Oh T, Kim H, Park B, Oh J (2009) Experimental infection of dogs with avian-origin canine influenza A virus (H3N2). Emerg Infect Dis 15:56–58

Songserm T, Amonsin A, Jam-on R, Sae-Heng N, Meemak N, Pariyothorn N, Payungporn S, Theamboonlers A, Poovorawan Y (2006) Avian influenza H5N1 in naturally infected domestic cat. Emerg Infect Dis 12:681–683

Sovinova O, Tumova B, Pouska F, Nemec J (1958) Isolation of a virus causing respiratory disease in horses. Acta Virol 2:52–61

Sponseller BA, Strait E, Jergens A, Trujillo J, Harmon K, Koster L, Jenkins-Moore M, Killian M, Swenson S, Bender H, Waller K, Miles K, Pearce T, Yoon KJ, Nara P (2010) Influenza A pandemic (H1N1) 2009 virus infection in domestic cat. Emerg Infect Dis 16:534–537

Suarez DL, Senne DA, Banks J, Brown IH, Essen SC, Lee CW, Manvell RJ, Mathieu-Benson C, Moreno V, Pedersen JC, Panigrahy B, Rojas H, Spackman E, Alexander DJ (2004) Recombination resulting in virulence shift in avian influenza outbreak, Chile. Emerg Infect Dis 10:693–699

Thanawongnuwech R, Amonsin A, Tantilertcharoen R, Damrongwatanapokin S, Theamboonlers A, Payungporn S, Nanthapornphiphat K, Ratanamungklanon S, Tunak E, Songserm T, Vivatthanavanich V, Lekdumrongsak T, Kesdangsakonwut S, Tunhikorn S, Poovorawan Y (2005) Probable tiger-to-tiger transmission of avian influenza H5N1. Emerg Infect Dis 11:699–701

Tong S, Zhu X, Li Y, Shi M, Zhang J, Bourgeois M, Yang H, Chen X, Recuenco S, Gomez J, Chen LM, Johnson A, Tao Y, Dreyfus C, Yu W, McBride R, Carney PJ, Gilbert AT, Chang J, Guo Z, Davis CT, Paulson JC, Stevens J, Rupprecht CE, Holmes EC, Wilson IA, Donis RO (2013) New world bats harbor diverse influenza A viruses. PLoS Pathog 9:e1003657

Tu J, Zhou H, Jiang T, Li C, Zhang A, Guo X, Zou W, Chen H, Jin M (2009) Isolation and molecular characterization of equine H3N8 influenza viruses from pigs in China. Arch Virol 154:887–890

United States Department of Agriculture ARS (2014) Animal influenza viruses: gap analysis workshop report. http://go.usa.gov/KpGP

Uyeki TM, Chong YH, Katz JM, Lim W, Ho YY, Wang SS, Tsang TH, Au WW, Chan SC, Rowe T, Hu-Primmer J, Bell JC, Thompson WW, Bridges CB, Cox NJ, Mak KH, Fukuda K (2002) Lack of evidence for human-to-human transmission of avian influenza A (H9N2) viruses in Hong Kong, China 1999. Emerg Infect Dis 8:154–159

van Gils JA, Munster VJ, Radersma R, Liefhebber D, Fouchier RA, Klaassen M (2007) Hampered foraging and migratory performance in swans infected with low-pathogenic avian influenza A virus. PLoS One 2:e184

Vijaykrishna D, Poon LL, Zhu HC, Ma SK, Li OT, Cheung CL, Smith GJ, Peiris JS, Guan Y (2010) Reassortment of pandemic H1N1/2009 influenza A virus in swine. Science 328:1529

Vijaykrishna D, Smith GJ, Pybus OG, Zhu H, Bhatt S, Poon LL, Riley S, Bahl J, Ma SK, Cheung CL, Perera RA, Chen H, Shortridge KF, Webby RJ, Webster RG, Guan Y, Peiris JS (2011) Long-term evolution and transmission dynamics of swine influenza A virus. Nature 473:519–522

Vincent A, Awada L, Brown I, Chen H, Claes F, Dauphin G, Donis R, Culhane M, Hamilton K, Lewis N, Mumford E, Nguyen T, Parchariyanon S, Pasick J, Pavade G, Pereda A, Peiris M, Saito T, Swenson S, Van Reeth K, Webby R, Wong F, Ciacci-Zanella J (2014) Review of influenza A virus in swine worldwide: a call for increased surveillance and research. Zoonoses Public Health 61:4–17

Virmani N, Bera BC, Singh BK, Shanmugasundaram K, Gulati BR, Barua S, Vaid RK, Gupta AK, Singh RK (2010) Equine influenza outbreak in India (2008–09): virus isolation, sero-epidemiology and phylogenetic analysis of HA gene. Vet Microbiol 143:224–237

Waddell GH, Teigland MB, Sigel MM (1963) A new influenza virus associated with equine respiratory disease. J Am Vet Med Assoc 143:587–590

Webster RG, Bean WJ, Gorman OT, Chambers TM, Kawaoka Y (1992) Evolution and ecology of influenza A viruses. Microbiol Rev 56:152–179

Webster RG, Hinshaw VS, Bean WJ, Van Wyke KL, Geraci JR, St Aubin DJ, Petursson G (1981) Characterization of an influenza A virus from seals. Virology 113:712–724

Webster RG, Peiris JS, Webby RJ, Guan Y (2013) H7N9 Influenza: what we learned from H5N1. Microbe 8:449–505

Webster RG, Yakhno M, Hinshaw VS, Bean WJ, Murti KG (1978) Intestinal influenza: replication and characterization of influenza viruses in ducks. Virology 84:268–278

WHO/OIE/FAO H5N1 Evolution Working Group (2014) Revised and updated nomenclature for highly pathogenic avian influenza A (H5N1) viruses. Influenza Other Respir Viruses 8(3):384–388

Zhang Y, Zhang Q, Kong H, Jiang Y, Gao Y, Deng G, Shi J, Tian G, Liu L, Liu J, Guan Y, Bu Z, Chen H (2013) H5N1 hybrid viruses bearing 2009/H1N1 virus genes transmit in guinea pigs by respiratory droplet. Science 340:1459–1463

Influenza A Virus Reassortment

John Steel and Anice C. Lowen

Abstract Reassortment is the process by which influenza viruses swap gene segments. This genetic exchange is possible due to the segmented nature of the viral genome and occurs when two differing influenza viruses co-infect a cell. The viral diversity generated through reassortment is vast and plays an important role in the evolution of influenza viruses. Herein we review recent insights into the contribution of reassortment to the natural history and epidemiology of influenza A viruses, gained through population scale phylogenic analyses. We describe methods currently used to study reassortment in the laboratory, and we summarize recent progress made using these experimental approaches to further our understanding of influenza virus reassortment and the contexts in which it occurs.

Abbreviations

Pan/99	A/Panama/2007/1999 (H3N2)
Finch/91	A/finch/England/2051/1991 (H5N2)
Moscow/99	A/Moscow/10/1999 (H3N2)
PR8	A/Puerto Rico/8/1934 (H1N1)
NL/09	A/Netherlands/602/2009 (H1N1)
Anhui/13	A/Anhui/1/2013 (H7N9)
RT	Reverse transcription
PCR	Polymerase chain reaction
HRM	High resolution melt analysis
Ct	Cycle threshold
cDNA	Complementary DNA
dsDNA	Double stranded DNA
vRNPs	Viral ribonucleoproteins
HPAI	Highly pathogenic avian influenza

J. Steel · A.C. Lowen (✉)
Department of Microbiology and Immunology,
Emory University School of Medicine, Atlanta, GA, USA
e-mail: anice.lowen@emory.edu

Current Topics in Microbiology and Immunology (2014) 385: 377–401
DOI: 10.1007/82_2014_395

Published Online: 10 July 2014

LPAI	Low pathogenic avian influenza
TRIG	Triple reassortant internal gene
pH1N1	2009 pandemic H1N1 influenza virus

Contents

1 Introduction

Reassortment is the process by which two differing influenza viruses exchange intact genome segments. It is possible due to the segmented nature of the viral genome and occurs when two distinct strains co-infect one cell. From a co-infection with two influenza A viruses, each carrying eight segments, $2^8 = 256$ different progeny genotypes are possible. Thus, gene swapping through reassortment has the potential to generate significant viral diversity. This diversity, in turn, facilitates rapid evolution under selective pressure, including adaptation to new host environments, evasion of host immune responses and acquisition of resistance to antiviral drugs.

A classical example that illustrates the potential for reassortment to give rise to more fit genotypes under both positive and negative selection is the generation of high-yield vaccine strains. Production of vaccine strains with a high growth phenotype in eggs is classically achieved through forced reassortment between the circulating strain of interest and the egg-adapted A/Puerto Rico/8/34 (PR8) virus. Eggs are co-infected with the two viruses and the major antigenic determinants of PR8 are negatively selected through the addition of antiserum. In this way, reassortant viruses possessing the HA and NA of the circulating strain and a varying number of internal proteins (which drive the high growth phenotype) of PR8 are isolated (Fulvini et al. 2011). The advent of this technology in 1969 revolutionized influenza vaccine production and the process is still standard practise today.

In nature, influenza viral evolution is fuelled by both abrupt genetic changes brought about by reassortment and the gradual acquisition of point mutations due to the error-prone viral polymerase (Scholtissek 1995). Clearly, an important interplay exists between these two processes. Reassortment may bring together two advantageous mutations that arose in different viruses (Ince et al. 2013) or allow the separation of an advantageous change in one segment from a deleterious mutation in another (Holmes et al. 2005). Genetic drift, in turn, allows the acquisition of post-reassortment mutations that ease fitness defects arising due to segment mismatch (Kaverin et al. 1998; Neverov et al. 2014).

A number of recent analyses of viral evolutionary dynamics within individual hosts and at the population level and have revealed the surprisingly high frequency of mixed influenza virus infections and underlined the importance of reassortment in generating epidemiologically significant influenza viruses. Studies in the lab have increased our understanding of how viral genomes mix within co-infected cells, what conditions of co-infection favor or limit reassortment, and, particularly for strains of public health importance, which segment combinations give rise to more fit reassortant genotypes. These recent gains in our understanding of reassortment and its contributions to influenza viral evolution are discussed below.

2 Prevalence and Impact of Reassortment Among Influenza Viruses Circulating in Nature

Reassortment among circulating influenza viruses is identified through phylogenic analysis of multiple genome segments. A separate phylogenic tree is generated for each vRNA segment and clades are identified based on the existence of common ancestors. If an influenza isolate has a reassortant genotype, this isolate will appear in two or more different clades within the set of eight phylogenic trees. A caveat of this approach is that reassortment between highly homologous viruses does not lead to 'clade-jumping' and is therefore difficult to differentiate from genetic drift. Even where reassortment events result in a large change, their robust detection requires an abundance of sequencing data: full genome sequences are needed and the characterization of a large number of isolates is critical for the definition of parental lineages. The Influenza Genome Sequencing Project, launched in 2004, has fully sequenced more than 13,000 influenza virus isolates to date, enabling a number of large-scale analyses of reassortment at the population level. Collectively, these studies have revealed that reassortment is highly prevalent in nature and a critical determinant of influenza virus population structure.

2.1 Reassortment in Wild Bird Reservoirs

The most diverse influenza virus gene pool is that found in avian species. Viruses of sixteen different HA subtypes and nine different NA subtypes co-circulate among multiple species of wild waterfowl and shorebirds (Webster et al. 1992; Munster et al. 2007). The internal genes of avian influenza viruses are markedly less diverse than the HA and NA and have recently been found to descend from the introduction of an equine H7N7 subtype strain into birds in the late 1800s (Worobey et al. 2014a). In the decades following this introduction, a large selective sweep, facilitated by reassortment between the equine-origin internal genes and avian influenza viruses of at least 16 different subtypes, displaced previously circulating internal genes to yield the current global diversity of avian influenza viruses (Worobey et al. 2014a). Within this reservoir, at least 112 of the 144 possible HA/NA combinations have been identified (Alexander 2007; Munster et al. 2007; Dugan et al. 2008; Olson et al. 2014). The prevalence of influenza viruses in birds varies widely with species, location and time of year but can be as high as 30 % (Hinshaw et al. 1985; Krauss et al. 2004; Munster et al. 2007). Migratory patterns of the birds furthermore allow diverse influenza viruses to be carried over large distances and brought together at common breeding grounds or stopover points (Krauss et al. 2004; Lam et al. 2012). One large surveillance study revealed that, among 167 wild bird samples analyzed, 26 % showed evidence of more than one HA and / or NA subtype, indicative of mixed infection (Dugan et al. 2008). The result is an abundance of reassortment and the generation of further viral genomic diversity (Dugan et al. 2008; Abolnik et al. 2010; Fusaro et al. 2011; Deng et al. 2013; Wille et al. 2013). Dugan et al. reported four different genotypes among five H4N6 isolates collected from mallard ducks in the same day and at the same location (Dugan et al. 2008). They furthermore found that reassortment was so common in their dataset that forcing the tree topology of one gene segment onto the sequence data of another segment resulted in a tree that was nearly as unlikely to be correct as were 500 trees generated at random (Dugan et al. 2008). Under these conditions, it can be difficult to determine which genome constellations are parental and which are the products of reassortment. Although genetic exchange among avian influenza viruses is prevalent, ecological barriers, including the migratory paths of avian hosts and spatial distance between these paths, limit gene flow (Lam et al. 2012). The most extreme example of this limited exchange is between avian influenza viruses circulating in the Western and Eastern hemispheres. Due to the separation of bird species between the two hemispheres, the influenza viruses that they carry form two distinct lineages, which mix only occasionally (Krauss et al. 2007; Koehler et al. 2008; Bahl et al. 2009; Lam et al. 2012; Worobey et al. 2014a).

Reassortment among avian influenza viruses and between avian and mammalian-adapted strains often precedes zoonotic transfer of avian influenza virus gene segments to humans. Each of the four pandemics of the last century can be taken as examples. The 1918 pandemic strain has been suggested to have arisen when reassortment brought together a human H1 subtype HA and avian N1 neuraminidase and internal genes (Worobey et al. 2014b), the six internal genes of this virus derive from an equine H7N7 lineage that was introduced into the avian reservoir in the late 1800s (Worobey et al. 2014). Circulation of these internal genes in horses may have conferred a degree of mammalian adaptation that later facilitated avian-to-human transmission of the 1918 virus. The 1957 and 1968 pandemic strains retained five internal gene segments from the 1918 lineage but took on HA, NA and PB1 (in 1957) or HA and PB1 (in 1968) segments from avian influenza viruses (Kilbourne 2006). Most recently, extensive reassortment in pigs brought together gene segments derived from avian, human and swine influenza virus lineages to produce the 2009 pandemic strain (Garten et al. 2009; Smith et al. 2009). Thus, diversity generated through reassortment can facilitate host species jumps leading to human pandemics. Domestic animals are frequently intermediaries in this process, acting as mixing vessels in which reassortment occurs (Scholtissek 1995; Lam et al. 2011), allowing host-adaptive changes which may also render viruses more fit in human hosts (Matsuoka et al. 2009), or amplifying novel reassortant strains and bringing them into proximity with humans (Reperant et al. 2012).

2.2 *Reassortment in Poultry*

Domestic birds, including chickens, turkeys, ducks, geese and a number of minor poultry species are susceptible to infection with a wide range of avian influenza A viruses (Capua et al. 2008; Lvov and Kaverin 2008; Werner et al. 2008; Pasick et al. 2012). Aside from turkeys, the signs of disease caused by low pathogenic avian influenza (LPAI) viruses in these species tend to be mild or subclinical (Wright et al. 2006); as a result, most information on the circulation of influenza viruses in poultry focuses on the H5 and H7 subtypes, which have the potential to become highly pathogenic. It is clear from this literature that the introduction of avian influenza viruses from wild birds into poultry is not uncommon. Furthermore, a number of such viruses have formed established lineages within poultry populations in various regions of the world. Reassortment among viruses enzootic in poultry, and between poultry viruses and strains introduced transiently from wild birds, occurs frequently and has had important consequences for zoonotic infection of humans (Li et al. 2004; Lam et al. 2013; Wu et al. 2013).

Since the 1997 H5N1 outbreak in Hong Kong, increased influenza virus surveillance in poultry has revealed that both H5N1 and H9N2 subtype strains are enzootic in Southeast Asia. These viruses co-circulate within multiple poultry species and frequent reassortment among them is largely responsible for the extensive diversification of H5N1 influenza viruses between 1997 and present (Duan et al. 2008; Lam et al. 2008; Zhao et al. 2008; Nguyen et al. 2012). While some of the reassortant genotypes have been detected only transiently, others have persisted in poultry and out-competed the parental A/Goose/Guangdong/1/96 (H5N1)-like viruses introduced from wild birds (Vijaykrishna et al. 2008).

A number of different reassortant H5N1 genotypes have caused zoonotic infection of humans (Zhao et al. 2008; Wan et al. 2011; Nguyen et al. 2012).

In addition to the predominant H9N2 and H5N1 subtype viruses, other subtypes are present in the poultry of Southeast Asia. As a result, reassortant viruses carrying predominant internal gene cassettes, similar to those of the enzootic H9N2 and H5N1 lineages, with a number of different HA and NA pairings have been detected (Chin et al. 2002; Lam et al. 2013; Wu et al. 2013; Chen et al. 2014; Garcia-Sastre and Schmolke 2014). The most notable example is that of the H7N9 subtype viruses that, between February 2013 and February 2014, have caused 360 human cases of severe disease and greater than 100 deaths in China (WHO 2014). The H7N9 outbreak lineage most likely arose through two separate reassortment events involving wild-bird viruses of H7N3 and H2N9 or H11N9 subtypes and an H9N2 virus enzootic in Chinese poultry (Lam et al. 2013; Wu et al. 2013). The potential for these shared internal genes to support human infection is clear: they have crossed from poultry to humans in the context of H9N2, H5N1, H7N9 and H10N8 subtype viruses (Guan et al. 1999; Lin et al. 2000; Lam et al. 2013; Wu et al. 2013; Chen et al. 2014; Garcia-Sastre and Schmolke 2014). In addition, parallels can be drawn between this avian internal gene cassette, the triple reassortant internal gene (TRIG) cassette in swine (see below and Vincent et al. 2008; Ma et al. 2010) and the historic internal genes derived from the equine influenza outbreak of 1872–1873 (Worobey et al. 2014). In all three cases, particular combinations of six internal genes appear to support efficient replication when coupled to diverse HA/NA glycoprotein genes.

2.3 Reassortment in Swine Reservoirs

Influenza viruses are widespread in swine and multiple lineages currently circulate in swine populations. Pigs are furthermore susceptible to infection with viruses adapted to humans and birds, at least in part due to the presence on the epithelia of their upper respiratory tracts of both α2,6- and α2,3-linked sialic acid receptors (Ito et al. 1998). The resultant potential for influenza viruses that normally circulate in three distinct host species to meet within pigs has led to their labeling as a "mixing vessel," ideal for the genesis of novel reassortants (Scholtissek 1995; Webster et al. 1995). The natural history of swine influenza viruses supports this moniker.

From 1918 until 1997, one major lineage circulated in North American swine populations: that introduced into pigs during the 1918 pandemic (Koen 1919; Shope 1931; Brockwell-Staats et al. 2009). This same lineage was detected in Europe most likely in the 1930s, in the 1950s and in 1976, when it spread to multiple European countries (Harnach et al. 1950; Kaplan and Payne 1959; Nardelli et al. 1978; Lange et al. 2013; Zell et al. 2013). As described briefly below, and in greater detail in a number of excellent reviews (Vincent et al. 2008; Brockwell-Staats et al. 2009; Zell et al. 2013), after 1979 in Europe and the late 1990s in North America and Asia, the picture became much more complicated, with inter-species transmission and reassortment leading to the establishment of numerous swine influenza virus lineages.

In 1979 an avian influenza virus of the H1N1 subtype entered European swine populations and spread widely, establishing the Eurasian avian-like swine lineage and replacing the classical swine lineage in Europe (Pensaert et al. 1981). Reassortment of an Eurasian avian-like swine virus with a human seasonal H3N2 strain led to the emergence of the human-like swine H3N2 lineage in the 1980s, while multiple reassortment events brought together the HA of the human seasonal H1N1 viruses, the NA of the human seasonal H3N2 viruses and the internal gene segments of the Eurasian avian-like swine H1N1 viruses to form the human-like swine H1N2 lineage in the 1990s (Castrucci et al. 1993; Brown et al. 1998; Zell et al. 2013). These three lineages, each carrying the internal genes of the Eurasian avian-like swine lineage, co-circulated and reassorted in European swine prior to 2009 (Van Reeth et al. 2008; Balint et al. 2009).

In North America in 1998, the classical swine lineage had remained relatively stable genetically for approximately 70 years. During this time period, human H3N2 viruses were also detected by serology, but did not become established in pigs (Chambers et al. 1991). In 1998, however, reassortant viruses carrying the H3 and N2 genes of contemporary human influenza viruses were detected in pigs. A double reassortant swine/human genotype and a triple reassortant swine/human/avian genotype were detected (Zhou et al. 1999; Karasin et al. 2000b). The triple reassortant virus spread efficiently among North American swine, forming a major lineage in this population (Webby et al. 2000). After the establishment of the triple reassortant H3N2 lineage, further reassortment gave rise to an array of additional lineages, each carrying the six internal gene segments of the triple reassortant virus (termed the triple reassortant internal gene [TRIG] cassette) with varying HA and NA genes (Karasin et al. 2000a, 2006; Webby et al. 2000, 2004; Vincent et al. 2008; Brockwell-Staats et al. 2009). Compared to the classical swine lineage, the triple reassortant swine lineage is prone to reassortment, possibly because the TRIG cassette supports efficient replication when combined with diverse HA and NA genes (Vincent et al. 2008; Ma et al. 2010).

Each of the major lineages of swine influenza prevalent in Europe and North America also circulate in Asia (Vijaykrishna et al. 2011). In addition, human H3N2 viruses have been detected repeatedly in Asian swine herds (Kundin 1970; Shortridge et al. 1977; Peiris et al. 2001). Reassortment among these strains is commonplace, resulting in novel genotypes that are present only transiently as well as a number of minor lineages that are limited geographically to Asia (Vijaykrishna et al. 2011). Importantly, reassortment between North American triple reassortant swine viruses and Eurasian avian-like swine viruses has been detected repeatedly in Asia, suggesting a possible source for the 2009 pandemic strain (pH1N1) (Vijaykrishna et al. 2011).

In the spring of 2009, a swine origin influenza virus spread to humans to cause the first influenza pandemic since 1968 (Eurosurveillance 2009; MMWR 2009). The genome of the pandemic strain comprised two gene segments from the Eurasian avian-like swine lineage (M and NA) and the remaining six segments from an H1 subtype strain of the triple reassortant swine lineage (Garten et al. 2009; Smith et al. 2009). Notably, the TRIG cassette was disrupted in the pandemic strain by the replacement of the M segment with that of the Eurasian avian-like swine lineage. The pandemic M segment has subsequently been found to confer fitness advantages both in the field and in the laboratory (Chou et al. 2011; Lakdawala et al. 2011; Nelson et al. 2012; Kitikoon et al. 2013; Campbell et al. 2014). Following the outbreak in humans, transmission of the 2009 pandemic strain back to pigs has resulted in further diversification in swine reservoirs through multiple reassortment events with several distinct swine influenza virus lineages (Vijaykrishna et al. 2010; Weingartl et al. 2010; Kitikoon et al. 2013). Zoonotic transmission of reassortant strains back to humans has followed, and the human isolates identified overwhelmingly carry the pandemic M segment in an H3N2 triple reassortant background (MMWR 2012; Kitikoon et al. 2013).

Although host barriers certainly exist, the frequency with which human adapted influenza viruses are detected in swine suggests the barriers are not high. Coupled with the repeated introduction of avian influenza viruses into pigs, the potential for viral diversification through reassortment in the swine reservoir is great. Such diversification over the past two to three decades resulted in a strain sufficiently transmissible and antigenically novel to cause a human pandemic.

2.4 Reassortment Among Human Seasonal Influenza Viruses

Phylogenic analysis of full genome sequences generated through the Influenza Genome Sequencing Project has revealed that human seasonal influenza viruses of both the H3N2 and H1N1 subtypes comprise multiple divergent clades co-circulating on a small spatial-temporal scale (Ghedin et al. 2005; Holmes et al. 2005; Nelson et al. 2006, 2008a; Rambaut et al. 2008; Westgeest et al. 2013). This local diversity appears to result from multiple introductions of influenza virus into a given region each winter season (Nelson et al. 2006, 2008a). Co-circulating lineages were furthermore found to reassort with high frequency, generating significant genomic diversity and epidemiologically important variants (Holmes et al. 2005; Nelson et al. 2006, 2008a, b; Rambaut et al. 2008). For example, the A/Fujian/411/2002-like viruses that caused the epidemic of 2003/2004 arose through reassortment between two co-circulating clades (Holmes et al. 2005; Westgeest et al. 2013). Namely, the HA segment of clade B H3N2 viruses, which had been circulating at a low level since 1999 and had recently acquired two key drift mutations in the HA head domain, swapped into the background of the more prevalent clade A viruses, probably in 2002. This is thought to have moved the antigenically variant HA into a more fit background, with the result that the reassortant strain spread efficiently and caused a large outbreak and vaccine mismatch in the 2003/2004 season (MMWR 2004; Holmes et al. 2005; Jin et al. 2005).

Similarly, phylogenic analysis of the 1918 lineage suggests that the unusually severe epidemics of 1947 and 1951 were the result of intra-subtype reassortment among H1N1 viruses (Nelson et al. 2008b). Another instance of intra-subtype reassortment was found to underlie the dramatic rise between 2004 and 2006 of adamantane resistance among H3N2 viruses. In this case, genetic exchange involving two clades of H3N2 subtype viruses appears to have coupled the S31N resistance marker to an advantageous mutation elsewhere in the viral genome, with the consequence that adamantane resistant viruses became widespread globally, despite only infrequent use of this class of drugs in much of the world (Simonsen et al. 2007).

The co-circulation of H3N2 and H1N1 subtype influenza viruses since 1977 has created the potential for inter-subtype reassortment within the human population. Such reassortment events do not appear to be commonplace but do occur, as evidenced by sporadic reports describing H1N2 clinical isolates prior to the 2009 pandemic (Nishikawa and Sugiyama 1983; Guo et al. 1992; Gregory et al. 2002; Xu et al. 2002; Chen et al. 2006; Al Faress et al. 2008). Furthermore, in the 2001/2002 flu season, an H1N2 virus carrying an A/Caledonia/20/1999 (H1N1)-like HA in an A/Panama/2007/1999 (H3N2)-like background circulated widely (Xu et al. 2002; Ellis et al. 2003). Reassortment in humans between H3N2 viruses and H1N1 strains derived from the 2009 pandemic has also been detected (Mukherjee et al. 2012).

3 Methods for Studying Reassortment in the Laboratory

3.1 Co-infection with Two Intact Viruses

The potential for reassortment between two influenza viruses can be evaluated experimentally by inoculating a laboratory substrate with a mixture of the viruses of interest and then determining the genotypes of progeny viruses arising from the co-infection (Gao and Palese 2009; Octaviani et al. 2010). The substrate used depends on the precise question being addressed, but high multiplicity (>3 PFU/cell) infection in cell culture is most common since this method ensures that all cells become co-infected. If one virus of the pairing is known to be less fit, the chances of reassortment may be improved by increasing the proportion of this virus in the inoculation mixture (Octaviani et al. 2010). Reassortant genotypes carrying segments predominantly from one parent can be favored by using UV exposure to partially inactivate the other parental strain (Banbura et al. 1991). In addition, selection pressure can be introduced, for example through the inclusion of an antibody in the culture medium, to disadvantage one parental strain (Banbura et al. 1991; Kiseleva et al. 2012; Kreibich et al. 2013). The generation of reassortants in this manner was an important means of mapping viral phenotypes prior to the advent of reverse genetics (Mayer et al. 1973; Palese and Schulman 1976b; Ritchey et al. 1976). Co-infection with two intact viruses remains essential when one wishes to evaluate the outcome of mixed infections in an unbiased fashion. That said, the detection of reassortants when using this method can be problematic if the two parental viruses are significantly more fit than the reassortant progeny (which is usually the case). For this reason, when the aim is to obtain and characterize particular reassortant viruses, or to evaluate which reassortants carrying, for example, a given HA segment are viable, reverse genetics-based approaches are now favored.

3.2 Reverse Genetics to Generate Specific Genotypes

Specific reassortant genotypes can be generated by reverse genetics (Fodor et al. 1999; Neumann et al. 1999) using a set of eight cDNAs that correspond to the viral RNA segments of interest (for example, Gao and Palese 2009; Li et al. 2010; Chou et al. 2011; Lakdawala et al. 2011; Octaviani et al. 2011; Imai et al. 2012; Zhang et al. 2013; Campbell et al. 2014). This is a very efficient means of generating specific reassortants or a set of reassortants; however, the successful recovery of less fit reassortant viruses will depend on the efficiency of the reverse genetics system. Unlike the co-infection method, this targeted approach does not, of course, allow estimation of which reassortants are most likely to arise out of a mixed infection in nature.

3.3 Reverse Genetics with Greater than Eight Viral cDNAs

Generating reassortant viruses through reverse genetics with greater than eight but fewer than sixteen cDNAs is a hybrid approach of the preceding two which has been taken up by a number of laboratories (Schrauwen et al. 2011, 2013; Essere et al. 2013; Kimble et al. 2014). For at least one segment, the cDNAs derived from both viruses of interest are included in the transfection. In this way, the problem of parental-type progeny swamping the system is avoided and intrinsic viral fitness is allowed to determine which reassortant strains are recovered.

3.4 Methods for Defining Reassortant Genotypes

Regardless of the genotyping method used, the identification of reassortant viruses arising from mixed infection requires the examination of clonal isolates. Clonal samples representing a single genotype can be obtained by performing limiting dilution or plaque assays with mixed viral populations. A range of different techniques can then be used to determine the source of each genome segment. The earliest method involved direct visualization of viral RNA using polyacrylamide gel electrophoresis (Palese and Schulman 1976a). This method allows genotyping based on electrophoretic mobility and therefore requires that the segments of the parental strains used differ in this property.

More recent technologies rely on reverse transcription of viral RNA and subsequent PCR amplification of cDNAs. Sequencing of partial or full viral genomes followed by comparison of sequences to those of the parental strains can be labor intensive but is a reliable method to detect reassortants (Ince et al. 2013). Another approach is to perform quantitative or standard PCR on viral cDNA with strain-specific primers targeting multiple gene segments (Gao and Palese 2009; Jackson et al. 2009; Dlugolenski et al. 2013; Stincarelli et al. 2013). Amplification products obtained by standard PCR can be visualized by gel electrophoresis, while Ct values obtained through qPCR can be interpreted directly. Each primer pair is designed, based on sequence differences between the parental strains, to amplify a cDNA of one parent but not the corresponding cDNA of the other parent. Thus, a set of 16 primer pairs allows full genotyping of influenza A viruses arising from mixed infection with two parental strains. A drawback of this approach is that it may not allow differentiation of segments that are well conserved between the parental strains.

We recently described a streamlined approach to genotyping reassortant viruses based on high resolution melt (HRM) analysis (Wittwer et al. 2003; Marshall et al. 2013). HRM technology exploits the fact that the melting properties of double stranded DNA (dsDNA) molecules are sequence dependent and allows the differentiation of PCR amplicons differing in sequence by as little as one nucleotide. HRM can be used to identify the parent of a given influenza virus segment by designing primers that (i) bind to regions of the segment that are conserved among the strains of interest and (ii) amplify a ~50–150 nucleotide region that contains one or more sequence differences. The full influenza A virus genome can be typed with a set of eight primer pairs, each corresponding to a different segment. One advantage of this approach is that very similar segments can be differentiated reliably (Marshall et al. 2013). A second is that, depending on the sequence properties of the strains of interest, a single primer pair per segment can be used to distinguish at least three parental viruses (Fig. 1). A drawback is that highly divergent segments, such as HA or NA segments of different subtypes, do not typically contain enough conserved sequence to allow for primers that anneal to identical regions of both parental strains. In this way, HRM and strain-specific PCR approaches are complementary and may work well when used in combination (Fig. 1).

4 Factors Affecting Reassortment Frequency

4.1 Under Unbiased Conditions, Reassortment Occurs with High Frequency

We recently addressed the question, how readily does reassortment occur under conditions unbiased by selective pressure, in cultured cells co-infected at high multiplicity and in guinea pigs co-infected at low or high doses (Marshall et al. 2013).

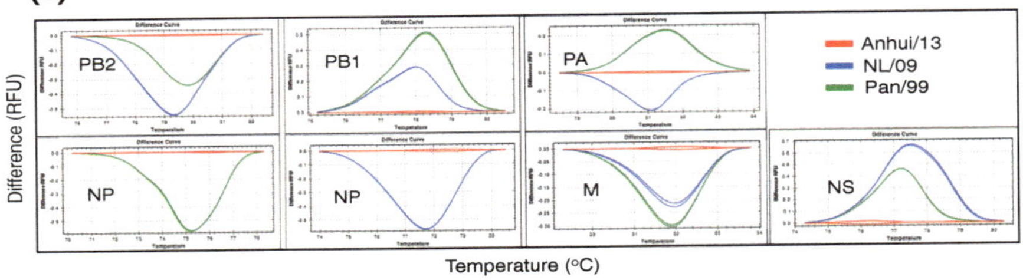

(b)

Identification of HA and NA segments by qPCR

	Primers		
HA	Anhui/13	Pan/99	NL/09
Anhui/13	9.2	>35	>35
Pan/99	>35	14.1	>35
NL/09	>35	34	12.2

	Primers		
NA	Anhui/13	Pan/99	NL/09
Anhui/13	21.2	>35	>35
Pan/99	>35	11.5	>35
NL/09	>35	>35	11.0

Fig. 1 Identification of influenza A/Anhui/1/2013 (H7N9), A/Panama/2007/1999 (H3N2) and A/Netherlands/602/2009 (H1N1) gene segments by high resolution melt or strain specific PCR analyses. **a** The results of high resolution melt analysis of PB2, PB1, PA, NP, M and NS gene segments derived from Anhui/13, Pan/99 and NL/09 viruses are shown. The melt curve of Anhui/13 is set as the reference in each graph and differences between strains in RFU detected at each temperature are plotted. A single primer pair was used for each of the PB2, PB1, PA, M and NS segments and separate primer pairs were used to differentiate NP segments of Anhui/13 and Pan/99 or Anhui/13 and NL/09. **b** Since HA and NA were too divergent to allow application of HRM, PCR using strain specific primers was used to type these segments. Tables show Ct values obtained using cDNA templates derived from each virus strain with primers specific for each strain

To achieve conditions where selective pressures do not act differentially on parental and reassortant progeny viruses and thereby obscure the baseline frequency of reassortment, we used parental viruses that differed only by silent mutations introduced into each genome segment. These mutations were shown not to attenuate the virus and allowed differentiation of the parental gene segments using HRM analysis.

Using these well-matched viruses, we found that reassortment was highly efficient in cell culture and prevalent in vivo. When MDCK cells were inoculated with a mixture containing 10 PFU/cell of each parental virus, 99 % of cells were co-infected and 88 % of progeny viruses carried a reassortant genotype. By titrating down the multiplicity of infection, we furthermore found that reassortment levels declined with co-infection levels. These results indicated that reassortment is not appreciably constrained within co-infected cells and therefore, when parental genomes are well matched, the extent of co-infection is the major determinant of reassortment efficiency.

When guinea pigs were inoculated intranasally with virus mixtures containing 1×10^3 PFU or 1×10^6 PFU of each parental virus, on average 30 and 59 % of viruses isolated on day 2 post-infection were reassortants, respectively (Marshall et al. 2013). When samples were collected at multiple time points after infection from guinea pigs co-infected with 2×10^3 PFU, viral genomic diversity was found to increase early in infection, as shedding titers increased, and then decline as the infection was cleared. In contrast, in guinea pigs infected at a high dose of 2×10^6 PFU, approximately 60 % of viruses shed throughout the course of infection were reassortant (Tao et al. 2014). Coupled with the results of our cell culture experiments, the robust reassortment observed in vivo indicated that co-infection is not rare in the guinea pig respiratory tract. Levels of reassortment were found to vary with dose and timing after inoculation. However, the persistence of parental genotypes at a frequency $\geq$22 % throughout the course of infection indicated that uniformly high multiplicity conditions were not achieved in vivo, irrespective of the day post-infection or inoculation dose.

4.2 Mixing of Genomes Within the Cell

Cell fusion studies and direct visualization of viral RNA molecules within infected cells have recently provided insight into where in the cell and at what stage of the viral life cycle reassortment may occur (Takizawa et al. 2010; Chou et al. 2013). The viral RNAs of an incoming virus particle were found to remain associated until they reached the nucleus (Chou et al. 2013). Segments were dispersed within the nucleus and upon delivery to the cytoplasm (Takizawa et al. 2010; Chou et al. 2013). Assembly of segments into full genome complexes occurred, not at the site of budding at the plasma membrane, but in the cytoplasm in association with Rab11-positive vesicles (Chou et al. 2013). Rab11-positive vesicles have been shown to mediate trafficking of viral ribonucleoproteins (vRNPs) to the plasma membrane (Amorim et al. 2011; Avilov et al. 2012). These results suggest that mixing of gene segments within co-infected cells is most likely to occur in the nucleus and in the cytoplasm following export of new vRNPs from the nucleus. The high baseline frequency of reassortment (88 %) seen when virtually all cells are co-infected furthermore suggests that this mixing occurs freely (Marshall et al. 2013).

4.3 Timing of Super-infection

Mixed influenza virus infections arise either through sequential, independent, infection events or through transmission of multiple variants from a host harbouring a mixed infection (Hughes et al. 2012; Murcia et al. 2012, 2013). Of course, the latter type of mixed infection requires that sequential infections occurred previously in the transmission chain. The importance of sequential transmission events prompts the question, how long after a primary influenza virus infection can a host be super-infected with a second influenza virus? We addressed this question in the guinea pig model using 10^3 PFU of our silently mutated variant of Pan/99 virus for primary infection and the same dose of Pan/99 wild-type virus for secondary infection (Marshall et al. 2013). We found that secondary inoculation within 24 h, but not at 48 h or 72 h, after primary infection lead to productive super-infection, as determined by detection of wild-type viral RNA at 2 d post-infection. The time window within which reassortant progeny were generated was somewhat more limited, however. Robust reassortment was seen with a 6 or 12 h delay between infections, while the frequency of reassortants was at the limit of detection with an 18 h delay and reassortants were not detected with a 24 h delay. Interestingly, reassortment levels were significantly higher in guinea pigs super-infected at 12 h compared to those co-infected at 0 h. Cell culture experiments performed under single vs. multiple cycle growth conditions suggested that allowing spread of the first virus prior to super-infection resulted in higher numbers of co-infected cells (Marshall et al. 2013). Thus, our data suggest that there is an optimal time window for super-infection to lead to co-infection and reassortment in vivo, between approximately 6 and 18 h after primary infection given the virus dosage used.

4.4 Genetic Incompatibility

Unlike the highly homologous model viruses used for the studies described above, genetically dissimilar influenza viruses do not reassort randomly. As was first described with the pairing of A/Puerto Rico/8/1934 (H1N1) and A/Hong Kong/1968 (H3N2) viruses, certain progeny genotypes arise more frequently than others (Lubeck et al. 1979). There are at least two reasons for this, both of which arise due to the evolutionary divergence among influenza virus lineages. The first is incompatibility among RNA packaging signals, which direct the incorporation of the eight gene segments into virions. The second comprises functional mismatches occurring throughout the viral life cycle.

4.4.1 Packaging Signal Mismatch

Segmentation of the viral genome is advantageous in that it allows reassortment and thereby facilitates rapid viral adaptation under selective pressure. However, segmentation also presents a disadvantage for the virus in that it complicates the process of virus assembly: to be infectious, a nascent influenza A virus must carry a complete set of eight segments. The precise mechanism by which influenza viruses ensure packaging of all segments into virions remains unclear. However, current evidence points to a selective mechanism mediated by RNA-RNA interactions among segment-specific packaging signals (Noda et al. 2006, 2012; Chou et al. 2012). Each packaging signal consists of the 3' and 5' untranslated regions plus a variable length of adjacent coding sequence at each end (Fujii et al. 2003, 2005; Gog et al. 2007; Marsh et al. 2007, 2008). These regions differ in sequence among the eight segments and, importantly for reassortment, the packaging signals of a given segment also differ among strains of influenza A virus. These strain differences in packaging signals were shown to dictate the genotypes of reassortant viruses obtained from mixed infection with the strains A/finch/England/2051/1991 (H5N2) [Finch/91] and A/Moscow/10/1999 (H3N2) [Moscow/99] (Essere et al. 2013). Incorporation of the Finch/91 HA segment into the Moscow/99 background was found to require either co-incorporation of the Finch/91 M segment or transfer of the packaging signals of the Moscow/99 HA segment to the Finch/91 HA segment. These results indicate a role for packaging signals in defining reassortment potential and suggest that the HA and M segments of Finch/91 interact during packaging. The network of interactions among viral segments has not been characterized in full, and which segments interact directly will impact reassortment (Greenbaum et al. 2012). Importantly, it also remains unclear whether the interaction network is conserved among strains. Thus, new evidence suggests that the spectrum of genotypes emerging from a co-infected cell is shaped by packaging signal compatibility, but further elucidation of packaging mechanisms is needed before the full impact on reassortment can be determined.

4.4.2 Functional Mismatch

While the formation of reassortant genotypes is governed in part by packaging signal compatibility, the subsequent fitness of reassortant viruses depends heavily on compatibility among viral components that function throughout the life cycle. For example, if physical interactions among viral polymerase subunits derived from different strains are suboptimal, the efficiency of viral replication and transcription will be affected (Li et al. 2008). Functional, rather than direct, interactions can also be subject to mismatch. Examples include the balance between HA binding and NA release activities (Kaverin et al. 2000; Mitnaul et al. 2000; Wagner et al. 2002) and the role of NS1 in modulating the race between the viral polymerase and cellular antiviral responses (Grimm et al. 2007). Such functional mismatches can lead to large fitness defects and therefore have important consequences for the detection of reassortant viruses in the laboratory and the potential for reassortant viruses to spread in nature.

The broad range of phenotypes generated through reassortment is illustrated by a number of risk assessment studies that have evaluated reassortants combining gene segments of highly pathogenic H5N1 avian influenza viruses and H1N1 or H3N2 subtype human strains (Maines et al. 2006; Chen et al. 2008; Jackson et al. 2009; Li et al. 2010; Octaviani et al. 2010; Cline et al. 2011; Zhang et al. 2013). Where large or comprehensive panels of reassortants were evaluated, as many as half of H5N1/H3N2 reassortants were found to have severe fitness defects (Chen et al. 2008; Li et al. 2010). In contrast, a panel of 126 H5N1/pH1N1 reassortants were reported to replicate efficiently in embryonated eggs (Zhang et al. 2013). Some studies identified reassortants with increased virulence relative to one or both parental viruses (Li et al. 2010; Cline et al. 2011; Zhang et al. 2013), while others found that all reassortants tested were less fit than the parental strains

(Maines et al. 2006; Chen et al. 2008; Jackson et al. 2009). Due to differences in the viral strains and host systems selected for study, there is little overlap among these publications in the genotypes found to confer the highest replication or virulence. In one study, the human H3N2 origin PB2 was found to be central to increased fitness of H5N1-based reassortants (Li et al. 2010); in another the M and NA segments derived from human strains were found to be incorporated most frequently into the H5 background (Schrauwen et al. 2013); in a third, a diverse set of genotypes supported efficient replication but those with highest virulence included the avian PB1, HA and NA segments (Chen et al. 2008); a fourth showed that the H5 HA segment alone was sufficient to increase fitness in a 2009 pandemic H1N1 background (Cline et al. 2011); a fifth found that the NS or PA segments of a pH1N1 strain supported guinea pig transmission of an H5N1 virus (Zhang et al. 2013). It is clear from these and other efforts to describe the likelihood of reassortment between two divergent influenza A viruses that outcomes are often specific to the strains under study, as well as the culture system or animal host used.

4.5 Selection Pressure

Clearly there is interplay between reassortment and selection; whether reassortant viruses arise only transiently or outcompete other variants depends not only on their intrinsic fitness, but also on the host environment and the selection pressures it presents. Indeed, the effects of mismatch between viral components may vary with host or environmental conditions (Chen et al. 2008). Experimental data demonstrating differential fitness of reassortant viruses in different contexts is limited. The concept is, however, validated by classical approaches to generating reassortant viruses, in which antibody or other pressures are applied to select against parental genotypes (Fulvini et al. 2011). Thus, while many reassortants carry fitness defects due to segment mismatch, these defects may be offset by selective advantages including escape from immune pressure and adaptation to new host species.

5 Conclusions and Perspectives

Much remains to be done to gain a comprehensive understanding of influenza virus reassortment and the conditions under which it occurs.

A key question that remains to be addressed is whether influenza A virus reassortment occurs more readily in some host species than others. The answer to this question is likely to be complex, with results depending on the viral strains studied and the relative susceptibility of various host species to those strains. Nevertheless, important patterns may emerge. One hypothesis is that reassortment is more efficient in contexts where influenza viruses are relatively limited in their tropism, such that co-infections are more likely to occur. Alternatively, some host species may harbor cell types that are broadly permissive for infection with a wide range of influenza viruses while, in other host species, divergent viruses may target distinct cell types. The identification of host environments conducive to reassortment is important in that it would allow prioritization of species for surveillance.

The impact of packaging signals on reassortment is another topic that should receive further research attention. It is clear that mismatch between the packaging signals of two viral strains can disrupt their reassortment. More remains to be learned, however, including the quantitative effect of such mismatches on the production of reassortant progeny, the segment pairings most affected, and the degree to which these effects are strain-specific.

Reassortment is a complex process: a wide range of factors comes into play in determining the outcome of mixed influenza virus infections. As viral genome sequencing data accumulates, population-based studies will gain even greater resolution and will continue to reveal novel insights into how reassortment, drift and selection interact to drive influenza virus evolution. At the same time, new approaches to studying reassortment in the lab will allow quantification of the impacts of virus, host and environmental factors on reassortment efficiency, thereby defining the conditions under which reassortant viruses are most likely to emerge in nature.

Acknowledgments Research in the authors' laboratories is supported by the NIH under R01 AI099000 (to AL) and the Center for Excellence in Influenza Research and Surveillance (CEIRS) contract number HHSN272201400004C (to JS and AL).

References

Abolnik C, Gerdes GH, Sinclair M et al (2010) Phylogenetic analysis of influenza A viruses (H6N8, H1N8, H4N2, H9N2, H10N7) isolated from wild birds, ducks, and ostriches in South Africa from 2007 to 2009. Avian Dis 54:313–322

Al Faress S, Ferraris O, Moules V, Valette M, Hay A, Lina B (2008) Identification and characterization of a late AH1N2 human reassortant in France during the 2002–2003 influenza season. Virus Res 132:33–41. doi:S0168-1702(07)00369-3, 10.1016/j.virusres.2007.10.007

Alexander DJ (2007) An overview of the epidemiology of avian influenza. Vaccine 25: 5637–5644

Amorim MJ, Bruce EA, Read EK, Foeglein A, Mahen R, Stuart AD, Digard P (2011) A Rab11 and microtubule dependent mechanism for cytoplasmic transport of influenza A virus vRNA. J Virol. doi:10.1128/JVI.02606-10

Avilov SV, Moisy D, Naffakh N, Cusack S (2012) Influenza A virus progeny vRNP trafficking in live infected cells studied with the virus-encoded fluorescently tagged PB2 protein. Vaccine 30:7411–7417. doi:10.1016/j.vaccine.2012.09.077, S0264-410X(12)01421-1

Bahl J, Vijaykrishna D, Holmes EC, Smith GJ, Guan Y (2009) Gene flow and competitive exclusion of avian influenza A virus in natural reservoir hosts. Virology 390:289–297. doi:S0042-6822(09)00287-6, 10.1016/j.virol.2009.05.002

Balint A, Metreveli G, Widen F et al (2009) The first Swedish H1N2 swine influenza virus isolate represents an uncommon reassortant. Virol J 6:180. doi:10.1186/1743-422X-6-180
Banbura MW, Kawaoka Y, Thomas TL, Webster RG (1991) Reassortants with equine 1 (H7N7) influenza virus hemagglutinin in an avian influenza virus genetic background are pathogenic in chickens. Virology 184:469–471
Brockwell-Staats C, Webster RG, Webby RJ (2009) Diversity of influenza viruses in swine and the emergence of a novel human pandemic influenza A (H1N1). Influenza Other Respir Viruses 3:207–213. doi:10.1111/j.1750-2659.2009.00096.x
Brown IH, Harris PA, McCauley JW, Alexander DJ (1998) Multiple genetic reassortment of avian and human influenza A viruses in European pigs, resulting in the emergence of an H1N2 virus of novel genotype. J Gen Virol 79(Pt 12):2947–2955
Campbell PJ, Danzy S, Kyriakis CS, Deymier MJ, Lowen AC, Steel J (2014) The M segment of the 2009 pandemic influenza virus confers increased NA activity, filamentous morphology and efficient contact transmissibility to A/Puerto Rico/8/1934-based reassortant viruses. J Virol. doi:10.1128/JVI.03607-13
Capua I, Cattoli G, Terregino C, Marangon S (2008) Avian influenza in Italy 1997-2006. In: Klenk HD, Matrosovich MN, Stech J (eds) Avian influenza. Karger, Basel, pp 59–70
Castrucci MR, Donatelli I, Sidoli L, Barigazzi G, Kawaoka Y, Webster RG (1993) Genetic reassortment between avian and human influenza A viruses in Italian pigs. Virology 193:503–506. doi:S0042-6822(83)71155-4, 10.1006/viro.1993.1155
Chambers TM, Hinshaw VS, Kawaoka Y, Easterday BC, Webster RG (1991) Influenza viral infection of swine in the United States 1988–1989. Arch Virol 116:261–265
Chen H, Yuan H, Gao R et al (2014) Clinical and epidemiological characteristics of a fatal case of avian influenza A H10N8 virus infection: a descriptive study. Lancet 383:714–721. doi:10.1016/S0140-6736(14)60111-2
Chen LM, Davis CT, Zhou H, Cox NJ, Donis RO (2008) Genetic compatibility and virulence of reassortants derived from contemporary avian H5N1 and human H3N2 influenza A viruses. PLoS Pathog 4:e1000072. doi:10.1371/journal.ppat.1000072
Chen MJ, La T, Zhao P, Tam JS, Rappaport R, Cheng SM (2006) Genetic and phylogenetic analysis of multi-continent human influenza A(H1N2) reassortant viruses isolated in 2001 through 2003. Virus Res 122:200–205. doi:S0168-1702(06)00245-0, 10.1016/j.virusres.2006.07.010
Chin PS, Hoffmann E, Webby R, Webster RG, Guan Y, Peiris M, Shortridge KF (2002) Molecular evolution of H6 influenza viruses from poultry in Southeastern China: prevalence of H6N1 influenza viruses possessing seven A/Hong Kong/156/97 (H5N1)-like genes in poultry. J Virol 76:507–516
Chou YY, Albrecht RA, Pica N et al (2011) The M segment of the 2009 new pandemic H1N1 influenza virus is critical for its high transmission efficiency in the guinea pig model. J Virol 85:11235–11241. doi:10.1128/JVI.05794-11
Chou YY, Heaton NS, Gao Q, Palese P, Singer R, Lionnet T (2013) Colocalization of different influenza viral RNA segments in the cytoplasm before viral budding as shown by single-molecule sensitivity FISH analysis. PLoS Pathog 9:e1003358. doi:10.1371/journal.ppat.1003358, PPATHOGENS-D-12-03007
Chou YY, Vafabakhsh R, Doganay S, Gao Q, Ha T, Palese P (2012) One influenza virus particle packages eight unique viral RNAs as shown by FISH analysis. Proc Natl Acad Sci USA 109:9101–9106. doi:10.1073/pnas.1206069109
Cline TD, Karlsson EA, Freiden P, Seufzer BJ, Rehg JE, Webby RJ, Schultz-Cherry S (2011) Increased pathogenicity of a reassortant 2009 pandemic H1N1 influenza virus containing an H5N1 hemagglutinin. J Virol 85:12262–12270. doi:10.1128/JVI.05582-11
Deng G, Tan D, Shi J et al (2013) Complex reassortment of multiple subtypes of avian influenza viruses in domestic ducks at the dongting lake region of china. J Virol 87:9452–9462. doi:10.1128/JVI.00776-13

Dlugolenski D, Jones L, Tompkins SM, Crameri G, Wang LF, Tripp RA (2013) Bat cells from Pteropus alecto are susceptible to influenza A virus infection and reassortment. Influenza Other Respi Viruses. doi:10.1111/irv.12128
Duan L, Bahl J, Smith GJ et al (2008) The development and genetic diversity of H5N1 influenza virus in China, 1996–2006. Virology 380:243–254. doi:S0042-6822(08)00485-6, 10.1016/j.virol.2008.07.038
Dugan VG, Chen R, Spiro DJ et al (2008) The evolutionary genetics and emergence of avian influenza viruses in wild birds. PLoS Pathog 4:e1000076. doi:10.1371/journal.ppat.1000076
Ellis JS, Alvarez-Aguero A, Gregory V, Lin YP, Hay A, Zambon MC (2003) Influenza AH1N2 viruses, United Kingdom, 2001–2002 influenza season. Emerg Infect Dis 9:304–310
Essere B, Yver M, Gavazzi C et al (2013) Critical role of segment-specific packaging signals in genetic reassortment of influenza A viruses. Proc Natl Acad Sci USA. doi:10.1073/pnas.1308649110
Eurosurveillance (2009) Pandemic alert level 6: scientific criteria for an influenza pandemic fulfilled. Euro Surveill 14:19237. doi: 19237
Fodor E, Devenish L, Engelhardt OG, Palese P, Brownlee GG, Garcia-Sastre A (1999) Rescue of influenza A virus from recombinant DNA. J Virol 73:9679–9682
Fujii K, Fujii Y, Noda T et al (2005) Importance of both the coding and the segment-specific noncoding regions of the influenza A virus NS segment for its efficient incorporation into virions. J Virol 79:3766–3774 doi:79/6/3766, 10.1128/JVI.79.6.3766-3774.2005
Fujii Y, Goto H, Watanabe T, Yoshida T, Kawaoka Y (2003) Selective incorporation of influenza virus RNA segments into virions. Proc Natl Acad Sci USA 100:2002–2007. doi:10.1073/pnas.0437772100
Fulvini AA, Ramanunninair M, Le J et al (2011) Gene constellation of influenza A virus reassortants with high growth phenotype prepared as seed candidates for vaccine production. PLoS One 6:e20823. doi:10.1371/journal.pone.0020823, PONE-D-11-01114
Fusaro A, Monne I, Salviato A et al (2011) Phylogeography and evolutionary history of reassortant H9N2 viruses with potential human health implications. J Virol 85:8413–8421. doi:10.1128/JVI.00219-11
Gao Q, Palese P (2009) Rewiring the RNAs of influenza virus to prevent reassortment. Proc Natl Acad Sci U S A 106:15891–15896. doi:10.1073/pnas.0908897106
Garcia-Sastre A, Schmolke M (2014) Avian influenza A H10N8—a virus on the verge? Lancet 383:676–677. doi:10.1016/S0140-6736(14)60163-X
Garten RJ, Davis CT, Russell CA et al (2009) Antigenic and genetic characteristics of swine-origin 2009 A(H1N1) influenza viruses circulating in humans. Science 325:197–201. doi:10.1126/science.1176225
Ghedin E, Sengamalay NA, Shumway M et al (2005) Large-scale sequencing of human influenza reveals the dynamic nature of viral genome evolution. Nature 437:1162–1166. doi:10.1038/nature04239
Gog JR, Afonso Edos S, Dalton RM et al (2007) Codon conservation in the influenza A virus genome defines RNA packaging signals. Nucleic Acids Res 35:1897–1907. doi:10.1093/nar/gkm087
Greenbaum BD, Li OT, Poon LL, Levine AJ, Rabadan R (2012) Viral reassortment as an information exchange between viral segments. Proc Natl Acad Sci USA 109:3341–3346. doi:10.1073/pnas.1113300109
Gregory V, Bennett M, Orkhan MH et al (2002) Emergence of influenza A H1N2 reassortant viruses in the human population during 2001. Virology 300:1–7. doi: S0042682202915138
Grimm D, Staeheli P, Hufbauer M et al (2007) Replication fitness determines high virulence of influenza A virus in mice carrying functional Mx1 resistance gene. Proc Natl Acad Sci USA 104:6806–6811. doi:10.1073/pnas.0701849104
Guan Y, Shortridge KF, Krauss S, Webster RG (1999) Molecular characterization of H9N2 influenza viruses: were they the donors of the "internal" genes of H5N1 viruses in Hong Kong? Proc Natl Acad Sci USA 96:9363–9367

Guo YJ, Xu XY, Cox NJ (1992) Human influenza A (H1N2) viruses isolated from China. J Gen Virol 73(Pt 2):383–387
Harnach R, Hubik R, Chivatal O (1950) Isolation of influenza virus in Czechoslovakia. Cas Cesk Vet 5:289
Hinshaw VS, Wood JM, Webster RG, Deibel R, Turner B (1985) Circulation of influenza viruses and paramyxoviruses in waterfowl originating from two different areas of North America. Bull World Health Organ 63:711–719
Holmes EC, Ghedin E, Miller N et al (2005) Whole-genome analysis of human influenza A virus reveals multiple persistent lineages and reassortment among recent H3N2 viruses. PLoS Biol 3:e300. doi:05-PLBI-RA-0438R1, 10.1371/journal.pbio.0030300
Hughes J, Allen RC, Baguelin M et al (2012) Transmission of equine influenza virus during an outbreak is characterized by frequent mixed infections and loose transmission bottlenecks. PLoS Pathog 8:e1003081. doi:10.1371/journal.ppat.1003081, PPATHOGENS-D-12-01833
Imai M, Watanabe T, Hatta M et al (2012) Experimental adaptation of an influenza H5 HA confers respiratory droplet transmission to a reassortant H5 HA/H1N1 virus in ferrets. Nature 486:420–428. doi:10.1038/nature10831
Ince WL, Gueye-Mbaye A, Bennink JR, Yewdell JW (2013) Reassortment complements spontaneous mutation in influenza A virus NP and M1 genes to accelerate adaptation to a new host. J Virol 87:4330–4338. doi:10.1128/JVI.02749-12
Ito T, Couceiro JN, Kelm S et al (1998) Molecular basis for the generation in pigs of influenza A viruses with pandemic potential. J Virol 72:7367–7373
Jackson S, Van Hoeven N, Chen LM, Maines TR, Cox NJ, Katz JM, Donis RO (2009) Reassortment between avian H5N1 and human H3N2 influenza viruses in ferrets: a public health risk assessment. J Virol 83:8131–8140. doi:10.1128/JVI.00534-09
Jin H, Zhou H, Liu H et al (2005) Two residues in the hemagglutinin of A/Fujian/411/02-like influenza viruses are responsible for antigenic drift from A/Panama/2007/99. Virology 336:113–119 doi:S0042-6822(05)00155-8, 10.1016/j.virol.2005.03.010
Kaplan MM, Payne AM (1959) Serological survey in animals for type A influenza in relation to the 1957 pandemic. Bull World Health Organ 20:465–488
Karasin AI, Carman S, Olsen CW (2006) Identification of human H1N2 and human-swine reassortant H1N2 and H1N1 influenza A viruses among pigs in Ontario, Canada (2003 to 2005). J Clin Microbiol 44:1123–1126. doi:44/3/1123, 10.1128/JCM.44.3.1123-1126.2006
Karasin AI, Olsen CW, Anderson GA (2000a) Genetic characterization of an H1N2 influenza virus isolated from a pig in Indiana. J Clin Microbiol 38:2453–2456
Karasin AI, Schutten MM, Cooper LA et al (2000b) Genetic characterization of H3N2 influenza viruses isolated from pigs in North America, 1977–1999: evidence for wholly human and reassortant virus genotypes. Virus Res 68:71–85. doi: S0168-1702(00)00154-4
Kaverin NV, Gambaryan AS, Bovin NV et al (1998) Postreassortment changes in influenza A virus hemagglutinin restoring HA-NA functional match. Virology 244:315–321 doi:S0042-6822(98)99119-X, 10.1006/viro.1998.9119
Kaverin NV, Matrosovich MN, Gambaryan AS et al (2000) Intergenic HA-NA interactions in influenza A virus: postreassortment substitutions of charged amino acid in the hemagglutinin of different subtypes. Virus Res 66:123–129. doi: S0168170299001318
Kilbourne ED (2006) Influenza pandemics of the 20th century. Emerg Infect Dis 12:9–14
Kimble JB, Angel M, Wan H, Sutton TC, Finch C, Perez DR (2014) Alternative reassortment events leading to transmissible H9N1 influenza viruses in the ferret model. J Virol 88:66–71. doi:10.1128/JVI.02677-13
Kiseleva I, Dubrovina I, Bazhenova E, Fedorova E, Larionova N, Rudenko L (2012) Possible outcomes of reassortment in vivo between wild type and live attenuated influenza vaccine strains. Vaccine 30:7395–7399. doi:10.1016/j.vaccine.2012.09.076, S0264-410X(12)01420-X
Kitikoon P, Nelson MI, Killian ML, Anderson TK, Koster L, Culhane MR, Vincent AL (2013) Genotype patterns of contemporary reassorted H3N2 virus in US swine. J Gen Virol 94:1236–1241. doi:10.1099/vir.0.51839-0, 94/Pt_6/1236

Koehler AV, Pearce JM, Flint PL, Franson JC, Ip HS (2008) Genetic evidence of intercontinental movement of avian influenza in a migratory bird: the northern pintail (Anas acuta). Mol Ecol 17:4754–4762. doi:MEC3953, 10.1111/j.1365-294X.2008.03953.x

Koen JS (1919) A practical method for field diagnosis of swine diseases. Am J Vet Med 14:468–470

Krauss S, Obert CA, Franks J et al (2007) Influenza in migratory birds and evidence of limited intercontinental virus exchange. PLoS Pathog 3:e167. doi:07-PLPA-RA-0309, 10.1371/journal.ppat.0030167

Krauss S, Walker D, Pryor SP, Niles L, Chenghong L, Hinshaw VS, Webster RG (2004) Influenza A viruses of migrating wild aquatic birds in North America. Vector Borne Zoonotic Dis 4:177–189

Kreibich A, Stech O, Hundt J, Ziller M, Mettenleiter TC, Stech J (2013) Avian influenza virus h3 hemagglutinin may enable high fitness of novel human virus reassortants. PLoS One 8:e79165. doi:10.1371/journal.pone.0079165, PONE-D-13-04453

Kundin WD (1970) Hong Kong A-2 influenza virus infection among swine during a human epidemic in Taiwan. Nature 228:857

Lakdawala SS, Lamirande EW, Suguitan AL Jr et al (2011) Eurasian-origin gene segments contribute to the transmissibility, aerosol release, and morphology of the 2009 pandemic H1N1 influenza virus. PLoS Pathog 7:e1002443. doi:10.1371/journal.ppat.1002443, PPATHOGENS-D-11-01769

Lam TT, Hon CC, Pybus OG et al (2008) Evolutionary and transmission dynamics of reassortant H5N1 influenza virus in Indonesia. PLoS Pathog 4:e1000130. doi:10.1371/journal.ppat.1000130

Lam TT, Ip HS, Ghedin E et al (2012) Migratory flyway and geographical distance are barriers to the gene flow of influenza virus among North American birds. Ecol Lett 15:24–33. doi:10.1111/j.1461-0248.2011.01703.x

Lam TT, Wang J, Shen Y et al (2013) The genesis and source of the H7N9 influenza viruses causing human infections in China. Nature. doi:10.1038/nature12515

Lam TT, Zhu H, Wang J et al (2011) Reassortment events among swine influenza A viruses in China: implications for the origin of the 2009 influenza pandemic. J Virol 85:10279–10285. doi:10.1128/JVI.05262-11

Lange J, Groth M, Kanrai P et al (2013) Circulation of classical swine influenza virus in Europe between the wars? Arch Virol. doi:10.1007/s00705-013-1950-x

Li C, Hatta M, Nidom CA, Muramoto Y, Watanabe S, Neumann G, Kawaoka Y (2010) Reassortment between avian H5N1 and human H3N2 influenza viruses creates hybrid viruses with substantial virulence. Proc Natl Acad Sci USA 107:4687–4692. doi:10.1073/pnas.0912807107

Li C, Hatta M, Watanabe S, Neumann G, Kawaoka Y (2008) Compatibility among polymerase subunit proteins is a restricting factor in reassortment between equine H7N7 and human H3N2 influenza viruses. J Virol 82:11880–11888. doi:10.1128/JVI.01445-08

Li KS, Guan Y, Wang J et al (2004) Genesis of a highly pathogenic and potentially pandemic H5N1 influenza virus in eastern Asia. Nature 430:209–213. doi:10.1038/nature02746

Lin YP, Shaw M, Gregory V et al (2000) Avian-to-human transmission of H9N2 subtype influenza A viruses: relationship between H9N2 and H5N1 human isolates. Proc Natl Acad Sci USA 97:9654–9658. doi:10.1073/pnas.160270697

Lubeck MD, Palese P, Schulman JL (1979) Nonrandom association of parental genes in influenza A virus recombinants. Virology 95:269–274

Lvov DK, Kaverin NV (2008) Avian influenza in Northern Eurasia. In: Klenk HD, Matrosovich MN, Stech J (eds) Avian influenza. Karger, Basel, pp 41–58

Ma W, Lager KM, Lekcharoensuk P et al (2010) Viral reassortment and transmission after co-infection of pigs with classical H1N1 and triple-reassortant H3N2 swine influenza viruses. J Gen Virol 91:2314–2321. doi:10.1099/vir.0.021402-0

Maines TR, Chen LM, Matsuoka Y et al (2006) Lack of transmission of H5N1 avian-human reassortant influenza viruses in a ferret model. Proc Natl Acad Sci USA 103:12121–12126. doi:10.1073/pnas.0605134103

Marsh GA, Hatami R, Palese P (2007) Specific residues of the influenza A virus hemagglutinin viral RNA are important for efficient packaging into budding virions. J Virol 81:9727–9736. doi:10.1128/JVI.01144-07

Marsh GA, Rabadan R, Levine AJ, Palese P (2008) Highly conserved regions of influenza a virus polymerase gene segments are critical for efficient viral RNA packaging. J Virol 82:2295–2304. doi:10.1128/JVI.02267-07

Marshall N, Priyamvada L, Ende Z, Steel J, Lowen AC (2013) Influenza virus reassortment occurs with high frequency in the absence of segment mismatch. PLoS Pathog 9:e1003421. doi:10.1371/journal.ppat.1003421, PPATHOGENS-D-12-03096

Matsuoka Y, Swayne DE, Thomas C et al (2009) Neuraminidase stalk length and additional glycosylation of the hemagglutinin influence the virulence of influenza H5N1 viruses for mice. J Virol 83:4704–4708. doi:10.1128/JVI.01987-08

Mayer V, Schulman JL, Kilbourne ED (1973) Nonlinkage of neurovirulence exclusively to viral hemagglutinin or neuraminidase in genetic recombinants of A-NWS (H0N1) influenza virus. J Virol 11:272–278

Mitnaul LJ, Matrosovich MN, Castrucci MR, Tuzikov AB, Bovin NV, Kobasa D, Kawaoka Y (2000) Balanced hemagglutinin and neuraminidase activities are critical for efficient replication of influenza A virus. J Virol 74:6015–6020

MMWR (2004) Preliminary assessment of the effectiveness of the 2003–2004 inactivated influenza vaccine–Colorado, December 2003. MMWR Morb Mortal Wkly Rep 53:8–11. doi:mm5301a

MMWR (2009) Outbreak of swine-origin influenza A (H1N1) virus infection–Mexico, March-April 2009. MMWR Morb Mortal Wkly Rep 58:467–470 doi:mm5817a5

MMWR (2012) Influenza A (H3N2) variant virus-related hospitalizations: Ohio, 2012. MMWR Morb Mortal Wkly Rep 61:764–767. doi:mm6138a3

Mukherjee TR, Agrawal AS, Chakrabarti S, Chawla-Sarkar M (2012) Full genomic analysis of an influenza A (H1N2) virus identified during 2009 pandemic in Eastern India: evidence of reassortment event between co-circulating A(H1N1)pdm09 and A/Brisbane/10/2007-like H3N2 strains. Virol J 9:233. doi:10.1186/1743-422X-9-233

Munster VJ, Baas C, Lexmond P et al (2007) Spatial, temporal, and species variation in prevalence of influenza A viruses in wild migratory birds. PLoS Pathog 3:e61. doi:06-PLPA-RA-0436R3, 10.1371/journal.ppat.0030061

Murcia PR, Baillie GJ, Stack JC et al (2013) Evolution of equine influenza virus in vaccinated horses. J Virol. doi:10.1128/JVI.03379-12

Murcia PR, Hughes J, Battista P et al (2012) Evolution of an Eurasian avian-like influenza virus in naive and vaccinated pigs. PLoS Pathog 8:e1002730. doi:10.1371/journal.ppat.1002730, PPATHOGENS-D-12-00315

Nardelli L, Pascucci S, Gualandi GL, Loda P (1978) Outbreaks of classical swine influenza in Italy in 1976. Zentralbl Veterinarmed B 25:853–857

Nelson MI, Edelman L, Spiro DJ et al (2008a) Molecular epidemiology of A/H3N2 and A/H1N1 influenza virus during a single epidemic season in the United States. PLoS Pathog 4:e1000133. doi:10.1371/journal.ppat.1000133

Nelson MI, Simonsen L, Viboud C et al (2006) Stochastic processes are key determinants of short-term evolution in influenza a virus. PLoS Pathog 2:e125. doi:06-PLPA-RA-0281R2, 10.1371/journal.ppat.0020125

Nelson MI, Viboud C, Simonsen L et al (2008b) Multiple reassortment events in the evolutionary history of H1N1 influenza A virus since 1918. PLoS Pathog 4:e1000012. doi:10.1371/journal.ppat.1000012

Nelson MI, Vincent AL, Kitikoon P, Holmes EC, Gramer MR (2012) Evolution of novel reassortant A/H3N2 influenza viruses in North American swine and humans, 2009-2011. J Virol 86:8872–8878. doi:10.1128/JVI.00259-12

Neumann G, Watanabe T, Ito H et al (1999) Generation of influenza A viruses entirely from cloned cDNAs. Proc Natl Acad Sci USA 96:9345–9350

Neverov AD, Lezhnina KV, Kondrashov AS, Bazykin GA (2014) Intrasubtype reassortments cause adaptive amino acid replacements in H3N2 influenza genes. PLoS Genet 10:e1004037. doi:10.1371/journal.pgen.1004037, PGENETICS-D-13-01606

Nguyen T, Rivailler P, Davis CT et al (2012) Evolution of highly pathogenic avian influenza (H5N1) virus populations in Vietnam between 2007 and 2010. Virology 432:405–416. doi:10.1016/j.virol.2012.06.021, S0042-6822(12)00323-6

Nishikawa F, Sugiyama T (1983) Direct isolation of H1N2 recombinant virus from a throat swab of a patient simultaneously infected with H1N1 and H3N2 influenza A viruses. J Clin Microbiol 18:425–427

Noda T, Sagara H, Yen A, Takada A, Kida H, Cheng RH, Kawaoka Y (2006) Architecture of ribonucleoprotein complexes in influenza A virus particles. Nature 439:490–492. doi:10.1038/nature04378

Noda T, Sugita Y, Aoyama K et al (2012) Three-dimensional analysis of ribonucleoprotein complexes in influenza A virus. Nat Commun 3:639. doi:10.1038/ncomms1647

Octaviani CP, Li C, Noda T, Kawaoka Y (2011) Reassortment between seasonal and swine-origin H1N1 influenza viruses generates viruses with enhanced growth capability in cell culture. Virus Res 156:147–150. doi:S0168-1702(10)00457-0, 10.1016/j.virusres.2010.12.014

Octaviani CP, Ozawa M, Yamada S, Goto H, Kawaoka Y (2010) High level of genetic compatibility between swine-origin H1N1 and highly pathogenic avian H5N1 influenza viruses. J Virol 84:10918–10922. doi:10.1128/JVI.01140-10

Olson SH, Parmley J, Soos C et al (2014) Sampling strategies and biodiversity of influenza a subtypes in wild birds. PLoS One 9:e90826. doi:10.1371/journal.pone.0090826, PONE-D-13-36482

Palese P, Schulman JL (1976a) Differences in RNA patterns of influenza A viruses. J Virol 17:876–884

Palese P, Schulman JL (1976b) Mapping of the influenza virus genome: identification of the hemagglutinin and the neuraminidase genes. Proc Natl Acad Sci USA 73:2142–2146

Pasick J, Pedersen J, Hernandez MS (2012) Avian influenza in North America, 2009–2011. Avian Dis 56:845–848

Peiris JS, Guan Y, Markwell D, Ghose P, Webster RG, Shortridge KF (2001) Cocirculation of avian H9N2 and contemporary "human" H3N2 influenza A viruses in pigs in southeastern China: potential for genetic reassortment? J Virol 75:9679–9686. doi:10.1128/JVI.75.20.9679-9686.2001

Pensaert M, Ottis K, Vandeputte J, Kaplan MM, Bachmann PA (1981) Evidence for the natural transmission of influenza A virus from wild ducks to swine and its potential importance for man. Bull World Health Organ 59:75–78

Rambaut A, Pybus OG, Nelson MI, Viboud C, Taubenberger JK, Holmes EC (2008) The genomic and epidemiological dynamics of human influenza A virus. Nature 453:615–619. doi:10.1038/nature06945

Reperant LA, Kuiken T, Osterhaus AD (2012) Adaptive pathways of zoonotic influenza viruses: from exposure to establishment in humans. Vaccine 30:4419–4434. doi:10.1016/j.vaccine.2012.04.049, S0264-410X(12)00585-3

Ritchey MB, Palese P, Schulman JL (1976) Mapping of the influenza virus genome. III. Identification of genes coding for nucleoprotein, membrane protein, and nonstructural protein. J Virol 20:307–313

Scholtissek C (1995) Molecular evolution of influenza viruses. Virus Genes 11:209–215

Schrauwen EJ, Bestebroer TM, Rimmelzwaan GF, Osterhaus AD, Fouchier RA, Herfst S (2013) Reassortment between Avian H5N1 and human influenza viruses is mainly restricted to the matrix and neuraminidase gene segments. PLoS One 8:e59889. doi:10.1371/journal.pone.0059889, PONE-D-12-35235

Schrauwen EJ, Herfst S, Chutinimitkul S et al (2011) Possible increased pathogenicity of pandemic (H1N1) 2009 influenza virus upon reassortment. Emerg Infect Dis 17:200–208

Shope RE (1931) Swine influenza: Iii. filtration experiments and etiology. J Exp Med 54:373–385
Shortridge KF, Webster RG, Butterfield WK, Campbell CH (1977) Persistence of Hong Kong influenza virus variants in pigs. Science 196:1454–1455
Simonsen L, Viboud C, Grenfell BT et al (2007) The genesis and spread of reassortment human influenza A/H3N2 viruses conferring adamantane resistance. Mol Biol Evol 24:1811–1820. doi:10.1093/molbev/msm103
Smith GJ, Vijaykrishna D, Bahl J et al (2009) Origins and evolutionary genomics of the 2009 swine-origin H1N1 influenza A epidemic. Nature 459:1122–1125. doi:10.1038/nature08182
Stincarelli M, Arvia R, De Marco MA et al (2013) Reassortment ability of the 2009 pandemic H1N1 influenza virus with circulating human and avian influenza viruses: public health risk implications. Virus Res 175:151–154. doi:10.1016/j.virusres.2013.04.012, S0168-1702(13)00129-9
Takizawa N, Kumakura M, Takeuchi K, Kobayashi N, Nagata K (2010) Sorting of influenza A virus RNA genome segments after nuclear export. Virology 401:248–256. doi:S0042-6822(10)00152-2, 10.1016/j.virol.2010.02.030
Tao H, Steel J, Lowen AC (2014) Intrahost dynamics of influenza virus reassortment. J Virol 88:7485–7492. doi:10.1128/JVI.00715-14
Van Reeth K, Brown IH, Durrwald R et al (2008) Seroprevalence of H1N1, H3N2 and H1N2 influenza viruses in pigs in seven European countries in 2002-2003. Influenza Other Respir Viruses 2:99–105 doi:10.1111/j.1750-2659.2008.00043.x, IRV043
Vijaykrishna D, Bahl J, Riley S et al (2008) Evolutionary dynamics and emergence of panzootic H5N1 influenza viruses. PLoS Pathog 4:e1000161. doi:10.1371/journal.ppat.1000161
Vijaykrishna D, Poon LL, Zhu HC et al (2010) Reassortment of pandemic H1N1/2009 influenza A virus in swine. Science 328:1529. doi:328/5985/1529, 10.1126/science.1189132
Vijaykrishna D, Smith GJ, Pybus OG et al (2011) Long-term evolution and transmission dynamics of swine influenza A virus. Nature 473:519–522. doi:10.1038/nature10004
Vincent AL, Ma W, Lager KM, Janke BH, Richt JA (2008) Swine influenza viruses a North American perspective. Adv Virus Res 72:127–154. doi:10.1016/S0065-3527(08)00403-X
Wagner R, Matrosovich M, Klenk HD (2002) Functional balance between haemagglutinin and neuraminidase in influenza virus infections. Rev Med Virol 12:159–166. doi:10.1002/rmv.352
Wan XF, Dong L, Lan Y et al (2011) Indications that live poultry markets are a major source of human H5N1 influenza virus infection in China. J Virol 85:13432–13438. doi:10.1128/JVI.05266-11
Webby RJ, Rossow K, Erickson G, Sims Y, Webster R (2004) Multiple lineages of antigenically and genetically diverse influenza A virus co-circulate in the United States swine population. Virus Res 103:67–73 doi:10.1016/j.virusres.2004.02.015, S0168170204001145
Webby RJ, Swenson SL, Krauss SL, Gerrish PJ, Goyal SM, Webster RG (2000) Evolution of swine H3N2 influenza viruses in the United States. J Virol 74:8243–8251
Webster RG, Bean WJ, Gorman OT, Chambers TM, Kawaoka Y (1992) Evolution and ecology of influenza A viruses. Microbiol Rev 56:152–179
Webster RG, Sharp GB, Claas EC (1995) Interspecies transmission of influenza viruses. Am J Respir Crit Care Med 152:S25–S30
Weingartl HM, Berhane Y, Hisanaga T et al (2010) Genetic and pathobiologic characterization of pandemic H1N1 2009 influenza viruses from a naturally infected swine herd. J Virol 84:2245–2256. doi:10.1128/JVI.02118-09
Werner O, Harder T, Veits J et al (2008) Avian influenza outbreaks in Germany—development of new avian vaccines. In: Klenk HD, Matrosovich MN, Stech J (eds) Avian influenza. Karger, Basel, pp 71–87
Westgeest KB, Russell CA, Lin X et al (2013) Genome-wide analysis of reassortment and evolution of human influenza A(H3N2) viruses circulating between 1968 and 2011. J Virol. doi:10.1128/jvi.02163-13
WHO (2014) Confirmed human cases of avian influenza A(H7N9) reported to WHO Report 13 WHO 2014

Wille M, Tolf C, Avril A, Latorre-Margalef N, Wallerstrom S, Olsen B, Waldenstrom J (2013) Frequency and patterns of reassortment in natural influenza A virus infection in a reservoir host. Virology 443:150–160. doi:10.1016/j.virol.2013.05.004, S0042-6822(13)00263-8
Wittwer CT, Reed GH, Gundry CN, Vandersteen JG, Pryor RJ (2003) High-resolution genotyping by amplicon melting analysis using LCGreen. Clin Chem 49:853–860
Worobey M, Han GZ, Rambaut A (2014a) A synchronized global sweep of the internal genes of modern avian influenza virus. Nature. doi:10.1038/nature13016
Worobey M, Han GZ, Rambaut A (2014b) Genesis and pathogenesis of the 1918 pandemic H1N1 influenza A virus. Proc Natl Acad Sci USA 111:8107–8112. doi:10.1073/pnas.1324197111
Wright PF, Neumann G, Kawaoka Y (2006) Orthomyxoviruses. In: Knipe DMH, Howley PM (eds) Fields virology. Lippincott-Raven, Philidelphia, pp 1691–1740
Wu A, Su C, Wang D, et al (2013) Sequential reassortments underlie diverse influenza H7N9 genotypes in China. Cell Host Microbe. doi:S1931-3128(13)00298-9, 10.1016/j.chom.2013.09.001
Xu X, Smith CB, Mungall BA, et al (2002) Intercontinental circulation of human influenza A(H1N2) reassortant viruses during the 2001-2002 influenza season. J Infect Dis 186:1490–1493. doi:JID020620, 10.1086/344738
Zell R, Scholtissek C, Ludwig S (2013) Genetics, evolution, and the zoonotic capacity of European swine influenza viruses. Curr Top Microbiol Immunol 370:29–55. doi:10.1007/82_2012_267
Zhang Y, Zhang Q, Kong H et al (2013) H5N1 hybrid viruses bearing 2009/H1N1 virus genes transmit in guinea pigs by respiratory droplet. Science 340:1459–1463. doi:10.1126/science.1229455
Zhao ZM, Shortridge KF, Garcia M, Guan Y, Wan XF (2008) Genotypic diversity of H5N1 highly pathogenic avian influenza viruses. J Gen Virol 89:2182-2193 doi: 89/9/2182, 10.1099/vir.0.2008/001875-0
Zhou NN, Senne DA, Landgraf JS et al (1999) Genetic reassortment of avian, swine, and human influenza A viruses in American pigs. J Virol 73:8851–8856

Antigenic Analyses of Highly Pathogenic Avian Influenza A Viruses

Ruben O. Donis

Abstract In response to the ongoing threat to animal and human health posed by HPAI endemic in poultry, Asia (H5N1) and North America (H7N3) have revived efforts to reduce pandemic risk by disease control at the source and improved pandemic vaccines. Discovery of conserved neutralization epitopes in the HA, which mediate broad protection within and across HA subtypes have changed the paradigm of "broadly reactive" or "universal" vaccine design. Development of such vaccines would benefit from comparative antigenic analysis of viruses with increasing divergence within (and between) HA subtypes. A review of recent work to define the antigenic properties of HPAI viruses revealed data generated through an array of experimental approaches. This information has supported diagnostics and vaccine development for animal and human health. Further harmonization of analytical methods is needed to determine the antigenic relationships among multiple lineages of rapidly evolving HPAI viruses.

Abbreviations

Aa	Amino acid
CVV	Candidate vaccine virus
ELISA	Enzyme linked immunosorbent assay
FAO	Food and agriculture organization
HA	Hemagglutinin
NA	Neuraminidase
HI	Hemagglutination inhibition
HPAI	High pathogenicity avian influenza
LPAI	Low pathogenicity avian influenza
Nt	Nucleotide
OIE	World organisation for animal health
RBC	Red blood cell

R.O. Donis (✉)
Influenza Division, Centers for Disease Control and Prevention,
1600 Clifton Road NE Mailstop A20, Atlanta, GA 30333, USA
e-mail: rdonis@cdc.gov

Current Topics in Microbiology and Immunology (2014) 385: 403–440 403
DOI: 10.1007/82_2014_422

Published Online: 5 September 2014

RBS	Receptor binding site
WHO	World health organization
gs/Gd/96 A/goose/Guangdong/1/1996-mAb	Monoclonal antibody

Contents

1 Introduction

The array of sporadic or endemic highly pathogenic avian influenza (HPAI) viruses poses a formidable challenge to the development of diagnostics and vaccines for animal health and pandemic preparedness. However, improved diagnostic technologies, including rapid tests and real-time PCR, have strengthened virologic surveillance for avian influenza in poultry and zoonotic infections in people, and have revealed the magnitude of the pandemic threat by these viruses. An understanding of the antigenic properties of potentially pandemic HPAI viruses is essential to reassess diagnostics, vaccines, and pandemic preparedness countermeasures. Genomic sequence analysis does not provide needed information on antigenic properties, but it remains the starting point for antigenic analysis as a valuable tool to select non-redundant viruses and to determine possible structure-function relationships.

Vaccination is the cornerstone of pandemic preparedness plans. Available pre-pandemic vaccines for avian influenza viruses were licensed without complications because they are produced using the same process as a licensed seasonal influenza vaccine (Fauci 2006; WHO 2009c). The latter achieve effectiveness levels not to exceed ~60 % on average (Osterholm et al. 2012; Treanor et al. 2012). Further improvements to influenza vaccine effectiveness will require elicitation of humoral and cellular memory immune responses with defined molecular specificity for protective epitopes, exceeding the achievements of natural infection (Nabel and Fauci 2010). Rational design of improved pre-pandemic vaccines would require a detailed understanding of influenza antigenicity, particularly to the HA, to develop vaccines that elicit host-protective immune responses and durable memory.

Specific and sensitive assays for serologic or cell-based diagnosis of zoonotic infections with avian influenza depend on a solid understanding of the performance characteristics of the assay, including the target influenza antigens and the immune response of the host. These assays could be a valuable tool for epidemiologic and public health studies. With the availability of detailed knowledge on the functional properties of viral epitopes, serologic/cellular diagnostic tests could be established to define correlates of host protection to guide development of vaccines and monitor vaccine efficacy. These tests could be invaluable in assessing the prevalence of immunity to emerging influenza viruses, which would be a critical parameter to establish their pandemic potential (Skowronski et al. 2013; Wu et al. 2014).

The ability of an influenza virus protein to bind specifically with the paratope of an antibody or a T cell receptor is described as its antigenicity, and is determined by primarily by non-covalent protein-protein interactions. The immunogenicity of an influenza virus protein is the ability to induce a humoral or cellular immune response in a competent host. Although antigenicity can be chemically defined, immunogenicity depends on complex interactions with the host immune system. In practice, the boundaries between the strict definitions of antigenicity and immunogenicity are blurred, especially when a polyclonal immune reagent is used as a tool to analyze and compare the properties of influenza viruses.

Influenza virus infection or (to a lesser extent) vaccination results in engagement of innate immune sensors (e.g., pattern recognition receptors) that respond by establishing a rapid but nonspecific and transient antiviral state. Most importantly, these signals trigger a more effective and durable response mediated by B and T lymphoid cell precursors, which engage with pathogen antigens and undergo rapid expansion, maturation, and differentiation into effectors. These cells carry immunoglobulin or T cell antigen receptors, respectively, on their surface. Therefore, the immunogenicity of a virus is defined by viral determinants with the capacity to engage immunoglobulin (B cell) and T cell receptors. The structural and functional properties of the virus, how it enters or is delivered to the host and presented to the immune system will determine which viral epitopes will elicit an adaptive immune response (antibodies and T cells), as well as its magnitude and longevity (memory). Therefore, a reductionist approach using monoclonal antibodies or T cells would enable analysis of the antigenic structure of an influenza antigen. In contrast, use of polyclonal preparations of antibodies or T cells provide a less detailed but practical delineation of the immune response and the antigenic and immunogenic determinants of the virus that are representative of the typical antiviral response associated with specific hosts. In this article, the term antigenicity is used to describe these inseparable attributes of influenza virus proteins in the context of relevant host species.

Although the intense research focus on structural analyses of antigen-antibody, or more specifically, epitope-paratope interactions has advanced the understanding of influenza antigenicity, the molecular determinants of viral immunogenicity, i.e., antigen-induced immune responses in vivo, are poorly understood. A more detailed understanding of influenza virus antigenicity in the natural hosts would help improve vaccines, diagnostics, and pandemic preparedness. In response to the ongoing threat to animal and human health posed by endemic HPAI in poultry, Asia (A(H5N1)) and North America (A(H7N3)) have revived efforts to improve pandemic preparedness through improved adjuvants and vaccines that provide broad protection within and across HA subtypes, often called "broadly reactive" or "universal" vaccines. Development of such vaccines would benefit from comparative antigenic analysis of viruses with increasing divergence within (and even between) HA subtypes.

2 Impact of Virus Diversity and Evolution on Antigenic Analysis

The segmented nature of the influenza genome facilitates genetic exchange between viruses by reassortment and contributes to viral diversity and evolution. In addition, the error-prone RNA-dependent RNA polymerase of the virus incorporates at least one mismatched nt per genome. Favorable mutations will become fixed in viruses emerging from a population bottleneck and enable "piggy-back" mutations that were present in the founder virus genome (Buonagurio et al. 1986). Genetic analyses of influenza viruses have shown that certain regions of the HA and NA genes evolve faster than the polymerase, NP or M genes, possibly due to fewer structural and functional constraints. Positive selection of antigenic variants in host populations that develop adaptive immunity following natural infection or vaccination has been demonstrated (Buonagurio et al. 1986; Fitch et al. 1991; Ina and Gojobori 1994) and is a considered a driver of antigenic drift.

HPAI viruses arise from the so-called "low-pathogenic" avian viruses as a result of genetic changes in the HA gene. These changes occur in the vicinity of the proteolytic cleavage site of the HA0, and result in the acquisition of susceptibility to furin-like proteases present in the secretory pathway of virtually all vertebrate cells (Bosch et al. 1981; Garten et al. 1981; Rott 1980; Rott et al. 1980). To date, only LPAI viruses with subtype H5 or H7 HA genes circulating in gallinaceous poultry or ratites have evolved toward a HPAI phenotype and the mechanisms involved remain elusive (Abolnik et al. 2009; Garcia et al. 1996; Rohm et al. 1996). Although we cannot rule out a role for other genes besides HA in the acquisition of HA cleavability by furin-like proteases, the HA alone imparts the HPAI phenotype. HPAI infections are lethal to a majority of gallinaceous poultry and some of these viruses also may cause lethal infections in other bird species. In contrast, HPAI infections in mammalian species are generally not highly virulent. Paradoxically, LPAI viruses that are characterized by low virulence for gallinaceous poultry can also cause fatal infections in some people; e.g., recent A(H7N9) infections reported in China (Zhou et al. 2013). Genomic and phylogenetic analyses of the Eurasian lineage of HPAI A(H5N1) circulating in Asia since 1996 and Africa since 2006 revealed persistence of the H5 HA lineage and complete replacement of the remaining genes by reassortment (Guan and Smith 2013; Vijaykrishna et al. 2008). Although detailed genomic analyses of all HPAI viruses have not been completed, the available evidence indicates that genes other than HA do not contribute necessary and sufficient functions for expression of the HPAI phenotype. Therefore, this article will be focused on the antigenic properties of HA of subtype H5 and H7 from HPAI viruses.

3 Approaches and Tools for Antigenic Analysis of HPAI

Antigenic analyses of the HPAI HAs have relied on several approaches that can be classified as follows: (i) statistical analyses of evolution, (ii) epitope characterization, and (iii) serologically defined antigenic relatedness. The first approach relies on codon use analysis to identify HA codons under positive selection. The second type of analysis uses mAbs to delineate epitopes on the HA 3-D structure. The third approach relies on panels of polyclonal antibody reagents and viruses to define antigenicity in terms of relationships to reference and field viruses. These methods generate complementary knowledge that is essential to developing models predictive of antigenic evolution to inform animal and public health interventions.

Implementation of antigenic analysis has benefited from relevant data collated into databases as well as computational tools to query the data and perform a variety of comparative analyses to relate antigenicity to genetic and structural data. These tools have become essential for research as well as for performing risk assessments and guiding interventions in response to outbreaks of HPAI.

3.1 Statistical Analysis of Evolution

Virologic surveillance programs have produced sequence information from viruses collected at regular intervals during extended periods of time. Codon usage analysis of the HA1 subunit of the H3 subtype revealed that antigenic sites exhibit significantly more non-synonymous than synonymous nucleotide substitutions, whereas the rest of the molecule undergoes primarily synonymous variation (Bush et al. 1999; Fitch et al. 1991, 1997). These findings indicated that the HA is undergoing positive selection leading to divergence. Continuous selective pressure on the antigenic sites of HA1 by host population immunity drive the emergence of novel functional variants and the extinction of previous populations, resulting in slender trees with extended trunks and relatively short branches by bootstrap analyses (Bahl et al. 2011). Therefore, codon use analysis has provided information on HA sites that may constitute important targets for neutralizing antibody responses. Similar studies exploit statistical analysis of sequons to identify the role of glycosylation in the evolution of antigenic sites (Bragstad et al. 2008; Cui et al. 2009; Zhang et al. 2004).

3.2 Epitope Characterization with Monoclonal Antibodies

Antigenic sites on the H3 and H1 HAs targeted by neutralizing antibodies were identified decades ago (Caton et al. 1982, 1983, 1986; Daniels et al. 1984; Skehel et al. 1983; Wiley et al. 1981; Yewdell et al. 1993). Initial studies of the H5 and H7 subtypes identified approximate epitope locations by sequence analysis of neutralization escape mutants. Precise structures of some epitopes were determined in short order by analysis of HA crystals in complex with mAbs (Cho et al. 2013; Dreyfus et al. 2013; Ekiert et al. 2009; Ekiert and Wilson 2012; Fleury et al. 1999; Han and Marasco 2011; Hong et al. 2013; Stewart and Nemerow 1997; Tsibane et al. 2012; Xu et al. 2013; Zhu et al. 2013). Although H5 and H7 HA structures in complex with antibody are still scarce, advances in high-throughput structural biology are likely to reverse this situation in the near future (Nabel 2012).

3.3 Serologically Defined Antigenic Relationships

Although the individually defined antigenic characteristics for each unique HA are of substantial value, there are many biological questions relevant to diagnostics and vaccines that require an understanding of the differences between the array of diverging viruses. The dominant role of humoral and mucosal antibody responses to animal and human influenza infections necessitates development of consensus antigenic profiles based on polyclonal antibody reagents. Therefore, antigenic analyses of influenza viruses are often conducted using panels of polyclonal antiserum and rationally selected viruses, which are the cornerstone of the process. The pairwise binding signals are often used to develop a fingerprint or cluster map for each virus relative to other members of the panel.

3.4 Databases, Bioinformatics, and Related Tools

The Immune Epitope Database and Analysis Resources (IEDB) (www.immuneepitope.org) was developed to capture epitope-related data for downstream to analyze and visualize existing epitope data (Bui et al. 2007a). Compared to human influenza, there is a limited number of epitopes reported for avian influenza strains/subtypes, and there is bias toward the hemagglutinin and nucleoprotein. However, nearly 500 peptides were identified as B or T cell epitopes from 91 H5N1 viruses (described in 84 publications). In contrast, a search for H7N7 yielded 10 peptide epitopes from 9 viruses, described in 12 publications. The IEDB provides two useful search and visualization tools; EpitopeViewer and Epitope conservancy. EpitopeViewer provides three-dimensional visualization of immune epitopes and analyzes their interactions with antigen-specific receptors of the immune system (antibodies, T cell receptors, and MHC molecules) for structures available in the Protein Data Bank (PDB). (Beaver et al. 2007). Epitope conservancy is an analysis tools implemented to address the issue of variability (or conservation) of epitopes, and to assist in the selection of epitopes with the desired pattern of conservation. (Bui et al. 2007b). The IEDB also provides tools to predict epitopes on the basis of antigen or antibody structures. DiscoTope is a tool to predict discontinuous epitopes on proteins using three-dimensional structure information in addition to sequence data. The method is based on aa and known epitope statistics, spatial information, and surface accessibility (Kringelum et al. 2012). Conversely, the Paratome tool predicts antigen-binding regions for Ig molecules; it accepts aa sequences or 3D structures (in PDB format) of antibodies as input to perform antigen site predictions (Kunik et al. 2012a, b).

Understanding the antigenic and genetic evolution of HPAI viruses requires integrating other knowledge on the epidemiology and geographic distribution of circulating viruses. The FAO and OIE collect animal disease outbreak information as well as animal population census data and store it in searchable databases. The WAHID interface provides access to the World Animal Health Information System epidemiologic information database on reported HPAI outbreaks globally. All countries, which are members of the World Organization of Animal Health (OIE), are bound by the statutory requirement to report all infections with H5 and H7 in poultry, including HPAI (Vallat et al. 2013). The OIE maintains a public database of all reported subtype H5 and H7 HPAI outbreaks (Ben Jebara 2007; Jebara et al. 2012). In addition, other databases, such as EMPRES-i, link the outbreak epidemiologic information to the genetic sequences of viruses associated with the corresponding outbreak (Claes et al. 2014).

The World Health Organization's (WHO) Global Influenza Surveillance and Response System (GISRS) coordinates a network of 108 National Influenza Centers and 6 Collaborating Centers working toward control and prevention of seasonal and pandemic influenza. HPAI viruses that cause human infections are available to the global animal and public health sectors under the terms of the

WHO Pandemic Influenza Preparedness Framework (Fidler and Gostin 2011; WHO 2009c). The viruses distributed under the PIP Framework can be found in the WHO Influenza Virus Traceability Mechanism (IVTM) database (URL: https://extranet.who.int/ivtm/). The WHO website also provides lists of reagents (URL: http://www.who.int). In addition, influenza reagents for public health applications are available from the Influenza Reagent Resource (URL: http://www.influenzareagentresource.org/WHOKitForms.aspx).

3.5 Laboratory Methods for Antigenic Characterization

The approach for antigenic analysis of HPAI viruses is determined by the information needs of the project. Subtle differences between two viruses are best detected using monoclonal antibody (mAb) panels that provide exquisite sensitivity and specificity. A more comprehensive understanding of the antigenic relationships within and between different viruses generally relies on polyclonal antisera. Regardless of the approach, it should be noted that detailed sequence analysis of all viruses used for laboratory assays or to generate antisera is essential to detect mixed virus populations; as the literature is replete with examples of antigenic differences due to mixed sequence populations (Deyde et al. 2009; Furuse et al. 2010; Ghedin et al. 2011; Hay et al. 2001; Ramakrishnan et al. 2009; Robertson et al. 1994).

The molecular composition (purity) and integrity of inactivated antigens should be established by appropriate biochemical methods and documented (Harvey et al. 2012; Li et al. 2014; Oxford et al. 1981). Accurate quantification of the viral components is necessary for consistent results (Luna et al. 2008; Pierce et al. 2011; Williams et al. 2008, 2012). Likewise, live virus preparation should be quantified to determine their infectivity with appropriate host systems and assessed for the presence of defective interfering particles as needed (Balish et al. 2013).

MAb panels to the Eurasian and American lineages of H5 and H7 HAs have been used for antigenic characterization. However, these studies addressed specific research questions rather than seek to understand the antigenic relationships among diverse HPAI virus lineages circulating in poultry. In principle, a large panel of carefully selected mAbs to well characterized epitopes on the surface of HA could provide an informative assessment of antigenic distances between divergent viruses. However, such mAb panels would require a large effort to fully characterize the mAbs and manufacture user-friendly assay kits. Unfortunately, demand for such products is limited to a few specialized research laboratories and would not support commercial development and distribution at a reasonable cost.

Antigenic analyses of HPAI have been generally conducted using polyclonal antisera. These reagents were raised by different methods and in diverse host species. Polyclonal antibodies to be used as a probe for antigenic analysis are the cornerstones of this approach. The molecular specificity and affinity of the antibody repertoire comprised in polyclonal antiserum are known to vary in accordance with immunization protocols; therefore, the following considerations must be noted. Generation of antibody reagents in chickens by live HPAI virus inoculation requires prophylactic interventions to prevent death; e.g., use of antiviral drugs, passive, or active immunization with inactivated antigens that would not interfere with the assays to be performed (e.g., NP) (Shahzad et al. 2008). Similar considerations apply to the use of ferrets, although only some HPAI are lethal to ferrets at the doses used for antiserum production (Belser et al. 2011; Belser and Tumpey 2013; Matsuoka et al. 2009). Animal health labs generally inoculate chickens with inactivated HPAI viruses (generally emulsified with adjuvant), in some cases with a booster immunization (with the same live or inactivated virus) to achieve higher titers. Public health laboratories rely on ferrets to raise antiserum by inoculation with infectious virus via the respiratory route, often followed by a parenteral boost with inactivated and adjuvanted antigen. Live or inactivated antigen delivered by mucosal infection or parenteral injection, respectively, requires standardization of doses, booster regimens, and blood collection timelines. In general, antisera collected within 2 weeks of a single immunization will have a narrow specificity (high discriminating power) for the immunizing antigen and closely related viruses. In contrast, hyperimmune antisera generated after two or more booster injections with the same adjuvanted virus have a propensity to react with conserved epitopes on divergent viruses; i.e., low discriminating power. The latter is the method currently used to produce HI-based subtyping antisera, which must recognize all viruses within a particular HA subtype as implemented in the WHO influenza subtyping kit (URL: http://www.influenzareagentresource.org/WHOKitForms.aspx).

Laboratory methods to analyze the antigenic characteristics of HPAI with specific antisera include microneutralization assays, hemagglutination inhibition (HI) assays, and ELISA, among others. The MN assay is considered the gold standard for the analysis of the antigenic properties of the HA because it is specific, sensitive and, in general, correlates with host protective immunity (Rowe et al. 1999). Despite efforts to automate and standardize this assay, it is still very labor intensive and prone to inter and intra-laboratory variability (Ampofo et al. 2012; Stephenson et al. 2009). The HI assay is a surrogate of the MN test, and detects antibodies that block the HA receptor binding site and allow even erythrocyte sedimentation. HI titers correlate quite well with MN titers in general (Grund et al. 2011). The RBC donor species can have a significant impact in the sensitivity and specificity of assays, and cells from different species; turkey, guinea pig or human, chicken, horse, and goose; are often evaluated. Horse is the source species of choice for RBC used in HI assays to measure antibody to HPAI in human sera (Jia et al. 2008; Kayali et al. 2008; Stephenson et al. 2003).

ELISA assays have been used to detect and quantify antibodies to the influenza NP, either directly or with a blocking assay format. They are also suitable to perform antigenic analysis with mAbs. However, this is not the case for polyclonal antibody reagents, as the relatively small antigenic differences between viruses detected by the HI or MN tests are often blurred in the ELISA test. It is thought that the polyclonal antibodies binding to HA epitopes that are not involved in neutralization and/or blocking receptor binding generate a substantial signal in ELISA readings. Because these sites are generally highly conserved in the HA, these signals tend to mask differences in antibody binding to variable (and discriminating) epitopes.

The diversity of approaches used to generate antisera and perform assays to determine the antigenic characteristics of HPAI viruses complicates the comparison of results from different laboratories. An international consortium was formed recently to standardize the methodology used in many laboratories around the world to measure antibody responses in humans (Van Kerkhove et al. 2012). Although the goals of this initiative focus on measuring antibody, the standards developed for HI and MN assays could be applied to the antigenic analysis of HPAI.

Restrictions in the international transfer of HPAI viruses represent another obstacle to achieve a comprehensive understanding of the antigenic relationships among HPAI viruses. The WHO has led a global effort to eliminate some of the barriers to the sharing of information on HPAI viruses, important for pandemic influenza preparedness (PIP) (WHO 2009c). The increased transparency and progressive benefit sharing has created an equitable environment conducive to sharing of materials for pandemic preparedness purposes.

3.6 Comparative Analysis of Antigenic Properties

Understanding antigenic differences between two viruses in a single laboratory is generally a straightforward process, involving generation of antibody reagents to the two viruses and performance of reciprocal MN and/or HI analyses. However, the global nature of the HPAI threat to animal and human health requires a comprehensive and consensus understanding of the antigenic relationships among viruses to support control and mitigation programs. In addition to the needed standardization of laboratory testing methods, the data should be analyzed, interpreted, and communicated as rigorously and effectively as possible. Antigenic relatedness among isolates belonging to different genetic clades within a subtype and between subtypes is generally evaluated with panels of antisera to viruses that represent each of the clades and subtypes of interest. HI tests with panels of antisera and viruses typically include the viruses used to raise the antisera (often called reference antigens) as well as other viruses whose properties need to be determined. HI data are difficult to interpret quantitatively. Comparing the titers of multiple viruses with different antisera is complicated by the fact that the range of titers of each antisera with the antigen used for immunization (the so-called homologous titer) can differ by an order of magnitude, and the same virus and antiserum may produce different values upon repeat testing, even in the same laboratory. To normalize these differences, Archetti and Horsfall calculated an antigenic relatedness (R value) as the ratio between homologous and heterologous titer as a percentage (Archetti and Horsfall 1950). This value is generally considered to indicate antigenic similarity between isolates having R values of more than 50 %. Although this is conceptually simple, it has been the most widely used method to interpret HI test results, often implicitly. However, R values are not sufficient to derive consensus results relating similar antisera and similar viruses. Despite these drawbacks, manual HI data analysis and interpretation have been considered reliable enough for detection of large antigenic differences, such as the differences that would support changes in WHO vaccine composition recommendation ($\geq$8-fold titer difference by HI test). Additional statistical and mathematical approaches implemented computationally to analyze and visualize HI test results can provide consensus antigenic distances among multiple virus groups. Square-root transformed HI titers can be used to derive a Bray–Curtis coefficient and construct similarity matrices. Hierarchical agglomerative clustering with group-average linking can be applied to generate dendrogram representations of antigenic relatedness among viruses (Smith et al. 2006). Non-metric multidimensional scaling also has been used to produce two- and three-dimensional maps to visualize the data (Smith et al. 2006).

Antigenic cartography was developed in 2001 to analyze and visualize the relationships among multiple antigen-antibody pairs (HI or other binding assay) data (Lapedes and Farber 2001; Perelson and Oster 1979). Since then, antigenic cartography has been applied by the WHO to supplement influenza surveillance activities (Ampofo et al. 2013; Smith et al. 2004). An advantage of using antigenic cartography over R values is the use of multiple HI values rather than one homologous titer to determine antigenic relationships. Therefore, antigenic relationships among test viruses and multiple reference viruses can be assessed simultaneously. Antigenic maps based on HI data do not always correlate with genetic phylogenies based on nt or aa distances. To this end, recent studies have attempted to integrate phylogenetic approaches with antigenic cartography (Bedford et al. 2014).

Antigenic cartography has also been used to evaluate antigenic properties HPAI A(H5N1) viruses on the basis of HI tests with ferret or chicken antisera (Fouchier and Smith 2010). The considerable antigenic variation within and between A(H5N1) clades suggested implications for the efficacy of poultry vaccines. This and other multidimensional scaling methods have been implemented as Web severs that users can access to analyze their data (Barnett et al. 2012; Cai et al. 2011). In summary, these promising methods have not been widely used and a theoretical framework to achieve a predictive understanding of HI data remains elusive.

3.7 Application of Antigenic Analysis to Animal and Human Health Vaccinology

Antigenic similarity between the virus antigen used to manufacture inactivated vaccines and the circulating viruses has been widely regarded as an important consideration for assessing vaccine efficacy and effectiveness in humans (Fiore et al. 2009; Lo et al. 2013; Skowronski et al. 2014). Likewise, antigenic similarity between circulating viruses and those used for preparation of live attenuated influenza vaccines (LAIV) is thought to contribute to their effectiveness (Belshe et al. 2004). Inactivated vaccines for use in horses and swine are also thought to provide better protection when the antigenic characteristics of the virus antigens are similar to those of the field viruses (Daly et al. 2004; de Jong et al. 2001). In contrast, inactivated vaccines developed for use in poultry have shown excellent protection and reduced shedding when the HA sequences differed by up to 13 % (Abbas et al. 2011; Swayne et al. 1999, 2000a, 2001). However, several studies have revealed that similarity between the antigenic properties of poultry vaccine and field viruses results in reduced virus shedding and to a lesser extent, reduced mortality (Lee et al. 2007b; Swayne et al. 2000a, b). Thus, the evidence seems to suggest that a closer antigenic relationship between vaccine antigens and field isolates may provide increased protection and decrease the risk of transmission by limiting virus shedding.

The World Health Organization's (WHO) Global Influenza Surveillance and Response System (GISRS) coordinates semiannual consultations to conduct zoonotic influenza risk assessments for pandemic preparedness (Trock et al. 2012). A critical element toward assessing pandemic risk is the immunologic status of the human population with regard to animal influenza virus being evaluated. Human susceptibility to infection and severity of illness by specific animal influenza viruses may depend on age and geographic area. To this end, the WHO monitors the circulation of animal viruses with pandemic potential, especially HPAI influenza viruses, due to their propensity to cause severe infections in people and their novel antigenic properties (different from those of human viruses that elicited population immunity). The antigenic properties of animal viruses with pandemic potential are a key consideration for the selection of appropriate viruses for development of candidate vaccine viruses. The ultimate goal of these consultations is to maintain a library of candidate vaccine viruses that could be used promptly for pandemic vaccine manufacture, if needed. Since the emergence of A(H5N1), a total of 29 candidate vaccine viruses have been developed for pandemic preparedness purposes (Table 4) (WHO 2006, 2007, 2009a, b, 2010a, b, 2011a, b, 2012a, b, 2013a, b, 2014a).

4 Genetic Diversity of the HA

The genetic diversity of H5 and H7 HA genes from HPAI viruses reflects that of their ancestor LPAI viruses circulating in aquatic birds. In other words, most of the major lineages of H5 and H7 HAs from LPAI viruses circulating in the Americas or Europe and Asia have occasionally been transmitted to terrestrial poultry and subsequently evolved into HPAI viruses (Tables 1 and 2). The evolution and emergence of HPAI viruses is poorly understood, but there is evidence indicating

Table 1 Lineage diversity of H5 subtype viruses

Lineage designation	Group[1]	Representative virus	Subtype	Avian habitat	HA cleavage site
North American	A	chicken/ Pennsylvania1/ 1983	H5N2	Terrestrial	PQKKKR/G
North American	B	chicken/Queretaro/ 14588-19/1995	H5N2	Terrestrial	PQRKRKTR/G
North American	C	chicken/Texas/ 298313/2004	H5N2	Terrestrial	PQRKKR/G
Eurasian/ Africa	E	chicken/Scotland/ 1959	H5N1	Terrestrial	PQRKKR/G
Eurasian	F	turkey/England/ N50-92/1991	H5N1	Terrestrial	PQRKRKTR/G
Eurasian	G	mallard/Italy/1980/ 1993	H5N2	Aquatic birds	PQRETR/G
Eurasian	H	duck/Primorie/2633/ 2001	H5N3	Aquatic only	PQRETR/G
Eurasian	I	duck/Altai/1285/1991	H5N3	Aquatic	PQRETR/G
Eurasian, African	J	goose/Guangdong/ 1/1996	H5N1*	Terrestrial	PQRERRRKKR/G
Eurasian	K	chicken/Italy/312/ 1997	H5N2	Terrestrial	PQRRRKKR/G

* Other subtypes reported include: H5N5, H5N6, H5N8 (Lee et al. 2014b; WHO-WPRO 2014; Zhao et al. 2013)
[1] Based on database designations (Lu et al. 2007)

Table 2 Genetic diversity of H7 subtype HPAI viruses

Lineage designation	Group[1]	Representative virus	Subtype	Avian habitat	HA cleavage site
Current Eurasian avian	A	turkey/ England/1963	H7N3	Terrestrial	PETPKRRRR/G
Ancestral Eurasian avian	B	fowl/Dobson/ 1927	H7N7	Terrestrial	PELPKKRRKR/G
Australian	C	chicken/ Victoria/1985	H7N7	Terrestrial	PEIPKKREKR/G
Equine 1	D	equine/Prague/ 1/1956	H7N7	Equine	PEAPAHKQLTHHMRKKR/G
South American	E	chicken/Chile/ 4977/2002	H7N3	Terrestrial	PEKPKTCSPLSRCRETR/G
North American	F	turkey/Oregon/ 1971	H7N3	Terrestrial	PENPKTR/G

[1] Based on database designations (Lu et al. 2007)

that HPAI virus can arise in the chicken population of a single farm, after LPAI virus introduction from aquatic birds (Bonfanti et al. 2014). However, in some cases LPAI viruses can circulate for extended periods in terrestrial poultry without acquisition of the HPAI phenotype; for example LPAI A(H7N9) viruses have circulated in the chicken population of China between March 2013 and May 2014 without evidence of changes in the HA cleavage site (Zhou et al. 2013).

5 Antigenic Diversity of HPAI Viruses

5.1 Antigenic Characteristics of the A/goose/Guangdong/1/96-Lineage A(H5N1) Viruses

Subtype H5 HPAI viruses have been detected in Europe since 1959. However, this outbreak and subsequent ones in Europe and other continents were contained promptly or became extinct due to host population losses (Swayne and Suarez 2000). The H5 HA gene of the A(H5N1) gs/Gd/96 virus is the closest known ancestor of the HPAI viruses that caused outbreaks in Chinese poultry in 1996 (Guan et al. 2002). This gs/GD/96-like (H5N1) virus caused unprecedented zoonotic infections during the following year, 1997(Claas et al. 1998; Subbarao et al. 1998). A total of 18 human infections with 9 fatalities were reported. Unlike all previous HPAI outbreaks in poultry, which were self-limiting and became extinct within months, this virus spread throughout Asia, and entered Africa and Europe by 2006, with over 60 countries reporting its detection (Guan et al. 2009; Liu et al. 2005; Prosser et al. 2011). Despite control efforts, HPAI viruses with the gs/GD/96-like HA became endemic in poultry from parts of China, Indonesia, Vietnam, Bangladesh, and Egypt, among others (FAO-DAH 2011). As expected, the vast geographic separation and large poultry populations in these countries supported extensive evolution of the A/goose/Guangdong/1/1996-like HA since its emergence in 1996.

5.1.1 Antigenic Characteristics Inferred from Statistical Analysis of Coding Sequences

Analysis of the HA coding region to determine statistically significant bias is an indicator for positive selection. However, establishing that positively selected codons are under immune selection generally relies on indirect evidence, such as independently mapped epitopes within the same subtype. In general, this type of information would be difficult to interpret without evidence that the sites under positive selection are indeed targets of immunity. Cattoli et al. evaluated gene and site-specific selection pressures for the HA gene segment of H5N1 viruses from Egypt, Indonesia, Nigeria, Turkey, and Thailand by calculating the ratio of non-synonymous (dN) to synonymous (dS) nt substitutions per site (ratio dN/dS). A total of 25 positively selected sites were identified in the HA genes analyzed; 12 of these were located at putative antigenic sites as follows. Six on site B: residues 137, 170, 171, 172, 201, 205; 5 on site A: 136, 145, 152, 156, 157; and 1 on site E:

104, based on the H3 convention (Cattoli et al. 2011). Similar studies conducted by Wei et al evaluated 240 HPAI H5N1 viruses representing four major clusters based on geographic location in agreement with their antigenic divergence (Wei et al. 2012). The authors identified six sites apparently evolving under positive selection, five of which persist in the population. Although some positively selected sites located either within or flanking the receptor-binding sites may increase the affinity to human-type receptors, other sites may be involved in glycosylation and antigenic drift (Wei et al. 2012).

5.1.2 Antigenic Characteristics Defined by Epitope Mapping

Two neutralization epitopes on the HPAI H5 have been defined by X-ray crystallography. The structure of a conserved neutralization epitope on the stem of the A/Vietnam/1203/2004 (H5N1) HA was defined by mAb F10 (Sui et al. 2009). Zhu et al determined the crystal structure of mAb Fab H5m9 in complex with the gs/GD/96 (H5N1) HA. The epitope defined by this mAb is conserved among H5N1 viruses (Zhu et al. 2013).

Early mAb escape mutant studies on the H5 HA of A/turkey/Ontario/7732/66 (H5N9) identified five distinct neutralization regions, termed groups 1 through 5. Group 1 mutations in the globular head (140–145) analogous to site A in H3 subtype and site Ca2 in H1 subtype; group 2 (156–157) equivalent to site B in H3 HA; and group 3 (129–133) corresponding to the Sa site in H1 subtype (Stevens et al. 2006; Wiley et al. 1981); group 4 overlapping the vestigial esterase domain; group 5 within the 130-loop structure. Multiple studies have subsequently reported neutralization escape mutants selected with mAbs. Although these studies helped define the antigenic evolution of H5 viruses, and are useful for detecting changes that can lead to the emergence of antigenic variants, the reported data has only so far led to a fragmented understanding of the overall antigenic structure of the H5 HA (Gerhard et al. 1981).

Velkov et al. proposed three broad epitope clusters based on their location on the HA spike in a recently published comprehensive review of the H5 epitope mapping data (Velkov et al. 2013). These comprised:

(1) HA2 stalk highly conserved epitopes, comprising nine potential overlapping sites (Corti et al. 2011; De Marco et al. 2012; Ekiert et al. 2009; Okuno et al. 1993, 1994; Prabhu et al. 2008; Smirnov et al. 2000; Sui et al. 2009; Throsby et al. 2008; Vareckova et al. 2008; Wang et al. 2010).
(2) Non-RBS globular domain moderately conserved epitopes, comprising ten potential sites (Cao et al. 2012; Du et al. 2009; Han et al. 2011; Hu et al. 2012; Li et al. 2011; Oh et al. 2010; Prabakaran et al. 2009a; Rockman et al. 2013).
(3) RBS variable epitopes; comprising a total of twenty potential epitopes (Chen et al. 2009; Ferreira et al. 2010; Hanson et al. 2006; Ho et al. 2009; Hoffmann et al. 2005; Kaverin et al. 2002, 2007; Khurana et al. 2009; Li et al. 2009; Masalova et al. 2011; Ohkura et al. 2012; Prabakaran et al. 2009a, b, 2010;

Prabhu et al. 2009; Rudneva et al. 2010; Sun et al. 2009; Wang et al. 2009; Yang et al. 2007; Yoshida et al. 2009).

(1) HA2 stalk highly conserved epitopes

Two independently identified mAbs, mAbsCR6261 (Ekiert et al. 2009) and F10 (Sui et al. 2009), engaged with almost identical neutralization epitopes on the stalk region of the H5 HA that were shared by an unprecedented number of subtypes; all the group 1 HAs (i.e., H1, H2, H6, H8, H9, H11, H12, H13, and H16) had conserved structures. These antibodies neutralized virus infectivity in vitro and protected from lethal influenza challenge in laboratory animal models by locking the fusion peptide and blocking the conformational change required for HA fusion activity. The identification of conserved neutralization epitopes in the stalk region of the H5 HA triggered a paradigm shift in seasonal and pandemic influenza vaccine development.

Human mAbs (CR8020 and CR8043) specifically recognizing group 2 influenza viruses and binding to highly conserved epitopes at the base of the HA stalk with neutralizing activity against H7N7 HPAI, as well as H3, and H10 viruses were reported (Ekiert et al. 2011; Friesen et al. 2013). The crystal structures of each mAb in complex with HA revealed similar epitopes, although the mAbs use different angles of approach and distinct atomic interactions. Binding profiles of mAbs against the group 2 HA stem also indicate that the group 2 viruses have a conserved site of vulnerability for vaccine design similar to the group 1 viruses (Ekiert et al. 2011). Following the identification of broadly neutralizing mAbs for groups 1 or group 2 viruses, a mAb with neutralizing activity against all 16 subtypes of influenza A viruses was identified. Crystal structures of mAb F16 in complex with HA subtypes H1 and H3 showed that the epitope is located in the HA2 stem region and binds to the highly conserved fusion peptide to block fusion. In addition, this mAb reacted with A(H5N1) HPAI viruses from clades 0, 1, 2.2, and 2.3.4, and A(H7Nx) viruses from the Eurasian lineage. The crystal structures of mAbs F10 and CR6261 in complex with the group 1 H5 HA compared to the that of F16 revealed that the latter binds to a larger region comprising helix A and the fusion peptide of the neighboring HA monomer; therefore, infectivity neutralization also may be mediated by binding to the unprocessed HA0 and thus blocking cleavage into HA1 and HA2.

(2) Non-RBS globular domain moderately conserved epitopes

A mouse mAb (termed HA-7) that neutralized H5N1 clades 0, 1, 2.2, 2.3.4, and 2.3.2.1 bound to the globular head of the H5 HA without blocking the receptor-binding step (Du et al. 2012). The epitope is located on residues 81–83 and 117–122 of HA1 (H5 numbering) near the vestigial esterase domain . Sequence comparisons revealed a high level of epitope conservation within the H5N1 viruses.

(3) RBS variable epitopes

A mAb (C05) that recognizes conserved elements of the receptor-binding site on the HA and neutralized H1, H2, and H3 subtype viruses challenged the prevailing hypothesis stating that the high variability of the receptor binding site of HA would preclude heterosubtypic reactivity (Ekiert et al. 2012). The crystal structure showed that a single heavy-chain complementarity-determining region 3 loop dominated recognition of the receptor-binding pocket, with minor contacts from heavy-chain complementarity-determining region 1. These interactions were sufficient to achieve nanomolar binding and allowed targeting of a small conserved functional site without contacts on adjacent hypervariable residues of the HA (Ekiert et al. 2012). Some mAbs that bind epitopes in the H5 HA RBS were not affected by aa changes associated with viral adaptation to droplet transmission (e.g., N158D/N224K/Q226L in the A/Vietnam/1203/2004 HA and T160A/Q226L/G228S in the A/Indonesia/5/2005 HA)(Thornburg et al. 2013).

5.1.3 Serologically Defined Antigenic Characterization

The extensive genetic evolution of gs/GD/96-like HA lineage is evidenced by frequent updates to the H5 clade nomenclature, which were necessary to assign names to newly emerged clades. A total of 38 clades have been described since the first classification of this HA into 10 first order clades; clade 0–9 (Table 3) (WHO-OIE-FAO 2008, 2009, 2012, 2014). Extensive sequence divergence has resulted in the recent designation of several fifth-order clades, by virtue of exceeding the 1.5 % average nt distance threshold in the three nomenclature revisions since 2008. Genetic divergence of the HA genes has resulted in extensive evolution of the aa sequences, particularly in the HA1 subunit, which determines antigenic drift. Although a few aa changes within critical antigenic sites are sufficient to mediate substantial antigenic difference, the gs/Gd/96 lineage has accumulated up to 60 aa differences between the HA1 of most divergent viruses (Table 5).

Antigenic analysis of gs/Gd/96 H5N1 viruses have been performed using antisera raised in chickens or ferrets. In addition, some studies were performed with hyper-immune goat antiserum or with monoclonal antibodies. For example, studies on the antigenic characteristics of isolates from the 1997 outbreak in Hong Kong with monoclonal antibodies could discriminate between two groups of H5N1 viruses from the outbreak based on the presence or absence of glycosylation at Asn158 (H3 numbering, 154 in H5), but could not further differentiate these viruses from those from North American lineage (Claas et al. 1998; Shortridge et al. 1998). Studies with ferret antisera confirmed the impact Asn158 glycosylation on the antigenic characteristics of the virus; antisera to non-glycosylated viruses lost significant levels of binding to glycosylated viruses (Bender et al. 1999).

Ducatez *et al.* evaluated the antigenic cross-reactivity of a panel of HPAI H5N1 viruses from nine different genetic clades using four ferret antisera in HI assays using either chicken or horse RBCs as well as MN tests in MDCK cells (Ducatez et al. 2011). Ferret antisera were raised against five divergent viruses representing clade 1 (A/Vietnam/1203/04), clade 2.2 (A/whooper swan/Mongolia/244/05),

Table 3 H5 HA clade emergence

Clade[1]	Year	Geographical location	Description/prototype virus
0	1996–2002	China, HK SAR*	Gs/Guangdong/1/1996
3	2000–2001	China, HK SAR, Vietnam	Chicken/Hong Kong/YU562/2001
4	2002/2003	China. HK SAR	Gs/Guiyang/337/2006
5	2000–2003	China, Vietnam	Gs/Guangxi/914/2004
6	2002/2004	China	Chicken/Hunan/01/2004
7	2002/2004	China: Yunnan, Hubei, Shanxi	Chicken/Shanxi/2/2006
7.1	2007	China	Chicken/Vietnam/NCVD-16/2008
7.2	2008	China	Chicken/Hebei/A-8/2009
8	2001–2004	HK SAR, China	Chicken/Hong Kong/YU777/2002
9	2003–2005	China	Duck/Guangxi/2775/2005
1	2002/2006	HK SAR, China Vietnam, Cambodia, Thailand, Laos, Malaysia	Vietnam/1203/2004
1.1	2004	Vietnam	Duck/Vietnam/1233/2005
1.1.1	2007	Vietnam	Duck/Vietnam/NCVD-010/2008
1.1.2	2008	Vietnam	Duck/Vietnam/NCVD-366/2009
2.1.1	2003–2005	Eastern Indonesia	Chicken/Indonesia/BL/2003
2.1.2	2005–2006	Western Indonesia	Indonesia/538H/2006
2.1.3	2004–2007	Eastern and Western	Indonesia/5/2005
2.1.3.1	2004	Indonesia	Chicken/Gunung Kidal/BBVW/2005
2.1.3.2	2004	Indonesia	Indonesia/CDC292N/2005
2.1.3.2a	2006	Indonesia	Chicken/Indonesia/Tanggerang/1/2007
2.2	2005	China, Mongolia, Europe, Middle East, Africa	Bar headed goose/Qinghai/1/2005
2.2.1	2005	China, Middle East, Africa	Chicken/Egypt/06207-NLQP/2006
2.2.1.1	2007	Egypt	Chicken/Egypt/07701S-NLQP/2007
2.2.1.1a	2009	Egypt	Chicken/Egypt/F10/2009
2.2.2	2006	Bangladesh, India	Chicken/Bangladesh/364/2007
2.2.2.1	2008	Bangladesh	Chicken/Bangladesh/CD-48-BL-165/2008
2.3.1	2003–2005	Hunan, Guangdong (China)	Duck/Hunan/303/2004

(continued)

Table 3 (continued)

Clade[1]	Year	Geographical location	Description/prototype virus
2.3.2	2004–2006	HK SAR, Southern China	Chicken/Guangxi/2461/2004
2.3.2.1	2006	China	Peregrine falcon/Hong_Kong/1143/2007
2.3.2.1a	2008	China, Vietnam, Bangladesh	Environment/Chang_Sh25/2009
2.3.2.1b	2008	China, Vietnam	Feral pigeon/Hong_Kong/3409/2009
2.3.2.1c	2008	China, Indonesia, Vietnam	Bar-headed goose/MongoliX53/2009
2.3.3	2004	Guiyang and Hunan (China)	Chicken/Guiyang/3055/2005
2.3.4	2005–2006	HK SAR, China, Thailand, Laos, Malaysia	Duck/Fujian/1734/2005
2.3.4.1	2008	China	Guizhou/1/2009
2.3.4.2	2006	China	Duck/Yunnan/6490/2006
2.4	2002–2005	Yunnan and Guangxi (China)	Chicken/Yunnan/115/2004
2.5	2003/2004	China, Korea, Japan Shantou	Chicken/Korea/ES/2003

[*] Hong Kong Special Administrative Region

[1] Based on previously publications; (WHO-OIE-FAO 2008, 2009, 2012, 2014)

clade 2.1 (A/duck/Hunan/795/02), clade 2.3.4 (A/duck/Laos/3295/06 and A/Japanese white-eye/Hong Kong/1038/06), and clade 4 (A/goose/Guiyang/337/06). There was general agreement between results from MN tests and HI tests with chicken RBC. However, HI tests with horse RBC resulted in reduced antigenic distances between viruses. The MN data showed that clade 1 viruses were very similar to each other. Likewise, clade 2.2 showed similar reactivity, whereas clade 2.3.2 and 2.3.4 viruses remained within 4-fold differences in the assay endpoints. The antigenic distances between viruses belonging to clade 1 and those in clade 2 were less clear than anticipated, with some viruses showing significant reactivity with viruses from the heterologous clade (i.e., clade 1 with clade 2 and vice versa). These finding highlighted the limitations of using phylogeny-based clustering to compare viruses. Despite the complexity of the data, the pairwise antigenic distance between viruses of different clades was approximately 4–8 assay units (HI or MN dilution endpoints) on average.

The WHO GISRS consultations have reviewed the antigenic characteristics of H5N1 viruses from the gs/Gd/96 lineage since 2004. The semi-annual reviews resulted in the identification of at least 25 antigenically distinct H5N1 viruses that merit consideration as candidate vaccine antigens for pandemic preparedness purposes (Table 4). It should be noted that the candidate vaccine viruses recommended by the WHO do not necessarily represent the entire spectrum of gs/Gd/96 lineage

Table 4 H5 antigenic clusters: WHO candidate vaccine viruses and OFFLU surveillance data

Virus	Geneticclade	Population status
Cluster 1		
A/Vietnam/1203/2004 (CDC-RG; SJRG-161052)	1	Replaced
A/Vietnam/1194/2004 (NIBRG-14)	1	Replaced
A/Cambodia/R0405050/2007 (NIBRG-88)	1.1	Replaced
A/Cambodia/W0526301/2012-like	1.1.2	Circulating
Cluster 2.1		
A/duck/Hunan/795/2002 (SJRG-166614)	2.1	Replaced
A/Indonesia/5/2005 (CDC-RG2)	2.1.3.2	Replaced
A/Indonesia/NIHRD11771/2011-like (NIIDRG-9)	2.1.3.2a	Circulating
Cluster 2.2		
A/bar-headed goose/Qinghai/1A/2005 (SJRG-163222)	2.2	Replaced
A/whooper swan/Mongolia/244/2005 (SJRG-163243)	2.2	Replaced
A/chicken/India/NIV33487/2006 (IBCDC-RG7)	2.2	Circulating
A/chicken/Bangladesh/CD-48-BL-165/2008*	2.2.2.1	Replaced
A/turkey/Turkey/1/2005 (NIBRG-23)	2.2.1	Replaced
A/Egypt/321/2007 (IDCDC-RG11)	2.2.1	Replaced
A/Egypt/N03072/2010 (IDCDC-RG29)	2.2.1	Circulating
A/Egypt/3300-NAMRU3/2008 (IDCDC-RG13)	2.2.1.1	Extinct
Cluster 2.3.2		
A/common magpie/Hong Kong/5052/2007 (SJRG-166615)	2.3.2.1	Replaced
A/Hubei/1/2010 (IDCDC-RG30)	2.3.2.1a	Circulating
A/duck/Bangladesh/19097/2013-like	2.3.2.1a	Circulating
A/Barn-Swallow/HK/D10-1161/2010 (SJ-003)	2.3.2.1b	Circulating
A/duck/Viet Nam/NCVD-1584/2012-like	2.3.2.1c	Circulating
Cluster 2.3.4		
A/Anhui/1/2005 (IBCDC-RG6)	2.3.4	Replaced
A/japanese white-eye/HongKong/1038/2006 (SJRG-164281)	2.3.4	Replaced
A/duck/Laos/3295/2006 (CBER-RG1)	2.3.4	Replaced
A/chicken/Hong Kong/AP156/2008 (SJ-002)	2.3.4	Replaced
A/chicken/Bangladesh/11rs1984-30/2011-like	2.3.4.2	Circulating
A/Guizhou/1/2013-like	2.3.4.2	Circulating
A/duck/Korea/Buan2/2014 (H5N8)*	2.3.4.6¶	Circulating
Cluster 4		
A/goose/Guiyang/337/2006 (SJRG-165396)	4	Extinct
Cluster 7		
A/chicken/Vietnam/NCVD-016/2008 (IDCDC-RG12)	7.1	Replaced
A/chicken/Vietnam/NCVD-03/2008 (IDCDC-RG25A)	7.1	Replaced
A/environment/Hubei/950/2013-like	7.2	Circulating

* *Source* OFFLU Surveillance

¶ Clade designation proposed, pending confirmation by WHO-OIE-FAO H5 Evolution Working Group

H5N1 viruses because only the antigenically distinct virus groups that pose a significant threat to human health merit selection for candidate vaccine virus development to meet pandemic preparedness goals. These include four first order clades 1, 2, 4, and 7 and higher order clades from clades 1, 2, and 7. Nevertheless, it is clear that this lineage of H5N1 viruses have achieved extensive antigenic divergence from the five ancestral clades described by (Ducatez et al. 2011). The early clade 1 antigenic cluster represented by A/Vietnam/1203/04 and A/Cambodia/405050/07 has diverged and the currently circulating viruses represented by A/Cambodia/W0526301/2012-like viruses are antigenically distinct from the ancestral viruses from 2007 and earlier (WHO 2013a). Viruses from the clade 2.1 antigenic cluster have continued to evolve in Indonesia and the circulating viruses, represented by A/Indonesia/NIHRD11771/2011, are antigenically distinct from their ancestors (2012b). Viruses from the clade 2.2 antigenic cluster have diverged into three distinct groups in South Asia and Africa, represented by A/chicken/India/NIV33487/2006, A/Egypt/N03072/2010 from genetic clade 2.2.1, and A/Egypt/3300-NAMRU3/2008 (from genetic clade 2.2.1.1); although the latter has become extinct (2011a). Viruses from antigenic cluster 2.3 have diverged extensively, giving rise to three groups that can be distinguished from the A/common magpie/HongKong/5052/2007-like ancestor virus. The first is represented by A/Hubei/1/2010, which more recently gave rise to an antigenically distinct A/duck/Bangladesh/19097/2013-like group. The second is represented by A/Barn-Swallow/HK/D10-1161/2010-like viruses, which continue to circulate in China. The third group is represented by A/duck/Viet Nam/NCVD-1584/2012-like viruses, which have spread recently to Indonesia and Lao. Antigenic cluster 2.4 has diverged into four distinct antigenic groups; two in China, one in Korea and one in Bangladesh. The Chinese antigenic groups are represented by A/Guizhou/1/2013-like viruses and by A/chicken/Hong Kong/AP156/2008-related viruses (WHO 2013b). The group represented by A/duck/Korea/Buan2/2014 (H5N8) are antigenically distinct reassortant viruses that emerged recently in China and have spread to Korea, Japan, and Laos, possibly by wild birds (Lee et al. 2014b; WHO-WPRO 2014). The fourth group is represented by A/chicken/Bangladesh/11rs1984-30/2011-like viruses (2011b) (Table 5).

Antigenic cluster 4 viruses have not been detected in recent years and appear to be extinct. Antigenic cluster 7 viruses have evolved extensively in China, and the representative virus A/environment/Hubei/950/2013 is antigenically distinct (2014a).

There is no consensus on the drivers of antigenic divergence among these virus groups. Vaccine-induced immunity has been postulated to mediate selection of antigenic variants based on studies, which show a possible correlation between the use of poultry vaccines and emergence of antigenic variant viruses (Cattoli et al. 2011; Lee and Suarez 2005). Immunity acquired by natural exposure may play a role in the evolution of viruses circulating in ducks, which often experience non-lethal infections with H5N1. Regardless of the forces that shape the evolution of the gs/Gd/96 lineage of H5N1 viruses, there is no doubt that the relentless antigenic evolution of the HA will continue to pose a challenge to pandemic vaccine preparedness efforts.

5.2 *Antigenic Characteristics of Eurasian H5 Viruses Distinct from A/goose/Guangdong/1/96-Like Virus*

Early studies (1997–1998) on the antigenic characteristics of European lineage H5N2 HPAI viruses (Donatelli et al. 2001) showed that the Italian H5N2 isolates were antigenically similar to, although distinguishable from, A/HK/156/97, isolated from the first outbreak of human influenza H5N1 virus in Hong Kong in 1997. Phylogenetic analysis of the hemagglutinin (HA) genes showed that the highly pathogenic Italian viruses clustered with the Hong Kong strains, and were more distantly related to the HPAI A/turkey/England/91 (H5N1) strain. However, when the antigenic properties of Italian HPAI (European lineage) were tested by Beato *et al.* in HI assays using chicken antisera, these viruses were antigenically distinct from viruses from the gs/Gd/96 lineage (Beato et al. 2010). In contrast, the Italian viruses were antigenically similar to each other.

5.3 *Antigenic Characteristics of North American H5*

HPAI H5N2 emerged in Mexico in 1994–1995, but control measures, including massive vaccination, were successful, and it has not been reported since 1996. However, the ancestral LPAI viruses remain endemic in Mexico as well as El Salvador and Guatemala. Antigenic analyses indicated that most viruses detected after the introduction of vaccine belonged to divergent genetic groups, and showed major antigenic differences relative to the vaccine viruses. These findings are consistent with the postulated role of vaccination in the selection of antigenically divergent viruses (Lee et al. 2004). The antigenic properties of the Mexican H5N2 viruses are similar to those of viruses that caused an outbreak in Pennsylvania, USA, in 1983. A recombinant fowlpox vectored vaccine expressing the HA of A/turkey/Ireland/83 (H5N8) protected chickens from challenge with HPAI A/Hong Kong/156/97 (H5N1), from the Eurasian lineage, as well as from challenge with A/chicken/Queretaro/14588/95(H5N2) and A/chicken/Pennsylvania/1370/83 (H5N2) from the North American lineage (Swayne et al. 2000a). The latter shared only 87–89 % aa sequence homology with the vaccine virus. Similarly, the Mexican lineage H5N2-based vaccine induced protection against challenge with Eurasian lineage H5N1 in chickens (Terregino et al. 2010).

Viruses closely related to the H5N2 from Mexico have caused outbreaks in Japan and Taiwan. These viruses were eradicated from Japan, but became endemic in Taiwan. Extensive reassortments have replaced the internal genes and NA, but the HA genes descend from an ancestor that resembles a virus that was used for vaccine manufacturing in Mexico. It has been postulated that incompletely inactivated poultry vaccines may have resulted in the introduction of the exotic virus into Japan and Taiwan (Lee et al. 2014a).

Table 5 Relative amino acid sequence differences among H5N1 candidate vaccine viruses selected by WHO

Virus	Clade	VN/1203/04	CB/W0526301/12	Indo/5/05	Indo/N-11771/11	bgh/Qinghai/1/05	Eg/N03072/10	Eg/3300-N3/08	ck/Bangl/-BL-165/08	cm/HK/5052/07	Hubei/1/10	barn-s/HK/1161/10	dk/VN/N-1584/12	dk/Bangl/19097/13	Anhui/1/2005	ck/Bangl/1984-30/11	Guizhou/1/2013	ck/HK/AP156/2008	br dk/Korea/2/14	gs/Guiyang/337/06	ck/VN/N-016/08	ck/VN/NCVD-03/08	env/Hubei/950/13
		1	1.1.2	2.1.3.2	2.1.3.2a	2.2	2.2.1	2.2.1.1	2.2.2.1	2.3.2.1	2.3.2.1a	2.3.2.1b	2.3.2.1c	2.3.2.1a	2.3.4.2	2.3.4.2	2.3.4.2	2.3.4	2.3.4.6	4	7.1	7.1	7.2
Viet Nam/1203/2004	1		3.1[a]	4.0	5.9	4.6	5.6	8.4	5.3	6.8	8.0	10.2	10.2	9.0	5.6	8.0	10.2	9.3	11.5	5.0	11.8	12.4	15.5
Cambodia/W0526301/2012	1.1.2	10[b]		5.9	7.7	6.5	7.5	9.6	7.1	8.7	9.6	10.2	10.8	10.2	7.4	9.9	11.1	10.5	12.1	6.5	13.3	13.7	17.7
Indonesia/5/2005	2.1.3.2	13	19		1.9	5.0	6.3	8.1	5.3	6.5	8.0	10.5	9.9	8.7	5.0	7.1	9.3	8.7	12.4	5.9	10.8	11.5	14.3
Indonesia/NIHRD11771/2011	2.1.3.2a	19	25	6		6.8	8.1	9.6	7.1	7.4	8.7	11.5	9.9	9.0	6.2	7.4	10.2	9.6	12.4	7.7	11.5	11.5	14.6
bar-headed goose/Qinghai/1/2005	2.2	15	21	16	22		1.6	5.6	1.5	6.2	7.1	9.6	8.7	8.4	5.0	8.0	10.2	7.7	11.5	7.1	11.5	12.4	14.6
Egypt/N03072/2010	2.2.1	18	24	20	26	5		5.6	2.5	6.9	7.5	10.0	9.1	8.8	6.3	9.1	11.3	9.1	12.5	7.8	12.2	13.2	15.0
Egypt/3300-NAMRU3/2008	2.2.1.1	27	31	26	31	18	18		6.2	9.0	9.9	12.1	11.5	10.6	8.7	10.2	12.1	10.2	14.3	10.6	12.4	13.4	15.0
chicken/Bangladesh/CD-48-BL-165/2008[c]	2.2.2.1	17	23	17	23	5	8	20		6.8	8.0	10.5	9.6	9.3	6.2	8.7	10.8	9.0	11.5	7.7	12.1	13.0	15.8
common magpie/Hong Kong/5052/2007	2.3.2.1	22	28	21	24	20	22	29	22		2.5	4.3	3.7	3.1	6.5	8.0	10.2	8.7	10.8	9.3	13.3	14.3	15.2
Hubei/1/2010	2.3.2.1a	26	31	26	28	23	24	32	26	8		5.6	2.5	2.5	7.4	9.3	10.5	9.0	10.5	10.8	13.9	14.9	15.8
barn swallow/Hong/Kong/1161/2010	2.3.2.1b	33	33	34	37	31	32	39	34	14	18		5.3	5.6	10.5	12.1	12.7	10.8	12.1	12.4	15.5	16.5	16.8
duck/Vietnam/NCVD-1584/2012	2.3.2.1c	33	35	32	32	28	29	37	31	12	8	17		3.1	8.7	10.5	11.8	9.9	11.1	12.7	14.9	15.8	16.8
duck/Bangladesh/19097/2013	2.3.2.1a	29	33	28	29	27	28	34	30	10	8	18	10		7.4	9.3	10.5	9.0	10.2	11.5	13.9	14.9	15.8
Anhui/1/2005	2.3.4.2	18	24	16	20	16	20	28	20	21	24	34	28	24		4.6	5.9	5.9	9.6	7.7	12.7	13.0	15.2
chicken/Bangladesh/11rs1984-30/2011	2.3.4.2	26	32	23	24	26	29	33	28	26	30	39	34	30	15		6.2	7.7	11.5	8.7	13.6	13.4	15.8
Guizhou/1/2013	2.3.4.2	33	36	30	33	33	36	39	35	33	34	41	38	34	19	20		7.7	10.5	10.5	14.6	15.2	17.4
chicken/Hong Kong/AP156/2008	2.3.4	30	34	28	31	25	29	33	29	28	29	35	32	29	19	25	25		11	11	14	15	15
broiler duck/KoreBuan2/2014(H5N8)[c]	2.3.4.6	37	39	40	40	37	40	46	37	35	34	39	36	33	31	37	34	36		12.4	17.6	17.7	18.6
goose/Guiyang/337/2006	4	16	21	19	25	23	25	34	25	30	35	40	41	37	25	28	34	35	40		10.5	11.5	14.9
chicken/Vietnam/NCVD-016/2008	7.1	38	43	35	37	37	39	40	39	43	45	50	48	45	41	44	47	45	57	34		4.0	11.8
chicken/Vietnam/NCVD-03/2008	7.1	40	44	37	37	40	42	43	42	46	48	53	51	48	42	43	49	47	57	37	13		11.8
enviroment/Hubei/950/2013	7.2	50	57	46	47	47	48	48	51	49	51	54	54	51	49	51	56	48	60	48	38	38	

[a] The upper right side of the table denotes pairwise aa distances (as %) between the HA1 region of the HA from the viruses indicated
[b] The lower left side of the table denotes the number of aa differences between the HA1 regions of the HA for indicated viruses
[c] A/broiler duck/KoreBuan2/2014(H5N8) and A/chicken/Bangladesh/CD-48-BL-165/2008 are not WHO candidate vaccine viruses

5.4 *Antigenic Characteristics of Eurasian and American H7 Viruses*

The antigenic characteristics of H7 HPAI viruses have not been analyzed as extensively as their H5 counterparts. However, this may change as a result of the H7N9 outbreak in China, which resulted in over 440 human infections (WHO 2014b). The live bird markets in China appear to support amplification of the virus in poultry (Fuller et al. 2014). Chickens appear to be the most permissive host for H7N9 replication and transmission (Pantin-Jackwood et al. 2014). Continued virus circulation in gallinaceous poultry may facilitate the emergence of HPAI H7N9 viruses.

5.4.1 Eurasian H7 Viruses

Epitope Mapping

The structure of an epitope on the H7 HA of HPAI A/Netherlands/219/2003 virus as defined by a neutralizing antibody was reported by (Dreyfus et al. 2012). This conserved epitope in the HA stem mediates protection against lethal challenge with type A and B viruses in animal models. Two mAbs, 213/2 and 253/1, developed against the HA of A/duck/Hokkaido/Vac-2/04 (H7N7) were found to recognize non-overlapping epitopes conserved among 13 different subtype H7 viruses isolated in Europe and Asia between 1963 and 2004 (Manzoor et al. 2008). Epitope mapping of the mAbs by escape mutant selection and sequencing indicated that the epitope for mAb 213/2 comprised Ser145 and Ser152, whereas the mAb 253/1 epitope included Ala169 (aa numbering from Met start codon of the HA). Both epitopes are thus located in the globular head of the HA molecule (Manzoor et al. 2008).

Serologically Defined Antigenic Relationships

The antigenic relationships of the HPAI H7N3 viruses from Pakistan with reference H7 viruses from American or Eurasian origin were examined by HI with chicken, ferret, and goat antisera (Aamir et al. 2009). The Pakistani H7N3 viruses were antigenically closely related and showed minimal antigenic drift between 1995 and 2002. Goat antisera to A/FPV/Rostock/34 (H7N1) showed antigenic similarity to Pakistan H7N3 viruses, but some differences from the American H7N3 viruses from 2003. In contrast, the chicken antisera to H7N3 isolates from Pakistan showed antigenic differences from A/Netherlands/219/03 (H7N7).

(Abbas et al. 2011) analyzed the antigenic relationships among HPAI and LPAI H7 viruses from the Americas (A/chicken/British Columbia/314514-2/04 H7N3, A/mallard/OH/421/87 H7N8 and A/chicken/Chile/176822/02 H7N3)), Europe (A/mallard/Netherlands/12/00 H7N3, A/mallard/Netherlands/9/05 H7N7, A/turkey/Italy/4580/99 H7N1), and several isolates from Pakistan using panels of chicken antisera produced with adjuvanted antigen boost. HI assay results indicated that titers of 2003 and 2004 isolates from Pakistan were ≤4-fold different from each other. In contrast, four H7 viruses from Pakistan did not cluster together: NARC-1/95, SPVC-6/04, and NARC-46/03 showed eightfold difference in titers with the main Pakistan cluster. The viruses from Chile, British Columbia, and the Netherlands showed HI titer differences at or below twofold. In contrast, isolates from Ohio and Italy were divergent with >eightfold titer differences.

Antigenic analysis of the LPAI H7N9 viruses that emerged in China in March 2013 indicated a clear relationship to the HPAI viruses isolated in Europe (WHO 2013a). Krammer et al. used serum from people vaccinated with a live cold-adapted, attenuated vaccine expressing HA of A/Netherlands/219/03 (H7N7) twice (28 days apart) and then boosted with A/mallard/Netherlands/12/00 (H7N7) inactivated vaccine, to analyze the antigenic characteristics of the homologous viruses with heterologous H7N9 LPAI from the 2013 outbreak in China (A/Shanghai/2/2013) and several H7 viruses from North America (Krammer et al. 2014). No antigenic differences were noted between these antigens in the HI test. Similar results were observed when hyperimmune sera were used in HI assays.

5.4.2 American H7 Viruses

Epitope Mapping

Early studies by Kida et al. characterized four operational epitopes (groups I–IV) on the HA of A/seal/Massachusetts/1/80 (H7N7), all mediating neutralization (Kida et al. 1982). Group I mAbs neutralized the virus and inhibited hemagglutination by binding residues Gly100, Thr156, and Gly196 (H7 numbering) near the RBS (Imai et al. 1998), while mAbs to group II had similar properties but recognized Ala151. In contrast, mAbs from groups III to IV failed to inhibit hemagglutination of the virus, yet effectively neutralized viral infectivity by binding epitopes Gly52 and Ala138.

Serologically Defined Antigenic Relationships

Antigenic characterization of representative North American (Canada and US) and South American (Chile) HPAI H7 viruses as well as an European isolate by HI analysis with ferret antisera, indicated large antigenic differences between the North American viruses and the rest of the viruses in one-way antigenic analysis (Pappas et al. 2007). Viruses from the 2012 HPAI H7N3 outbreak in laying hens in Mexico, such as A/chicken/Jalisco/CPA1/2012 (H7N3), or a related human isolate A/Mexico/INDRE/7218/2012 were shown to be antigenically similar to the H7 viruses isolated in 2004 from the HPAI H7N3 outbreak in British Columbia by one way antigenic analysis (2012b). These studies highlight the impact of experimental methods on the antigenic relationships between viruses.

6 Perspectives

Sequence similarity is insufficient to predict antigenic properties (Durviaux et al. 2014; Lee et al. 2007a; Lees et al. 2010; Liao et al. 2008). Structural similarity and physical chemistry attributes of the HA provide additional information to estimate the degree of antigenic similarity between related viruses. These principles have been implemented as bioinformatics tools, structural modeling, and statistical and mathematical algorithms, which are providing insights into the evolution of influenza viruses and their antigenic properties. Despite considerable progress, *in silico* determination of antigenic characteristics remains elusive. However, statistical algorithms including antigenic cartography are useful in facilitating compilation of large datasets and in aiding visualization of antigenic relatedness. Although the HI assay is likely to remain the first choice for the antigenic characterization of viruses in the foreseeable future, alternative assays such as micro-neutralization provide an important adjunct for this purpose. Advanced recombinant methods to produce mAbs for epitope mapping should allow analysis of the repertoire of specificities of a polyclonal response. These data would inform correlations between serology and protective immune response, and aid the development of laboratory tools for measuring immunity to HPAI influenza viruses. Standardized laboratory methods coupled with production and distribution of reference reagents (viruses, mAbs, and polyclonal antisera) would help to establish a global understanding of the antigenic relatedness of viruses circulating in different continents. Recent collaboration between the WHO GISRS, OFFLU, and academic partners has resulted in the increased availability of viruses and antisera for antigenic characterization. However, the available data remain insufficient to establish a comprehensive understanding of the antigenic characteristics of the global pool of HPAI and their precursors. Despite the focus on the HA as a critical determinant of host protection, there is a need to expand our understanding of the role of immunity to the NA and other influenza proteins, which are undoubtedly an integral part of host responses to influenza infections. The discovery of broadly neutralizing mAbs has changed the paradigm for influenza vaccine development; the HA is now considered a viable target for universal (heterosubtypic) vaccines for seasonal and pandemic influenza. On the opposite end of the spectrum, a few aa differences can have a significant impact on vaccine effectiveness. These observations emphasize the importance of both nuanced and comprehensive analyses of interactions between influenza HA (and other proteins) with antibodies (and T cell receptors) and ultimately, their role in host protection.

Acknowledgments The author thanks Drs. Giovanni Cattoli and Catherine Bigger, for critical reading of the manuscript, and colleagues in the Influenza Division, Centers for Disease Control and Prevention for valuable information and insight. The contributions of the WHO GISRS member laboratories, including National Influenza Centers, H5N1 Reference Laboratories, and Essential Regulatory Laboratories are greatly appreciated. We also thank the OIE/FAO Network of Expertise on Avian Influenza (OFFLU) for contributing viruses, reagents, and surveillance data.

References

Aamir UB, Naeem K, Ahmed Z, Obert CA, Franks J, Krauss S, Seiler P, Webster RG (2009) Zoonotic potential of highly pathogenic avian H7N3 influenza viruses from Pakistan. Virology 390:212–220
Abbas MA, Spackman E, Fouchier R, Smith D, Ahmed Z, Siddique N, Sarmento L, Naeem K, McKinley ET, Hameed A, Rehmani S, Swayne DE (2011) H7 avian influenza virus vaccines protect chickens against challenge with antigenically diverse isolates. Vaccine 29:7424–7429. doi:10.1016/j.vaccine.2011.07.064 S0264-410X(11)01109-1
Abolnik C, Londt BZ, Manvell RJ, Shell W, Banks J, Gerdes GH, Akol G, Brown IH (2009) Characterisation of a highly pathogenic influenza A virus of subtype H5N2 isolated from ostriches in South Africa in 2004. Influenza Other Respir Viruses 3:63–68
Ampofo WK, Baylor N, Cobey S, Cox NJ, Daves S, Edwards S, Ferguson N, Grohmann G, Hay A, Katz J, Kullabutr K, Lambert L, Levandowski R, Mishra AC, Monto A, Siqueira M, Tashiro M, Waddell AL, Wairagkar N, Wood J, Zambon M, Zhang W (2012) Improving influenza vaccine virus selection: report of a WHO informal consultation held at WHO headquarters, Geneva, Switzerland, 14–16 June 2010. Influenza Other Respir Viruses 6(142–52):e1–e5. doi:10.1111/j.1750-2659.2011.00277.x
Ampofo WK, Baylor N, Cobey S, Cox NJ, Daves S, Edwards S, Ferguson N, Grohmann G, Hay A, Katz J, Kullabutr K, Lambert L, Levandowski R, Mishra AC, Monto A, Siqueira M, Tashiro M, Waddell AL, Wairagkar N, Wood J, Zambon M, Zhang W (2013) Improving influenza vaccine virus selection: report of a WHO informal consultation held at WHO headquarters, Geneva, Switzerland, 14–16 June 2010. Influenza Other Respir Viruses 7(Suppl 2):52–53. doi:10.1111/irv.12081
Archetti I, Horsfall FL Jr (1950) Persistent antigenic variation of influenza A viruses after incomplete neutralization in ovo with heterologous immune serum. J Exp Med 92:441–462
Bahl J, Nelson MI, Chan KH, Chen R, Vijaykrishna D, Halpin RA, Stockwell TB, Lin X, Wentworth DE, Ghedin E, Guan Y, Peiris JS, Riley S, Rambaut A, Holmes EC, Smith GJ (2011) Temporally structured metapopulation dynamics and persistence of influenza A H3N2 virus in humans. Proc Natl Acad Sci U S A 108:19359–19364. doi:10.1073/pnas.1109314108 1109314108
Balish AL, Katz JM, Klimov AI (2013) Influenza: propagation, quantification, and storage. Curr Protoc Microbiol Chapter 15:Unit15G 1. doi:10.1002/9780471729259.mc15g01s29
Barnett JL, Yang J, Cai Z, Zhang T, Wan XF (2012) AntigenMap 3D: an online antigenic cartography resource. Bioinformatics 28:1292–1293. doi:10.1093/bioinformatics/bts105 bts105
Beato MS, Monne I, Mancin M, Bertoli E, Capua I (2010) A proof-of-principle study to identify suitable vaccine seed candidates to combat introductions of Eurasian lineage H5 and H7 subtype avian influenza viruses. Avian Pathol 39:375–382. doi:10.1080/03079457.2010.513376
Beaver JE, Bourne PE, Ponomarenko JV (2007) Immunome Res 3:3. doi:10.1186/1745-7580-3-3
Bedford T, Suchard MA, Lemey P, Dudas G, Gregory V, Hay AJ, McCauley JW, Russell CA, Smith DJ, Rambaut A (2014) Integrating influenza antigenic dynamics with molecular evolution. Elife 3:e01914. doi:10.7554/eLife.01914
Belser JA, Katz JM, Tumpey TM (2011) The ferret as a model organism to study influenza a virus infection. Dis Model Mech 4:575–579. doi:10.1242/dmm.007823 dmm.007823
Belser JA, Tumpey TM (2013) H5N1 pathogenesis studies in mammalian models. Virus Res. doi:10.1016/j.virusres.2013.02.003
Belshe R, Lee MS, Walker RE, Stoddard J, Mendelman PM (2004) Safety, immunogenicity and efficacy of intranasal, live attenuated influenza vaccine. Expert Rev Vaccines 3:643–654
Ben Jebara K (2007) WAHIS and the role of the OIE's reference laboratories and collaborating centres. Dev Biol (Basel) 128:69–72

Bender C, Hall H, Huang J, Klimov A, Cox N, Hay A, Gregory V, Cameron K, Lim W, Subbarao K (1999) Characterization of the surface proteins of influenza A (H5N1) viruses isolated from humans in 1997–1998. Virology 254:115–123

Bonfanti L, Monne I, Tamba M, Santucci U, Massi P, Patregnani T, Loli Piccolomini L, Natalini S, Ferri G, Cattoli G, Marangon S (2014) Highly pathogenic H7N7 avian influenza in Italy. Vet Rec. doi:10.1136/vr.102202

Bosch FX, Garten W, Klenk HD, Rott R (1981) Proteolytic cleavage of influenza virus hemagglutinins: primary structure of the connecting peptide between HA1 and HA2 determines proteolytic cleavability and pathogenicity of Avian influenza viruses. Virology 113:725–735

Bragstad K, Nielsen LP, Fomsgaard A (2008) The evolution of human influenza A viruses from 1999 to 2006: a complete genome study. Virol J 5:40. doi:10.1186/1743-422X-5-40

Bui HH, Peters B, Assarsson E, Mbawuike I, Sette A (2007a) Ab and T cell epitopes of influenza A virus, knowledge and opportunities. Proc Natl Acad Sci U S A 104:246–251

Bui HH, Sidney J, Li W, Fusseder N, Sette A (2007b) Development of an epitope conservancy analysis tool to facilitate the design of epitope-based diagnostics and vaccines. BMC Bioinformatics 8:361. doi:10.1186/1471-2105-8-361

Buonagurio DA, Nakada S, Parvin JD, Krystal M, Palese P, Fitch WM (1986) Evolution of human influenza A viruses over 50 years: rapid, uniform rate of change in NS gene. Science 232:980–982

Bush RM, Fitch WM, Bender CA, Cox NJ (1999) Positive selection on the H3 hemagglutinin gene of human influenza virus A. Mol Biol Evol 16:1457–1465

Cai Z, Zhang T, Wan XF (2011) Concepts and applications for influenza antigenic cartography. Influenza Other Respir Viruses 5(Suppl 1):204–207

Cao Z, Meng J, Li X, Wu R, Huang Y, He Y (2012) The epitope and neutralization mechanism of AVFluIgG01, a broad-reactive human monoclonal antibody against H5N1 influenza virus. PLoS One 7:e38126. doi:10.1371/journal.pone.0038126

Caton AJ, Brownlee GG, Staudt LM, Gerhard W (1986) Structural and functional implications of a restricted antibody response to a defined antigenic region on the influenza virus hemagglutinin. Embo J 5:1577–1587

Caton AJ, Brownlee GG, Yewdell JW, Gerhard W (1982) The antigenic structure of the influenza virus A/PR/8/34 hemagglutinin (H1 subtype). Cell 31:417–427

Caton AJ, Raymond FL, Brownlee GG, Yewdell JW, Gerhard W (1983) Antigenic variation in influenza virus. Biochem Soc Trans 11:435–441

Cattoli G, Fusaro A, Monne I, Coven F, Joannis T, El-Hamid HS, Hussein AA, Cornelius C, Amarin NM, Mancin M, Holmes EC, Capua I (2011) Evidence for differing evolutionary dynamics of A/H5N1 viruses among countries applying or not applying avian influenza vaccination in poultry. Vaccine 29:9368–9375. doi:10.1016/j.vaccine.2011.09.127 S0264-410X(11)01584-2

Chen Y, Qin K, Wu WL, Li G, Zhang J, Du H, Ng MH, Shih JW, Peiris JS, Guan Y, Chen H, Xia N (2009) Broad cross-protection against H5N1 avian influenza virus infection by means of monoclonal antibodies that map to conserved viral epitopes. J Infect Dis 199:49–58. doi:10.1086/594374

Cho KJ, Lee JH, Hong KW, Kim SH, Park Y, Lee JY, Kang S, Kim S, Yang JH, Kim EK, Seok JH, Unzai S, Park SY, Saelens X, Kim CJ, Lee JY, Kang C, Oh HB, Chung MS, Kim KH (2013) Insight into structural diversity of influenza virus haemagglutinin. J Gen Virol 94:1712–1722. doi:10.1099/vir.0.051136-0

Claas EC, Osterhaus AD, van Beek R, De Jong JC, Rimmelzwaan GF, Senne DA, Krauss S, Shortridge KF, Webster RG (1998) Human influenza A H5N1 virus related to a highly pathogenic avian influenza virus. Lancet 351:472–477

Claes F, Kuznetsov D, Liechti R, Von Dobschuetz S, Dinh Truong B, Gleizes A, Conversa D, Colonna A, Demaio E, Ramazzotto S, Larfaoui F, Pinto J, Le Mercier P, Xenarios I, Dauphin G (2014) The EMPRES-i genetic module: a novel tool linking epidemiological outbreak information and genetic characteristics of influenza viruses. Database (Oxford) 2014: bau008. doi:10.1093/database/bau008

Corti D, Voss J, Gamblin SJ, Codoni G, Macagno A, Jarrossay D, Vachieri SG, Pinna D, Minola A, Vanzetta F, Silacci C, Fernandez-Rodriguez BM, Agatic G, Bianchi S, Giacchetto-Sasselli I, Calder L, Sallusto F, Collins P, Haire LF, Temperton N, Langedijk JP, Skehel JJ, Lanzavecchia A (2011) A neutralizing antibody selected from plasma cells that binds to group 1 and group 2 influenza A hemagglutinins. Science 333:850–856. doi:10.1126/science.1205669

Cui J, Smith T, Robbins PW, Samuelson J (2009) Darwinian selection for sites of Asn-linked glycosylation in phylogenetically disparate eukaryotes and viruses. Proc Natl Acad Sci U S A 106:13421-13426. doi:10.1073/pnas.0905818106 0905818106

Daly JM, Newton JR, Mumford JA (2004) Current perspectives on control of equine influenza. Vet Res 35:411–423

Daniels RS, Douglas AR, Skehel JJ, Wiley DC, Naeve CW, Webster RG, Rogers GN, Paulson JC (1984) Antigenic analyses of influenza virus haemagglutinins with different receptor-binding specificities. Virology 138:174–177

de Jong JC, Heinen PP, Loeffen WL, van Nieuwstadt AP, Claas EC, Bestebroer TM, Bijlsma K, Verweij C, Osterhaus AD, Rimmelzwaan GF, Fouchier RA, Kimman TG (2001) Antigenic and molecular heterogeneity in recent swine influenza A(H1N1) virus isolates with possible implications for vaccination policy. Vaccine 19:4452–4464

De Marco D, Clementi N, Mancini N, Solforosi L, Moreno GJ, Sun X, Tumpey TM, Gubareva LV, Mishin V, Clementi M, Burioni R (2012) A non-VH1-69 heterosubtypic neutralizing human monoclonal antibody protects mice against H1N1 and H5N1 viruses. PLoS One 7:e34415. doi:10.1371/journal.pone.0034415 PONE-D-11-19380

Deyde VM, Okomo-Adhiambo M, Sheu TG, Wallis TR, Fry A, Dharan N, Klimov AI, Gubareva LV (2009) Pyrosequencing as a tool to detect molecular markers of resistance to neuraminidase inhibitors in seasonal influenza A viruses. Antiviral Res 81:16–24

Donatelli I, Campitelli L, Di Trani L, Puzelli S, Selli L, Fioretti A, Alexander DJ, Tollis M, Krauss S, Webster RG (2001) Characterization of H5N2 influenza viruses from Italian poultry. J Gen Virol 82:623–630

Dreyfus C, Ekiert DC, Wilson IA (2013) Structure of a classical broadly neutralizing stem antibody in complex with a pandemic H2 hemagglutinin. J Virol. doi:10.1128/JVI.02975-12

Dreyfus C, Laursen NS, Kwaks T, Zuijdgeest D, Khayat R, Ekiert DC, Lee JH, Metlagel Z, Bujny MV, Jongeneelen M, van der Vlugt R, Lamrani M, Korse HJ, Geelen E, Sahin O, Sieuwerts M, Brakenhoff JP, Vogels R, Li OT, Poon LL, Peiris M, Koudstaal W, Ward AB, Wilson IA, Goudsmit J, Friesen RH (2012) Highly conserved protective epitopes on influenza B viruses. Science 337:1343–1348. doi:10.1126/science.1222908

Du A, Daidoji T, Koma T, Ibrahim MS, Nakamura S, de Silva UC, Ueda M, Yang CS, Yasunaga T, Ikuta K, Nakaya T (2009) Detection of circulating Asian H5N1 viruses by a newly established monoclonal antibody. Biochem Biophys Res Commun 378:197–202

Du L, Jin L, Zhao G, Sun S, Li J, Yu H, Li Y, Zheng BJ, Liddington RC, Zhou Y, Jiang S (2012) Identification and structural characterization of a broadly neutralizing antibody targeting a novel conserved epitope on influenza H5N1 hemagglutinin. J Virol. doi:10.1128/JVI.02344-12 JVI.02344-12

Ducatez MF, Cai Z, Peiris M, Guan Y, Ye Z, Wan XF, Webby RJ (2011) Extent of antigenic cross-reactivity among highly pathogenic H5N1 influenza viruses. J Clin Microbiol 49:3531–3536. doi:10.1128/JCM.01279-11 JCM.01279-11

Durviaux S, Treanor J, Beran J, Duval X, Esen M, Feldman G, Frey SE, Launay O, Leroux-Roels G, McElhaney JE, Nowakowski A, Ruiz-Palacios GM, van Essen GA, Oostvogels L, Devaster JM, Walravens K (2014) Genetic and antigenic typing of seasonal influenza virus breakthrough cases from a 2008–2009 vaccine efficacy trial. Clin Vaccine Immunol 21:271–279. doi:10.1128/CVI.00544-13 CVI.00544-13

Ekiert DC, Bhabha G, Elsliger MA, Friesen RH, Jongeneelen M, Throsby M, Goudsmit J, Wilson IA (2009) Antibody recognition of a highly conserved influenza virus epitope. Science 324:246–251. doi:10.1126/science.1171491

Ekiert DC, Friesen RH, Bhabha G, Kwaks T, Jongeneelen M, Yu W, Ophorst C, Cox F, Korse HJ, Brandenburg B, Vogels R, Brakenhoff JP, Kompier R, Koldijk MH, Cornelissen LA, Poon LL, Peiris M, Koudstaal W, Wilson IA, Goudsmit J (2011) A highly conserved neutralizing epitope on group 2 influenza A viruses. Science 333:843–850. doi:10.1126/science.1204839
Ekiert DC, Kashyap AK, Steel J, Rubrum A, Bhabha G, Khayat R, Lee JH, Dillon MA, O'Neil RE, Faynboym AM, Horowitz M, Horowitz L, Ward AB, Palese P, Webby R, Lerner RA, Bhatt RR, Wilson IA (2012) Cross-neutralization of influenza A viruses mediated by a single antibody loop. Nature 489:526–532. doi:10.1038/nature11414
Ekiert DC, Wilson IA (2012) Broadly neutralizing antibodies against influenza virus and prospects for universal therapies. Curr Opin Virol 2:134–141. doi:10.1016/j.coviro.2012.02.005
FAO-DAH (2011) Stamping out H5N1 avian influenza could take decades. http://www.fao.org/news/story/en/item/66118/icode/. Accessed 15 Jul 2012
Fauci AS (2006) Pandemic influenza threat and preparedness. Emerg Infect Dis 12:73–77
Ferreira HL, Lambrecht B, van Borm S, Torrieri-Dramard L, Klatzmann D, Bellier B, van den Berg T (2010) Identification of a dominant epitope in the hemagglutinin of an Asian highly pathogenic avian influenza H5N1 clade 1 virus by selection of escape mutants. Avian Dis 54:565–571
Fidler DP, Gostin LO (2011) The WHO pandemic influenza preparedness framework: a milestone in global governance for health. JAMA 306:200–201. doi: 10.1001/jama.2011.960 306/2/200
Fiore AE, Bridges CB, Cox NJ (2009) Seasonal influenza vaccines. Curr Top Microbiol Immunol 333:43–82. doi:10.1007/978-3-540-92165-3_3
Fitch WM, Bush RM, Bender CA, Cox NJ (1997) Long term trends in the evolution of H(3) HA1 human influenza type A. Proc Natl Acad Sci U S A 94:7712–7718
Fitch WM, Leiter JM, Li XQ, Palese P (1991) Positive Darwinian evolution in human influenza A viruses. Proc Natl Acad Sci U S A 88:4270–4274
Fleury D, Barrere B, Bizebard T, Daniels RS, Skehel JJ, Knossow M (1999) A complex of influenza hemagglutinin with a neutralizing antibody that binds outside the virus receptor binding site. Nat Struct Biol 6:530–534
Fouchier RAM, Smith DJ (2010) Use of antigenic cartography in vaccine seed strain selection. Avian Dis 54:220–223. doi:10.1637/8740-032509-ResNote.1
Friesen RHE, Lee PS, Stoop EJM, Hoffman RMB, Ekiert DC, Bhabha G, Yu W, Juraszek J, Koudstaal W, Jongeneelen M, Korse HJWM, Ophorst C, Brinkman-van der Linden ECM, Throsby M, Kwakkenbos MJ, Bakker AQ, Beaumont T, Spits H, Kwaks T, Vogels R, Ward AB, Goudsmit J, Wilson IA (2013) A common solution to group 2 influenza virus neutralization. In: Proceedings of the National Academy of Sciences. doi:10.1073/pnas.1319058110
Fuller T, Havers F, Xu C, Fang LQ, Cao WC, Shu Y, Widdowson MA, Smith TB (2014) Identifying areas with a high risk of human infection with the avian influenza A (H7N9) virus in East Asia. J Infect. doi:10.1016/j.jinf.2014.03.006
Furuse Y, Suzuki A, Kishi M, Nukiwa N, Shimizu M, Sawayama R, Fuji N, Oshitani H (2010) Occurrence of mixed populations of influenza A viruses that can be maintained through transmission in a single host and potential for reassortment. J Clin Microbiol 48:369–374. doi:10.1128/JCM.01795-09 JCM.01795-09
Garcia M, Crawford JM, Latimer JW, Rivera-Cruz E, Perdue ML (1996) Heterogeneity in the haemagglutinin gene and emergence of the highly pathogenic phenotype among recent H5N2 avian influenza viruses from Mexico. J Gen Virol 77(Pt 7):1493–1504
Garten W, Bosch FX, Linder D, Rott R, Klenk HD (1981) Proteolytic activation of the influenza virus hemagglutinin: The structure of the cleavage site and the enzymes involved in cleavage. Virology 115:361–374
Gerhard W, Yewdell J, Frankel ME, Webster R (1981) Antigenic structure of influenza virus haemagglutinin defined by hybridoma antibodies. Nature 290:713–717
Ghedin E, Laplante J, DePasse J, Wentworth DE, Santos RP, Lepow ML, Porter J, Stellrecht K, Lin X, Operario D, Griesemer S, Fitch A, Halpin RA, Stockwell TB, Spiro DJ, Holmes EC, St George K (2011) Deep sequencing reveals mixed infection with 2009 pandemic influenza A

(H1N1) virus strains and the emergence of oseltamivir resistance. J Infect Dis 203:168–174. doi:10.1093/infdis/jiq040 jiq040

Grund S, Adams O, Wahlisch S, Schweiger B (2011) Comparison of hemagglutination inhibition assay, an ELISA-based micro-neutralization assay and colorimetric microneutralization assay to detect antibody responses to vaccination against influenza A H1N1 2009 virus. J Virol Methods 171:369–373. doi:10.1016/j.jviromet.2010.11.024

Guan Y, Peiris JS, Lipatov AS, Ellis TM, Dyrting KC, Krauss S, Zhang LJ, Webster RG, Shortridge KF (2002) Emergence of multiple genotypes of H5N1 avian influenza viruses in Hong Kong SAR. Proc Natl Acad Sci U S A 99:8950–8955

Guan Y, Smith GJ (2013) The emergence and diversification of panzootic H5N1 influenza viruses. Virus Res. doi:10.1016/j.virusres.2013.05.012

Guan Y, Smith GJ, Webby R, Webster RG (2009) Molecular epidemiology of H5N1 avian influenza. Rev Sci Tech 28:39–47

Han T, Marasco WA (2011) Structural basis of influenza virus neutralization. Ann N Y Acad Sci 1217:178–190. doi:10.1111/j.1749-6632.2010.05829.x

Han T, Sui J, Bennett AS, Liddington RC, Donis RO, Zhu Q, Marasco WA (2011) Fine epitope mapping of monoclonal antibodies against hemagglutinin of a highly pathogenic H5N1 influenza virus using yeast surface display. Biochem Biophys Res Commun 409:253–259. doi:10.1016/j.bbrc.2011.04.139

Hanson BJ, Boon AC, Lim AP, Webb A, Ooi EE, Webby RJ (2006) Passive immunoprophylaxis and therapy with humanized monoclonal antibody specific for influenza A H5 hemagglutinin in mice. Respir Res 7:126

Harvey R, Hamill M, Robertson JS, Minor PD, Vodeiko GM, Weir JP, Takahashi H, Harada Y, Itamura S, Bamford P, Dalla Pozza T, Engelhardt OG (2012) Application of deglycosylation to SDS PAGE analysis improves calibration of influenza antigen standards. Biologicals 40:96–99. doi:10.1016/j.biologicals.2011.12.009 S1045-1056(11)00201-6

Hay AJ, Gregory V, Douglas AR, Lin YP (2001) The evolution of human influenza viruses. Philos Trans R Soc Lond B Biol Sci 356:1861–1870

Ho HT, Qiang HL, He F, Meng T, Szyporta M, Prabhu N, Prabakaran M, Chan KP, Kwang J (2009) Rapid detection of H5N1 subtype influenza viruses by antigen-capture enzyme-linked immunosorbent assay using H5 and N1 specific monoclonal antibodies. Clin Vaccine Immunol 16(5):726–732

Hoffmann E, Lipatov AS, Webby RJ, Govorkova EA, Webster RG (2005) Role of specific hemagglutinin amino acids in the immunogenicity and protection of H5N1 influenza virus vaccines. Proc Natl Acad Sci U S A 102:12915–12920

Hong M, Lee PS, Hoffman RM, Zhu X, Krause JC, Laursen NS, Yoon SI, Song L, Tussey L, Crowe JE Jr, Ward AB, Wilson IA (2013) Antibody recognition of the pandemic H1N1 Influenza virus hemagglutinin receptor binding site. J Virol 87:12471–12480. doi:10.1128/JVI.01388-13

Hu H, Voss J, Zhang G, Buchy P, Zuo T, Wang L, Wang F, Zhou F, Wang G, Tsai C, Calder L, Gamblin SJ, Zhang L, Deubel V, Zhou B, Skehel JJ, Zhou P (2012) A human antibody recognizing a conserved epitope of H5 hemagglutinin broadly neutralizes highly pathogenic avian influenza H5N1 viruses. J Virol 86:2978–2989. doi:10.1128/JVI.06665-11 JVI.06665-11

Imai M, Sugimoto K, Okazaki K, Kida H (1998) Fusion of influenza virus with the endosomal membrane is inhibited by monoclonal antibodies to defined epitopes on the hemagglutinin. Virus Res 53:129–139

Ina Y, Gojobori T (1994) Statistical analysis of nucleotide sequences of the hemagglutinin gene of human influenza A viruses. Proc Natl Acad Sci U S A 91:8388–8392

Jebara KB, Caceres P, Berlingieri F, Weber-Vintzel L (2012) Ten years' work on the world organisation for animal health (OIE) worldwide animal disease notification system. Prev Vet Med 107:149–159. doi:10.1016/j.prevetmed.2012.08.008

Jia N, Wang SX, Liu YX, Zhang PH, Zuo SQ, Lin Z, Dang RL, Ma YH, Zhang C, Zhang L, Lu S, Cao WC (2008) Increased sensitivity for detecting avian influenza-specific antibodies by a modified hemagglutination inhibition assay using horse erythrocytes. J Virol Methods 153:43–48

Kaverin NV, Rudneva IA, Govorkova EA, Timofeeva TA, Shilov AA, Kochergin-Nikitsky KS, Krylov PS, Webster RG (2007) Epitope mapping of the hemagglutinin molecule of a highly pathogenic H5N1 influenza virus by using monoclonal antibodies. J Virol 81:12911–12917. doi:10.1128/JVI.01522-07

Kaverin NV, Rudneva IA, Ilyushina NA, Varich NL, Lipatov AS, Smirnov YA, Govorkova EA, Gitelman AK, Lvov DK, Webster RG (2002) Structure of antigenic sites on the haemagglutinin molecule of H5 avian influenza virus and phenotypic variation of escape mutants. J Gen Virol 83:2497–2505

Kayali G, Setterquist SF, Capuano AW, Myers KP, Gill JS, Gray GC (2008) Testing human sera for antibodies against avian influenza viruses: horse RBC hemagglutination inhibition versus microneutralization assays. J Clin Virol 43:73–78

Khurana S, Suguitan AL Jr, Rivera Y, Simmons CP, Lanzavecchia A, Sallusto F, Manischewitz J, King LR, Subbarao K, Golding H (2009) Antigenic fingerprinting of H5N1 avian influenza using convalescent sera and monoclonal antibodies reveals potential vaccine and diagnostic targets. PLoS Med 6:e1000049. doi:10.1371/journal.pmed.1000049

Kida H, Brown LE, Webster RG (1982) Biological activity of monoclonal antibodies to operationally defined antigenic regions on the hemagglutinin molecule of A/Seal/Massachusetts/1/80 (H7N7) influenza virus. Virology 122:38–47

Krammer F, Albrecht RA, Tan GS, Margine I, Hai R, Schmolke M, Runstadler J, Andrews SF, Wilson PC, Cox RJ, Treanor JJ, Garcia-Sastre A, Palese P (2014) Divergent H7 immunogens offer protection from H7N9 virus challenge. J Virol 88:3976–3985. doi:10.1128/JVI.03095-13

Kringelum JV, Lundegaard C, Lund O, Nielsen M (2012) Reliable B cell epitope predictions: impacts of method development and improved benchmarking. PLoS Comput Biol 8:e1002829. doi:10.1371/journal.pcbi.1002829

Kunik V, Ashkenazi S, Ofran Y (2012a) Paratome: an online tool for systematic identification of antigen-binding regions in antibodies based on sequence or structure. Nucleic Acids Res 40:W521–W524. doi:10.1093/nar/gks480

Kunik V, Peters B, Ofran Y (2012b) Structural consensus among antibodies defines the antigen binding site. PLoS Comput Biol 8:e1002388. doi:10.1371/journal.pcbi.1002388

Lapedes A, Farber R (2001) The geometry of shape space: application to influenza. J Theor Biol 212:57–69

Lee CC, Zhu H, Huang PY, Peng L, Chang YC, Yip CH, Li YT, Cheung CL, Compans R, Yang C, Smith DK, Lam TT, King CC, Guan Y (2014a) The emergence and evolution of avian H5N2 influenza viruses in chickens in Taiwan. J Virol. doi:10.1128/JVI.00139-14

Lee CW, Senne DA, Suarez DL (2004) Effect of vaccine use in the evolution of Mexican lineage H5N2 avian influenza virus. J Virol 78:8372–8381

Lee CW, Suarez DL (2005) Avian influenza virus: prospects for prevention and control by vaccination. Anim Health Res Rev 6:1–15

Lee MS, Chen MC, Liao YC, Hsiung CA (2007a) Identifying potential immunodominant positions and predicting antigenic variants of influenza A/H3N2 viruses. Vaccine 25:8133–8139

Lee YJ, Kang HM, Lee EK, Song BM, Jeong J, Kwon YK, Kim HR, Lee KJ, Hong MS, Jang I, Choi KS, Kim JY, Lee HJ, Kang MS, Jeong OM, Baek JH, Joo YS, Park YH, Lee HS (2014b) Novel reassortant influenza A(H5N8) viruses, South Korea, 2014. Emerg Infect Dis 20:1086–1089. doi:10.3201/eid2006.140233

Lee YJ, Sung HW, Choi JG, Lee EK, Jeong OM, Kwon YK, Kwon JH, Song CS, Kimd JH (2007b) Effects of homologous and heterologous neuraminidase vaccines in chickens against H5N1 highly pathogenic avian influenza. Avian Dis 51:476–478

Lees WD, Moss DS, Shepherd AJ (2010) A computational analysis of the antigenic properties of haemagglutinin in influenza A H3N2. Bioinformatics. doi:10.1093/bioinformatics/btq160 btq160

Li J, Wang Y, Liang Y, Ni B, Wan Y, Liao Z, Chan KH, Yuen KY, Fu X, Shang X, Wang S, Yi D, Guo B, Di B, Wang M, Che X, Wu Y (2009) Fine antigenic variation within H5N1 influenza virus hemagglutinin's antigenic sites defined by yeast cell surface display. Eur J Immunol 39:3498–3510. doi:10.1002/eji.200939532

Li Z, Liu Z, Ma C, Zhang L, Su Y, Gao GF, Cui L, He W (2011) Identification of amino acids in highly pathogenic avian influenza H5N1 virus hemagglutinin that determine avian influenza species specificity. Arch Virol 156:1803–1812. doi:10.1007/s00705-011-1056-2

Li Z, Sun W, Wu D, Gao X, Sun N, Liu N (2014) Mass spectrometry analysis coupled with de novo sequencing reveals amino acid substitutions in nucleocapsid protein from influenza A virus. Int J Mol Sci 15:2465–2474. doi:10.3390/ijms15022465

Liao YC, Lee MS, Ko CY, Hsiung CA (2008) Bioinformatics models for predicting antigenic variants of influenza A/H3N2 virus. Bioinformatics 24:505–512

Liu J, Xiao H, Lei F, Zhu Q, Qin K, Zhang XW, Zhang XL, Zhao D, Wang G, Feng Y, Ma J, Liu W, Wang J, Gao GF (2005) Highly pathogenic H5N1 influenza virus infection in migratory birds. Science 309:1206

Lo YC, Chuang JH, Kuo HW, Huang WT, Hsu YF, Liu MT, Chen CH, Huang HH, Chang CH, Chou JH, Chang FY, Lin TY, Chiu WT (2013) Surveillance and vaccine effectiveness of an influenza epidemic predominated by vaccine-mismatched influenza B/Yamagata-lineage viruses in Taiwan, 2011–2012 season. PLoS One 8:e58222. doi:10.1371/journal.pone.0058222

Lu G, Rowley T, Garten R, Donis RO (2007) FluGenome: a web tool for genotyping influenza A virus. Nucleic Acids Res 35:W275–W279. doi:10.1093/nar/gkm365

Luna LG, Williams TL, Pirkle JL, Barr JR (2008) Ultra performance liquid chromatography isotope dilution tandem mass spectrometry for the absolute quantification of proteins and peptides. Anal Chem 80:2688–2693

Manzoor R, Sakoda Y, Sakabe S, Mochizuki T, Namba Y, Tsuda Y, Kida H (2008) Development of a pen-site test kit for the rapid diagnosis of H7 highly pathogenic avian influenza. J Vet Med Sci 70:557–562

Masalova OV, Klimova RR, Chichev EV, Fediakina IT, Loginova SY, Borisevich SV, Bondarev VP, Deryabin PG, Lvov DK, Kushch AA (2011) Development of monoclonal antibodies to highly pathogenic avian influenza H5N1 virus and their application to diagnostics, prophylaxis, and therapy. Acta Virol 55:3–14

Matsuoka Y, Lamirande EW, Subbarao K (2009) The ferret model for influenza. Curr Protoc Microbiol Chapter 15:Unit 15G 2. doi:10.1002/9780471729259.mc15g02s13

Nabel GJ (2012) Rational design of vaccines for AIDS and influenza. Trans Am Clin Climatol Assoc 123:9–16

Nabel GJ, Fauci AS (2010) Induction of unnatural immunity: prospects for a broadly protective universal influenza vaccine. Nat Med 16:1389–1391. doi:10.1038/nm1210-1389 nm1210-1389

Oh HL, Akerstrom S, Shen S, Bereczky S, Karlberg H, Klingstrom J, Lal SK, Mirazimi A, Tan YJ (2010) An antibody against a novel and conserved epitope in the hemagglutinin 1 subunit neutralizes numerous H5N1 influenza viruses. J Virol 84:8275–8286. doi:10.1128/JVI.02593-09

Ohkura T, Kikuchi Y, Kono N, Itamura S, Komase K, Momose F, Morikawa Y (2012) Epitope mapping of neutralizing monoclonal antibody in avian influenza A H5N1 virus hemagglutinin. Biochem Biophys Res Commun 418:38–43. doi:10.1016/j.bbrc.2011.12.108 S0006-291X(11)02317-5

Okuno Y, Isegawa Y, Sasao F, Ueda S (1993) A common neutralizing epitope conserved between the hemagglutinins of influenza A virus H1 and H2 strains. J Virol 67:2552–2558

Okuno Y, Matsumoto K, Isegawa Y, Ueda S (1994) Protection against the mouse-adapted A/FM/1/47 strain of influenza A virus in mice by a monoclonal antibody with cross-neutralizing activity among H1 and H2 strains. J Virol 68:517–520

Osterholm MT, Kelley NS, Sommer A, Belongia EA (2012) Efficacy and effectiveness of influenza vaccines: a systematic review and meta-analysis. Lancet Infect Dis 12:36–44. doi:10.1016/S1473-3099(11)70295-X

Oxford JS, Corcoran T, Hugentobler AL (1981) Quantitative analysis of the protein composition of influenza A and B viruses using high resolution SDS polyacrylamide gels. J Biol Stand 9:483–491

Pantin-Jackwood MJ, Miller PJ, Spackman E, Swayne DE, Susta L, Costa-Hurtado M, Suarez DL (2014) Role of poultry in spread of novel H7N9 influenza virus in China. J Virol. doi:10.1128/JVI.03689-13

Pappas C, Matsuoka Y, Swayne DE, Donis RO (2007) Development and evaluation of an influenza virus subtype H7N2 vaccine candidate for pandemic preparedness. Clin Vaccine Immunol 14:1425–1432

Perelson AS, Oster GF (1979) Theoretical studies of clonal selection: minimal antibody repertoire size and reliability of self-non-self discrimination. J Theor Biol 81:645–670. doi:http://dx.doi.org/10.1016/0022-5193(79)90275-3

Pierce CL, Williams TL, Moura H, Pirkle JL, Cox NJ, Stevens J, Donis RO, Barr JR (2011) Quantification of immunoreactive viral influenza proteins by immunoaffinity capture and isotope-dilution liquid chromatography-tandem mass spectrometry. Anal Chem. doi:10.1021/ac2006526

Prabakaran M, He F, Meng T, Madhan S, Yunrui T, Jia Q, Kwang J (2010) Neutralizing epitopes of influenza virus hemagglutinin: Target for the development of universal vaccine against H5N1 lineages. J Virol. doi:10.1128/JVI.00891-10 JVI.00891-10

Prabakaran M, Ho HT, Prabhu N, Velumani S, Szyporta M, He F, Chan KP, Chen LM, Matsuoka Y, Donis RO, Kwang J (2009a) Development of epitope-blocking ELISA for universal detection of antibodies to human H5N1 influenza viruses. PLoS One 4:e4566. doi:10.1371/journal.pone.0004566

Prabakaran M, Prabhu N, He F, Hongliang Q, Ho HT, Qiang J, Meng T, Goutama M, Kwang J (2009b) Combination therapy using chimeric monoclonal antibodies protects mice from lethal H5N1 infection and prevents formation of escape mutants. PLoS One 4:e5672

Prabhu N, Prabakaran M, Ho HT, Velumani S, Qiang J, Goutama M, Kwang J (2008) Monoclonal antibodies against the fusion peptide of Hemagglutinin protect mice from lethal Influenza A H5N1 Infection. J Virol 83(6):2553–2562

Prabhu N, Prabakaran M, Hongliang Q, He F, Ho HT, Qiang J, Goutama M, Lim AP, Hanson BJ, Kwang J (2009) Prophylactic and therapeutic efficacy of a chimeric monoclonal antibody specific for H5 haemagglutinin against lethal H5N1 influenza. Antivir Ther 14:911–921. doi:10.3851/IMP1413

Prosser DJ, Cui P, Takekawa JY, Tang M, Hou Y, Collins BM, Yan B, Hill NJ, Li T, Li Y, Lei F, Guo S, Xing Z, He Y, Zhou Y, Douglas DC, Perry WM, Newman SH (2011) Wild bird migration across the Qinghai-Tibetan plateau: a transmission route for highly pathogenic H5N1. PLoS One 6:e17622. doi:10.1371/journal.pone.0017622

Ramakrishnan MA, Tu ZJ, Singh S, Chockalingam AK, Gramer MR, Wang P, Goyal SM, Yang M, Halvorson DA, Sreevatsan S (2009) The feasibility of using high resolution genome sequencing of influenza A viruses to detect mixed infections and quasispecies. PLoS One 4:e7105. doi:10.1371/journal.pone.0007105

Robertson JS, Cook P, Nicolson C, Newman R, Wood JM (1994) Mixed populations in influenza virus vaccine strains. Vaccine 12:1317–1322

Rockman S, Camuglia S, Vandenberg K, Ong C, Baker MA, Nation RL, Li J, Velkov T (2013) Reverse engineering the antigenic architecture of the haemagglutinin from influenza H5N1 clade 1 and 2.2 viruses with fine epitope mapping using monoclonal antibodies. Mol Immunol 53:435–442. doi:10.1016/j.molimm.2012.10.001

Rohm C, Suss J, Pohle V, Webster RG (1996) Different hemagglutinin cleavage site variants of H7N7 in an influenza outbreak in chickens in Leipzig, Germany. Virology 218:253–257

Rott R (1980) Genetic determinants for infectivity and pathogenicity of influenza viruses. Philos Trans R Soc Lond B Biol Sci 288:393–399

Rott R, Reinacher M, Orlich M, Klenk HD (1980) Cleavability of hemagglutinin determines spread of avian influenza viruses in the chorioallantoic membrane of chicken embryo. Arch Virol 65:123–133

Rowe T, Abernathy RA, Hu-Primmer J, Thompson WW, Lu X, Lim W, Fukuda K, Cox NJ, Katz JM (1999) Detection of antibody to avian influenza A (H5N1) virus in human serum by using a combination of serologic assays. J Clin Microbiol 37:937–943

Rudneva IA, Kushch AA, Masalova OV, Timofeeva TA, Klimova RR, Shilov AA, Ignatieva AV, Krylov PS, Kaverin NV (2010) Antigenic epitopes in the hemagglutinin of Qinghai-type influenza H5N1 virus. Viral Immunol 23:181–187. doi:10.1089/vim.2009.0086

Shahzad MI, Naeem K, Mukhtar M, Khanum A (2008) Passive immunization against highly pathogenic Avian Influenza Virus (AIV) strain H7N3 with antiserum generated from viral polypeptides protect poultry birds from lethal viral infection. Virol J 5:144

Shortridge KF, Zhou NN, Guan Y, Gao P, Ito T, Kawaoka Y, Kodihalli S, Krauss S, Markwell D, Murti KG, Norwood M, Senne D, Sims L, Takada A, Webster RG (1998) Characterization of avian H5N1 influenza viruses from poultry in Hong Kong. Virology 252:331–342

Skehel JJ, Daniels RS, Douglas AR, Wiley DC (1983) Antigenic and amino acid sequence variations in the haemagglutinins of type A influenza viruses recently isolated from human subjects. Bull World Health Organ 61:671–676

Skowronski DM, Janjua NZ, Sabaiduc S, De Serres G, Winter AL, Gubbay JB, Dickinson JA, Fonseca K, Charest H, Bastien N, Li Y, Kwindt TL, Mahmud SM, Van Caeseele P, Krajden M, Petric M (2014) Influenza A/subtype and B/lineage effectiveness estimates for the 2011–2012 trivalent vaccine: cross-season and cross-lineage protection with unchanged vaccine. J Infect Dis. doi:10.1093/infdis/jiu048

Skowronski DM, Moser FS, Janjua NZ, Davoudi B, English KM, Purych D, Petric M, Pourbohloul B (2013) H3N2v and other influenza epidemic risk based on age-specific estimates of sero-protection and contact network interactions. PLoS One 8:e54015. doi:10.1371/journal.pone.0054015

Smirnov YA, Lipatov AS, Gitelman AK, Claas EC, Osterhaus AD (2000) Prevention and treatment of bronchopneumonia in mice caused by mouse-adapted variant of avian H5N2 influenza A virus using monoclonal antibody against conserved epitope in the HA stem region. Arch Virol 145:1733–1741

Smith DJ, Lapedes AS, de Jong JC, Bestebroer TM, Rimmelzwaan GF, Osterhaus AD, Fouchier RA (2004) Mapping the antigenic and genetic evolution of influenza virus. Science 305:371–376. doi:10.1126/science.1097211

Smith GJ, Fan XH, Wang J, Li KS, Qin K, Zhang JX, Vijaykrishna D, Cheung CL, Huang K, Rayner JM, Peiris JS, Chen H, Webster RG, Guan Y (2006) Emergence and predominance of an H5N1 influenza variant in China. Proc Natl Acad Sci U S A 103:16936–16941

Stephenson I, Heath A, Major D, Newman RW, Hoschler K, Junzi W, Katz JM, Weir JP, Zambon MC, Wood JM (2009) Reproducibility of serologic assays for influenza virus A (H5N1). Emerg Infect Dis 15:1252–1259

Stephenson I, Wood JM, Nicholson KG, Zambon MC (2003) Sialic acid receptor specificity on erythrocytes affects detection of antibody to avian influenza haemagglutinin. J Med Virol 70:391–398

Stevens J, Blixt O, Tumpey TM, Taubenberger JK, Paulson JC, Wilson IA (2006) Structure and receptor specificity of the hemagglutinin from an H5N1 influenza virus. Science 312:404–410

Stewart PL, Nemerow GR (1997) Recent structural solutions for antibody neutralization of viruses. Trends Microbiol 5:229–233

Subbarao K, Klimov A, Katz J, Regnery H, Lim W, Hall H, Perdue M, Swayne D, Bender C, Huang J, Hemphill M, Rowe T, Shaw M, Xu X, Fukuda K, Cox N (1998) Characterization of an avian influenza A (H5N1) virus isolated from a child with a fatal respiratory illness. Science 279:393–396

Sui J, Hwang WC, Perez S, Wei G, Aird D, Chen LM, Santelli E, Stec B, Cadwell G, Ali M, Wan H, Murakami A, Yammanuru A, Han T, Cox NJ, Bankston LA, Donis RO, Liddington RC, Marasco WA (2009) Structural and functional bases for broad-spectrum neutralization of avian and human influenza A viruses. Nat Struct Mol Biol 16:265–273. doi:10.1038/nsmb.1566

Sun L, Lu X, Li C, Wang M, Liu Q, Li Z, Hu X, Li J, Liu F, Li Q, Belser JA, Hancock K, Shu Y, Katz JM, Liang M, Li D (2009) Generation, characterization and epitope mapping of two neutralizing and protective human recombinant antibodies against influenza A H5N1 viruses. PLoS One 4:e5476. doi:10.1371/journal.pone.0005476

Swayne DE, Beck JR, Garcia M, Stone HD (1999) Influence of virus strain and antigen mass on efficacy of H5 avian influenza inactivated vaccines. Avian Pathol 28:245–255. doi:10.1080/03079459994731

Swayne DE, Beck JR, Perdue ML, Beard CW (2001) Efficacy of vaccines in chickens against highly pathogenic Hong Kong H5N1 avian influenza. Avian Dis 45:355–365

Swayne DE, Garcia M, Beck JR, Kinney N, Suarez DL (2000a) Protection against diverse highly pathogenic H5 avian influenza viruses in chickens immunized with a recombinant fowlpox vaccine containing an H5 avian influenza hemagglutinin gene insert. Vaccine 18:1088–1095

Swayne DE, Perdue ML, Beck JR, Garcia M, Suarez DL (2000b) Vaccines protect chickens against H5 highly pathogenic avian influenza in the face of genetic changes in field viruses over multiple years. Vet Microbiol 74:165–172

Swayne DE, Suarez DL (2000) Highly pathogenic avian influenza. Rev Sci Tech 19:463–482

Terregino C, Toffan A, Cilloni F, Monne I, Bertoli E, Castellanos L, Amarin N, Mancin M, Capua I (2010) Evaluation of the protection induced by avian influenza vaccines containing a 1994 Mexican H5N2 LPAI seed strain against a 2008 Egyptian H5N1 HPAI virus belonging to clade 2.2.1 by means of serological and in vivo tests. Avian Pathol 39:215–222. doi:10.1080/03079451003781858

Thornburg NJ, Nannemann DP, Blum DL, Belser JA, Tumpey TM, Deshpande S, Fritz GA, Sapparapu G, Krause JC, Lee JH, Ward AB, Lee DE, Li S, Winarski KL, Spiller BW, Meiler J, Crowe JE Jr (2013) Human antibodies that neutralize respiratory droplet transmissible H5N1 influenza viruses. J Clin Invest 123:4405–4409. doi:10.1172/JCI69377

Throsby M, van den Brink E, Jongeneelen M, Poon LL, Alard P, Cornelissen L, Bakker A, Cox F, van Deventer E, Guan Y, Cinatl J, ter Meulen J, Lasters I, Carsetti R, Peiris M, de Kruif J, Goudsmit J (2008) Heterosubtypic neutralizing monoclonal antibodies cross-protective against H5N1 and H1N1 recovered from human IgM+ memory B cells. PLoS One 3:e3942. doi:10.1371/journal.pone.0003942

Treanor JJ, Talbot HK, Ohmit SE, Coleman LA, Thompson MG, Cheng PY, Petrie JG, Lofthus G, Meece JK, Williams JV, Berman L, Breese Hall C, Monto AS, Griffin MR, Belongia E, Shay DK (2012) Effectiveness of seasonal influenza vaccines in the United States during a season with circulation of all three vaccine strains. Clin Infect Dis 55:951–959. doi:10.1093/cid/cis574 cis574

Trock SC, Burke SA, Cox NJ (2012) Development of an influenza virologic risk assessment tool. Avian Dis 56:1058–1061

Tsibane T, Ekiert DC, Krause JC, Martinez O, Crowe JE Jr, Wilson IA, Basler CF (2012) Influenza human monoclonal antibody 1F1 interacts with three major antigenic sites and residues mediating human receptor specificity in H1N1 viruses. PLoS Pathog 8:e1003067. doi:10.1371/journal.ppat.1003067

Vallat B, Thiermann A, Ben Jebara K, Dehove A (2013) Notification of animal and human diseases: the global legal basis. Rev Sci Tech 32:331–335

Van Kerkhove MD, Broberg E, Engelhardt OG, Wood J, Nicoll A (2012) The consortium for the standardization of influenza seroepidemiology (CONSISE): a global partnership to standardize influenza seroepidemiology and develop influenza investigation protocols to inform public health policy. Influenza Other Respir Viruses. doi:10.1111/irv.12068

Vareckova E, Mucha V, Kostolansky F, Gubareva LV, Klimov A (2008) HA2-specific monoclonal antibodies as tools for differential recognition of influenza A virus antigenic subtypes. Virus Res 132:181–186. doi:10.1016/j.virusres.2007.10.004

Velkov T, Ong C, Baker MA, Kim H, Li J, Nation RL, Huang JX, Cooper MA, Rockman S (2013) The antigenic architecture of the hemagglutinin of influenza H5N1 viruses. Mol Immunol 56:705–719. doi:10.1016/j.molimm.2013.07.010

Vijaykrishna D, Bahl J, Riley S, Duan L, Zhang JX, Chen H, Peiris JS, Smith GJ, Guan Y (2008) Evolutionary dynamics and emergence of panzootic H5N1 influenza viruses. PLoS Pathog 4:e1000161

Wang SF, Chen KH, Thitithanyanont A, Yao L, Lee YM, Chan YJ, Liu SJ, Chong P, Liu WT, Huang JC, Chen YM (2009) Generating and characterizing monoclonal and polyclonal antibodies against avian H5N1 hemagglutinin protein. Biochem Biophys Res Commun 382:691–696. doi:10.1016/j.bbrc.2009.03.119

Wang TT, Tan GS, Hai R, Pica N, Ngai L, Ekiert DC, Wilson IA, Garcia-Sastre A, Moran TM, Palese P (2010) Vaccination with a synthetic peptide from the influenza virus hemagglutinin provides protection against distinct viral subtypes. Proc Natl Acad Sci U S A 107:18979–18984. doi:10.1073/pnas.1013387107

Wei K, Chen Y, Chen J, Wu L, Xie D (2012) Evolution and adaptation of hemagglutinin gene of human H5N1 influenza virus. Virus Genes 44:450–458. doi:10.1007/s11262-012-0717-x

WHO-OIE-FAO HNEWG (2008) Toward a unified nomenclature system for highly pathogenic avian influenza virus (H5N1). Emerg Infect Dis 14:e1

WHO-OIE-FAO HNEWG (2009) Continuing progress towards a unified nomenclature for the highly pathogenic H5N1 avian influenza viruses: divergence of clade 2.2 viruses. Influenza Other Respir Viruses 3:59–62. doi:10.1111/j.1750-2659.2009.00078.x

WHO-OIE-FAO HNEWG (2012) Continued evolution of highly pathogenic avian influenza A (H5N1): updated nomenclature. Influenza Other Respir Viruses 6:1–5. doi:10.1111/j. 1750-2659.2011.00298.x

WHO-OIE-FAO HNEWG (2014) Revised and updated nomenclature for highly pathogenic avian influenza A (H5N1) viruses. Influenza Other Respir Viruses 8:384–388. doi:10.1111/irv.12230

WHO-WPRO (2014) Avian influenza weekly update number 432. Accessed 9 May 2014

WHO (2006) Antigenic and genetic characteristics of H5N1 viruses and candidate H5N1 vaccine viruses developed for potential use as pre-pandemic vaccines. Wkly Epidemiol Rec 81:328–330

WHO (2007) Antigenic and genetic characteristics of H5N1 viruses and candidate H5N1 vaccine viruses developed for potential use as pre-pandemic vaccines, March 2007. Wkly Epidemiol Rec 82:164–167

WHO (2009a) Antigenic and genetic characteristics of H5N1 viruses and candidate vaccine viruses developed for potential use in human vaccines, February 2009. Wkly Epidemiol Rec 84:72–76

WHO (2009b) Antigenic and genetic characteristics of influenza A(H5N1) viruses and candidate vaccine viruses developed for potential use in human vaccines. Wkly Epidemiol Rec 84:432–436

WHO (2009c) Pandemic influenza preparedness and response: a who guidance document. Geneva

WHO (2010a) Antigenic and genetic characteristics of influenza A(H5N1) and influenza A(H9N2) viruses and candidate vaccine viruses developed for potential use in human vaccines—February 2010. Wkly Epidemiol Rec 85:100–107

WHO (2010b) Antigenic and genetic characteristics of influenza A(H5N1)and influenza A(H9N2) viruses and candidate vaccine viruses developed for potential use in human vaccines. Wkly Epidemiol Rec 85:418–424

WHO (2011a) Antigenic and genetic characteristics of influenza A(H5N1) and influenza A(H9N2) viruses for development of candidate vaccines viruses for pandemic preparedness—February 2011. Wkly Epidemiol Rec 86:93–100

WHO (2011b) Antigenic and genetic characteristics of zoonotic influenza viruses and development of candidate vaccine viruses for pandemic preparedness. Wkly Epidemiol Rec 86:469–480

WHO (2012a) Antigenic and genetic characteristics of zoonotic influenza viruses and development of candidate vaccine viruses for pandemic preparedness. Wkly Epidemiol Rec 87:97–108

WHO (2012b) Antigenic and genetic characteristics of zoonotic influenza viruses and development of candidate vaccine viruses for pandemic preparedness. Wkly Epidemiol Rec 87:401–412

WHO (2013a) Antigenic and genetic characteristics of zoonotic influenza viruses and development of candidate vaccine viruses for pandemic preparedness. Wkly Epidemiol Rec 88:449–463

WHO (2013b) Antigenic and genetic characteristics of zoonotic influenza viruses and development of candidate vaccine viruses for pandemic preparedness. Wkly Epidemiol Rec 88:117–125
WHO (2014a) Antigenic and genetic characteristics of zoonotic influenza viruses and development of candidate vaccine viruses for pandemic preparedness. Wkly Epidemiol Rec 89:105–115
WHO (2014b) Confirmed human cases of avian influenza A(H7N9) reported to WHO. Report 18, data in WHO/HQ as of 14 July 2014. http://www.who.int/influenza/human_animal_interface/influenza_h7n9/18_reportwebh7n9number_20140714.pdf?ua=1Accessed 14 July 2014
Wiley DC, Wilson IA, Skehel JJ (1981) Structural identification of the antibody-binding sites of Hong Kong influenza haemagglutinin and their involvement in antigenic variation. Nature 289:373–378
Williams TL, Luna L, Guo Z, Cox NJ, Pirkle JL, Donis RO, Barr JR (2008) Quantification of influenza virus hemagglutinins in complex mixtures using isotope dilution tandem mass spectrometry. Vaccine 26:2510–2520. doi:10.1016/j.vaccine.2008.03.014
Williams TL, Pirkle JL, Barr JR (2012) Simultaneous quantification of hemagglutinin and neuraminidase of influenza virus using isotope dilution mass spectrometry. Vaccine 30:2475–2482. doi:10.1016/j.vaccine.2011.12.056 S0264-410X(11)01991-8
Wu JT, Leung K, Perera RA, Chu DK, Lee CK, Hung IF, Lin CK, Lo SV, Lau YL, Leung GM, Cowling BJ, Peiris JS (2014) Inferring influenza infection attack rate from seroprevalence data. PLoS Pathog 10:e1004054. doi:10.1371/journal.ppat.1004054
Xu R, Krause JC, McBride R, Paulson JC, Crowe JE Jr, Wilson IA (2013) A recurring motif for antibody recognition of the receptor-binding site of influenza hemagglutinin. Nat Struct Mol Biol 20:363–370. doi:10.1038/nsmb.2500
Yang ZY, Wei CJ, Kong WP, Wu L, Xu L, Smith DF, Nabel GJ (2007) Immunization by avian H5 influenza hemagglutinin mutants with altered receptor binding specificity. Science 317:825–828. doi:10.1126/science.1135165
Yewdell JW, Taylor A, Yellen A, Caton A, Gerhard W, Bachi T (1993) Mutations in or near the fusion peptide of the influenza virus hemagglutinin affect an antigenic site in the globular region. J Virol 67:933–942
Yoshida R, Igarashi M, Ozaki H, Kishida N, Tomabechi D, Kida H, Ito K, Takada A (2009) Cross-protective potential of a novel monoclonal antibody directed against antigenic site B of the hemagglutinin of influenza A viruses. PLoS Pathog 5:e1000350. doi:10.1371/journal.ppat.1000350
Zhang M, Gaschen B, Blay W, Foley B, Haigwood N, Kuiken C, Korber B (2004) Tracking global patterns of N-linked glycosylation site variation in highly variable viral glycoproteins: HIV, SIV, and HCV envelopes and influenza hemagglutinin. Glycobiology 14:1229–1246. doi:10.1093/glycob/cwh106
Zhao K, Gu M, Zhong L, Duan Z, Zhang Y, Zhu Y, Zhao G, Zhao M, Chen Z, Hu S, Liu W, Liu X, Peng D, Liu X (2013) Characterization of three H5N5 and one H5N8 highly pathogenic avian influenza viruses in China. Vet Microbiol 163:351–357. doi:10.1016/j.vetmic.2012.12.025
Zhou J, Wang D, Gao R, Zhao B, Song J, Qi X, Zhang Y, Shi Y, Yang L, Zhu W, Bai T, Qin K, Lan Y, Zou S, Guo J, Dong J, Dong L, Zhang Y, Wei H, Li X, Lu J, Liu L, Zhao X, Li X, Huang W, Wen L, Bo H, Xin L, Chen Y, Xu C, Pei Y, Yang Y, Zhang X, Wang S, Feng Z, Han J, Yang W, Gao GF, Wu G, Li D, Wang Y, Shu Y (2013) Biological features of novel avian influenza A (H7N9) virus. Nature 499:500–503. doi:10.1038/nature12379
Zhu X, Guo YH, Jiang T, Wang YD, Chan KH, Li XF, Yu W, McBride R, Paulson JC, Yuen KY, Qin CF, Che XY, Wilson IA (2013) A unique and conserved neutralization epitope in H5N1 influenza viruses identified by an antibody against the A/Goose/Guangdong/1/96 hemagglutinin. J Virol 87:12619–12635. doi:10.1128/JVI.01577-13

Index

Current Topics in Microbiology and Immunology (2014) 385: 441–444
DOI: 10.1007/978-3-319-11155-1